Soil Fertility and Fertilizers

SOIL FERTILITY

AND FERTILIZERS

THIRD EDITION

Samuel L. Tisdale
Vice President
The Sulphur Institute
FORMERLY PROFESSOR OF SOILS,
NORTH CAROLINA STATE UNIVERSITY

Werner L. Nelson
Senior Vice President
Potash Institute of North America
FORMERLY PROFESSOR OF AGRONOMY,
NORTH CAROLINA STATE UNIVERSITY

Macmillan Publishing Co., Inc.
NEW YORK

Collier Macmillan Publishers
LONDON

Copyright © 1975, Macmillan Publishing Co., Inc.

Printed in the United States of America

All rights reserved. No part of this book may be reproduced or transmitted in any form or
by any means, electronic or mechanical, including photocopying, recording, or any information storage and retrieval system, without permission in writing from the Publisher.

Earlier editions © 1956 and copyright © 1966 by Macmillan Publishing Co., Inc.

Macmillan Publishing Co., Inc.
866 Third Avenue, New York, New York 10022

Collier-Macmillan Canada, Ltd.

Library of Congress Cataloging in Publication Data

Tisdale, Samuel L
    Soil fertility and fertilizers.

    Includes bibliographies.
    1.   Soil fertility.   2.   Fertilizers and manures.
3.   Plants – Nutrition.   I.   Nelson, Werner L., joint author.   II.   Title.
S633.T66   1974   631.4'2   74–78
ISBN 0-02-420860-4

Printing: 2 3 4 5 6 7 8   Year: 5 6 7 8 9 0

Preface

The purpose of the third edition of *Soil Fertility and Fertilizers* is the same as that of the first two; to present the fundamental principles of soil fertility and fertilizer manufacture and use in a manner suitable for students of agriculture at the junior and senior levels in college. As with the first edition, this text will be of greatest value to those who have first completed a fairly comprehensive beginning course in soils.

The sequence of the chapters and the organization of each is essentially the same as in the second edition. However, there has been extensive updating of the material covered with numerous references being made to literature published since 1966. The chapter on growth and the factors affecting it has been reorganized, with the material dealing with growth expressions following that in which the growth factors themselves are discussed.

Basic soil-plant relationships, Chapter 4, are discussed in greater depth in keeping with the current ideas relating to effective cation exchange capacity and such factors as mass flow, diffusion, and contact exchange and the impact they have on ion absorption by plant roots.

Chapter 11, "Liming," has been significantly reworked, bringing into focus the influence of Al, Mn, and Fe on soil acidity. The most recent ideas concerning the liming of soils in the warm humid regions of the country, as well as humid, cool areas, are discussed in detail.

The latest ideas relating to soil fertility evaluation, fertilizer application and the economics of lime, fertilizer use, and good soil management are presented in Chapters 12 through 16. The material in these chapters has been updated and revised, citing the latest published experimental data relating to these topics.

A summary and list of questions are included at the end of each chapter as was the case with the second edition.

The cost:return figures in this text are used largely for illustrative purposes. The world agricultural picture is in a state of flux brought about by a number of factors, and prices have fluctuated dramatically. The physical relations between plants and their environment, however, remain the same regardless of prices, but to determine whether a given practice is economically sound, current prices must be applied to the physical crop response data.

World demand for food and fiber has increased tremendously since 1965. With it there has been a corresponding increase in the production and consumption of fertilizers. However, in the 1970's an energy crisis developed which seriously threatens to curtail the production of ammonia, the basis of almost all chemical nitrogen fertilizer. If this crisis continues, the impact on food production could be serious, and alternative methods of supplying nitrogen for plant growth may have to be considered. Among the alternatives are the greater use of sewage sludge, legume crops, and farm manures, practices which a decade ago appeared to be on the way out.

A discussion of the factors affecting agricultural prices and energy supplies is beyond the scope of a text such as this, but an understanding of the principles of soil fertility and plant growth set forth in this book can be of considerable help in meeting the production problems that confront the farmer in an energy-short world. To successfully manipulate input:output economics in a fluid agricultural price situation, an understanding of the relation between plant growth and the factors affecting it, including soil fertility, is a necessity. It is that central theme around which this text has been built.

We have drawn liberally from the published work of our colleagues in North America and in other countries. Many of the illustrations contained in previously published reports were supplied by the authors of these papers. To these people, too numerous to list here, we are especially grateful. Many of our colleagues have made valuable suggestions for changes and to them we extend our thanks.

Special acknowledgment is made to the following publishers for their generosity in permitting the reproduction of tables and figures: The American Society of Agronomy, The American Chemical Society, Academic Press, Inc., The David McKay Publishing Co., The Williams & Wilkins Co., John Wiley & Sons, The Soil Science Society of America, The Potash Institute of North America, Rheinhold Publishing Co., and Martinus Nijhoff.

Last but by no means least, our thanks to Mrs. Allyne Tisdale, Mrs. Jeanette Nelson, and Mrs. Jean Watkins for their invaluable assistance in the typing and preparation of the manuscript.

S. L. T.
W. L. N.

Contents

INTRODUCTION: FERTILIZERS IN A CHANGING WORLD

Now more than ever the importance of an adequate supply of plant nutrients to ensure efficient crop production is being recognized. Growers are continually striving to overcome nutrient deficiencies as well as use improved management practices in order that yields may more nearly approach the genetic limit of crop plants. As a result of this effort, great progress in fertilizer technology and in the use of plant nutrients has been made in the United States, as well as in other countries, especially since 1960. A wider understanding of plant and soil chemistry has led to improved fertilization and cultural practices. Improved technology has led to the production of more efficient forms of fertilizer.

Until about 1900 the demands for greater agricultural yields in the United States were met primarily by bringing new land into cultivation. Although some new land can still be made arable by irrigation, drainage, or clearing of forested areas, the relative increases in acreage of crop land by these means will be quite small. Actually the annual loss of land in farms from 1962 to 1972 was 6,000,000 acres or about 0.5% per year. This loss to urban expansion, roads, and recreational areas will continue. It is certain that any substantial improvement in agricultural production must come from larger yields on land already in cultivation. Higher yields per acre mean a higher net profit per acre and lower unit production costs.

The tremendous increase in plant-nutrient consumption from 1950 to 1973 indicates that the importance of fertilizers to crop production in the United States is being accepted. During this period the respective tonnage increases in the use of nitrogen, phosphorus, and potassium were as follows:

Tons Used (USDA)

Years	N	P_2O_5	K_2O
1949–1950	956,000	1,930,000	1,070,000
1964–1965	4,605,442	3,785,230	2,828,458
1972–1973	8,339,000	5,072,000	4,412,000

TABLE I–1. Fertilizer Consumption in Relation to Arable Land and Per Capita

	Kilograms of fertilizer consumed per hectare of arable land*			Per capita total
	N	P_2O_5	K_2O	N, P_2O_5, K_2O
United States	38.3	22.6	20.4	75.5
USSR	22.3	10.5	12.1	42.6
Belgium	198.3	176.9	202.8	50.1
Denmark	115.4	50.1	71.5	127.0
France	79.8	101.1	78.7	96.8
East Germany	119.8	80.1	132.7	100.4
West Germany	139.9	115.7	152.6	55.8
Greece	56.7	34.1	5.0	38.8
Netherlands	440.6	119.4	148.9	45.6
United Kingdom	128.7	70.8	68.7	34.7
Canada	7.7	7.7	4.4	39.8
Brazil	9.3	15.0	11.8	11.2
Peru	27.3	5.0	1.2	7.1
China	30.1	8.2	1.3	5.6
India	10.7	3.4	1.8	4.8
Japan	160.8	122.0	107.0	20.1
Turkey	10.2	6.8	0.3	13.0
Algeria	11.6	13.2	4.4	13.7
South Africa	17.3	24.5	9.0	29.7
Egypt	122.7	16.7	0.7	11.5
Zaire	0.4	0.3	0.2	0.4
Australia	2.8	17.4	1.7	76.5
New Zealand	11.9	416.0	139.5	165.9
World	23.2	14.5	12.0	19.6

* Annual Fertilizer Review, Food and Agriculture Organization of the United Nations, 1972.

In spite of the progress made, the rates of application in the United States, as shown in Table I-1, are much less than those of many European countries. However, the use per capita in the United States is among the top six countries and this helps to increase opportunities for export of agricultural products. In Europe the land is old in terms of agricultural use, and it must be maintained at a high level of production because of the high population density. In comparison with European soils, the soils of the United States are relatively new agriculturally and much dependence is still placed on native fertility in some areas. Then, too, many millions of acres in this country are in dry-land or desert areas where water rather than plant nutrients is the limiting factor. The information given in Table I-1, however, is an indication of the importance of and the need for fertilizers in various parts of the world.

World use of plant nutrients is spreading rapidly but future requirements, as estimated by the Food and Agricultural Organization of the
United Nations, Tennessee Valley Authority, and other organizations,
will be 50 per cent greater in 1980 than in 1972.

The relation between fertilizer use and the value index of crop production per arable acre for thirty countries is shown in Figure I-1. Fertilizers are more than just an index to modern agricultural methods; they are
also a powerful factor that motivates the improvement of other cultural
practices. The whole of the increase in yields shown in Figure I-1 is, of
course, not the result of the use of fertilizers, but these increases do indicate what can be achieved when fertilizers are adapted to improved
cropping practices. It is of interest that the lower use countries have
made considerable strides in fertilizer use and value index. FAO reports
the following $N : P_2O_5 : K_2O$ ratios in fertilizer use for 1972 as follows:

	N	P_2O_5	K_2O
Developed countries	100	76	72
Developing countries	100	41	18

A significant development in the 1940's in the United States was the
increase in use of liming materials. The application of limestone reached
a peak during 1946 and 1947, but since that time the demand has
declined. This is unfortunate, for greater use of nitrogen and more intensive cropping add to the need for lime. It has been estimated that in the
United States more than 88 million tons of limestone are required annually, but only 24 million tons are being applied. The effect of limestone is both direct and indirect, and its use on acid soils is essential if
maximum returns are to be obtained from fertilization.

The gradually increasing yields of the principal crops in the United
States (see table) are the result of improved varieties, cultural practices,
and pest control as well as the use of fertilizers. However, the actual fertility of many of our soils is decreasing, for greater quantities of plant nutrients are being removed than are being added.

It is estimated that two thirds of the four billion people in the world
exist on an inadequate diet. This condition, coupled with the prediction
that the population will reach seven to eight billions by the year 2000,
presents an imposing challenge to the world's producers of food and
fiber.

Year	Corn (bu.)	Wheat (bu.)	Soybeans (bu.)	Cotton (lb. lint)	Alfalfa (tons)
1950	37.6	14.3	21.7	269	2.1
1964	62.1	26.2	22.8	524	2.4
1972	96.9	32.7	28.0	495	2.9

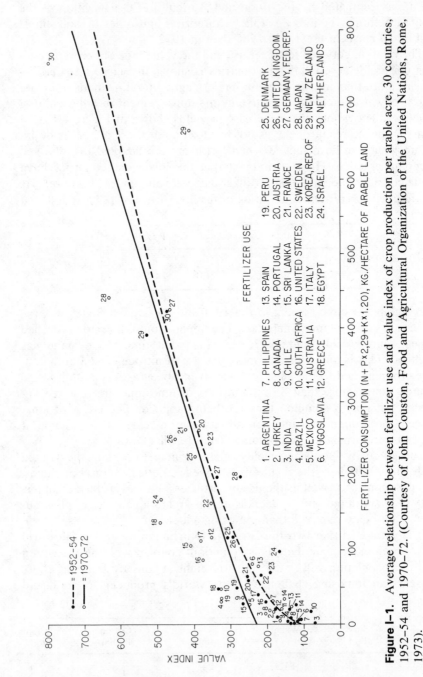

Figure I–1. Average relationship between fertilizer use and value index of crop production per arable acre, 30 countries, 1952–54 and 1970–72. (Courtesy of John Couston, Food and Agricultural Organization of the United Nations, Rome, 1973).

4

1. SOIL FERTILITY—
PAST AND PRESENT

The period in the development of the human race during which man began the cultivation of plants marks the dawn of agriculture. The exact time, of course, is not known, but it was certainly several thousand years before the birth of Christ. Until then man hunted almost exclusively for his food and was nomadic in his habits.

ANCIENT RECORDS

As the years went by man became less of a wanderer and more of a settler. Families, clans, and villages developed, and with them came the development of the skill we call agriculture. It is generally agreed that one area in the world that shows evidence of a very early civilization is Mesopotamia, situated between the Tigris and Euphrates rivers in what is now Iraq. Writings dating back to 2500 B.C. mention the fertility of the land. It is recorded that the yield of barley was 86-fold and even 300-fold in some areas, which means, of course, that for every unit of seed planted 86 to 300 units were harvested.

Herodotus, the Greek historian, reporting on his travels through Mesopotamia some 2,000 years later, mentions the phenomenal yields obtained by the inhabitants of this land. The high production was probably the result of a well-developed irrigation system and soil of high fertility, attributable in part to annual flooding by the river. Theophrastus, writing around 300 B.C., referred to the richness of the Tigris alluvium and stated that the water was allowed to remain on the land as long as possible so that a large amount of silt might be deposited.

In time man learned that certain soils would fail to produce satisfactory yields when cropped continuously. The practice of adding animal and vegetable manures to the soil to restore fertility probably developed from such observations, but how or when fertilization actually began is not known. Greek mythology, however, offers one picturesque explanation: Augeas, a legendary king of Elis, was famous for his stable, which contained 3,000 oxen. This stable had not been cleaned for thirty years. King Augeas contracted with Hercules to clean the stable out and agreed to give him 10 per cent of the cattle in return. Hercules is said to have accomplished this task by turning the River Alpheus through the stable, thus carrying away the accumulated filth and presumably depositing it on the adjacent land. Augeas then refused payment for this service, whereupon a war ensued, and Augeas was put to death by Hercules.

In the Greek epic poem *The Odyssey,* attributed to the blind poet Homer, who is thought to have lived between 900 and 700 B.C., the manuring of vineyards by the father of Odysseus is mentioned. The manure heap, which would suggest its systematic collection and storage, is also referred to. Argos, the faithful hound of Odysseus, was described as lying on such a heap when his master returned after an absence of twenty years. Having recognized his master, Argos wagged his tail feebly and "went down into the blackness of death." These writings suggest that manuring was an agricultural practice in Greece nine centuries before the birth of Christ.

Xenophon, who lived between 434 and 355 B.C., observed that "the estate has gone to ruin" because "someone didn't know it was well to manure the land." And again, ". . . there is nothing so good as manure."

Theophrastus (372–287 B.C.) recommended the abundant manuring of thin soils but suggested that rich soils be manured sparingly. He also endorsed a practice considered good today—the use of bedding in the stall. He pointed out that this would conserve the urine and bulk and that the humus value of the manure would be increased. It is interesting to note that Theophrastus suggested that plants with high nutrient requirements also had a high water requirement.

The truck gardens and olive groves around Athens were enriched by sewage from the city. A canal system was used, and there is evidence of a device for regulating the flow. It is believed that the sewage was sold to farmers. The ancients also fertilized their vineyards and groves with water that contained dissolved manure.

Manures were classified according to their richness or concentration. Theophrastus, for example, listed them in the following order of decreasing value: human, swine, goat, sheep, cow, oxen, and horse. Later Varro, an early writer on Roman agriculture, developed a similar list but rated bird and fowl manure as superior to human excrement. Columella recommended the feeding of snail clover to cattle because he felt that it enriched the excrement.

Not only did the ancients recognize the merits of manure, but they also observed the effect that dead bodies had on increasing the growth of crops. Archilochus made such an observation around 700 B.C., and the Old Testament records are even earlier. In Deuteronomy it is directed that the blood of animals should be poured on the ground. The increase in the fertility of land that has received the bodies of the dead has been acknowledged down through the years, but probably most poetically by Omar Khayyam, the astronomer-poet of Persia, who around the end of the eleventh century wrote

> I sometime think that never blows so red
> The rose as where some buried Caesar bled;
> That every hyacinth the garden wears
> Dropt in her lap from some once lovely head.
>
> And this delightful herb whose tender green
> Fledges the rivers lip on which we lean—
> Ah, lean upon it lightly! for who knows
> From what once lovely lip it springs unseen.

The value of green-manure crops, particularly legumes, was also soon recognized. Theophrastus noted that a bean crop (*Vicia faba*) was plowed under by the farmers of Thessaly and Macedonia. He observed that even when thickly sown and large amounts of seed were produced the crop enriched the soil.

Xenophon, around 400 B.C., recommended spring plowing because "the land is more friable then" and "the grass turned up is long enough at that season to serve as manure, but not having shed seed, it will not grow." He also pointed out that "every kind of vegetation, every kind of soil in stagnant water turns into manure."

Cato (234–149 B.C.) suggested that poor vineyard land be interplanted with a crop of *acinum*. It is not known what this crop is, but it was not allowed to go to seed, and the implication is that it was turned under. He also said that the best legumes for enriching the soil were field beans, lupines, and vetch.

Lupine was quite popular with the ancients. Columella listed numerous legume crops, including lupines, vetch, lentils, chick peas, clover, and alfalfa, that were satisfactory for soil improvement. Many of the early writers agreed, however, that lupine was the best general-purpose green-manure crop because it grew well under a wide range of soil conditions, furnished food for man and beast, was easy to seed, and quick to grow.

Virgil (70–19 B.C.) advocated the application of legumes, as indicated in the following passage:

Or, changing the season, you will sow there yellow wheat, whence before you have taken up the joyful pulse, with rustling pods, or the vetch's slender offspring and the bitter lupine's brittle stalks, and rustling grove.

The use of what might now be called mineral fertilizers or soil amendments was not entirely unknown to the ancients. Theophrastus suggested the mixing of different soils as a means of "remedying defects and adding heart to the soil." This practice may have been beneficial from several standpoints. The addition of fertile soil to infertile soil could lead to increased fertility, and the practice of mixing one soil with another may have provided better inoculation of legume seed on some fields. Again, the mixing of coarse-textured soils with those of fine texture or vice versa may have caused an improvement in the water and air relations in the soils of the fields so treated.

The value of marl was also recognized. The early dwellers of Aegina dug up marl and applied it to their land. The Romans, who learned this practice from the Greeks and Gauls, even classified the various liming materials and recommended that one type be applied to grain and another to meadow. Pliny (62–113) stated that lime should be spread thinly on the ground and that one treatment was "sufficient for many years, though not 50." Columella also recommended the spreading of marl on a gravelly soil and the mixing of gravel with a dense calcareous soil.

The Bible records the value of wood ashes in its reference to the burning of briars and bushes by the Jews, and Xenophon and Virgil both report the burning of stubble to clear fields and destroy weeds. Cato advised the vine keeper to burn prunings on the spot and to plow in the ashes to enrich the soil. Pliny states that the use of lime from lime kilns was excellent for olive groves, and some farmers burned manure and applied the ashes to their fields. Columella also suggested the spreading of ashes or lime on lowland soils to destroy acidity.

Saltpeter, or potassium nitrate, was mentioned both by Theophrastus and Pliny as useful for fertilizing plants and is referred to in the Bible in the book of Luke. Brine was mentioned by Theophrastus. Apparently recognizing that palm trees required large quantities of salt, early farmers poured brine around the roots of their trees.

Even as soil scientists today are searching for methods of predicting the fitness of soil for the production of crops, so did the minds of the early agricultural philosophers and writers turn to such matters. Virgil, for example, believed that soil that was "blackish and fat under the deep pressed share, and whose mold is loose and crumbling is generally best for corn."

Virgil wrote on the soil characteristic known today as bulk density. His advice on determining this property was

. . . first, you shall mark out a place with your eye, and order a pit to be sunk deep in solid ground, and again return all the mold into its place, and level with your feet the sands at top. If they prove deficient, the soil is loose and more fit for cattle and bounteous vines; but if they deny the possibility of returning to their places, and there be an over plus of mold after the pit is filled up, it is a dense soil; expect reluctant clods and stiff ridges, and give the first plowing to the land with sturdy bullocks.

Pliny apparently held a different view of the merits of the test proposed by Virgil. He stated that soil thrown back into the hole from which it was dug will never fill the hole, and that by this method it is impossible to form any opinion concerning its density or thinness.

Later, Columella, writing on ways of testing the soil to determine its physical suitability for agricultural use, set down an approach different from that of Virgil. He suggested that a trench be dug and that the soil removed therefrom be returned to it. If the hole were then not filled completely, the soil was lean, but if some soil were left out the soil was fat.

Virgil describes another means which might today be considered the prototype of a chemical soil test.

> But saltish ground, and that which is accounted bitter, where corn can never thrive, will give proof of this effect. Snatch from the smoky roofs bushels of close woven twigs and the strainers of the wine press. Hither let some of that vicious mold, and sweet water from the spring, be pressed brimful; be sure that all the water will strain out and bid drops pass through the twigs. But the taste will clearly make discovery; and in its bitterness will distort the wry faces of the tasters with the sensation.

Columella also suggested a taste test to measure the degree of acidity and salinity of soils, and Pliny stated that the bitterness of soils might be detected by the presence of black and underground herbs.

Pliny wrote that "among the proofs of the goodness of soil is the comparative thickness of the stem in corn," and Columella stated simply that the best test for the suitability of land for a given crop was whether it would grow.

Many of the early writers (and, for that matter, many people today) believed that the color of the soil was a criterion of its fertility. The general idea was that black soils were fertile and light or gray soils infertile. Columella disagreed with this viewpoint, pointing to the infertility of the black marshland soils and the high fertility of the light-colored soils of Libya. He felt that such factors as structure, texture, and acidity were far better guides to an estimation of soil fertility.

Much of the early writing regarding soil fertility consisted largely of descriptions of farm practices. There seems to be little evidence of an experimental approach to farm problems, but many of these manuscripts do reflect a rather keen comprehension of certain of the factors now known to affect plant growth. Some of the practices described in these early writings are as sound today as they were then. Many more are not, but it was centuries before man began to work intelligently toward a solution of the agricultural problems with which he was confronted.

The age of the Greeks from perhaps 800 to 200 B.C. was indeed a Golden Age. Many of the men of this period reflected genius that was unequaled, or at least not permitted expression, for centuries to come. Their writings, their culture, their agriculture were copied by the Romans, and the philosophy of many of the Greeks of this period dominated the thinking of men for more than 2,000 years.

SOIL FERTILITY DURING THE FIRST EIGHTEEN CENTURIES A.D.

After the decline of Rome there were few contributions to the development of agriculture until the publication of *Opus ruralium commodorum,* a collection of local agricultural practices, by Pietro de Crescenzi (1230–1307). De Crescenzi is referred to by some as the founder of modern agronomy, but his manuscript seems to be confined to the work of writers from the time of Homer. His contribution was largely that of summarizing the material. He did, however, suggest an increase in the rate of manuring over that in use at the time.

After the appearance of de Crescenzi's work little was added to agricultural knowledge for many years, although Palissy in 1563 is credited with the observation that the ash content of plants represented the material they had removed from the soil.

Around the beginning of the seventeenth century Francis Bacon (1561–1624) suggested that the principal nourishment of plants was water. He believed that the main purpose of the soil was to keep the plants erect and to protect them from heat and cold and that each plant drew from the soil a substance unique for its own particular nourishment. Bacon maintained further that continued production of the same type of plant on a soil would impoverish it for that particular species.

During this same period Jan Baptiste van Helmont (1577–1644), a Flemish physician and chemist, reported the results of an experiment which he believed proved that water was the sole nutrient of plants. He placed 200 lb. of soil into an earthen container, moistened the soil, and planted a willow shoot weighing 5 lb. He carefully shielded the soil in the crock from dust, and only rain or distilled water was added. After a period of five years, van Helmont terminated the experiment. The tree weighed 169 lb. and about 3 oz. He could account for all but about 2 oz. of the 200 lb. of soil originally used. Because he had added only water, his conclusion was that water was the sole nutrient of the plant, for he attributed the loss of the 2 oz. of soil to experimental error.

Of course, it is *now* known that both carbon dioxide (CO_2) and minerals from the soil are required for the nutrition of plants. It should be remembered, however, that this work was done at a time *before* anything was known about mineral nutrition or photosynthesis. Van Helmont's work and his erroneous conclusions were actually valuable contributions to our knowledge, for, even though they were wrong, his conclusions stimulated later investigations, the results of which have led to a much better understanding of plant nutrition.

The work of van Helmont was repeated several years later by no less a figure than Robert Boyle (1627–1691) of England. Boyle is probably best known for expressing the relation of the volume of a gas to its pressure. He was also interested in biology and a great believer in the experimental approach to the solution of problems dealing with science. He believed that observation was the only road to truth. Boyle confirmed the findings of van Helmont, but he went one step farther. As a result of the chemical analyses he performed on plant samples, he stated that plants contained salts, spirits, earth, and oil, all of which were formed from water.

About this same time J. R. Glauber (1604–1668), a German chemist, suggested that saltpeter (KNO_3) and not water was the "principle of vegetation." He collected the salt from soil under the pens of cattle and argued that it must have come from the droppings of these animals. He further stated that because the animals ate forage the saltpeter must have come originally from the plants. When he applied this salt to plants and observed the large increases in growth it produced, he was convinced that soil fertility and the value of manure were due entirely to saltpeter.

John Mayow (1643–1679), an English chemist, supported the viewpoint of Glauber. Mayow estimated the quantities of niter in the soil at various times during the year and found it in its greatest concentration in the spring. Failing to find any during the summer, he concluded that the saltpeter had been absorbed, or sucked up, as he put it, by the plant during its period of rapid growth.

Experimental techniques were quite crude at this time. Even with the contributions made by Mayow, Glauber, Boyle, and Bacon, there was really little work that could, by present-day standards, be termed good research. About 1700, however, a study was made which was outstanding and which represented a considerable advance in the progress of agricultural science. An Englishman by the name of John Woodward, who was acquainted with the work of Boyle and van Helmont, grew spearmint in samples of water he had obtained from various sources: rain water, river water, sewage water, and sewage water plus garden mold. He carefully measured the quantity of water transpired by the plants and recorded the weight of the plants at the beginning and end of the experiment. He found that the growth of the spearmint was proportional to the amount of impurities in the water and concluded that terrestrial matter, or earth, rather than water, was the principle of vegetation. Although his conclusion is not correct in its entirety, it represents an advance in knowledge, and his experimental technique was considerably better than any that had been used before.

There was much understandable ignorance concerning the nutrition of plants during this period. Many quaint ideas came into being and "bode their hour or two and went their way." Not the least of these ideas were introduced by another enterprising Englishman, Jethro Tull (1674–1741). Tull was educated at Oxford, which is generally considered a bit unusual for a person of agricultural leanings. He appears to have been interested in politics, but ill health forced his retirement to the farm. There he carried out numerous experiments, most of which dealt with cultural practices. He believed that the soil should be finely pulverized to provide the "proper pabulum" for the growing plant. According to Tull, the soil particles were actually ingested through openings in the plant roots. The pressure caused by the swelling of the growing roots was thought to force this finely divided soil into "the lacteal mouths of the roots," after which it entered the "circulatory system" of the plant.

Tull's ideas about plant nutrition were, to say the least, a bit odd. However, his experiments led to the development of two valuable pieces of farm equipment, the drill and the horse-drawn cultivator. His published book, *Horse Hoeing Husbandry,* was long considered an authoritative text in English agricultural circles.

Around 1762 John Wynn Baker, a Tull adherent, established an experimental farm in England, the purpose of which was the public exhibition of the results of experiments in agriculture. Baker's work was praised later by Arthur Young who, however, admonished his readers to beware of giving too much credit to calculations based on the results of only a few years' work, an admonition that is as timely today as it was when originally made.

One of the more famous of the eighteenth-century English agriculturists was Arthur Young (1741–1820). Young conducted pot tests to find those substances that would improve the yield of crops. He grew barley in sand to which he added such materials as charcoal, train oil, poultry dung, spirits of wine, niter, gunpowder, pitch, oyster shells, and numerous other materials. Some of the materials produced plant growth, others did not. Young, a prolific writer, published a work entitled *Annals of Agriculture,* in forty-six volumes, which was highly regarded and made a considerable impact on English agriculture.

Many of the agricultural writings of the seventeenth and eighteenth centuries reflected the idea that plants were composed of one substance, and most of the workers during this period were searching for this *principle of vegetation.* Around 1775, however, Francis Home stated that there was not only one principle but probably many, among which he included air, water, earth, salts, oil, and fire in a fixed state. Home felt that the problems of agriculture were essentially those of the nutrition of plants. He carried out pot experiments to measure the effects of different substances on plant growth and made chemical analyses of plant materials. His work was considered to be a valuable stepping stone in the progress of scientific agriculture.

The idea that plants contained fire in a fixed state, or phlogiston, as it was called, lingered in the minds of men for many years. There was also the belief that organic material or humus was taken in directly by plants and that it constituted their principal nutrient. This idea persisted down through the years. It was difficult to dispel because the results of chemical analyses had shown that plants and humus contained essentially the same elements. Too, the process of photosynthesis had not yet been discovered. Joseph Priestley's observation around 1775 that sprigs of mint "purified" air led him to suggest that plants reversed the effect of breathing. At the time of this observation oxygen had not been discovered. Later, when he did discover this gas, he failed to recognize its relation to plants.

The discovery of oxygen by Priestley was the keystone to a number of other discoveries that went far toward unlocking the mystery of plant life. Jan Ingenhousz (1730–1799) showed that the purification of air took place in the presence of light, but in the dark the air was not purified. Coupled with this discovery was the statement by Jean Senebier (1742–1809), a Swiss natural philosopher and historian, that the increase in the weight of van Helmont's willow tree was the result of air!

PROGRESS DURING THE NINETEENTH CENTURY

These discoveries stimulated the thinking of Theodore de Saussure, whose father was acquainted with the work of Senebier. He attacked two of the problems on which Senebier had worked—the effect of air on plants and the origin of salts in plants. As a result, de Saussure was able to demonstrate that plants absorbed oxygen and liberated carbon dioxide, the central theme of respiration. In addition, he found that plants would absorb carbon dioxide with the release of oxygen in the presence of light. If, however, plants were kept in an environment free of carbon dioxide, they died.

DeSaussure concluded that the soil furnishes only a small fraction of the nutrients needed by plants, but he demonstrated that it does supply both ash and nitrogen. He effectively dispelled the idea that plants spontaneously generate potash and stated further that the plant root does not behave as a mere filter. Rather the membranes are selectively permeable, allowing for a more rapid entrance of water than of salts. He also showed the differential absorption of salts and the inconstancy of plant composition, which varies with the nature of the soil and the age of the plant.

De Saussure's conclusion that the carbon contained by plants was derived from the air did not meet with immediate acceptance by his colleagues. No less a figure than Sir Humphrey Davy, who published his work *The Elements of Agricultural Chemistry* about 1813, stated that although some plants may have received their carbon from the air the major portion was taken in through the roots. Davy was so enthusiastic in this belief that he recommended the use of oil as a fertilizer because of its carbon and hydrogen content.

The middle of the nineteenth to the beginning of the twentieth century was a time during which much progress was made in the understanding of plant nutrition and crop fertilization. Among the men of this period whose contributions loom large was Jean Baptiste Boussingault (1802–1882), a widely traveled French chemist who established a farm in Alsace on which he carried out field-plot experiments. Boussingault employed the careful techniques of de Saussure in weighing and analyzing the manures he added to his plots and the crops he harvested. He maintained a balance sheet which showed how much of the various plant-nutrient elements came from rain, soil, and air, analyzed the composition of his crops during various stages of growth, and determined that the best rotation was that which produced the largest amount of organic matter in addition to that added in the manure. Boussingault is called by some the father of the field-plot method of experimentation.

Although several of the agricultural scientists of this period recognized the value of de Saussure's observations, the old humus theory still had many adherents. It was such a natural theory that it was difficult to dispel. Many must have felt then, as some do today, that the decay of plant and animal material gives rise to products that are best suited for the nutrition of growing plants. Justus von Liebig (1803–1873), a German chemist, very effectively deposed the humus myth. The presentation of his paper at a prominent scientific meeting jarred the conservative thinkers of the day to such an extent that only a few scientists since that time have dared to suggest that the carbon contained in plants comes from any source other than carbon dioxide. Liebig made the following statements:

1. Most of the carbon in plants comes from the carbon dioxide of the atmosphere.
2. Hydrogen and oxygen come from water.
3. The alkaline metals are needed for the neutralization of acids formed by plants as a result of their metabolic activities.
4. Phosphates are necessary for seed formation.
5. Plants absorb everything indiscriminately from the soil but excrete from their roots those materials that are nonessential.

Not all of Liebig's ideas, of course, were correct. He thought that acetic acid was excreted by the roots. He also believed that the NH_4^+ form of nitrogen was the one absorbed and that plants might obtain this compound from soil, manure, or air.

Liebig firmly believed that by analyzing the plant and studying the elements it contained one could formulate a set of fertilizer recommendations based on these analyses. It was also his opinion that the growth of plants was proportional to the amount of mineral substances available in the fertilizer. He eventually developed the law of the minimum, which, in effect, says that the growth of plants is limited by the plant-nutrient element present in the smallest quantity, all others being present in adequate amounts. This concept dominated the thinking of agricultural workers for a long time thereafter.

Liebig manufactured a fertilizer based on his ideas of plant nutrition. The formulation of the mixture was perfectly sound, but he made the mistake of fusing the phosphate and potash salts with lime. As a result the fertilizer was a complete failure. Nonetheless, the contributions that Liebig made to the advancement of agriculture were monumental, and he is perhaps quite rightly recognized as the father of agricultural chemistry.

Following on the heels of Liebig's now famous paper was the establishment in 1843 of an agricultural experiment station at Rothamsted, England. The founders of this institution were J. B. Lawes and J. H. Gilbert. Work here was conducted along the same lines as that carried out earlier by Boussingault in France.

Lawes and Gilbert did not believe that all of the maxims set down by Liebig were correct. Twelve years after the founding of the station they settled the following points:

1. Crops require both phosphorus and potash, but the composition of the plant ash is no measurement of the amounts of these constituents required by the plant.
2. Nonlegume crops require a supply of nitrogen. Without this element, no growth will be obtained, regardless of the quantities of phosphorus and potassium present. The amount of ammonia nitrogen contributed by the atmosphere is insufficient for the needs of crops.
3. Soil fertility could be maintained for some years by means of chemical fertilizers.
4. The beneficial effect of fallow lies in the increase in the availability of nitrogen compounds in the soil.

For a long time, and even today in some sections, farmers were reluctant to believe that fertility could be maintained by the use of chemical fertilizers alone. The early work at Rothamsted proved conclusively, however, that it can be done. When soils become depleted in spite of the use of commercial fertilizers, it is largely because of inadequate amounts of these materials and improper soil management rather than the mineral fertilizer *per se*. Work at the Rothamsted station has continued on through the years, and today this station is recognized as one of the world's leading centers of agricultural research.

A discovery of considerable importance was made in England in 1852. Thomas Way, observing that a Yorkshire farmer was able to reduce ammonia losses from manure by the addition of soil, first demonstrated the phenomenon known as cation exchange. However, its tremendous significance was not immediately recognized, even by the great Liebig.

The value of some of the early results from the Rothamsted experiments was recognized by Georges Ville, a Frenchman of Vincennes. Ville maintained that the use of chemical fertilizers was the only method of supporting soil fertility. He made recommendations for the fertilization of crops based on the results of field trials and drew up a simple scheme of plot tests that could be used by farmers to determine for themselves just what fertilizers were needed for their crops.

The problem of soil and plant nitrogen remained unsolved. Several workers had observed the unusual behavior of legumes. In some instances they grew well in the absence of added nitrogen, whereas in others no growth was obtained. Nonlegumes, on the other hand, always failed to grow when there was insufficient nitrogen in the soil.

In 1878 some light was thrown on the confusion by two French bacteriologists, Theodore Schloessing and Alfred Müntz. These scientists purified sewage water by passing it through a filter made of sand and limestone. They analyzed the filtrate periodically, and for twenty-eight days only ammonia nitrogen was detected. At the end of this time nitrates began to appear in the filtrate. Schloessing and Müntz found that the production of nitrates could be stopped by adding chloroform and that it could be started again by adding a little fresh sewage water. They concluded that nitrification was the result of bacterial action.

The results of these experiments were applied to the soils by Robert Warrington of England. He showed that nitrification could be stopped by carbon disulfide and chloroform and that it could be started again by adding a small amount of unsterilized soil. He also demonstrated that the reaction was a two-step phenomenon, the ammonia first being converted to nitrites and the nitrites subsequently to nitrates.

Warrington was unable to isolate the organisms responsible for nitrification. This task remained for S. Winogradsky, who effected the isolation by the use of a silica-gel plate, rather than the usual agar medium, because the organisms are autotrophic and obtain their carbon from the carbon dioxide of the atmosphere.

As to the erratic behavior of legume plants with respect to nitrogen, two Germans, Hellriegel and Wilfarth, in 1886 concluded that bacteria must be present in the nodules attached to legume roots. Further, these organisms were believed to assimilate gaseous nitrogen from the atmosphere and to convert it to a form that could be used by higher plants. This was the first specific information regarding nitrogen fixation by legumes. Hellriegel and Wilfarth based their arguments on observations made in certain of their experiments. They did not, however, isolate the responsible organisms. This was later done by M. W. Beijerinck, who called the organism *Bacillus radicicola*.

THE DEVELOPMENT OF SOIL FERTILITY IN THE UNITED STATES

Although most of the advances made in agriculture during the eighteenth century were accomplished on the Continent, a few early American contributions were sufficiently significant to mention. In 1733 James E. Oglethorpe established an experimental garden on the bluffs of the Savannah River, which is the present site of the city of Savannah, Georgia. This garden was devoted to the production of exotic food crops and is said to have been a place of beauty while it was maintained. Interest was lost in it, however, and it soon ceased to exist, but because it was largely the result of British interests it probably cannot be truly considered an "American" undertaking.

Benjamin Franklin demonstrated the value of gypsum. On a prominent hillside he applied gypsum to the land in a pattern which outlined the words, "This land has been plastered." The increased growth of pasture in the area to which the gypsum had been applied served as an effective demonstration of its fertilizer value.

In 1785 a society was formed in South Carolina which had among its objectives the setting up of an experimental farm. Eleven years later President Washington, in his annual message to the Congress, pleaded for the establishment of a national board of agriculture. Some of the most important contributions to early American agriculture were made by Edmond Ruffin of Virginia from about 1825 to 1845. He is believed to have been one of the first to use lime on humid-region soils to replace fertility elements lost by crop removal and leaching. Ruffin was a careful observer, a studious reader, and possessed of a keen and inquiring mind. Although his use of lime to bolster the yields of crops was known to the ancients, it was apparently a new experience in a new America.

It was not until 1862 that the Department of Agriculture was established, and, in the same year, the Morrill Act provided for state colleges of agriculture and mechanical arts. The first organized agricultural experiment station, set up in 1875 at Middletown, Connecticut, was supported by state funds. In 1877 North Carolina established a similar station, followed closely by New Jersey, New York, Ohio, and Massachusetts. In 1888 the Hatch Act called for the setting up of state experiment stations to be operated in conjunction with the Land-Grant colleges, and an annual grant of $15,000 was made available to each state for their support. Although much of the early experimental work was largely demonstration, a "scientific" approach to agricultural problems was gradually developed in this country.

The idea that plants excreted weak acids was firmly fixed in the minds of many, and a method for determining the fertility status of soils by weak acid extraction came into being. Although such treatment produced results that were of little absolute value, it was found, when large numbers of soils were studied, that the amounts of nutrients extracted generally were correlated with crop yields.

The suggestion was advanced that mineral elements were present in the soil in two rather well-defined forms: in the undecomposed primary minerals and in secondary minerals known as zeolites. These elements were thought to be held in available form by the zeolites.

The idea of extracting soils with acids to determine their fertility status persisted, and E. W. Hilgard (1833–1916) found that the maximum solubility of soil minerals in hydrochloric acid (HCl) was obtained when the acid was at a specific gravity of 1.115, which happens to correspond to the concentration of the acid obtained on prolonged boiling. Hilgard attached particular significance to this fact. The strong acid digestion became quite popular, and numerous analyses of soils were made by this method. It was later shown that there was little foundation for assuming that this technique would give data of predictive value, and its use was discontinued.

Two early workers who contributed much to the development of interest in soil fertility in the United States were Milton Whitney and C. G. Hopkins. Around the beginning of the twentieth century these two men engaged in a controversy which attracted nationwide attention and which, in fact, became quite bitter. Whitney maintained that the total supply of nutrients in soils was inexhaustible and that the important factor from the standpoint of plant nutrition was the rate at which these nutrients entered into the soil solution. Hopkins, on the other hand, felt that this philosophy would lead to soil depletion and a serious decline in crop production. He made a survey of the soils of Illinois and reduced soil fertility to a system of bookkeeping. As a result of these exhaustive studies, he concluded that Illinois soils required the addition of only lime and phosphate. So effectively did he preach this doctrine that the use of lime and rock phosphate in a corn, oats, and clover rotation was a continuous practice in this state for many years.

The controversy between Whitney and Hopkins finally waned. Whitney's ideas were shown to be at least partly incorrect, but the argument, though bitter, did much to stimulate the thinking of agricultural scientists of the period.

Just after the turn of the twentieth century most experiment stations had established field plots which showed the remarkable benefits of fertilization. As a result of these experiments, the major problem of soil fertility could be delimited in a broad way. It was shown, for example, that there was a widespread need for phosphatic fertilizers, that potassium was generally lacking in the coastal plains regions, and that nitrogen was particularly deficient in the soils of the South. The soils east of the Mississippi were generally acid and needed lime, whereas those west of the river were as a rule fairly well supplied with calcium. Even though the broad outline of the fertility status of soils in the United States had been fairly well defined, it was soon apparent that blanket fertilizer recommendations based on such knowledge could not be made. Each farm required individual attention, as did each field within that farm. The interest in soil tests sprang up anew.

During the last thirty years much headway has been made toward an understanding of the problems of soil fertility. To enumerate the men whose contributions have advanced our knowledge would require far more space than is available here. These advances have not been the work of scientists of any one country. The English, who began around 1600, have continued to make great strides. The agricultural workers of France, Germany, Scandinavia, Russia, Canada, Australia, New Zealand, as well as the United States and other countries, have unraveled many problems which have held back the progress of the science. The fruits of these studies are everywhere apparent, for agricultural production in advanced countries is higher today than it has ever been, and the free world as a whole is generally better fed, clothed, and housed than at any time in the past. This could not be so if the production of crops today were at the level of Europe in the Dark Ages, when the average yield of grain was 6 to 10 bu./A.

Progress in agriculture depends on research of a high caliber. For every problem solved by the scientist today, many more are raised. Agricultural scientists must delve into questions of a fundamental nature, questions that deal more with the *why* of things than with the *what.*

It is not the purpose of this chapter to cover all of the significant events in the development of the science of soil fertility. Much has been omitted, and much more could have been written. Certainly the advances made toward the end of the nineteenth century and in the twentieth have been largely responsible for the present state of our learning. These advances were covered only superficially in this chapter, but the remaining chapters, in the information they contain, confirm the importance of these events to progress in soil fertility. Though brief, it is hoped that this sketch will give the student some idea of the time, effort, and thought that has been devoted in the last 4,500 years to accumulating what is still insufficient knowledge.

Selected References

Diagnostic Techniques for Soils and Crops. Washington D.C.: Pot. Inst. North Amer., 1948.

Fussell, G. E., "John Wynne Baker: an improver in eighteenth-century agriculture." *Agr. Hist.,* **5:**151–161 (1931).

Gruber, J. W., "Irrigation and land use in ancient Mesopotamia." *Agr. Hist.,* **22:**69–77 (1948).

Holland, J. W., "The beginnings of public agricultural experimentation in America: the trustee's garden in Georgia." *Agr. Hist.,* **12:**271–298 (1938).

Jacob, K. D., "Fertilizers in retrospect and prospect." *Proc. Fertiliser Soc. (London)* (April 1963).

Lucretius Carus Titus, *De Rerum Naturum.* Translated by William Ellery Leonard. New York: Dutton, 1921.

Miller, A. C., Jr., "Jefferson as an agriculturist." *Agr. Hist.,* **16:**65–78 (1942).

Olson, Lois, "Columella and the beginnings of soil science." *Agr. Hist.,* **17:**65–72 (1943).

Olson, Lois, "Pietro de Crescenzi: the founder of modern agronomy." *Agr. Hist.,* **18:**35–40 (1944).

Olson, Lois, "Cato's views on the farmer's obligation to the land. *Agr. Hist.,* **19:**129–133 (1945).

Plinius Secundus Caius, *Natural History,* Book XVII. Translated by John Bostock and H. T. Riley. London: Henry G. Bohn (1857).

Russell, E. W., *Soil Conditions and Plant Growth,* 9th Ed. London: Longmans, Green, 1961.

Semple, Ellen Churchill, "Ancient Mediterranean agriculture." *Agr. Hist.,* **2:**61 (1928).

Virgil *(Publius Vergilius Maro), Georgicus.* Translated by Theodore Williams. Cambridge: Harvard University Press, 1915.

Xenophon, *Oeconomicus.* Translated by E. C. Marchant. New York: Putnam.

2. GROWTH AND THE FACTORS AFFECTING IT

Stated in its most elemental terms, the success of the farming operation depends largely on the growth of the crop. If plant growth and yield of the harvest are good, the farmer will have been successful; and, barring an unfavorable economy, he will receive a return on his investment in labor and capital in keeping with the quality and quantity of the crop. On the other hand, if plant growth has been poor and the yield of the harvest is low, the farmer will receive a correspondingly lower return, if any at all.

From the practical standpoint of a profitable agriculture, growth and the factors affecting it occupy a place of importance second to none in the farming enterprise. Because of this, some of these factors, and the effect they may have on limiting plant growth responses to an adequate supply of plant nutrients, are discussed briefly.

GROWTH

Growth is defined by Webster as the *progressive development of an organism*. There are, however, several ways in which this development can be expressed. It may refer to the development of some specific organ or organs or to the plant as a whole. Growth may be expressed in terms of dry weight, length, height, or diameter. The various ways in which growth has been expressed will be treated later in this chapter. From an agricultural point, however, the factors which influence plant growth are of the greatest importance, for a knowledge of these factors and how they affect crop growth can make it possible for the farm operator to manipulate them to his advantage in maximizing the return he gets on his farming operation.

FACTORS AFFECTING PLANT GROWTH

All of the factors that influence plant growth may not have been identified. Those that are known, however, can be classified as genetic or environmental.

Genetics. The importance of genetics in the growth of agricultural crops is shown by the large increases in yields which have come about through the use of hybrid corn and other improved crop plants. The high-yielding potential and other characteristics (such as quality, disease resistance, and drought hardiness) are related to the genetic constitution of the plant. Just why one variety or hybrid outperforms another is not known, but research in recent years has demonstrated the influence that the genes exert on physiological processes by a controlling mechanism on the synthesis of enzymes. Research in this and related fields will doubtless provide the answers to many of the puzzling questions concerning plant behavior.

VARIETY AND PLANT NUTRIENT NEEDS. It is obvious that a hybrid corn producing 200 bu./A. will require more plant nutrients than one producing 100 bushels. This important fact has sometimes been overlooked when a shift to the higher-yielding variety is made. Under low-fertility conditions a given variety may not be allowed to develop the full potential of its yielding capacity. In fertile soils the new variety will deplete the soil more rapidly and eventually yields will be depressed if nutrient supplements are not provided.

An example of the importance of genetics as a yield-limiting factor is shown in Figure 2-1. Notice especially the inability of the Arksoy variety to produce the high yields obtained with the other varities on the moderate and high-fertility soils. One of the first steps in a successful crop farming enterprise is the selection of hybrids or varieties that are genetically capable of producing high crop yields and of utilizing to the fullest extent the supply of plant nutrients that will be made available to them.

In research leading to the development of new hybrids or varieties, as well as in the use of these new lines in the farming enterprise, pest control measures must be adopted whenever necessary. Frequently, greater vegetative growth results from the use of these newer materials under higher fertility conditions, and insects, disease, and weeds may be encouraged. If such pests are not controlled, many otherwise promising high-yielding materials will be discarded in a research program or fail to perform adequately on the farm. This is another example of the limiting factor concept.

VARIETY-FERTILITY INTERACTIONS. The differential response of various crops to applied plant nutrients has been demonstrated on numerous occasions. In general, varieties that have a small range of adaptation tend to show significant variety-fertilizer interactions, whereas

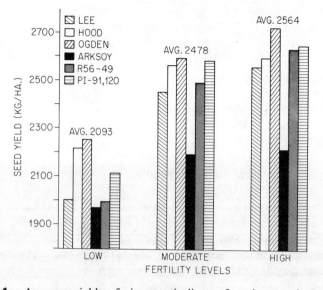

Figure 2–1. Average yields of six genetic lines of soybeans when grown at three levels of soil fertility for six years, Stuttgart, Arkansas. [Caviness and Hardy, *Agron. J.*, **62**:236 (1970).]

those with a wide range of adaptation do not. As early as 1922 the Tennessee Experiment Station was recommending varieties of corn on the basis of the fertility level of the soil. The varieties selected for poor soils were entirely different from those suggested for the more fertile soils.

In the United States the present trend is to supply adequate plant nutrients on the low-fertility soils. The recommendations for varieties or hybrids thus do not need to be made on the basis of the fertility level of the soil but rather on their ability to withstand insects, diseases, or unfavorable conditions of moisture or temperature. Soil fertility need no longer be a limiting factor.

IMPORTANCE OF PROGRESSIVE RESEARCH IN PLANT GENETICS. The genetic constitution of a given plant species limits the extent to which that plant may develop. No environmental conditions, no matter how favorable, can further extend these limits. It is imperative that there be a forward-moving plant-breeding program that will produce new varieties or hybrids capable of achieving maximum growth under specified environmental conditions. What has been done in the area of plant breeding and genetics is amply attested to by the crop production records kept by the better growers in this and other countries. What can be done will be determined by tomorrow's needs and the imagination and skill of geneticists and plant breeders the world over.

Environmental Factors. Environment is defined as the *aggregate of all the external conditions and influences affecting the life and develop-*

ment of an organism. Among the environmental factors known to influence plant growth, the following are probably the most important:

1. Temperature.
2. Moisture supply.
3. Radiant energy.
4. Composition of the atmosphere.
5. Gas content of the soil.
6. Soil reaction.
7. Biotic factors.
8. Supply of mineral nutrient elements.

Many environmental factors do not behave independently. An example is the inverse relation that exists between soil air and soil moisture or between the content of oxygen and carbon dioxide in the soil atmosphere. As the soil moisture increases, the soil air decreases, and as the carbon dioxide content of the soil air increases, the oxygen content decreases.

Another example is the relation between the diffusion rate of oxygen in the soil and soil temperature. It will be shown in a subsequent section of this chapter that the partial pressure of oxygen in the root environment is extremely important to the growth of plants. The maintenance of this pressure is related to the diffusion rate of oxygen to the root surface which in turn is influenced by the soil temperature.

These examples are offered simply to illustrate the dependence on other factors of so-called independent variables of plant environment. Even though these factors are not independent of one another, they will be separately treated for simplicity of discussion.

TEMPERATURE. Temperature is a measure of the intensity of heat. Physicists consider that the temperature of our universe ranges from a low of $-273°C$. to a high of several million degrees near the center of the sun. In terms of biological life as we know it, this is an almost unbelievably wide range. The limit of survival of those living organisms on this planet has generally been reported to be between -35 and $75°C$. The range of growth for most agricultural plants, however, is usually much narrower—perhaps between 15 and $40°C$. At temperatures much below or above these limits growth decreases rapidly. Consequently, the range of temperature in which earth life may continue is startlingly small when contrasted with the known range of temperatures.

Temperature directly affects the plant functions of photosynthesis, respiration, cell-wall permeability, absorption of water and nutrients, transpiration, enzyme activity, and protein coagulation. This influence is reflected in the growth of the plant.

The effect of temperature on photosynthesis is complex and differs with plants of various species as well as with the carbon dioxide content of the atmosphere, the intensity of light, and the duration of light of a given intensity. The present consensus among physiologists is that if light is limiting temperature has little effect on photosynthesis rate, but if carbon dioxide is limiting and light intensity is not, photosynthesis is increased by an increase in temperature. The complexity of these relationships is illustrated by the data arranged graphically in Figure 2-2.

Neither space nor purpose permits a more thorough treatment of this topic, one that is more appropriately treated in a course on plant physiology. However, because temperature is a factor related to photosynthesis and under certain conditions may limit crop responses to applied fertilizer, its bearing on the topic of soil fertility is of no small importance.

Respiration is also affected by changes in temperature. Barring a discussion of the physiological aspects of this process, it should be stated that, in general, respiration takes place more slowly at low temperatures and increases as the temperature rises. At very high temperatures the rate of respiration is initially great but is not maintained. After a few hours at elevated temperatures, respiration rates for at least some plants drop off rather rapidly.

For many crop plants of the temperate zone the temperature optimum for photosynthesis is lower than that for respiration. This has been

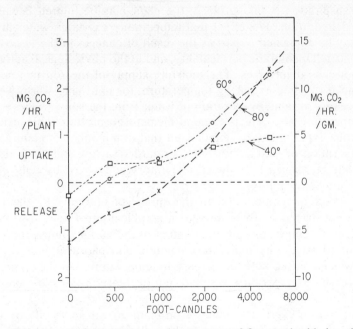

Figure 2-2. Net carbon dioxide exchange rates of five-week-old rice plants as related to temperature and light intensity. [Ormrod, *Agron. J.*, **53**:94 (1961).]

suggested as one reason for the higher yields of starchy crops, such as corn and potatoes, in cool climates as contrasted with the yield of these crops in warmer regions. It must be emphasized that the genetic constitution of a plant will determine its adaptability to climate, but for a species of given adaptation the general relation between the climate and type of crop may be expected to hold. It is possible that under conditions of prolonged temperatures above the optimum, a plant may literally suffer from starvation simply because respiration is taking place more rapidly than photosynthesis.

Transpiration, or the loss of water vapor from the stomata of leaves, is influenced by temperature. Transpiration rates as a rule are low at low temperatures and increase with rising temperatures. Under conditions of excessive transpiration water losses may exceed the water intake by the plant and wilting soon follows.

The absorption of water by plant roots is affected by temperature. Again, the influence of temperature is modified by species, but with a number of plants adapted to conditions of the temperate zone, absorption increases with a rise in temperature of the rooting medium from 0°C. to about 60 or 70°C. Above this point there is a leveling off of the rate of absorption.

Low soil temperature may adversely affect the growth of plants by its effect on the absorption of water. If soil temperatures are low, yet excessive transpiration is taking place, the plant may be injured because of tissue dehydration. The effect that temperature exerts on water absorption may be explained in part as the result of changes in the viscosity of water, in cell membrane permeability, and in the physiological activity of the root cells themselves. The moisture supply of the soil may also be influenced to some extent by temperature, for unusually warm weather produces more rapid evaporation of water from the soil surface.

Temperature also affects mineral element absorption. The results of numerous experiments have indicated that in a number of plant species the absorption of solutes by roots is retarded at low soil temperatures. This may be caused by reduced respiratory activity or by reduced cell membrane permeability.

The effect of temperature on the uptake of nutrients by the potato plant is illustrated by the data shown graphically in Figure 2-3. Notice the effect of temperature on the content of the various nutrients in both tops and roots. For example, the content of phosphorus in both tops and roots was increased with an increase in temperature. On the other hand, an increase in temperature resulted in a decrease in the root content of potassium.

Temperature exerts its influence on plant growth indirectly by its effect on the microbial population of the soil. The activity of the nitrobacteria as well as of most heterotrophic organisms increases with a rise in

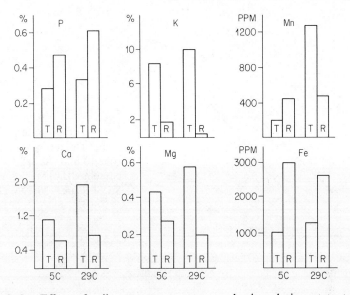

Figure 2–3. Effect of soil temperature on several minerals in potato tops (T) and roots (R). [Epstein, *Agron. J.*, **63**:664 (1971).]

temperature. The soil pH may change with temperature, which, in turn, may affect plant growth. It has been observed that the soil pH increases in winter and decreases in summer and is generally considered to be related to the activities of microorganisms, since microbial activity is accompanied by a release of carbon dioxide which combines with water to form carbonic and other acids. In soils that are only slightly acid this small change in pH may influence the availability of such micronutrient elements as manganese, zinc, or iron.

Numerous studies on the direct relationship between yield or dry-matter production and temperature have been made. This relation varies for different plant species and the effect of temperature may also be modified by soil type. This is well illustrated by the data shown graphically in Figure 2-4.

These data emphasize once again the importance of the limiting factor concept stated earlier in this chapter. The temperature-plant function relations illustrated here point up the need for a working knowledge of this concept, for the planting of a crop or variety not adapted to prevailing temperatures will result in lowered yields and a lower return from inputs of labor and capital.

Temperature may also alter the composition of the soil air; this again is the result of increases or decreases in the activity of microorganisms. When the activity of the micropopulation is great, there will be higher partial pressure of the carbon dioxide of the soil atmosphere as the ox-

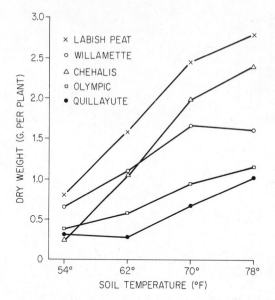

Figure 2-4. Effects of soil temperature on dry weights of bean plants grown in five soils (0 and 70 pp2m-P rates combined). [Mack et al., *SSSA Proc.*, **30**:236 (1966).]

ygen content decreases. Under conditions restricting the diffusion of gases into and out of the soil, a decrease in the O_2 pressure resulting from such activity might influence the rate of respiration of the plant roots, hence their ability to absorb nutrients.

Practical application has been made of the plant growth-temperature relationship to develop what is termed the heat unit concept. Heat units have been described in various ways: degree days, optimum days, and effective degrees. All of the terms express the amount of heat energy that has been absorbed by the soil over a given period of time. The number of units required to bring a crop to maturity (or some particular stage of growth) has been determined for various crop plants. Some commercial farmers, particularly the growers of canning and fresh-frozen crops, make use of this technique for determining planting and harvesting dates with a degree of precision never before obtainable. Use of this technique has considerably improved the efficiency of canning and freezing operations, for it permits planting dates to be staggered so that harvests will provide a steady flow of high quality produce to the canner or freezer. For a more complete discussion of this phase of temperature-plant growth relationship the reader is referred to Katz, and Gilmore and Rogers in the suggested reading list at the end of this chapter.

MOISTURE SUPPLY. The growth of many plants is proportional to the amounts of water present, for growth is restricted both at very low and very high levels of soil moisture. Water is required by plants for the manufacture of carbohydrates, to maintain hydration of protoplasm, and

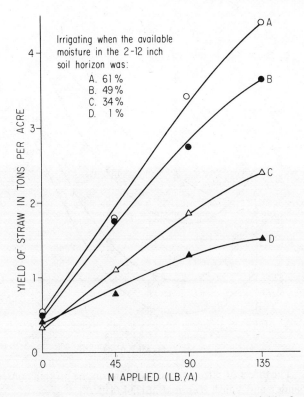

Figure 2–5. The effect of nitrogen and moisture on the yield of wheat straw. [Fernandez and Laird, *Agron. J.*, **51**:35 (1959).]

as a vehicle for the translocation of foods and mineral elements. Internal moisture stress causes reduction both in cell division and cell elongation, hence in growth.

The effect of several levels of soil moisture and increasing rates of applied nitrogen on the yield of wheat straw is shown in Figure 2-5. At any given level of applied nitrogen an increase in the amount of available water raises the yield of straw, which is greatest at the highest rate of applied nitrogen. Conversely, an increase in the rate of nitrogen at any given level of applied water increases the yield of wheat, which is greatest at the highest level of applied water. This illustrates very nicely the limiting effect that too little moisture can have on plant responses to applied fertilizer. It is necessary also to apply sufficient fertilizer to make the greatest use of available water. This can be seen by comparing the yield of moisture treatments A and B at the 45-lb. rate of applied nitrogen.

Yield is not the only plant property affected by soil moisture. Protein content of grain is frequently influenced by the degree of available water.

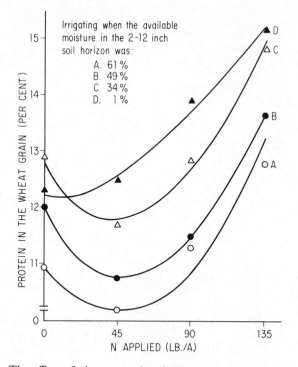

Figure 2–6. The effect of nitrogen and moisture on the protein content of wheat grain. [Fernandez and Laird, *Agron. J.,* **51**:34 (1959).]

Higher percentages of protein are generally associated with lower levels of soil moisture, as described by the curves in Figure 2-6.

Cotton is another important crop that responds to irrigation. Most of the irrigated cotton in the United States is grown in California, Arizona, and Texas, where irrigation, high fertility, and excellent management practices have been combined to produce perhaps the highest cotton yields in this country. However, it has been found that even in the humid regions of the South, the original cotton-producing center in the U.S., irrigation is profitable. It makes possible a greater use of fertilizer in an area that is generally characterized by soils of low fertility. The impact of the effect of water and high rates of applied fertilizer nitrogen on the yield of lint cotton in Alabama is illustrated in Figure 2-7.

Soil moisture level also has a pronounced effect on the uptake of plant nutrients. As a general rule, there is an increase in the uptake of cations and anions as soil moisture tension is decreased from the permanent wilting percentage to field capacity. When the pores become flooded with water, however, root respiration is affected and ion uptake is decreased.

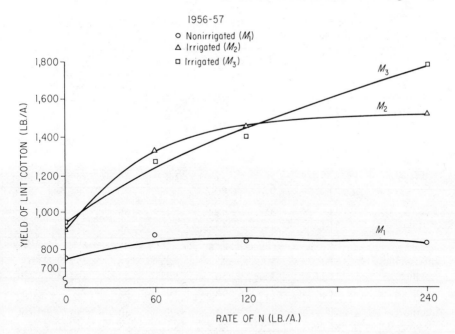

Figure 2–7. The effect of moisture and nitrogen levels on the yield of lint cotton. [Scarsbrook et al., *Agron. J.*, **51**:719 (1959).]

The effect of decreasing soil moisture tension on the uptake of nutrients by several forage plants is given in Table 2-1.

An increasing supply of moisture permits better nutrient uptake by crops. When moisture supplies are adequate, an improvement in nutrient supply increases and water-use efficiency of plants. Water-use efficiency, or WUE, is the amount of dry matter that can be produced from a given quantity of water. It is sometimes expressed as pounds of water required to produce a pound of dry matter. It is also described as the weight of dry matter derived from an acre inch of water. The effect of applied phosphate fertilizer on the WUE of alfalfa at three levels of soil moisture is illustrated by the data in Table 2-2. In this table Y is the yield in total tons per acre for the three-year period and ET is the evapotranspiration measured in inches. Y/ET in tons per acre inch is a measure of the WUE. It is obvious that under all three moisture regimes a rise in the rate of applied phosphate caused an increase in the WUE by alfalfa.

Phosphorus is a relatively immobile element in the soil and does not move with the soil water. If crops are to be grown under nonirrigated conditions and there is considerable drying of the soil profile, placement of the phosphorus fertilizer in a soil zone that is moist will enhance its

TABLE 2-1. Relative Uptake of Ions by Forage Plants (Third Harvest) Grown at Four Soil Moisture Levels*

Plant	Moisture level†	Relative uptake with moisture level M-3 = 100					
		N	P	K	Ca	Mg	S
Atlantic	M-1	55	44	52	50	62	60
alfalfa	M-U	70	59	72	59	62	60
	M-2	92	84	82	90	100	80
African	M-1	72	53	72	74	86	60
alfalfa	M-U	55	52	60	48	57	60
	M-2	92	90	83	80	86	80
Red	M-1	58	49	54	59	64	50
clover	M-U	74	63	77	76	79	72
	M-2	91	85	94	98	100	100
Intermediate	M-1	30	27	28	26	27	25
white	M-U	53	50	50	49	54	50
clover	M-2	72	68	70	72	72	75
Ladino	M-1	40	34	45	40	50	50
clover	M-U	72	66	83	65	66	75
	M-2	94	89	94	82	83	75
Fescue	M-1	74	52	71	64	61	67
	M-U	69	59	76	64	61	67
	M-2	98	94	107	109	92	100
Reed	M-1	105	70	95	100	80	100
canary-	M-U	94	73	88	86	80	100
grass	M-2	127	108	127	128	110	133
Orchard-	M-1	82	54	70	78	75	75
grass	M-U	102	78	90	100	88	100
	M-2	113	104	109	100	100	125

* Kilmer et al., *Agron. J.*, **52**:284 (1960).
† Inches of water supplied were M-1 = 2.64, M-U = 3.99, M-2 = 5.75 and M-3 = 7.24.

uptake. It is obvious, however, that there are limits to the placement of phosphate fertilizer. Fertilizer placement is discussed at length in Chapter 13.

In the arid regions in which irrigation is practiced moisture supply is assured and is conveniently controlled. Because of this control there is little uncertainty about responses to fertilizer application because of too little or too much water. In humid or subhumid regions the uncertain pattern of rainfall distribution makes the use of a given amount of fertilizer less profitable in some years than in others. If a grower could determine the probability of getting a profitable return from a given fertilizer

TABLE 2–2. Yield, Evapotranspiration, and Water-Use Efficiency for Alfalfa as Affected by Irrigation Treatment and Phosphate Fertilization*

| | Irrigation treatment | | | | | |
| | Dry | | Medium | | Wet | |
P_2O_5 applied (lb./A.)	Y† (tons/A.)	Y/ET† (tons/A.-in.)	Y (tons/A.)	Y/ET (tons/A.-in.)	Y (tons/A.)	Y/ET (tons/A.-in.)
100	25.77	0.120	28.74	0.129	33.29	0.130
200	28.62	0.133	30.73	0.138	37.72	0.148
400	34.52	0.160	37.65	0.169	44.79	0.175
600	34.94	0.162	40.57	0.182	47.20	0.185
ET (in.)	215.3		223.3		255.3	

* Viets, *Advan. Agron.*, **14**:244 (1962).
† Y = yield, tons per acre.
‡ ET = evapotranspiration, inches of water.

investment, he would be in a better position to plan his cropping program for maximum net profit. It is necessary first that crop responses to fertilizer inputs under different moisture regimes be known. It is also necessary to know the frequency of drought occurrence before an expression can be developed to make such input-output calculations possible. Workers at the Tennessee Experiment Station have addressed themselves to this problem. By using crop response data to inputs of nitrogen applied under different moisture regimes they have developed a quadratic equation which relates expected yields of corn to the rate of nitrogen fertilizer applied and to a drought index value.

$$Y = b_0 + b_1 N + b_2 N^2 + b_3 D + b_4 D^2 + b_5 ND \qquad (1)$$

where Y is the yield, N the pounds of applied nitrogen per acre, D, the drought days or drought index, and $b_0 \cdots b_5$, the constants. The yields corresponding to various inputs of N and D are calculated. On the basis of their knowledge of drought occurrence in the area in which the yield data were obtained, these workers developed tables of probabilities for obtaining a certain yield when a given amount of nitrogen fertilizer was applied. Table 2-3 is an example. These data may be converted to net profit per acre when the value of the crop and production costs are known.

Use of this table can be illustrated in the following example. Assume that a farmer wants to realize at least 88 bu. of corn, seven years out of ten, in the Johnsonville, Tennessee, area. He can do this by using 100 lb. of nitrogen with the expectation that on the average his yield will be 88 bu. or better seven years out of ten.

Soil moisture level also influences plant growth indirectly by its effect

TABLE 2-3.   Estimated Yields of Corn at Several Rates of Nitrogen Application and Different Drought Conditions with Equal Probability of One in Ten at Johnsonville, Tennessee*

Nitrogen (lb./A.)	Expected average drought index values with equal probability of 1 in 10									
	103	74	59	49	40	35	32	27	12	−5
0	31	49	58	64	69	72	74	77	86	96
25	36	55	65	72	78	81	83	88	96	108
50	39	60	71	78	85	89	91	95	106	118
75	40	64	76	84	91	95	98	102	114	128
100	41	66	79	88	96	101	103	108	121	136
125	40	68	82	91	100	105	107	112	126	142
150	38	68	83	93	102	107	110	115	131	148
175	35	66	83	94	103	109	112	118	134	152
200	30	64	81	93	103	109	113	118	136	155

* Parks and Knetsch, *Agron. J.*, **51**:364 (1959).

on the behavior of soil microorganisms. At extremely low or extremely high moisture levels the activity of the nitrifying organisms is inhibited, with the result that plants may have at their disposal a reduced supply of available nitrogen.

The points discussed are of particular importance in planning fertilizer use in any balanced farm program, for the application of fertilizer at rates inconsistent with the supply of available moisture is unwise and uneconomical, a consequence of the principle of limiting factors enunciated so frequently in this chapter.

RADIANT ENERGY.   Radiant energy is a significant factor in plant growth and development. The quality, intensity, and duration of light are all important. Studies have been made of the effect of light quality on plant growth, but such experiments are difficult to conduct because of the necessity for controlling simultaneously the wavelength and the intensity of the radiation. Even though the results of these studies have suggested that the full spectrum of sunlight is generally most satisfactory for plant growth, light quality is also an influence. The impact of this variable has not been studied to the same extent as the effect of light intensity and photoperiod. Even though light quality is known to affect plant growth, it is not likely in the foreseeable future that this factor can be controlled on a large-scale field basis, though it may be quite feasible for small acreages and for specialty crops with high acre value and in areas in which the production of such crops is not too extensive. An example of the effect of light quality on the relative growth of two alfalfa varieties is given in Figure 2-8.

The intensity of light as a factor in plant development has also been investigated. It has been shown that most plants are generally able to

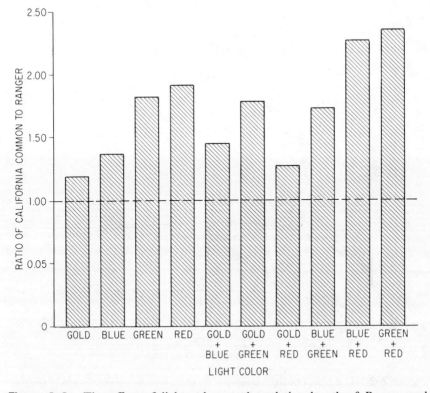

Figure 2–8. The effect of light color on the relative length of Ranger and California common alfalfa stems grown with photoperiods of eight hours. The total length of the bars represents the length of California common alfalfa. The length of the Ranger alfalfa is equal to 1.00. [Nittler et al., *Agron. J.*, **51**:728 (1959).]

make good growth at light intensities of less than full daylight. However, plants do differ in their response to light of varying intensity, as illustrated by the curves shown in Figure 2-9.

Note that the dogwood (*Cornus florida*) was the least responsive to increasing radiation intensities. This species is found in the understory of deciduous or mixed deciduous-coniferous forests of the eastern states, especially on the Appalachian plateau, a habitat in which the light intensity is considerably less than that of full daylight.

It has been shown that corn continues to respond to increasing insolation. This explains why corn breeders are developing hybrids with a more upright leaf to intercept more light.

Studies by Japanese workers, who used wheat as the test plant, indicate that the absorption of the ammonia form of nitrogen, sulfate, and water was increased with increasing light intensity but that absorption of

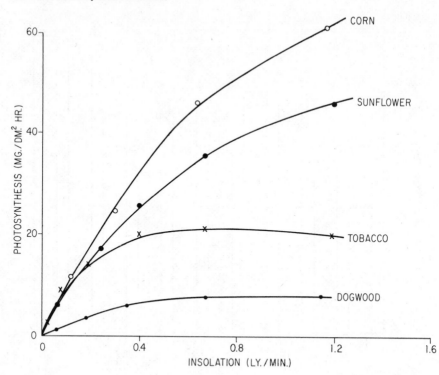

Figure 2–9. The effect of light intensity on photosynthesis in four plant species. [Waggoner et al., *Agron. J.*, **55**:38 (1963).]

Ca and Mg was little affected. The effect of light intensity on the uptake of phosphate and potassium was marked. It was also observed that oxygen uptake by the roots increased with increasing light intensity.

The influence of light intensity on crop growth is not simply an academic question. Its effect on yield and plant growth under field conditions and what can be done about it have not been completely evaluated. However, some information on this topic has been accumulated, and brief mention is made of one or two examples that deal with changes in light intensity which result from shading and its subsequent effect on plant growth. Inadequate plant population can limit crop yields. However, there is also a point beyond which an increase in plant population does not produce greater yields because of the competition among the plants for nutrients, water, and light. The effect of shading caused by increased plant population on the yield of several corn hybrids has recently been studied. It was found that some of the hybrids were shade-sensitive and that yields were reduced considerably more than those of the other hybrids which were similarly treated. The hybrids were arbitrarily classed as tolerant and intolerant, depending on their yielding

TABLE 2–4. The Effect of Shade on the Yield of Shade-Tolerant (T) and Intolerant (I) Corn Hybrids*

Variety	Classi-fication	Yield in sun (bu./A.)	Percentage of reduction of yield in shade
P158 × C103	I	93	45
HY × C103	I	113	40
Ohio M15	I	84	43
Connecticut 870	I	116	34
WF9 × 38–11	I	102	45
P334 × P367	T	91	22
P334 × C103	T	116	18
Pennsylvania 602	T	100	14
Eastern states 800	T	109	20
Pennsylvania 807	T	112	29
C103 × D14	T	98	11

* Stinson and Moss, *Agron. J.*, **52**:483 (1960).

ability under shaded conditions. A portion of the yield results is shown in Table 2-4.

Shading of crops can also occur when two different species are grown in a mixture, such as a grass-clover pasture. This problem has been studied by several workers in Australia. Balanced growth between the grass and clover is an important problem in good pasture management. It is known, for example, that excessive nitrogen fertilization of grass-clover pastures will drive the clover out of the mixture. Professor C. M. Donald and his associates in Australia studied this problem with the following conclusions:

1. Increased rates of nitrogen application give increased yields of grass.
2. Increased yields of grass give higher leaf areas of grass disposed above the clover-leaf canopy.
3. Higher leaf areas above the clover reduce the light density at the clover-leaf canopy.
4. Reduced light density at the clover leaf canopy causes reduced growth of clover.

The loss of clover from clover-grass pastures heavily fertilized with nitrogen is frequently observed in many areas of the United States. This has been attributed largely to competition for moisture and nutrients, particularly potassium, but Donald's work suggests the possible limiting effect of too low a light intensity.

Even though light quality and intensity may be of limited significance from the standpoint of field-grown crops, the duration of the light period is important. The behavior of the plant in relation to day length is termed photoperiodism. On the basis of their reaction to the photoperiod, plants have been classed as short-day, long-day, or indeterminate. Short-day plants are those that will flower only when the photoperiod is as short or shorter than some critical period of time. If the time of exposure to light is longer than this critical period, the plants will develop vegetatively without completing their reproductive cycle. Mammoth tobacco and coleus are examples.

Long-day plants are those that will bloom only if the period of time during which they are exposed to light is as long or longer than some critical period. If the plants are exposed to light for periods shorter than this critical time, they develop only vegetatively. Grains and clovers are members of this group.

Plants that flower and complete their reproductive cycle over a wide range of day lengths are classed as indeterminate. Cotton and buckwheat are representative of this group.

The phenomenon of photoperiodism was first described by Garner and Allard. They observed that a variety of burley tobacco known as Maryland Mammoth failed to bloom in the field during the summer months. When one of these plants was transferred to the greenhouse for the winter, however, it flowered profusely. Subsequent work showed this behavior to be related to the length of time that the plant was exposed to light. Since this discovery numerous other plants have been found to exhibit photoperiodism. From the standpoint of agriculture this property is of obvious importance, and the failure of crops to bloom or to set seed may frequently be related to day length.

Since the earlier work of Garner and Allard, the failure of some crops to flower in certain geographical areas has been traced to this property. Knowledge of the growth-photoperiod relationship has been a powerful tool in research and has led to the development of better-adapted varieties and hybrids for specific areas. The commercial exploitation of this knowledge is perhaps best illustrated by the production of chrysanthemums. This flower crop can be made to bloom at any predetermined time under greenhouse conditions simply by controlling the photoperiod. Suitable means for controlling illumination have been developed, and commercial florists can now depend on a uniform supply of these flowers.

COMPOSITION OF THE ATMOSPHERE. Carbon dioxide content of the atmosphere, usually about 0.03 per cent by volume, plays a role of great importance in the biological world. Through photosynthetic activity, the carbon dioxide is chemically bound in organic molecules in the plant.

Carbon dioxide is continually being returned to the atmosphere as a product of respiration of animals and plants. The decomposition of organic residues by the microorganisms is an important source of this gas, and one of the beneficial effects of manure or plant residues may be the result of carbon dioxide released.

Although the normal value of carbon dioxide is 0.03 per cent, the concentration may range from one half to several times this figure. In a field of corn on a quiet day the carbon dioxide content may become measurably less during daylight hours when there is a high rate of photosynthesis. Similarly, in a dense forest the content may drop considerably. In a closed greenhouse the carbon dioxide level may drop appreciably during the daylight hours, as indicated by the data given in Figure 2-10.

The impact on plant growth of increasing the carbon dioxide content of the atmosphere has been investigated and sometimes with conflicting results. Much of the early research on this problem was carried out by workers in northern Europe and later by investigators in the United States. In these early studies some workers reported impressive responses from increasing the carbon dioxide concentration, whereas

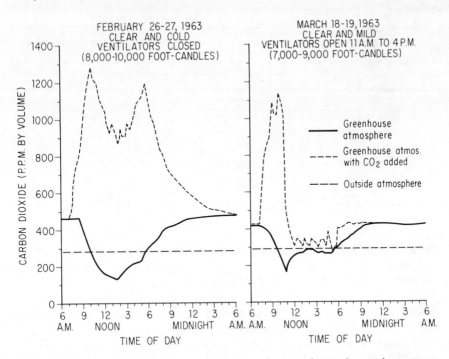

Figure 2–10. Carbon dioxide concentrations in experimental greenhouses, on the outside on a clear cold day with the ventilators closed, and on a clear mild day on which the vents were open after 10:00 A.M. [Wittwer et al., *Econ. Bot.,* **18:**334 (1964).]

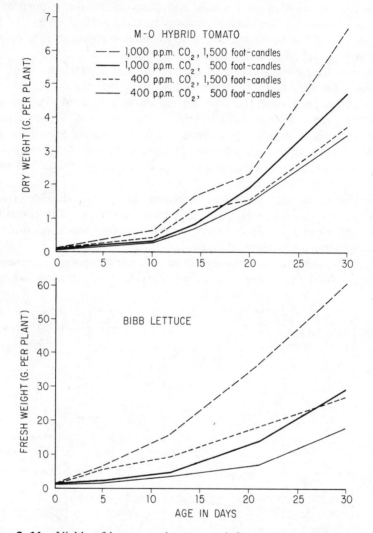

Figure 2–11. Yields of lettuce and tomatoes influenced by light intensity and carbon dioxide level of the atmosphere in controlled environment growth chambers. [Wittwer et al., *Econ. Bot.*, **18**:334 (1964).]

others observed depressions from such treatment. Subsequent studies showed that such growth depressions were caused by toxic impurities in the carbon dioxide gas rather than by the increased concentration of the carbon dioxide itself. It is generally recognized today that increasing carbon dioxide concentrations several times can have a pronounced positive impact on growth and crop yields.

In the United States some excellent work on this subject has been carried out by Wittwer and Robb at the University of Michigan. From the standpoint of present commercial application to food production, use of additional carbon dioxide seems to have its greatest potential in the growth of greenhouse crops. The effect of increasing light intensity and carbon dioxide content on the growth of tomatoes and lettuce is shown in Figure 2-11. Table 2-5 deals with the effect of increases of carbon dioxide on the yield of lettuce in terms of the weight of ten heads of the harvested product; its influence on the cumulative yield of tomatoes is analyzed in Figure 2-12. Not only was there an increase in the yield of tomatoes, but improvement was observed in the quality of the harvest. This is indicated by the data in Table 2-6, which gives the percentages of number one fruit harvested in areas treated with two levels of carbon dioxide.

Other plants, among which are cucumbers, flower crops, foliage crops (greens), peas, beans, and potatoes, have also responded to increased carbon dioxide concentrations.

When the level of carbon dioxide is increased, the light requirement may be also. Maximum carbon dioxide fixation at normal atmospheric levels occurs at relatively low light intensities of 1,500 to 2,000 foot-candles for some plants. Some recent studies, however, have shown that higher light intensities are required for certain crops. As Wittwer and Robb point out, the greatest potential benefits to be derived from carbon dioxide enrichment of a greenhouse atmosphere come when normal light intensities are in excess of the saturation value for a normal (0.03%) carbon dioxide atmosphere. In this connection it should also be noted that as carbon dioxide levels of the atmosphere are increased photosynthesis becomes more sensitive to temperature.

Although most of the work on increasing atmospheric carbon dioxide has been carried out under growth-chamber and greenhouse conditions,

TABLE 2-5.   Yields of Lettuce Varieties Influenced by Carbon Dioxide Added to the Greenhouse Atmosphere*

Lettuce variety	Yields† in pounds per ten heads	
	$-CO_2$	$+CO_2$‡
Bibb	1.2	1.9
Cheshunt No. 5B	1.6	2.8
Grand Rapids H-54	1.3	2.6
Mean (all varieties)	1.4	2.4

* Wittwer et al., *Econ. Botany*, **18**:334 (1964).
† Yields of each variety, and the mean of all varieties, significantly greater at $+CO_2$ (odds 99:1).
‡ Levels of carbon dioxide ranged from 125–500 ppm. in $-CO_2$ and from 800 to 2000 ppm. in $+CO_2$ during the daylight hours, except when ventilators were open.

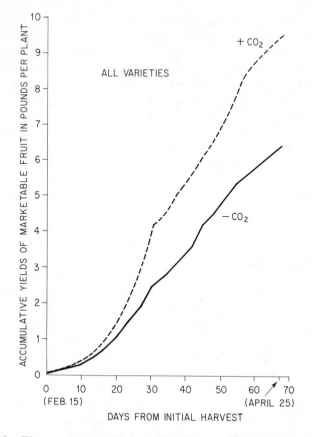

Figure 2–12. The mean accumulative yields of twenty-two varieties and selections of tomatoes subjected to 125 to 500 ppm. ($-CO_2$) and 800 to 2000 ppm. ($+CO_2$) of carbon dioxide during daylight hours in a greenhouse atmosphere. [Wittwer et al., *Econ. Bot.*, **18**:334 (1964).]

some research has been carried out under field conditions in the southern United States by workers at the USDA. A technique was developed for metering CO_2 into a field plot of cotton so that a concentration of this gas of 450 to 500 ppm. was maintained at three-fourths plant height. The CO_2 profile of the atmosphere in the field is shown in Figure 2-13. This technique was found to increase the net production of photosynthate by about 35 per cent.

The foregoing discussion illustrates that plant growth can be increased by the supplemental use of CO_2. Its use under greenhouse conditions appears quite feasible, but whether its use under field conditions can be developed on a commercial scale remains to be seen.

The quality of the atmosphere surrounding aboveground parts of plants may under certain conditions influence growth. Certain gases, such as sulfur dioxide (SO_2), carbon monoxide (CO), and hydrofluoric

TABLE 2–6. Grade of Fruit of Michigan-Ohio Hybrid Tomato Influenced by Carbon Dioxide Added to the Greenhouse Atmosphere[*]

Date of Harvest, 1963	Percentage of U.S. No. 1 fruit	
	$-CO_2$	$+CO_2$
March 15	36	54
March 18	44	48
March 22	62	72
March 25	63	73
March 29	68	86
April 1	69	75
April 4	78	86
April 8	85	89
April 11	90	88
April 15	88	98
April 18	86	98
April 22	97	99
April 25	95	99
May 2	84	93

[*] Wittwer et al., *Econ. Botany*, **18**:334 (1964).

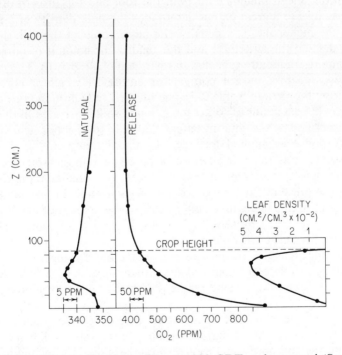

Figure 2–13. Carbon dioxide profiles at 1030 CDT under natural (September 16) and under release conditions (September 18, 1969), and leaf area density vs. height. [Harper et al., *Agron. J.*, **65**:7 (1973).]

acid (HF), when released into the air in sufficient quantities, are toxic to plants. Though the exception rather than the rule, isolated cases of injury from these gases have been reported.

What has not received adequate attention is that, while SO_2 in abnormally high concentrations is injurious to plants, at lower concentrations it is taken in through the leaves and fills a portion of the plants need for sulfur. Studies carried out in Germany have shown that virtually all of the plants sulfur requirement can be met using atmospherically applied SO_2. It is unlikely that this practice will ever be adopted, but it should be mentioned that reducing the atmospheric levels of SO_2 will in many instances mean that sulfur supplied to agricultural crops will have to be increased proportionately through fertilizer additions of this nutrient.

Injury to vegetation has also been reported in the vicinity of factories manufacturing metallic aluminum as well as those producing various phosphates. In both cases the fluorine released to the atmosphere as a result of these operations has been responsible for the damage. The destruction of the plants, however, may not be so important as the toxicity to grazing cattle.

SOIL STRUCTURE AND COMPOSITION OF SOIL AIR. The structure of soils, particularly those containing appreciable quantities of silt and clay, has a pronounced influence on both the root and top growth of plants. Soil structure to a great extent determines the bulk density of a soil. As a rule, the higher the bulk density, the more compact the soil, the more poorly defined the structure, and the smaller the amount of pore space. This is quite frequently reflected in restricted plant growth. The effect of fertilizer, moisture, and soil compaction on the dry weight of corn plants is given by the data in Table 2-7. The marked reduction in both top and root growth as a result of soil compaction is well illustrated in this table. Also described is the limiting effect that soil compaction can have on plant response to applied fertilizer. Compare treatments 1, 2, and 5.

Bulk density is really a measure of pore space in the soil; the higher the bulk density for a given textural class, the smaller the amount of pore space present. Pore space, of course, is occupied by air and water, the amount of one being inversely related to the amount of other. The effect of the air space present in a soil to the growth of tomato plants is shown by the curves in Figure 2-14.

High bulk densities inhibit the emergence of seedlings. Anthocyanin tends to accumulate in tomato plants grown on soils with bulk densities of 1.4 to 1.7, and these same plants tend to be high in proteins and low in sugar.

High bulk densities offer increased mechanical resistance to root penetration. They almost certainly influence the rate of diffusion of oxygen into the soil pores, and root respiration is directly related to a continuing and adequate supply of this gas. The effect of increasing the per-

Table 2–7.   The Effect of Moisture, Fertility Level, and Degree of Soil Compaction on the Growth of Corn Plants[*]

Treatments	Weight of tops (g.)	Weight of roots (g.)	Top : root ratio (g.)	Weight of total plant (g.)
Loose, wet, fertilized	39.4	14.8	1 : 0.38	54.2
Loose, wet, unfertilized	23.5	10.1	1 : 0.43	33.7
Loose, dry, fertilized	27.5	9.3	1 : 0.34	36.8
Loose, dry, unfertilized	20.3	9.3	1 : 0.46	29.6
Compact, wet, fertilized	16.0	6.5	1 : 0.40	22.5
Compact, wet, unfertilized	17.0	7.7	1 : 0.45	24.7
Compact, dry, fertilized	20.1	11.3	1 : 0.56	31.4
Compact, dry, unfertilized	19.3	9.9	1 : 0.51	29.2

[*] Bertrand and Kohnke, *SSSA Proc.*, **21**:137 (1957).

centage of oxygen on growth of snapdragon plants is illustrated in Figure 2-15.

Although bulk density is an important influence on plant growth, the effect of the amount of oxygen in the soil air is of equal or greater importance, as illustrated in Figure 2-16.

Air with the oxygen percentages indicated in the tables was maintained over the surface of each of the containers throughout the growth period of the snapdragons, which were the test plants. The soil in the containers had a bulk density of 1.2. The effect on root growth of the increasing oxygen percentage is obvious.

Under field conditions oxygen diffusion into the soil is determined largely by the moisture level of the soil if bulk density is not a limiting factor. On well-drained soils with good structure oxygen content is not likely to retard plant growth except for possible infrequent periods of flooding, during which it may become a serious consideration because of the impact that oxygen supply has on ion uptake. The increasing importance of oxygen with decreasing moisture tension is illustrated in Figure 2-17. Note that raising the oxygen supply at low moisture tensions results in a continued increase in ion uptake by the roots to an oxygen percentage of 8 to 10. At higher moisture tensions ion uptake is reduced

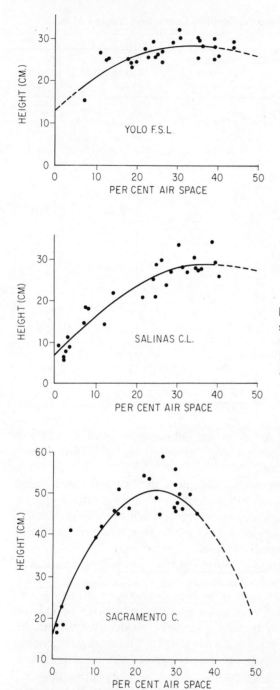

Figure 2–14. The effect of soil air space on the growth of tomato plants in three California soils. [Flocker et al., *SSSA Proc.,* **23:**191 (1959).]

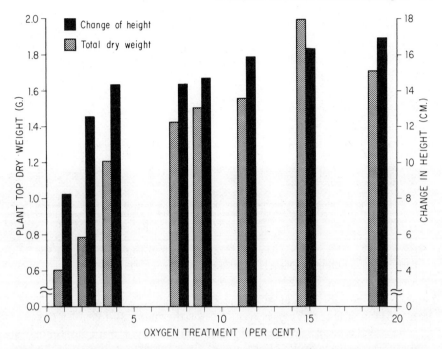

Figure 2–15. The effect of oxygen concentration of soil air on the growth of snapdragon plants. [Letey et al., *SSSA Proc.*, **25**:184 (1961).]

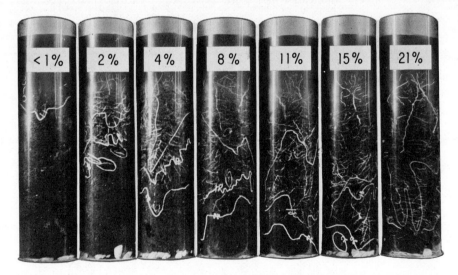

Figure 2–16. The effect of oxygen concentration maintained at the soil surface on root development. [Stolzy et al., *SSSA Proc.*, **25**:464 (1961).]

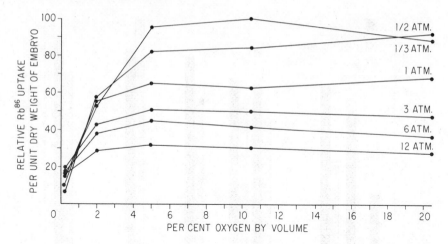

Figure 2–17. The effect of oxygen level and soil moisture tension on Rb^{86} uptake by corn seedlings. [Danielson et al., *SSSA Proc.*, **21**:5 (1957).]

and, quite possibly because of the limiting effect of the high moisture tension, much less oxygen is required for the maximum.

The oxygen supply at the root absorbing surface is critical. Hence not only the gross oxygen level of the soil air is important, but so is the rate at which oxygen diffuses through the soil to maintain an adequate partial pressure at the root surface. The influence of various rates of oxygen diffusion on the growth of pea plants on soils of three different fertility levels is nicely illustrated in Figure 2-18. At low rates of oxygen dif-

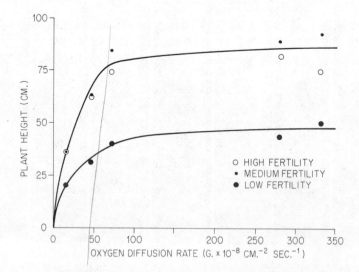

Figure 2–18. The effect of oxygen diffusion rate and fertility level on the growth of pea plants. [Cline and Erickson, *SSSA Proc.*, **23**:334 (1959).]

fusion small increases in diffusion rate have a much greater impact on corn grown on soils of medium or high fertility levels than the same increase on soils of low fertility level.

Agricultural plants differ widely in their sensitivity to soil oxygen supply. Paddy rice is grown under conditions of complete soil submergence. Tobacco is so sensitive to poor soil aeration that flooding a field for only a few hours causes serious damage or complete loss of the crop. Some pasture species are more tolerant of poorly aerated conditions than others, and species adapted to the conditions peculiar to the area should be selected. Good soil structure and aeration are imperative for maximum yields of most agricultural crops, and the limiting effect that inadequate root oxygen supply can have on growth must be considered in sound crop production programs.

SOIL REACTION. Soil reaction (soil acidity, pH) may affect plant development by its influence on the availability of certain elements required for growth. Examples are the reduced availability of phosphates in acid soils high in iron and aluminum and of manganese in high organic matter soils, which also have a high pH, and a decline in the availability of molybdenum with decreases in soil pH. Acid mineral soils are frequently high in aluminum, manganese, and iron, and excessive amounts of these elements, especially aluminum, are toxic to plants. When ammoniacal forms of nitrogen are applied to the surface of soils with pH values greater than 7.00, ammonia may be lost by volatilization. Expected responses to the applied nitrogen will not be obtained.

Certain soil-borne diseases are influenced by soil pH. Potato scab and black root rot of tobacco are favored by neutral-to-alkaline conditions, but both diseases can be almost completely controlled by lowering the soil pH to 5.5 or less.

The importance of soil acidity to crop growth and the availability of plant nutrients are treated in detail in Chapter 11. Soil acidity is a property of the greatest importance to the grower and one that is easily and economically altered.

BIOTIC FACTORS. The many biotic factors that can limit plant growth present a constant hazard to farming operations and pose a potential threat of reduced crop yields, if not of crop failure.

Heavier fertilization may encourage greater vegetative growth and better environmental conditions for certain disease organisms. The imbalance of nutrients available to plants may also be the reason for increased incidence of disease. Both conditions are illustrated by the data in Table 2-8. Leaf blight of corn was increased by liming the soil. However, the application of potassium decreased the incidence of the disease. When no lime or potassium was added, the addition of nitrogen tended to increase corn blight. Phosphate, however, tended to overcome the effect of nitrogen on the disease, particularly when no lime was added. Even though no yield data were presented in this particular case,

TABLE 2-8.   The Effect of Applications of Nitrogen, Phosphorus, Potassium, and Lime on the Incidence of Corn Leaf Blight*

Lb. K/A.	Lime	No phosphate (lb. N/A.)			Rock phosphate (lb. N/A.)			Superphosphate (lb. N/A.)		
		0	40	80	0	40	80	0	40	80
0	Yes	4.2†	3.2	4.0	4.2	4.5	4.2	4.8	4.8	4.2
	No	1.2	1.5	2.8	1.2	2.8	2.2	0.8	2.0	1.8
41.5	Yes	1.8	1.8	2.8	1.2	2.2	2.8	3.2	3.0	2.8
	No	0.8	1.0	1.5	1.5	1.5	1.8	1.2	2.0	1.5
83.0	Yes	1.5	2.0	2.0	1.0	2.5	2.2	1.5	3.2	2.2
	No	1.0	0.8	1.5	1.2	1.0	1.5	0.8	1.0	1.2

* Hooker et al., *Agron. J.*, 55:411 (1963).
† The leaf blight ratings are the average of two observations and range from 0.5 (slight infection) to 5.0 (severe infection).

plant diseases are almost always accompanied by yield decreases. Controlling disease is essential for maximum crop yields and the greatest response to applied fertilizer.

Certain pests may impose an added fertilizer requirement. Root knot and other nematodes, for example, attack the roots of certain crops and reduce absorption. It is then necessary to supply a greater concentration of nutrient elements in the soil to provide reasonable growth. These pests attack many species of crop plants and cause serious reductions in yield. Fortunately, they can be controlled both by crop rotation and by chemical treatment of the soil. One of numerous examples of the beneficial effects on crop yield of controlling nematodes is given in Table 2-9.

TABLE 2-9.   The Effect of Nematode Control on the Yield and Nutrient Uptake of Corn*

Treatment†		Yield (bu./A.)	Total pounds of elements in grain and stover per acre			
N rate (lb./A.)	Fumigation with EDB		P	K	Ca	Mg
50	5 gal./A.	35.1	6.00	16.11	5.55	5.03
	Unfumigated	25.6	3.29	9.67	3.93	3.17
100	5 gal./A.	48.4	6.00	18.58	7.57	6.73
	Unfumigated	38.6	5.53	14.74	5.89	4.88
100‡	5 gal./A.	49.5	8.41	21.03	7.32	7.91
	Unfumigated	40.5	6.38	16.32	5.99	5.62
LSD(p = 5%)		8.9				

* Wilcox et al., *Agron. J.*, 51:20 (1959).
† In addition to the nitrogen, the corn received 50 lb. P_2O_5 and 50 lb. K_2O/A. in the row.
‡ Plus 1½ tons/A. of dolomitic limestone.

Note the reduction in yield, particularly in the response to applied nitrogen, when no soil treatment was used.

Closely allied with disease is the problem of insects. Any infestation may seriously limit plant growth, and uncontrolled it may be responsible for the failure of a farming enterprise. Numerous examples can be cited. Heavier fertilization may encourage certain insects, such as the cotton boll weevil, by greater vegetative growth. Definite advances have been made in breeding insect-resistant strains of certain crops, and great strides have been made in the development of insecticides over the last two decades.

Weeds are another serious deterrent to efficient crop production, for they compete for moisture, nutrients, and in many instances light. Weeds can be controlled by adequate cultivation and by chemicals. Chemical weed control is rapidly becoming an established practice in many areas, particularly with high-value crops. The use of chemicals reduces root injury, which frequently results from mechanical cultivation, and lessens the structural damage caused by heavy equipment on many fine-textured soils.

The subjects discussed in this chapter are more appropriately treated in other texts. They could not be passed over without mention, however, *because of the limiting effect they may have on the efficient use of plant nutrients.*

GROWTH EXPRESSIONS

The manner in which plant growth responds to various inputs of plant nutrients has been the special study of a group of scientists who might be termed biometricians. These scientists have used various mathematical models to describe or define plant growth, and several of these concepts will be briefly presented in the following sections. If such mathematical models can be developed, then they can be useful in predicting what crop yields can be expected if the supply of elements available to the crop are known.

Growth as Related to Time. Whether growth is given as the increase in dry weight or in height of the plant, there is a fairly constant relationship between the measure of growth employed and time. The general pattern is one of initially small increases in size, followed by rapid growth, and then by a period during which the size of the plant increases slowly or not at all. This pattern is illustrated by the generalized growth curve shown in Figure 2-19.

Growth Related to the Factors Affecting It. Growth curves are helpful to an understanding of the general pattern of plant development. They indicate nothing, however, of the factors affecting growth, such as the supply of mineral nutrient elements, light, carbon dioxide, and water. The plant is a product both of its genetic constitution and its environment. The genetic pattern is a fixed quantity for a given plant and deter-

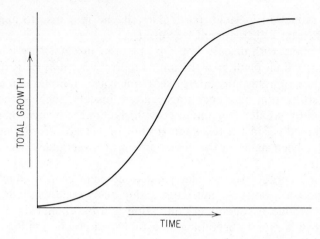

Figure 2-19. A generalized curve illustrating the growth pattern of an annual plant.

mines, so to speak, its potential for maximum growth under an environment favorable to its development. From the standpoint of that particular plant, attainment of this maximum development is dependent on a favorable environment. In other words, plant growth is a function of various environmental or growth factors, which may be considered as variables whose magnitude and combination determine the amount of growth that will be made. Symbolically, this may be expressed as

$$G = f(x_1, x_2, x_3 \ldots x_n)$$

where

$$G = \text{some measure of plant growth}$$

and

$$x_1, x_2, x_3 \cdots x_n = \text{the various growth factors}$$

Further, if all but one of the growth factors are present in adequate amounts, an increase in the quantity of this limiting factor will generally result in increases in plant growth. Symbolically,

$$G = f(x_1)_{x_2, x_3 \ldots x_n}$$

This, however, is not a simple linear relationship. It has been shown repeatedly that the addition of each successive increment of a growth factor results in an increase in growth that is less than that obtained from additions of the preceding increment.

Mitscherlich's Equation. The growth equation discussed in the preceding paragraph is a generalized expression relating growth to all of the factors involved. In 1909 E. A. Mitscherlich developed an equation which related growth to the supply of plant nutrients. He observed that when plants were supplied with adequate amounts of all but one nutrient their growth was proportional to the amount of this one limiting element which was supplied to the soil. Plant growth increased as more of this element was added, but not in direct proportion to the amount of the growth factor added. The increase in growth with each successive addition of the element in question was progressively smaller. Mitscherlich expressed this mathematically as

$$dy/dx = (A - y)C \qquad (2)$$

where dy is the increase in yield resulting from an increment of the growth factor dx, dx is an increment of the growth factor x, A is the maximum possible yield obtained by supplying all growth factors in optimum amounts, y is the yield obtained after any given quantity of the factor x has been applied, and C is a proportionality constant which depends on the nature of growth factor.

The value of C was found to be 0.122 for nitrogen, 0.60 for P_2O_5, and 0.40 for K_2O. Numerous workers since Mitscherlich have found that C in fact is not a constant value, but instead a value that varies rather widely for different crops grown under different climatic conditions. It is the lack of constancy of C that has caused much of the criticism of the Mitscherlich growth equation. The same critiscism may be applied to the Spillman equation, for it reduces itself to the same form.

Spillman's Equation. As is so often the case in scientific investigations, this same principle was independently developed several years later by W. J. Spillman. Spillman expressed the relation as

$$y = M(1 - R^x) \qquad (3)$$

where y is the amount of growth produced by a given quantity of the growth factor x, x is the quantity of the growth factor, M is the maximum yield possible when all growth factors are present in optimum amounts, and R is a constant.

Further work by Spillman showed that, though of a different form, his equation and that of Mitscherlich could both be reduced to

$$y = A(1 - 10^{-cx}) \qquad (4)$$

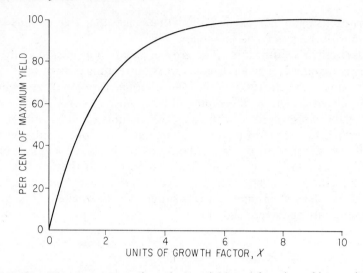

Figure 2-20. The percentage of maximum yield as a function of increasing additions of a growth factor *x*.

where *y* is the yield produced by a given quantity of *x*, *x* is the growth factor, *A* is the maximum yield possible, and *c* is a constant dependent on the nature of the growth factor. None of these expressions is conveniently handled as written, but they may be stated in the equivalent form

$$\log (A - y) = \log A - 0.301(x) \qquad (5)$$

The symbols used are the same as those in equation 4.

If the function is graphed, the curve obtained appears as shown in Figure 2-20.

Calculation of Relative Yields from Addition of Increasing Amounts of a Growth Factor. The value 0.301 replaces the constant *c* when yields are expressed on a relative basis of *A* = 100. When conventional units of yield are employed, the value of *c*, as already mentioned, varies with the growth factor.

If *A*, the maximum yield, is considered to be 100 per cent, Equation 5 reduces to

$$\log (100 - y) = \log 100 - 0.301(x) \qquad (6)$$

It is possible to determine the relative yield expected from the addition of a given number of units of *x*. It will be helpful if the student observes how these calculations are made.

If none of the growth factor is available, that is, *x* = 0, then *y* = 0; but suppose one unit of *x* is present. Then

$$\log (100 - y) = \log 100 - 0.301(1)$$
$$\log (100 - y) = 2 - 0.301$$
$$\log (100 - y) = 1.699$$
$$100 - y = 50$$
$$y = 50$$

the addition of one unit of the growth factor x results in a yield that is 50 per cent of the maximum.

Assume, however, that two units of the growth factor were present. In this instance

$$\log (100 - y) = \log 100 - 0.301(2)$$
$$\log (100 - y) = 2.000 - 0.602$$
$$\log (100 - y) = 1.398$$
$$100 - y = 25$$
$$y = 75$$

The same operation may be repeated until 10 units of the growth factor have been added. The result of such a series of calculations is given in tabular form.

Units of growth factor (x)	Yield (%)	Increase in yield (%)
0	0	–
1	50	50
2	75	25
3	87.5	12.5
4	93.75	6.25
5	96.88	3.125
6	98.44	1.562
7	99.22	0.781
8	99.61	0.390
9	99.80	0.195
10	99.90	0.098

It is obvious that the successive increases of a growth factor result in a yield increase that is 50 per cent of that resulting from addition of the preceding unit until a point is reached at which further increases, for all intents and purposes, are of no consequence.

The Baule Unit. Reference has been made so far only to units of a growth factor. Certainly this is a rather meaningless term, and it is necessary to understand its relation to familiar terms before proceeding with the discussion. This unit has been designated as a baule after the German mathematician who collaborated with Mitscherlich. Baule suggested that the unit of fertilizer, or any other growth factor, be taken as that amount necessary to produce a yield that is 50 per cent of the maximum possible. As almost everyone knows, plants require different absolute amounts of nitrogen, phosphorus, and potassium, but the amount in pounds of each required to produce a yield that is 50 per cent of the maximum possible is termed one baule unit. Obviously, according to this concept, one baule of a growth factor is equivalent to one baule of any other growth factor in terms of growth-promoting ability. The values of the baule unit in pounds per acre of N, P_2O_5 and K_2O are 223, 45, and 76, respectively, as calculated from the results of Mitscherlich's work.

Yield Increases in Response to More Than One Growth Factor. An extension of the baule unit concept has been developed for cases in which two or more growth factors are limiting. It has been shown that when all growth factors but one, x, were present in optimum amounts the addition of one unit of this factor x would produce a yield that was 50 per cent of the maximum possible. Suppose, however, that all except two growth factors, x_1 and x_2, were present in optimum amounts. If one baule of each is added, the yield obtained will not be 50 per cent but 50 per cent times 50 per cent, or 25 per cent of the maximum. If all growth factors but three were present in optimum amounts, the simultaneous addition of one unit of each of the three would result in a yield that was $50 \times 50 \times 50$, or 12.5 per cent of the maximum. This relationship is expressed by the general equation

$$y = A(1 - 10^{-0.301x_1})(1 - 10^{-0.301x_2})(1 - 10^{-0.301x_3}) \qquad (7)$$

in which x_1, x_2, and x_3 are quantities of the growth factors to be added. The equation is very simply handled by determining the percentage yields that correspond to given values for x_1, x_2, and x_3 and then multiplying these figures together.

It is of interest to note that the rule of decreasing increments applies even though all growth factors are not present in optimum amounts. This makes possible a calculation of the maximum possible yield under a given set of climatic and genetic conditions when fertilizers are added in increasing amounts.

It also makes possible the calculation of the greatest possible crop production from a limited amount of fertilizer, for the greatest increases in yield result from the addition of the first unit increment. The German people made extensive use of this concept during World War II. It is probably fair to say that the practical application of Mitscherlich's ideas was largely responsible for preventing a greater degree of starvation in Germany than actually occurred during the war years.

The Agrobiology of O. W. Wilcox. O. W. Wilcox has employed the

Mitscherlich equation to extend a branch of science that he calls quantitative agrobiology. This science, he states, "Comprises the general and specific *quantitative* relations between plants and the other factors of their growth and yield." Wilcox assumes that the Mitscherlich equation is valid, an opinion that is by no means shared by many agricultural scientists. He further develops from the Mitscherlich equation the inverse nitrogen:yield concept which says in effect that the yield of a crop is inversely proportional to its nitrogen content. Symbolically,

$$Y = \frac{k}{n} \tag{8}$$

where Y is the yield, n is the percentage of nitrogen in the crop, and k is a constant. To support this contention, Wilcox cites a number of examples of crop yields and nitrogen percentages that follow this general rule.

From the Mitscherlich equation, Wilcox further extends the inverse yield:nitrogen concept and evaluated the constant in Equation 8. He finds that the value of k is 318 lb./A. and states that this is the maximum amount of nitrogen that can be absorbed in one season by an annual crop growing on an acre of land. This presumably would occur with the production of a maximum yield which could happen only if none of the various growth factors was limiting. It is obvious that crop yields would increase ad infinitum if only the nitrogen content of the crop on a percentage basis could be made to decrease.

Although there are numerous examples reported in which the Mitscherlich-Baule-Wilcox equations fail to describe plant growth adequately as a function of plant nutrient inputs, it is felt that these concepts should be considered, for they represent an original attempt to develop a unified theory of plant growth as a function of mineral nutrition. The ideas of Mitscherlich, Baule, and Wilcox are not completely without foundation. Plant growth as a function of nutrient inputs *is* logarithmic and generally follows a pattern of diminishing increases, as expressed in the Mitscherlich equation. The growth of annual plants *does* tend to reach a maximum with increasing inputs of nutrients under a particular set of environmental conditions, and often the plants that produce the highest yield of dry matter have the lowest percentage of nitrogen in their tissues. However, it remains for posterity to determine whether a single expression may be developed that will universally predict the amount of growth that can be produced from the input of a given quantity of plant nutrients when environmental and genetic growth factors are adequately described.

An example of the use to which the Mitscherlich concept can be put in soil fertility research was developed by workers at the Tennessee

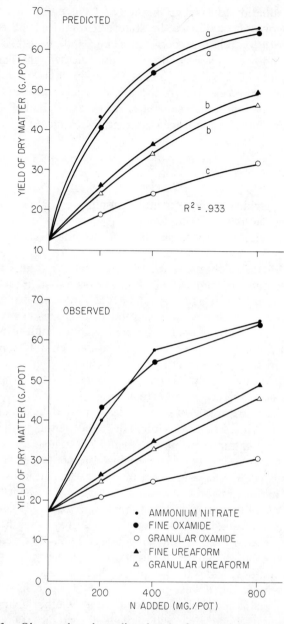

Figure 2–21. Observed and predicted corn forage yield as affected by rate, source, and granule size of fertilizer N. [Engelstad and Khasawnch, *Agron. J.,* **61**:473 (1969).]

Valley Authority at Muscle Shoals. Alabama. The effectiveness of various fertilizer materials as sources of nitrogen for corn was determined. The yield data obtained followed the Mitscherlich diminishing yield response pattern. The model so developed provided for a common limiting yield and a common intercept. It can be fitted by standard regression techniques, and the coefficients obtained can be readily compared by statistical techniques. A comparison of the observed and values predicted by this technique is shown in Figure 2-21.

Factorial Experiments and Regression Equations. Much fertilizer research has made use of what is known as the factorial experiment. In such studies the effect on crop yield of several levels of different input factors is evaluated simultaneously in one experiment. For example, the effect of several levels of nitrogen and phosphorus on the yield of a crop can be studied simultaneously in one experimental layout. If three levels of nitrogen and three levels of phosphorus are examined, the experiment is said to be a 3^2 factorial and requires nine treatments to evaluate the effect on yield of all combinations of nitrogen and phosphorus. The results are usually subjected to various statistical treatments. An equation, normally termed a regression equation, is developed in which the yield is functionally related to the inputs of the fertilizer variables.

Research workers at the Iowa and North Carolina Experiment Stations were among the first to make extensive use of this type of experiment, which is now popular in America. When climate, soil type, plant population, fertilizer placement, and other factors are uniform and constant, such studies help to predict fertilizer requirements, but the equations developed from them are not universally applicable. An example of this type of equation was given earlier in this chapter in the discussion dealing with the effect of moisture on plant growth. Such factorial studies are an outgrowth of the Mitscherlich concept, and their equations are generally similar in form to the Mitscherlich equation.

Bray's Nutrient Mobility Concept. A modification of the Mitscherlich-Baule-Spillman concept has been proposed by the late Dr. Roger Bray and his co-workers at the University of Illinois. In brief, it is claimed that crop yields obey the percentage sufficiency concept of Mitscherlich for such elements as phosphorus and potassium, which are relatively immobile in the soil. This concept, in turn, is based on Bray's nutrient mobility concept which states that

> as the mobility of a nutrient in the soil decreases, the amount of that nutrient needed in the soil to produce a maximum yield (the soil nutrient requirement) increases from a variable net value, determined principally by the magnitude of the yield and the optimum percentage composition of the crop, to an amount whose value tends to be a constant.

The magnitude of this constant is independent of the magnitude of the yield of the crop, provided that the kind of plant, planting pattern and rate, and fertility pattern remain constant and that similar soil and seasonal conditions prevail. He further states that for a mobile element such as nitrate nitrogen Liebig's law of the minimum best expresses the growth of a crop. Bray has modified the Mitscherlich equation to

$$\log (A - Y) = \log A - C_1 b - Cx \tag{9}$$

where A, Y, and x have the connotation already given, C_1 is a constant representing the efficiency of b for yields in which b represents the amount of an immobile but available form of nutrient, such as phosphorus or potassium, measurable by some suitable soil test, and C represents the efficiency factor for x, which is the added fertilizer form of the nutrient b.

On the basis of work done in Illinois, Bray has shown that the values for C_1 and C are specific and fairly constant over a wide area in the state, regardless of yield and season for each of the following crops: corn, wheat, and soybeans. The factors that will alter the values, however, are wide differences in soil type, plant population and planting patterns, and the form and distribution in the soil of the immobile nutrient element under study. Bray's concepts form a useful basis for making fertilizer recommendations in those situations in which the previously noted conditions are uniform. An excellent discussion of the use of this concept may be found in the paper by Vavra and Bray in the references listed at the end of this chapter.

Limited Applications of Growth Expressions.   It is obvious that there is a lack of agreement among the various concepts developed to describe the relation between plant growth and nutrient element input. Steenbjerg and Jakobsen of Denmark, in commenting on the variability among growth response curves, point out that "the constants in formulas are not constants because the variables in the formulas are not independent variables." Factors other than nutrient interactions obviously affect the shape of yield curves. They include other environmental factors which were discussed in the preceding sections. The change in shape and position of yield-plant nutrient input curves with changes in environmental conditions is of the greatest importance to the practical agriculturalist. The understanding of the interplay among these factors and their successful manipulation by the farm operator to bring about maximum crop production at a minimum production cost determines in the final analysis the degree of success of the farm operation.

The term *limiting growth factor* or more simply *limiting factor,* used frequently throughout this book, has been clearly illustrated by the variable nature of the response curves and surfaces previously discussed. If, for example, a crop has inadequate moisture, the application of a given amount of fertilizer will provide a lower yield than if moisture were adequate. Another example, and an important one, is the application of fertilizer to a crop growing on a soil that is too acid for maximum growth, regardless of the amount of fertilizer added. If lime is not applied, acidity becomes the limiting factor that keeps yield responses to the added fertilizer low and reduces the farmer's return on his investment. The importance to practical farm operations of the concept of a growth pattern and how it may be altered by various "limiting factors" cannot be overstated.

An adequate supply of mineral nutrient elements is required for maximum agricultural production, but alone they are no guarantee of a bumper crop because of the possible limiting effect of the numerous items that influence plant growth. The remaining chapters in this book deal largely with the role that mineral elements play in the production of agricultural crops.

Summary

1. Plant growth as a function of time, genetic make-up of the plant, and environmental factors was discussed. The importance of selecting crops that are genetically capable of making maximum utilization of the supply of available plant nutrients was pointed out.
2. The environmental factors were considered in relation to their effect on plant growth as well as their impact on limiting crop response to applied plant nutrients. The concept of limiting growth factors was discussed and the need for recognizing the importance of this concept in practical farming operations pointed out.
3. Growth of annual plants follows a well-defined pattern. Plant responses to environmental conditions, including the supply of plant nutrients, also follow a set pattern. When growth is plotted as a function of increasing amounts of applied nutrients, successive increments of fertilizer give successively smaller increases in plant growth. Such curves, known as response curves, have been studied by numerous investigators and various mathematical formulas have been developed to describe them.

Questions

1. Equations that express plant growth as a function of inputs of a plant nutrient differ from one another to varying degrees. To what do you attribute these differences?
2. Despite the differences among the various equations referred to in Question 1, they all seem to have one thing in common. What is it?
3. Crop yields have been increasing over the years because of improvements in tillage, varieties, pest control, fertilization, and so

on. From a *theoretical* standpoint, what do you consider to be the factor that will *ultimately* limit further increases in plant growth? From a *practical* standpoint, what do you think this factor will be?

4. Roger Bray of Illinois has had unusually good results in predicting crop responses, especially by corn, to applications of potassium and phosphate fertilizers from the equations and soil tests that he has developed. What are some of the factors that may have contributed to this success?

5. Plant yields in terms of forage, grain, or fruit are the criteria of the effectiveness of various fertilizer inputs. If one is interested in the effect of the imposed treatment on the plants ability to convert radiant energy to a usable form, are these yield figures in your opinion the best criteria of treatment effect? Why? (*Hint:* Compare the highest yields of soybeans with the highest yields of corn.) What criterion would you use?

6. What growth factor in the past has been frequently overlooked by plant breeders in developing new crop varieties?

7. According to the Mitscherlich-Baule concept, what percentage of maximim growth would be produced if only three plant nutrients were the factors limiting growth and they were supplied to the extent of 6, 4, and 2 baules, respectively?

8. Among environmental factors limiting crop response to applied nutrients, which is probably the most easily and inexpensively overcome?

9. What are some of the environmental factors affecting plant response to applied nutrients that are more easily overcome or controlled in the greenhouse than in the field? If you were planning to control these factors in a greenhouse operation set up for the commercial production of crops, what items would you consider before instituting control?

10. In your opinion could the limited or partial control of the carbon dioxide content of the aboveground atmosphere under field conditions be successfully undertaken? Under what conditions would you expect such control to be successful?

11. Of *all* of the growth factors limiting crop production, which is most easily corrected in commercial farming operations?

12. Study the various growth curves discussed in this chapter. Should a commercial grower always attempt to produce maximum yields? Why? At what point along the yield-fertilizer input curve do you feel that he should operate? Elaborate.

13. Why is soil structure so important in influencing crop responses to applied plant nutrients? Be specific. What can be done about soil structure? On which type of soil is structure the more important — loamy sands or silty clay loams? Why?

14. Light or radiant energy was listed as an environmental factor affecting growth and response to applied plant nutrients. In what ways does light influence growth under field conditions? Under controlled conditions? What can be done about the effect of light on growth?
15. Though not listed as such, man is the one biotic factor influencing crop growth to the greatest extent. Why is this statement made, especially in relation to commercial farming?
16. In the strictest sense, is it correct to refer to the various environmental factors as independent variables? Why?

Selected References

Adams, J. E., "Effect of soil temperature on grain sorghum growth and yield." *Agron. J.*, **54**:257 (1962).

Bertrand, A. R., and H. Kohnke, "Subsoil conditions and their effects on oxygen supply and the growth of corn roots." *SSSA Proc.*, **21**:135 (1957).

Black, C. A., and O. Kempthorne, "Willcox's agrobiology: I. Theory of the nitrogen constant 318." *Agron. J.*, **46**:303 (1954).

Black, C. A., and O. Kempthorne, "Willcox's agrobiology: II. Application of the nitrogen constant 318." *Agron. J.*, **46**:307 (1954).

Black, C. A., *Soil-Plant Relationships*. New York: Wiley, 1968.

Bray, R. H., "Confirmation of the nutrient mobility concept of soil-plant relationships." *Soil Sci.*, **95**:124 (1963).

Brown, D. M., "Soybean ecology: I. Development-temperature relationships from controlled environment studies." *Agron. J.*, **52**:493 (1960).

Brown, D. M., and L. J. Chapman, "Soybean ecology: II. Development-temperature-moisture relationships from field studies." *Agron. J.*, **52**:496 (1960).

Brown, D. A., G. A. Place, and J. V. Pettiet, "The effect of soil moisture upon cation exchange in soils and nutrient uptake by plants." *Trans. 7th Int. Congr. Soil Sci.*, **3**:443 (1960).

Caviness, C. E., and G. W. Hardy, "Response of six diverse genetic lines of soybeans to different levels of soil fertility." *Agron. J.*, **62**:236 (1970).

Connell, G. H., F. T. Bingham, J. C. Lingle, and M. J. Garber, "Yield and nutrient composition of tomatoes in relation to soil temperature, moisture, and phosphorus levels." *SSSA Proc.*, **27**:560 (1963).

Cline, R. H., and A. E. Erickson, "The effect of oxygen diffusion rate and applied fertilizer on the growth, yield, and chemical composition of peas." *SSSA Proc.*, **23**:333 (1959).

Danielson, R. E., and M. B. Russell, "Ion absorption by corn roots as influenced by moisture and aeration." *SSSA Proc.*, **21**:3 (1957).

Donald, C. M., "The interaction of competition for light and for nutrients." *Australian J. Agr. Res.*, **9**:421 (1958).

Engelstad, O. P., and F. E. Khasawnch, "Use of a concurrent Mitscherlich model in fertilizer evaluation." *Agron. J.*, **61**:473 (1969).

Epstein, E., "Effect of soil temperature on mineral element composition and morphology of the potato plant." *Agron. J.*, **63**:664 (1971).

Flocker, W. J., and R. C. Menary, "Some physiological responses in two tomato varieties associated with levels of soil bulk density." *Hilgardia*, **30**:101 (1960).

Flocker, W. J., and D. R. Nielson, "The absorption of nutrient elements by tomatoes associated with levels of bulk density." *SSSA Proc.,* **26:**183 (1962).

Flocker, W. J., J. A. Vomocil, and F. D. Howard, "Some growth responses of tomatoes to soil compaction." *SSSA Proc.,* **23:**188 (1959).

Gilmore, E. C., and J. S. Rogers, "Heat units as a method of measuring maturity in corn." *Agron. J.,* **50:**611 (1958).

Harper, L. A., D. N. Baker, J. E. Box, Jr., and J. D. Hesketh, "Carbon dioxide and the photosynthesis of field crops: A metered carbon dioxide release in cotton under field conditions." *Agron. J.,* **65:**7 (1973).

Hooker, A. L., P. E. Johnson, M. C. Shurtleff, and W. D. Pardee, "Soil fertility and northern corn leaf blight infection." *Agron. J.,* **55:**411 (1963).

Jones, M. B., C. M. McKell, and S. S. Winans, "Effect of soil temperature and nitrogen fertilization on the growth of soft chess (*Bromus mollis*) at two elevations." *Agron. J.,* **55:**44 (1963).

Katz, Y. H., "The relationship between heat unit accumulation and the planting and harvesting of canning peas." *Agron. J.,* **44:**74 (1952).

Kilmer, V. J., O. L. Bennett, J. F. Stahly, and D. R. Timmons, "Yield and mineral composition of eight forage species grown at four levels of soil moisture." *Agron. J.,* **52:**282 (1960).

Kramer, P. J., "Water stress and plant growth." *Agron. J.,* **55:**31 (1963).

Letey, J., and G. B. Blank, "Influence of environment on the vegetative growth of plants watered at various soil moisture suctions." *Agron. J.,* **53:**151 (1961).

Letey, J., O. R. Lunt, L. N. Stolzy, and T. E. Szuszkiewicz, "Plant growth, water use, and nutritional response to rhizosphere differentials of oxygen concentration." *SSSA Proc.,* **25:**183 (1961).

Letey, J., L. H. Stolzy, G. B. Blank, and O. R. Lunt, "Effect of temperature on oxygen diffusion rates and subsequent shoot growth, root growth, and mineral content of two plant species." *Soil Sci.,* **92:**314 (1961).

Letey, J., L. H. Stolzy, N. Valoras, and T. E. Szuszkiewicz, "Influence of oxygen diffusion rate on sunflower growth at various soil and air temperatures." *Agron. J.,* **54:**316 (1962).

Letey, J., L. H. Stolzy, N. Valoras, and T. E. Szuszkiewicz, "Influence of soil oxygen on growth and mineral content of barley." *Agron. J.,* **54:**538 (1962).

Mack, H. J., S. C. Fang, and S. B. Apple, Jr., "Response of snap beans (*Phaseolus vulgaris L.*) to soil temperature and phosphorus fertilization on five western Oregon soils." *SSSA Proc.,* **30:**236 (1966).

Nielsen, K. F., R. L. Halstead, A. J. McLean, S. J. Bourget, and R. M. Holmes, "The influence of soil temperature on the growth and mineral composition of corn, bromegrass, and potatoes." *SSSA Proc.,* **25:**369 (1961).

Nielsen, K. F., R. L. Halstead, A. J. McLean, R. M. Holmes, and S. J. Bourget, "The influence of soil temperature on the growth and mineral composition of oats." *Can. J. Soil Sci.,* **40:**255 (1960).

Nittler, L. W., and G. H. Gibbs, "The response of alfalfa varieties to photoperiod, color of light, and temperature." *Agron. J.,* **51:**727 (1959).

Ormrod, D. P., "Photosynthesis rates of young rice plants as affected by light intensity and temperature." *Agron. J.,* **53:**93 (1961).

Parks, W. L., and J. L. Knetsch, "Corn yields as influenced by nitrogen level and drought intensity." *Agron. J.,* **51:**363 (1959).

Parks, W. L., and J. L. Knetsch, "Utilizing drought days in evaluating irrigation and fertility response studies." *SSSA Proc.,* **24:**289 (1960).

Phillips, R. E., and D. Kirkham, "Soil compaction in the field and corn growth." *Agron. J.,* **54:**29 (1962).

Rhykerd, C. L., R. Langston, and G. O. Mott, "Influence of light on the foliar growth of alfalfa, red clover, and bird's-foot trefoil." *Agron. J.,* **51:**199 (1959).

Rhykerd, C. L., R. Langston, and G. O. Mott, "Effect of intensity and quantity of light on growth of alfalfa, red clover, and bird's-foot trefoil." *Agron. J.,* **52:**115 (1960).

Scarsbrook, C. E., O. L. Bennett, and R. W. Pearson, "The interaction of nitrogen and moisture on cotton yields and other characteristics." *Agron. J.,* **51:**718 (1959).

Steenbjerg, F., and S. T. Jakobsen, "Plant nutrition and yield curves." *Soil Sci.,* **95:**69 (1963).

Stern, W. R., and C. M. Donald, "Light relationships in grass-clover swards." *Australian J. Agr. Res.,* **13:**599 (1962).

Stern, W. R. and C. M. Donald, "The influence of leaf area and radiation on the growth of clover in swards." *Australian J. Agr. Res.,* **13:**615 (1962).

Stinson, H. T., Jr., and D. N. Moss, "Some effects of shade upon corn hybrids tolerant and intolerant of dense planting." *Agron. J.,* **52:**482 (1960).

Stolzy, L. H., J. Letey, T. E. Szuszkiewicz, and O. R. Lunt, "Root growth and diffusion rates as functions of oxygen concentration." *SSSA Proc.,* **25:**463 (1961).

Van der Paauw, F., "Critical remarks concerning the validity of the Mitscherlich effect law." *Plant Soil,* **4:**97 (1952).

Vavra, J. P., and R. H. Bray, "Yield and composition response of wheat to soluble phosphate drilled in the row." *Agron. J.* **15:**326 (1959).

Viets, F. G., Jr., "Fertilizers and the efficient use of water." *Advan. Agron.,* **14:**223 (1962).

Waggoner, P. E., D. N. Moss, and J. D. Hesketh, "Radiation in the plant environment and photosynthesis." *Agron. J.,* **55:**36 (1963).

Weiking, R. M., "Growth of ryegrass as influenced by temperature and solar radiation," *Agron. J.,* **55:**519 (1963).

White, W. C., and C. A. Black, "Willcox's agrobiology: III. The inverse yield-nitrogen law." *Agron. J.,* **46:**310 (1954).

Wiersum, L. K., "Uptake of nitrogen and phosphorus in relation to soil structure and nutrient mobility." *Plant Soil,* **16:**62 (1962).

Wilcox, G. E., J. P. Hollis, M. J. Fielding, L. D. Newsom, and D. A. Russel, "The effect of nematode control on the growth and nutrition of certain agronomic crops." *Agron. J.,* **51:**17 (1959).

Willcox, O. W., *The ABC of Agrobiology.* New York: Norton, 1937.

Wittwer, S. H., and W. Robb, "Carbon dioxide enrichment of greenhouse atmospheres for vegetable crop production." *Econ. Botany,* **18:**343 (1964).

Worley, R. E., R. E. Blaser, and G. W. Thomas, "Temperature effect on potassium uptake and respiration by warm and cool season grasses and legumes." *Crop Sci.,* **3:**13 (1963).

3. ELEMENTS REQUIRED IN PLANT NUTRITION

Protoplasm is common to plant and animal life. Animals, however, must rely on other animals or plants as a source of food so that they may continue to thrive. In the final analysis animals are completely dependent on the plant kingdom for life. The protoplasm of green plants, however, can exist and increase independently of any animal life. All that is required is a supply of water, carbon dioxide, and several mineral elements to make the green plant in light a completely self-sufficient organism. The raw materials that the plant consumes in the manufacture of its own tissues then assume a role of no small importance to persons engaged in the production of crop plants.

ESSENTIALITY OF ELEMENTS IN PLANT NUTRITION

A strict definition of essentiality is by no means simple, though attempts have been made to develop such a definition. Because the requirements for essentiality imposed by these definitions have frequently been found to be too rigid from a practical standpoint, a more useful definition of "essentiality" proposed by D. J. Nicholas of the Long Ashton Research Station is employed in this text. Nicholas has suggested that the term "functional or metabolism nutrient" be used to include any mineral element that functions in plant metabolism, whether or not its action is specific. This definition avoids the confusion that sometimes occurs when more rigid criteria of essentiality are imposed, which say, among other things, that an element is essential *only* if symptoms of its deficiency can be prevented or overcome by supplying that particular element. If this latter definition is adherred to, molybdenum would not be classed as essential for in some species of *Azotobacter* vanadium can substitute completely for Mo. In higher plants, chlorine has been shown to be necessary for growth and is generally classed as "essential." However, by the rigid definition it would not be so classed for bromine may substitute for the chlorine, though at higher concentrations.

One last illustration supporting the advantage of the looser definition of plant nutrient "essentiality" is the case of sodium. Using the rigid definition, this element would not be considered essential. However, it has been known for years to increase the yields of several crops, such as sugar beets, celery, table beets, and turnips. To the grower seeking to maximize the income from his farming operation from the production of these crops, sodium would most certainly have to be considered essential. It is in this light that the essentiality of plant nutrients is regarded in this text.

ELEMENTS REQUIRED IN PLANT NUTRITION

Carbon, hydrogen, oxygen, nitrogen, phosphorus, and sulfur are the elements of which proteins, hence protoplasm, are composed. In addition to these six, there are fourteen other elements which are essential to the growth of some plant or plants: calcium, magnesium, potassium, iron, manganese, molybdenum, copper, boron, zinc, chlorine, sodium, cobalt, vanadium, and silicon. Not *all* are required for *all* plants but *all* have been found to be essential to *some*. These mineral elements, in addition to phosphorus and sulfur, usually constitute what is known as the *plant ash,* or the mineral remaining after the "burning off" of carbon, hydrogen, oxygen, and nitrogen. Each of the twenty plays a role in the growth and development of plants, and when present in insufficient quantities can reduce growth and yields.

The carbon, hydrogen, and oxygen contained in plants are obtained from carbon dioxide and water. They are converted to simple carbohydrates by photosynthesis and ultimately elaborated into amino acids, proteins, and protoplasm. These elements are not considered to be mineral nutrients, and, with the exception of the control man exerts over water and, to a lesser extent, carbon dioxide, there is little of practical importance that can be done to alter the supply to plants.

Plant content of mineral elements is affected by a host of factors. Their percentage composition in crops therefore varies considerably and should be kept in mind when consulting tables of data showing plant composition of the various elements. Some of the older data still in use may list figures that are too low because of the inadequacy of earlier analytical methods. A case in point is sulfur. Recent studies have shown that some of the earlier percentage figures for the plant content of this element are too low. As pointed out by Venema, values of two to a hundredfold higher for sulfur are found today with modern analytical methods.

Plant composition data have in some cases been mistakenly used as the sole basis for formulating a fertilizer program, the idea being that the quantities of the elements removed by the crop should be the quantities replaced by the fertilizer. This approach ignores such important factors as losses by leaching, fixation by the soil in an unavailable form of certain elements, efficiency of various plants in absorbing certain elements, and so on. When considered with these factors, however, such data can be a helpful guide to the formulation of a sound fertility program.

Shown in Table 13-1 are the contents of some of the mineral elements in the more common field and horticultural crop plants. The most recent data available have been used, but the figures should nonetheless be considered only as averages. Soil, climate, crop variety, and management factors will have a considerable impact on plant composition and in individual cases may cause appreciable variation from the figures in the table.

The roles of the various mineral elements in plant growth are covered briefly in the following sections.

Nitrogen. Nitrogen is a vitally important plant nutrient, the supply of which can be controlled by man. It is absorbed by plants primarily in the form of nitrates, though smaller amounts of other forms can be absorbed including the ammonium ion and urea. In moist, warm, well-aerated soils most of the nitrogen compounds will be converted to the NO_3^- form.

Once in the plant the nitrate is reduced to ammonium N using the energy provided by photosynthesis. The NH_4^+-N so produced is combined with so-called "carbon skeletons" to form glutamic acid, which is in turn elaborated into over 100 different amino acids.

Twenty or twenty-one of the different amino acids are then joined together through peptide linkages to form proteins. The order in which these amino acids are linked together is controlled by the genetic makeup of the plant. As a result, though environmental factors, and especially the amount of N supplied to the crops, may affect the amount of protein produced, the kind of protein is largely controlled by genetic factors.

The proteins formed in plant cells are largely functional rather than structural; that is, they are enzymes. As such, they control the metabolic processes that take place in plants including those involved in the reduction of NO_3^- and the synthesis of protein. So, the products of protein are essential to the synthesis process itself. These functional proteins are not stable entities, of course, for they are continually being degraded and resynthesized.

In addition to its role in the formation of proteins, nitrogen is an integral part of the chlorophyll molecule. This molecule consists essentially of a central magnesium atom around which are coordinated four pyrrol rings, each of which contains one nitrogen and four carbon atoms. The nitrogen bonds are shared with those of the carbon and magnesium atoms.

An adequate supply of nitrogen is associated with vigorous vegetative growth and a deep green color. Excessive quantities of nitrogen can, under some conditions, prolong the growing period and delay crop maturity. This is most likely to occur when adequate supplies of the other plant nutrients are not present. The effect of nitrogen on delaying maturity is not so important as it was once considered. This is illustrated by the data in Table 3-1.

The proportion of cotton harvested at the first picking is a measure of the maturity of the crop—the greater the proportion picked at that time, the more mature the crop. The data in Table 3-1 indicate that (1) all treatments increased the yield of cotton and (2) all treatments delayed maturity. However, when phosphorus and potassium were supplied in addition to nitrogen, maturity was not so seriously affected as when nitrogen was applied alone. Adequate nitrogen also speeds the maturity of corn.

The supply of nitrogen is related to carbohydrate utilization. When nitrogen supplies are insufficient, carbohydrates will be deposited in vegetative cells which will cause them to thicken. When nitrogen supplies are adequate, and conditions are favorable for growth, proteins are formed from the manufactured carbohydrates. Less carbohydrate is thus deposited in the vegetative portion, more protoplasm is formed, and, because protoplasm is highly hydrated, a more succulent plant results.

Excessive succulence in some crops may have a harmful effect. With crops such as cotton, a weakening of the fiber may result. With grain crops, lodging may occur, particularly when potassium supplies are inadequate or when varieties not adapted to high levels of nitrogen fertilization are used. Excessive nitrogen fertilization will also reduce the sugar content of sugar beets. In some cases excessive succulence may make a plant more susceptible to disease or insect attack.

It is not intended to imply that nitrogen fertilization is detrimental. Quite the contrary. It is only under the unusual conditions of excessive

TABLE 3-1. The Effect of Nitrogen on Yield and Maturity of Cotton Grown on an Olivier Silt Loam*

			Yield		
Treatment lbs/A			First picking		Total
N	+ P_2O_5	+ K_2O	Pounds of seed cotton per acre	Percentage of total	Pounds of seed cotton per acre
0	0	0	1034	71	1465
60	0	0	866	56	1536
60	60	60	1121	63	1793
90	60	60	1042	58	1802

* From *Arkansas Agr. Exp. Sta. Mimeo. Ser. No. 92* (February 1960).

applications of nitrogen combined with inadequate supplies of the other elements or the use of unadapted varieties that these harmful effects are observed. One of the greatest boons to the development of a strong and efficient agriculture has been the production of inexpensive synthetic nitrogen fertilizers. When used in conjunction with other plant nutrients in a sound crop-management program, nitrogen fertilizers greatly increase crop yields and the grower's net income.

When plants are deficient in nitrogen, they become stunted and yellow in appearance. This yellowing, or chlorosis, usually appears first on the lower leaves; the upper leaves remain green. In cases of severe nitrogen shortage the leaves will turn brown and die. In grasses the lower leaves usually "fire," or turn brown, beginning at the leaf tip and progressing along the midrib until the entire leaf is dead. The appearance of normal and nitrogen-deficient corn plants is shown in Figure 3-1.

The tendency of the young upper leaves to remain green as the lower leaves yellow or die is an indication of the mobility of nitrogen in the plant. When the roots are unable to absorb sufficient amounts of this element to meet the growing requirement, nitrogen compounds in the older

Figure 3–1. Corn plants receiving adequate and inadequate nitrogen. Observe the thin spindly appearance and the light color of the nitrogen-deficient plants in the foreground, contrasted with the vigorous growth and dark color of the plants receiving nitrogen in the background. (Courtesy of Professor H. V. Jordan, ARS, U.S. Department of Agriculture, Starkville, Mississippi.)

plant parts will undergo lysis. The protein nitrogen thus is converted to a soluble form, translocated to the active meristematic regions, and reused in the synthesis of new protoplasm.

Phosphorus. Phosphorus, with nitrogen and potassium, is classed as a major nutrient element. It occurs in most plants, however, in quantities that are much smaller than those of nitrogen and potassium. It is generally considered that plants absorb most of their phosphorus as the primary orthophosphate ion, $H_2PO_4^-$. Smaller amounts of the secondary orthophosphate ion, HPO_4^{2-}, are absorbed. In fact, there are about ten times as many absorption sites on plant roots for $H_2PO_4^-$ as there are for HPO_4^{2-}. The relative amounts of these two ions, which will be absorbed by plants, are affected by the pH of the medium surrounding the roots. Lower pH values will increase the absorption of the $H_2PO_4^-$ ion, whereas, higher pH values will increase absorption of the HPO_4^{2-} form.

Other forms of phosphorus, among which are pyrophosphates and metaphosphates, may possibly be absorbed by plant roots. Both ionic forms are found in certain fertilizer phosphates. Potassium and calcium metaphosphates have been produced in limited amounts in the United States for several years by the Tennessee Valley Authority (TVA) and used in experimental sales and educational programs. TVA has developed a series of calcium, potassium, and ammonium polyphosphates in which the pyrophosphate ion predominates. Because both meta- and pyrophosphates hydrolyze in aqueous solutions, their absorption as such is limited. Regardless of the form of phosphate actually absorbed by the plant, pyrophosphates and metaphosphates are good sources of fertilizer phosphorus.

Plants may also absorb certain soluble organic phosphates. Nucleic acid and phytin are taken in by plants from sterile sand or solution cultures. Both compounds may occur as degradation products of the decomposition of soil organic matter and as such could be utilized directly by growing plants. Because of their instability in the presence of an active microbial population, however, their importance as sources of phosphorus as such for higher plants under field conditions is limited.

Phosphorus has for some time been recognized as a constituent of nucleic acid, phytin, and phospholipids. An adequate supply early in the life of the plant is important in laying down the primordia for its reproductive parts. Phosphorus has also been associated with early maturity of crops, particularly the cereals, and a shortage is accompanied by a marked reduction in plant growth. It is considered essential to seed formation and is found in large quantities in seed and fruit.

A good supply of phosphorus has been associated historically with increased root growth. Ohlrogge and his associates at Purdue University have shown that when a soluble phosphate and ammonium nitrogen are applied together in a band, plant roots proliferate extensively in that

ammonium N helps in uptake of P as opposed to NO₃-N

area. There is also a greatly increased uptake of phosphorus, which is not observed if nitrate nitrogen is used instead of the ammonium form. This phenomenon has not been fully explained. The larger uptake of phosphorus coupled with the increase in root proliferation may, however, lend support to the long-held view that phosphorus increases root growth.

Several other gross quantitative effects on plant growth are attributed to phosphate fertilization. A good supply of phosphorus is said to hasten plant maturity. This statement is reasonably true, for one of the striking observations frequently made of grain crops fertilized with increasing quantities of phosphate fertilizer is the shorter period required for ripening the grain on those plots receiving the higher rates of phosphate, as illustrated in Figure 3-2.

An adequate supply of phosphorus is associated with greater strength of cereal straw. The quality of certain fruit, forage, vegetable, and grain crops is said to be improved and disease resistance increased when they are adequately supplied with this element.

Phosphorus is readily mobilized in plants, and, when a deficiency occurs, the element contained in the older tissues is transferred to the active meristematic regions. However, because of the marked effect that

Figure 3–2. The effect of phosphate fertilization on the maturity of small grains. Notice the more advanced maturity of the small grains receiving the phosphorus (*left*) in contrast to those that have received no phosphorus (*right*). (Courtesy of Dr. O. H. Long, University of Tennessee.)

a deficiency of this element has on retarding over-all growth, the striking foliar symptoms which are evidence of a deficiency in certain other nutrients, such as nitrogen or potassium, are seldom observed.

During the last two decades a great deal has been learned about the specific role of phosphorus in plant growth and development. Very briefly, phosphate or phosphoryl radicals in plant cells are transported to an acceptor group by a transfer reaction, called phosphorylation, and the reactivity of a compound is increased. Phosphorylation results in a lowering of activation energy barriers and overcomes otherwise unfavorable thermodynamic conditions within the plant system. As a result of more favorable conditions, the number of reactions chemically possible in biological systems is enormously increased.

Phosphate turnover in plants includes three distinct phases. In the first phase inorganic phosphate is absorbed and combined with organic molecules or radicals. In the second step these primarily phosphorylated compounds transfer the phosphoryl group to other molecules, a step known as transphosphorylation. In the third and last phase phosphate or pyrophosphate is split from the phosphorylated intermediates, either by hydrolytic cleavage or by substitution of an organic radical. The main source of energy for the incorporation of phosphate into organic combinations is the oxidation-reduction potential energy set free in oxidative metabolism.

The phosphate-organic bonds of several biologically important compounds were initially classified as high energy or low energy on the basis of the free energy change on hydrolysis. This classification is somewhat arbitrary, however, for the transition from one category to the other is continuous. It has been proposed that the *energy wealth* of the various phosphate bonds be referred to as the phosphoryl transfer potential (Ptp). Its value in kilocalories corresponds to the amount of biochemical work that can be obtained from a given phosphate bond.

Table 3-2 shows the Ptp free energy changes (ΔF) on hydrolysis of several phosphorylated metabolites. The abbreviations AMP, ADP, and ATP refer to adenosine mono-, di-, and triphosphate, respectively. These and a host of other organic-phosphate compounds are responsible in one way or another for most of the energy changes in aerobic and anaerobic life processes, whether plant or animal. To list but a few, these phosphate compounds have been shown to be essential for photosynthesis, the interconversion of carbohydrates and related compounds, glycolysis, amino-acid metabolism, fat metabolism, sulfur metabolism, biological oxidations, and a host of other life process reactions too numerous to list.

Phosphorus is indeed the ubiquitous and essential element in the energy transfer processes so vital to life and growth.

Potassium. The third so-called major element required for plant

K - imp. in animal nutrition

TABLE 3-2. Free Energy Changes on the Hydrolysis of Some Phosphorylated Metabolites*

Metabolite	Temperature (°C.)	pH	Free energy change in kilocalories at specified temperature and pH
ATP ($\longrightarrow$ ADP + inorganic phosphate)	30	7	− 7.0
ATP ($\longrightarrow$ AMP + pyrophosphate)	37	7.5	− 8.6
ADP ($\longrightarrow$ AMP + inorganic phosphate)	25	7.0	− 6.4
Pyrophosphate	30	7.0	− 6.5
Glucose-1-phosphate	30	7.0	− 5.0
Galactose-1-phosphate	25	7.0	− 5.0
Glucose-6-phosphate	25	7.0	− 3.3
Fructose-6-phosphate	30	7.0	− 3.8
Fructose-1-phosphate	38	5.8	− 3.5
Glycerol-phosphate	25	7	− 2.1
Glyceraldehyde-1,3-diphosphate	25	6.9	−11.8
Phosphoenolpyruvate	30	7.4	−12.7
Acetylphosphate	29	7.0	−10.1

* Marre, *Ann. Rev. Plant Physiol.*, **12**:200 (1961). Reprinted with the permission of the author and the publisher. Annual Reviews, Inc., Palo Alto, California.

growth is potassium. It is absorbed as the potassium ion K^+ and is found in soils in varying amounts, but the fraction of the total potassium in the exchangeable or plant available form is usually small. Potassium equilibria in soils are discussed in a later chapter.

Fertilizer potassium is added to soils in the form of such soluble salts as potassium chloride, potassium sulfate, potassium nitrate, and potassium-magnesium sulfate.

Plant requirements for this element are quite high. When potassium is present in short supply, characteristic deficiency symptoms appear in the plant. Typical potassium deficiency symptoms of alfalfa are shown in Figure 3-3.

Potassium is a mobile element which is translocated to the younger, meristematic tissues if a shortage occurs. As a result, the deficiency symptoms usually appear first on the lower leaves of annual plants, progressing toward the top as the severity of the deficiency increases.

Unlike nitrogen, sulfur, phosphorus, and several others, potassium apparently does not form an integral part of such plant components as protoplasm, fats, and cellulose. Its function appears rather to be catalytic in nature. Despite this fact, it is nonetheless essential to the following physiological functions:

Figure 3–3. Potassium-deficient alfalfa. (*Western Potash News Letter,* **W-29,** April 1963.)

" *Essentiality* of K :

1. Carbohydrate metabolism or formation and breakdown and translocation of starch.
2. Nitrogen metabolism and synthesis of proteins.
3. Control and regulation of activities of various essential mineral elements.
4. Neutralization of physiologically important organic acids.
5. Activation of various enzymes.
6. Promotion of the growth of meristematic tissue.
7. Adjustment of stomatal movement and water relations.
   8. Role in photosynthesis

The gross impact on crop production of these effects is reflected in several ways. Perhaps the first visible indication of a potassium deficiency is the development of the leaf characteristics. In addition, potassium shortage is frequently accompanied by a weakening of the straw of grain crops, which results in lodging of small grains, illustrated in Figure 3-4, and stalk breakage in corn and sorghum, listed in Table 3-3.

Potassium deficiencies greatly reduce crop yields. In fact, serious yield reductions may result without the appearance of deficiency symptoms. This phenomenon has been termed hidden hunger and is not necessarily restricted to the element potassium. Crops may exhibit hidden hunger for other elements as well, a point that is discussed more fully in Chapter 12.

Figure 3–4. The effect of potassium fertilization on the lodging of small grains: The plants on the right received adequate potassium; those on the left did not. (*Eastern Potash News Letter,* **E-118,** December 1963.)

TABLE 3–3. The Effect of Nitrogen and Potassium Fertilization on the Percentages of Dead and Lodged Stalks of Dixie 29 Corn*

Fertilizer added†	Dead stalks (%)				Lodged plants (%)	
	1957	1958	1959	1960	1959	1960
60–35–0	55.6	48.9	78.5	51.4	21.8	19.2
120–35–0	42.9	60.3	78.7	55.6	25.4	24.6
60–35–42	35.2	58.6	58.4	18.7	5.1	1.7
120–35–42	37.8	52.2	54.6	33.5	6.0	6.7
60–35–83	25.1	42.2	43.4	22.8	1.6	1.9
120–35–83	29.5	55.2	41.0	23.0	1.7	3.3
60–35–166	19.9	48.8	32.6	17.7	2.0	3.2
120–35–166	22.9	44.1	34.8	27.6	1.4	2.4
LSD (0.05):						
K	10.4	NS	5.6	8.8	3.1	5.6
N	NS	NS	NS	3.7	NS	2.3

* Josephson, *Agron. J.,* **54:**179 (1962).
† Expressed as pounds of elements, N-P-K, per acre.

Potassium deficiency is associated with a decrease in resistance to certain plant diseases. A shortage of the element has been found to increase powdery mildew of wheat and in some cases wheat-stem sawfly damage. Root rot and winter killing of alfalfa are greater with an inadequate potassium supply. The winter hardiness, hence the survival, of Coastal Bermuda grass is greater with adequate potassium.

The quality of some crops, particularly fruits and vegetables, is decreased with insufficient potassium. This is illustrated by Figure 3-5.

The over-all effects of potassium deficiency on plant growth and quality are, of course, the result of the accompanying physiological disturbances within the plant system. A deficiency of potassium, for example, alters the activities of invertase, diastase, peptase, and catalase in sugar cane. It is also responsible for the activation of pyruvic kinase in some plants.

Photosynthesis is decreased with insufficient potassium, whereas at the same time respiration may be increased. This seriously reduces the supply of carbohydrates and consequently the growth of the plant. The translocation of sugar from the leaves of cane is greatly reduced in potassium-deficient plants. It was found, for example, that in normal cane plants the downward movement of sugars from the leaf proceeded at a rate of approximately 2.5 cm./min. In K-deficient plants the rate of translocation was reduced to less than half that of the K-sufficient plants.

The role of potassium in maintaining adequate water relations in plants is an important one. Maintenance of plant turgor is essential to the proper functioning of photosynthetic and metabolic processes. This is well illustrated by work done in Hawaii with sugar cane. Plants were grown in nutrient *solution* cultures for six weeks with an adequate supply of potassium. At that time, and for the six weeks thereafter, these same plants were transferred to nutrient solutions with different levels of

Figure 3–5. Potassium deficient (*upper*) and normal (*lower*) grapes. (Courtesy of Peter Christensen, *Farm Adviser,* Fresno County, California, and the Potash Institute of North America, formerly the American Potash Institute.)

potassium. The plants grown in solutions with continuously adequate potassium maintained high internal moisture levels. Those supplied with low levels of this element for the last six-week period experienced a decrease in the internal moisture level of the leaves on their lower halves. The two lowest leaves actually dried up and died, but not, of course, as the result of an inadequate *external* moisture supply, for the entire study was conducted with nutrient solutions.

The last important effect of potassium on plant metabolism to be considered here is its relation to protein synthesis. Studies with sugar cane have shown that in potassium-deficient plants nonprotein nitrogen accumulates in the leaves. Other studies have shown that free amino acids accumulate in the leaves of K-deficient barley plants and that in extremely deficient plants the concentration of these free acids decreases with an increase in the concentration of amides. Relatively recent work has shown that in certain grasses grown with inadequate potassium there is a buildup of amide nitrogen and conversion to protein is reduced. When adequate potassium was supplied, however, there was an increase in protein nitrogen and a corresponding decrease in amide nitrogen. Some of these relationships are illustrated by the data in Table 3-4.

The significance of the term b_{yx} is the degree to which the property Y (nonprotein nitrogen, α-amino N, or true protein) is related to the prop-

TABLE 3–4. Relationship Between the Nitrogen Content of Orchard Grass and the Concentration of Certain Nitrogenous Fractions*

Dependent variable (Y)	Potassium applied (lb./A.)	b_{yx}
Malate (mg./g.)	0	+0.257
	166	+0.885
	322	+0.727
Nonprotein nitrogen (%)	0	+0.851
	166	+0.606
	322	+0.331
α-amino N (%)	0	+0.183
	166	+0.128
	322	+0.083
True protein (%)	0	+0.148
	166	+0.389
	322	+0.676

* Teel, *Soil Sci.*, **39**:50 (1962). Reprinted with permission of The Williams & Wilkins Co., Baltimore.

erty X (the nitrogen content of the plant). The data show quite clearly that as the potassium level increases, the levels of nonprotein compounds in relation to total nitrogen in the plant decreases.

Studies have shown that the relation between potassium and protein synthesis in plants may have an important bearing on the nutrition of ruminant animals. It must be emphasized that at this writing the problem has not been resolved. It appears, however, that herbage high in nonprotein nitrogen may be injurious to animals because of the ease with which this fraction is deaminated. Deamination results in the rapid release of large amounts of ammonia (NH_3) in the rumen which may be injurious to the animal. Figure 3-6 shows lambs fed on diets both low and adequate in potassium.

It is not the purpose of this book to cover the requirements of potassium in animal physiology, but recent findings concerning potassium in animal metabolism are worth noting. Because animals, including man, derive a large proportion of their potassium from the consumption of plants, the adequate fertilization of crops with this plant nutrient assumes a position of even greater importance.

Figure 3–6. The effect of dietary level of potassium on the growth and appearance of lambs. The lamb on the left received a diet that contained 0.1 per cent K. The lamb on the right received a diet containing 0.6 per cent K. (Courtesy of Professors R. L. Preston and P. P. Telle, University of Missouri, and the Potash Institute of North America, formerly the American Potash Institute.)

A principal function of potassium in animal metabolism is the production of bioelectric currents. Animal cells use a portion of the energy resulting from oxidative metabolism to keep potassium on the inside and sodium on the outside of cell membranes. The potential difference created by the high concentration of potassium on the inside and sodium on the outside of cells generates an electric current which propagates nerve impulses. This is shown schematically in Figure 3-7. These impulses are transmitted along nerve fibers between the brain and the muscles in intervals of microseconds. The operation of this system is explained in its simplest terms. Cellular potassium is associated with the negatively charged organic molecules immediately under the cell membranes. The net internal charge is negative. A substance called acetylcholine, when stimulated, permits the sodium ions to enter the cells and the potassium ions to move out, thus generating electrical impulses. Another compound, an enzyme called acetylcholinestrase, appears to stop this flow of ions across the cell walls and prevents the further generation of electrical impulses. A similar reaction apparently activates a material known as actinomyosin which is responsible for muscle-fiber activity.

The importance of potassium in human nutrition was highlighted by the United States space program. It was found that a high level of dietary potassium was needed by the astronauts during their space voyages in order to prevent certain heart disturbances. Consequently, the diet of these astronauts was developed to ensure, among other things, that the daily intake of potassium was adequate to prevent the development of any of several physiological aberrations which can result from a K deficiency in the animal system.

Much more is still to be learned about potassium in plant and animal nutrition, but the foregoing suggests that its function is far more impor-

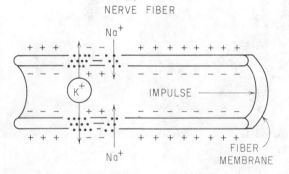

Figure 3-7. Schematic diagram of the function of potassium in the transmission of nerve impulses. (*Midwestern Potash News Letter,* **M-118,** November-December 1963.)

tant, especially in animal nutrition, than was supposed only a few years ago.

Calcium. Calcium is an element required by all higher plants; absorbed as the ion Ca^{2+}, it is found in abundant quantities in the leaves of plants and, in some species, in plant cells precipitated as calcium oxalate. It may also occur in the ionic form in cell sap. A deficiency of calcium manifests itself in the failure of the terminal buds of plants to develop. The same applies to the apical tips of roots. As a result of these two phenomena, plant growth ceases in the absence of an adequate supply. Calcium deficiency of tobacco is illustrated in Figure 3-8.

In corn a shortage of calcium prevents the emergence and unfolding of the new leaves, whose tips are almost colorless and are covered with a sticky gelatinous material which causes them to adhere to one another.

The specific physiological functions performed by calcium in plants are not clearly defined. Classically, calcium has been considered as necessary to the formation of the middle lamellae of cells because of its alleged role in the synthesis of calcium pectate. Some early research, which seems not to have received much attention, continues to throw doubt on the importance of calcium in this particular plant function. Fourteen types of plant tissue were analyzed for pectate (determined as pectic acid) after these tissues were handled in four different ways. One set was dried at 98°C. in a hot-air oven, another at 70°C. in a vacuum, a third sundried, and in the last group the fresh tissues were placed directly in the boiling alcohol. Pectic acid was absent in thirteen of the fourteen samples of tissue which were immersed in the boiling alcohol. It was present, however, in all samples subjected to the three drying treatments at elevated temperatures. The conclusion was that there is no pectate normally present in most living plant tissues and that the presence of this material in the samples subject to the three drying treatments was an artifact. In a comparatively recent paper it was stated that calcium is a constituent of the middle lamella, but it did not specify that the lamella is composed of Ca pectate.

Figure 3-8. Calcium deficiency of tobacco. (Courtesy of Professor R. R. Bennett, North Carolina Agricultural Extension Service, North Carolina State College, Raleigh.)

It has also been suggested that calcium favors the formation of and increases the protein content of mitochondria. If this is so, the role played by the mitochondria in aerobic respiration, hence salt uptake, indicates that there may be a direct relationship between calcium and ion uptake in general.

Calcium is related to protein synthesis by its enhancement of the uptake of nitrate nitrogen and is associated with the activity of certain enzyme systems.

An understanding of the part played by calcium in cell elongation, and therefore the development of meristematic tissue, has been complicated by the contradictory results of various research workers, apparently because of its interaction with certain plant growth regulators, with other cations, and, in plant roots, with the pH of the surrounding medium. Despite the confusion, it is a matter of common knowledge that plant roots and tips do not elongate in the absence of calcium.

Small amounts of calcium are required to produce normal mitosis and it has been suggested that this element has a specific function in the organization of chromatin or of the mitotic spindle. It has also been suggested that Ca is associated with the activation of certain enzyme systems and that it is directly involved in chromosome stability and that it is a constituent of chromosome structure. Calcium has been shown to have an effect on carbohydrate translocation in plants.

Finally, calcium is generally considered to be an immobile element, at least in herbaceous plants. It has been found, however, that in at least one forest species (Western white pine) previously deposited calcium moves from older to developing tissue. Whether this occurs in other forest species is not known.

Magnesium. Magnesium is absorbed in the form of the ion, Mg^{2+}. It is the only mineral constituent of the chlorophyll molecule and is located at its center, as described in the section on nitrogen. The importance of magnesium is obvious, for without chlorophyll the autotrophic green plant would fail to carry on photosynthesis. Although a large portion of plant magnesium resides in the chlorophyll, appreciable quantities are frequently found in seeds. It appears to be related to phosphorus metabolism and is considered to be specific in the activation of a number of plant enzyme systems.

Magnesium is a mobile element and is readily translocated from older to younger plant parts in the event of a deficiency. Consequently, the symptom often appears first on the lower leaves. In many species the deficiency results in an interveinal chlorosis of the leaf, in which only the veins remain green. In more advanced stages the leaf tissue becomes uniformly pale yellow, then brown and necrotic. In other species, notably cotton, the lower leaves may develop a reddish purple cast which gradually turns brown and finally necrotic. Magnesium deficiency of tobacco is illustrated in Figure 3-9.

Figure 3–9. Magnesium deficiency (sand drown) of tobacco. (Courtesy of Dr. J. E. McMurtrey, U.S. Department of Agriculture, Beltsville, Maryland.)

Magnesium is required for the activation of many enzymes concerned with carbohydrate metabolism and is prominent in the so-called citric acid cycle which is important to cell respiration. Numerous phosphorylation reactions relating to nitrogen metabolism in plants are catalyzed by this element. Although the role of magnesium in the structure of chlorophyll has been known for some time, it is only within comparatively recent years that its importance in enzyme activation has been recognized.

Magnesium is related to the synthesis of oil. With sulfur, it brings about significant increases in the oil content of several crops, a point that is well illustrated by the data on page 87 in the section dealing with sulfur as a plant nutrient.

Sulfur. Sulfur is absorbed by plant roots almost exclusively as the sulfate ion, SO_4^{2-}. Some SO_2 is absorbed through plant leaves and utilized by the plant; sulfur dioxide in anything but very small concentrations, however, is toxic. It has been found that elemental sulfur applied as a dust to fruit-tree leaves finds its way in small amounts to the internal plant system within a relatively short time, but the manner in which this water-insoluble material penetrates the plant is not known. It is of no practical importance.

Like nitrogen, much of the SO_4^{2-} is reduced in the plant, and the sulfur is found in the —S—S— and —SH form. Sulfate sulfur in large amounts may also be retained as such in the tissues and cell sap without apparent injury. It is present in equal or lesser amounts than phosphorus in such plants as wheat, corn, beans, and potatoes but in larger amounts in alfalfa, cabbage, and turnips.

A deficiency of sulfur, which has a pronounced retarding effect on plant growth, is characterized by uniformly chlorotic plants—stunted, thin-stemmed, and spindly. These symptoms in many plants resemble those of nitrogen and have undoubtedly led to many improper assessments of the cause of the trouble. Unlike nitrogen, however, sulfur does not appear to be easily translocated from older to younger plant parts under stress caused by its deficiency.

Crop deficiencies of sulfur have long been known in localized areas throughout the world. In comparatively recent years reported deficiencies of this element have shown a marked increase essentially because of four factors:

1. Increased use of sulfur-free fertilizer.
2. Decreased use of sulfur as an insecticide and fungicide.
3. Decreased concentration of sulfur compounds in the atmosphere and rainfall because of the decreased consumption of high-sulfur fuels.
4. Increased crop yields that require larger amounts of the essential plant nutrient elements.

Deficiencies of sulfur have been reported for a number of agricultural crops growing in the field, among which are alfalfa, clover, cotton, tobacco, grasses, corn, wheat, cruciferous crops, peanuts, and pome and stone fruit.

The effect of sulfur on the appearance of the leaves of pear trees is shown in Figure 3-10. The branch on the left was taken from a tree which had received no sulfur, whereas that on the right was taken from a tree which had been supplied with this element.

The specific functions of sulfur in plant growth and metabolism are numerous and important:

1. It is required for the synthesis of the sulfur-containing amino acids, cystine, cysteine, and methonine, and for protein synthesis.
2. It activates certain proteolytic enzymes such as the papainases, examples of which are papain, bromelin, and ficin.
3. It is a constituent of certain vitamins, of coenzyme A, and of glutathione.
4. It is present in the oils of plants of the mustard and onion families.

Figure 3-10. The effect of sulfur fertilization on the appearance of the leaves of pear trees grown in the area of Wenatchee, Washington. The branch on the left came from a tree that had received no sulfur and the branch on the right from a tree that had received this element. (Courtesy of Dr. Nels R. Benson and Dr. H. M. Reisenauer, Washington State University.)

5. It increases the oil content of crops such as flax and soybeans.
6. Disulfide linkages (—S—S—) have recently been associated with the structure of protoplasm, and the quantity of sulfhydryl groups (—SH) in plants has in some cases been related to increased cold resistance.
7. It is required for nitrogen fixation by leguminous plants and is a part of the nitrogenase enzyme system that is associated with this reaction.

Space does not permit a detailed discussion of all the functions of sulfur in plant nutrition, but the importance to man and agriculture of several of them, will be elaborated. The requirements of human and nonruminant animal nutrition for sulfur-containing amino acids cannot be overstated. The biological value of a protein, which is expressed by its EAA (essential amino acid) index, is determined by comparing its amino acid content with that of some standard high-quality animal protein. Many studies of the nutrititive value of proteins have shown that their lack of sulfur-containing amino acids is the factor that limits their biological value. In fact, workers investigating the nutritional value of food from various sources and countries have concluded not only that the content of sulfur-containing amino acids determines the biological value of the contained protein but that it is even more important than the lysine content of these foods. They have further concluded that a large segment of the world's population is living on a diet strongly deficient in methionine.

Within limits, the sulfur-containing amino acid content of plants can be altered by sulfur fertilization. This is illustrated by the data shown in Table 3-5.

The methionine, cystine, and total sulfur contents of the two strains of alfalfa increased with larger inputs of sulfur in the nutrient medium. The ability of the two experimental strains, C-3 and C-10, however, to synthesize methionine and cystine differed greatly. The ability of different plant strains or species to utilize effectively the supply of mineral nutrients in synthesizing various compounds is an important property, particularly from the standpoint of animal and human nutrition. It is one to which more attention could well be given in research programs.

A deficiency of sulfur can cause an accumulation of nonprotein nitrogen in plants, which can be detrimental to ruminant animals if it is not corrected with feeding supplements containing sulfur in either the organic or inorganic form. Ruminants are able to utilize sulfate, sulfide, and, to a lesser extent, elemental sulfur in the synthesis of proteins. Nonruminants cannot and must have methionine in their diets.

In nonleguminous plants that have been given liberal quantities of nitrogen fertilizers, nitrates as well as amides may accumulate in the tissues. Nitrates in large quantities are toxic to animals. If sulfur is limiting, nitrates accumulate in plant tissue. This problem was recently discussed by Norwegian workers who showed that the danger of high ni-

TABLE 3–5. The Effect of Sulfate Ion Concentration on the Methionine, Cystine, and Total Sulfur Contents of Two Strains of Alfalfa*

$SO_4 =$ ion concentration (ppm.)	Methionine (mg./g. of N)		Cystine (mg./g. of N)		Sulfur %	
	C-3	C-10	C-3	C-10	C-3	C-10
0	10.6	17.6	21.5	24.4	0.100	0.089
1	20.8	27.6	28.6	35.2	0.103	0.098
3	33.6	34.9	37.0	43.6	0.129	0.121
9	38.0	40.3	38.9	42.9	0.186	0.200
27	41.4	43.9	42.9	45.0	0.229	0.227
81	43.4	44.3	43.6	46.0	0.244	0.242

* Tisdale et al., *Agron. J.* **42**:221 (1950).

TABLE 3-6. The Effect of Sulfur Fertilization on the N:S Ratios of Barley, Oats, Timothy, and Turnip Foliage, Norway, 1961*

	N:S Ratio	
Crop	− S	+ S
Barley, grain	27.2	16.4
Barley, stover	45.7	6.5
Oat, grain	26.8	14.8
Oat, stover	49.3	6.0
Timothy, second crop	31.0	11.8
Turnip, second crop	42.9	19.1

* Ødelien, *Tidsskr, Norske Landbruk,* **70**:35 (1963).

trate levels in plants could be reduced with adequate sulfur fertilization. High nitrate levels were also related to wide N:S ratios (% total N:% total S) in plants; N:S ratios of 10:1 to 20:1 are usually considered suitable for plant material to be fed to ruminants. The effect of sulfur fertilization on four crops grown in Norway is shown in Table 3-6.

The impact of sulfur fertilization on narrowing the N:S ratio is apparent from these data. This property should also be considered in determining the suitability of forage for animal feed, though it is not widely taken into consideration at the present writing, at least in the United States.

Sulfur is also related to the formation of oil in crops such as flax, soybeans, and peanuts. This effect has only recently been generally recognized. It has been found that the addition of magnesium to the sulfur further enhances the production of oil. The effect of magnesium and sulfur on the oil content of soybeans is illustrated by the data in Table 3-7.

It should be stated that the reported effects of sulfur on the oil content of crops have been more numerous in Europe than in the United States, where workers have generally been more concerned with increasing the

TABLE 3-7. The Effect of Cl, S, and Mg on the Oil Content of Soybeans*

Treatment	Seed yield (kg./ha.)	Oil content (%)	Oil yield (kg./ha.)
N + P	566	22.9	129.6
N + P + K + Cl	645	24.0	154.8
N + P + K + SO$_4$	720	24.8	178.6
N + P + K + SO$_4$ + Mg	784	25.4	199.1

* K. C. W., Venema, *Potash Trop. Agric.,* **5** (October 1962).

oil content of crops by breeding and selection and little attention has been given to the impact that plant nutrition might have on this important plant function.

Boron. Boron is absorbed in one or more of its ionic forms, such as $B_4O_7^{2-}$, $H_2BO_3^-$, HBO_3^{2-}, or BO_3^{3-}. It is required in generally small quantities, for some plants, notably of the bean family, are quite sensitive to this element. Amounts applied to crops such as alfalfa are toxic to beans, a fact that can be confirmed by more than one farmer who has made the mistake of applying a borated alfalfa fertilizer to his bean crop.

Boron is not readily translocated from older to meristematic regions. When a deficiency occurs, the first visual symptom is the cessation of growth of the terminal bud, followed shortly thereafter by death of the young leaves. The youngest leaves become pale green, losing more color at the base than at the tip. The basal tissues break down, and, if growth continues, the leaves have a one-sided or twisted appearance. Usually the leaves die and terminal growth ceases.

Several root crops are affected by a deficiency of boron. The breaking down of internal tissues of the root gives rise to darkened areas described as brown or black heart. Boron-deficient sugar beets are shown in Figure 3-11.

Deficiencies of boron have been reported in deciduous and citrus fruit as well as in small fruit species. The well-known internal cork of apple is caused by a deficiency of this element. In citrus the peel is uneven in thickness, the fruit is lumpy, and there are gummy deposits in the albedo of the fruit. Just why a shortage of boron brings on the various symptoms described is not known. Several roles ascribed to this element will be reviewed briefly.

The borate ion has an unusual affinity for polyhydroxyl compounds with ortho configurations. Because of this, plus a wealth of experimental evidence that has accumulated, it has been suggested that boron can account for four distinct roles or regulatory mechanisms in many plants. These are:

1. Translocation of sugars across membranes.
2. Regulatory effects on oxidation by polyphenolase activity.
3. Modification of equilibrium in phosphate ester metabolism.
4. Influencing extent of the catalytic effects of O-diphenols in cell metabolism, including the inhibition of indoleacetic acid oxidation and possibly promoting pyridine nucleotide-quinone reductase activity, which is especially high in roots.

Boron is believed to influence cell development by the control it exerts on polysaccharide formation. The rate of cell division in plants has been listed as a function of boron content. Boron is also said to func-

Figure 3-11. Boron-deficient sugar beet roots. Note blackened heart tissue. (Courtesy of Professor R. L. Cook, Michigan State University, and reprinted from *Hunger Signs in Crops,* 3rd ed., 1964, with permission of the David McKay Co., New York.)

tion in pectin synthesis, a function, however, that should be viewed in light of the discussion on page 81 concerning the presence of pectin compounds in plants.

Boron is believed to inhibit the formation of starch by combining with the active site of phosphorylase. In this manner boron perhaps functions in a protective way by preventing the excessive polymerization of sugars at the sites of sugar synthesis.

It has been observed in experimental work that bean plants grown without boron transpired less than those grown with it, and its role in water relations was thereby suggested. This reduction in transpiration was attributed to a decrease in the rate of water absorption and to the abnormal morphology of the boron-deficient plants and their higher sugar and hydrophyllic colloid concentration.

Work carried out in the United Kingdom has shown that in the absence of boron there is a cessation of cell division but an enlargement of the apical cells of bean roots grown in solution cultures. Cell division apparently did not cease because of a lack of sugar, protein, or nucleic acid, and it was postulated that, in the absence of boron, abnormalities in the cell wall prevented the cell from becoming organized for mitosis.

Work carried out in Russia is reported to have shown that a shortage of boron decreased the content of ribonucleic acid (RNA) in the tips of stems and roots in bean and sunflower plants grown in nutrient solutions. The desoxyribonucleic acid (DNA) content of these organs was also reduced in the sunflowers. When boron was applied, the return to normal of the RNA and DNA content of the stems and roots led to the conclusion that a boron deficiency results in a disturbance of nucleic acid metabolism rather than carbohydrate metabolism.

Regardless of our inability to explain the precise role of boron in plants, it is essential in varying, but usually small, quantities to the growth of many important agricultural crops.

Iron. Iron may be absorbed by the roots of plants in ionic form or as complex organic salts. It is also absorbed by the leaves when iron sulfate and complex iron compounds, termed chelates, are applied as foliar sprays. Iron nails driven into the trunks of trees have been used to supply this element in certain species. Although ferric iron may be absorbed by plants, the metabolically active form seems to be the ferrous ion. Plant tissues containing large quantities of ferric iron may exhibit iron-deficiency symptoms.

Iron deficiency has been observed in many species. It is most frequently seen in crops growing on calcareous or alkaline soils, though the presence of high phosphate levels may also bring about this condition in acid soils in some species. Citrus and deciduous fruit not uncommonly exhibit iron chlorosis. It is also fairly common in blueberries (which are grown on very acid soils) and in sorghum, grown in neutral to alkaline soils. Other crops known to exhibit deficiencies of this element are soybeans, bush beans, strawberries, vegetable crops, and ornamentals.

A deficiency of iron shows up first in the young leaves of plants. It does not appear to be translocated from older tissues to the tip meristem, and as a result there is cessation of growth. The young leaves develop an interveinal chlorosis which progresses rapidly over the entire leaf. In severe cases the leaves turn completely white. This symptom is illustrated in Figure 3-12, which shows an iron-deficient apple shoot.

Iron functions specifically in the activation of several enzyme systems: fumaric hydrogenase, catalase, oxidase, and cytochromes. A shortage of iron also impairs the chlorophyll-producing mechanism, for it has been shown that the chlorophyll content of plants is related to a continuous supply of iron. In contrast, however, there seems to be no relation between quantities of iron supplied intermittently and the chlorophyll content of plants.

Iron is believed to be associated with the synthesis of chloroplastic protein. The amount of iron in relation to the amounts of other elements

Figure 3–12. Iron-deficient mature apple shoot. Note chlorosis of the younger leaves. (Courtesy of New York State Agricultural Experiment Station, Geneva, and reprinted from *Hunger Signs in Crops,* 3rd ed., 1964, with permission of the David McKay Co., New York.)

in many instances is as important or more so than the absolute quantities of this element present in tissue. The quantities of iron in relation to the amounts of molybdenum, phosphorus, manganese, and copper are of special importance.

Iron has been shown to be capable of partly replacing molybdenum as the metal cofactor necessary for the functioning of nitrate reductase in soybeans. Another function attributed to iron is its apparent necessity for the flavin enzyme which reduces cytochrome-c.

Manganese. Manganese is absorbed by plants as the manganous ion, Mn^{2+}, and in molecular combinations with certain organic complexing agents such as EDTA. It can also be absorbed in either form directly through the leaves and is commonly applied as a foliar spray to correct a deficiency.

Like iron, manganese is a relatively immobile element. The symptoms of deficiency usually show up first in the younger leaves. In broad-leaved plants the symptom is one of interveinal chlorosis, which also occurs in members of the grass family, but it is less conspicuous. Several soybean leaves with varying degrees of manganese deficiency are shown in Figure 3-13.

Manganese deficiency of several crops has been described by such terms as grey speck of oats, marsh spot of peas, and speckled yellows of sugar beets. These terms are not so widely used now as they were when the cause of the symptoms was not known.

Like iron and others in the heavy-metal group, manganese functions in the activation of numerous enzymes concerned with carbohydrate metabolism, phosphorylation reactions, and the citric acid cycle and with other metals in the activation of such enzymes as arginase, cysteine desulfhydrase, desoxyribonuclease, and yeast phosphatase. Apparently it is a specific activator of the enzymes prolidase and glutamyl transferase.

Attributed to manganese is its function in certain photochemical processes such as the Hill reaction. It has been shown by Australian workers that manganese is a constituent of tomato chloroplasts. A fraction containing this element cannot be removed from isolated chloroplasts even with repeated washing, by lysis, or by purification by density gradient centrifugation. The Hill reaction activity of isolated chloro-

Figure 3–13. Different stages of manganese deficiency of soybean leaves. (Courtesy of Department of Agronomy, Purdue University, and reprinted from *Hunger Signs in Crops,* 3rd ed., 1964, with permission of the David McKay Co., New York.)

plasts was found to be proportional both to the level of manganese supplied in the external nutrient and to the manganese content of the isolated chloroplasts.

It has also been shown that the manganese in chloroplasts was not removed by organic solvents but that it could be removed by metal complexing agents such as cyanide. A high concentration of KCN is required to expel manganese from the isolated chloroplasts and a high cyanide concentration is necessary to inhibit the Hill reaction. These results led to the postulate that manganese is directly involved in the Hill reaction system of chloroplasts.

Manganese is required by plants in only small quantities, large amounts being toxic. Crinkle leaf of cotton is a manganese toxicity that is sometimes observed on highly acid red and yellow soils of the old Cotton Belt in the southern United States. Manganese toxicity has also been detected in tobacco growing on extremely acid soils. In both cases the condition can be corrected by liming the soil to a pH of 5.5 or more.

Copper. Copper is absorbed by plants as the cupric ion, Cu^{2+}, and may be absorbed as a salt of an organic complex such as EDTA. Copper salts are also absorbed through the leaves, and deficiencies are often corrected or prevented by applications of this element in the biological control sprays applied to crops.

Copper deficiency has been reported in numerous plants, although it is more prevalent among crops growing on peat and muck soils. Crops responding to copper fertilization include red beets, carrots, clover, corn, oats, and fruit trees. Symptoms of the deficiency vary with the crop. In corn the youngest leaves become yellow and stunted, and as the deficiency becomes more severe the young leaves pale and the older leaves die back. In advanced stages dead tissue appears along the tips and edges of the leaves in a pattern similar to that of potassium deficiency. Copper-deficient small grain plants lose color in the younger leaves, which eventually break and the tips die. In many vegetable crops the leaves lack turgor. They develop a bluish-green cast, become chlorotic, and curl, and flower production fails to take place.

Copper is a metal activator of several enzymes, which include tyrosinase, laccase, ascorbic acid oxidase, and butyryl Co-A dehydrogenase. It has also been suggested that it may be one of the metals concerned with the light reaction in plants. Copper deficiency also causes iron accumulation in corn plants, especially in the nodes. Like iron and manganese, the amounts of copper present in a plant in relation to the amounts of the other heavy metals is perhaps of greater importance to proper plant functioning than the absolute amount of copper *per se.*

Zinc. Zinc is absorbed by plant roots as the ion, Zn^{2+}, and may also be absorbed as a molecular complex of such chelating agents as EDTA. Sprays containing soluble zinc salts or zinc complexes are applied to the foliage of plants to correct a deficiency of this element, for it is also capable of entering the plant system directly through the leaves.

Deficiencies of zinc have been observed in corn, sorghum, deciduous and citrus fruit, nut trees, tung trees, legumes, cotton, and several other vegetable crops. It makes its appearance first on the younger leaves, starting with an interveinal chlorosis followed by a great reduction in the rate of shoot growth. In many plants this produces a symptom known as *rosetting.* In corn and sorghum the symptom has been called *white bud,* though this term is objected to by some. A broad band of bleached tissue occurs on each side of the midrib, beginning at the base of the corn leaf, but the midrib and leaf margins remain green. Zinc deficiency in cotton results in a symptom called *little leaf.* Zinc-deficient peach twigs are shown in Figure 3-14.

Like most of the other so-called micronutrients, zinc is toxic to plants in any but small quantities. Its concentration in the plants in relation to the other heavy metals is of greater importance than the absolute amounts present. Zinc deficiencies have been discovered in a wide range of soil conditions, but they seem to be more frequent on calcareous soils and on those excessively high in phosphorus.

Zinc functions in plants largely as a metal activator of enzymes: enolase, yeast aldolase, oxalacetic decarboxylase, lecithinase, cysteine

Figure 3–14. Zinc-deficient peach shoots. The branch on the left was taken from a normal tree and the remaining twigs from trees that were progressively more zinc-deficient. (Courtesy of Dr. A. C. McClung, North Carolina Agricultural Experiment Station, Raleigh.)

desulfhydrase, histidine deaminase, carbonic anhydrase, dihydropeptidase, and glycylglycine dipeptidase.

Molybdenum. The essentiality of molybdenum has only recently been established. However, its importance to crop production in many areas has been quickly recognized, and in these areas it is included as an essential item in fertilizer formulas. Molybdenum is probably absorbed by plant roots as the ion MoO_4^{2-}. It is required in only small amounts, a point of considerable importance because an excess may cause toxicity to animals grazing in treated fields. The minute quantities required and the physical problems associated with its application are illustrated by the fact that in Australia only 24 oz. of molybdic oxide are mixed with a metric ton (2,240 lb.) of superphosphate before application to the soil. The problem of mixing 24 oz. uniformly in a mass of 2,240 lb. is obvious.

Molybdenum deficiencies have been reported for many crops which include clovers, alfalfa, grasses, tomatoes, sweet potatoes, soybeans, and other vegetables. Molybdenum deficiency of three grasses grown in Australia is shown in Figure 3-15.

Symptoms of the deficiency differ with various crops, but as a rule they are first observed as an interveinal chlorosis. With legumes the plants usually turn pale yellow and become stunted—symptoms characteristic of nitrogen deficiency. This, in fact, is what it is, for molybdenum

Figure 3–15. Molybdenum deficiency of three grass species grown in Australia. (Left to right, Phalaris, wheat, and perennial rye grass.) Three pots on left, *plus* Mo; three pots on right, *minus* Mo. (Courtesy of Dr. J. Lipsett, Div. Plant Ind., CSIRO, Canberra, ACT.)

is required by *Rhizobia* for nitrogen fixation. It is also required by nonlegumes for nitrate reduction.

Nitrogen fixation, assimilation, and reduction are, of course, prerequisite to amino acid and protein synthesis, and a deficiency of molybdenum has an adverse effect on these reactions. It also results in the accumulation of nitrates and an apparent lowering of the activity of ascorbic acid oxidase. Molybdenum is known to be specific for the activation of the enzymes nitrate reductase and xanthine oxidase.

Chlorine. Chlorine is absorbed as the chloride ion, Cl^-, and has only recently been classified as an essential element. Bromine at somewhat higher concentrations than chlorine may substitute for it in part at least, in a way similar to that in which sodium substitutes for potassium.

Most of the chlorine deficiencies have been obtained with nutrient culture studies in the greenhouse. However, a few responses have been observed in the field. Plants responding to this element include tobacco, tomatoes, buckwheat, peas, lettuce, cabbage, carrots, sugar beets, barley, corn, potatoes, and cotton.

Chlorine in excessive quantities has a detrimental effect on some plants, examples of which are tobacco and potatoes. In both species the leaves become thickened and tend to roll. The storage quality of the potato tuber and the smoking quality of tobacco are lowered by this excess.

The symptoms of chlorine deficiency are not easily identified. Plants

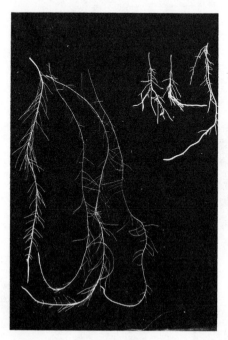

Figure 3–16. The roots of chlorine-sufficient (*left*) and chlorine-deficient (*right*) tomato plants. [Johnson et al., *Plant Soil,* **8**:337 (1957). Reprinted with permission of Martinus Nijhoff, The Hague.]

so affected are said to wilt, to become chlorotic and necrotic in some areas, and to exhibit leaf bronzing. In nutrient cultures it has been shown that chlorine deficiency is associated with reduced root growth, a condition illustrated in Figure 3-16. Little or nothing is known of the role of chlorine in nutrition other than that it is required for the growth and development of the plants listed.

Cobalt. It has not been definitely shown that cobalt (Co) is an element needed by the higher green plants, though several responses by nonleguminous crops to applications of this element have been reported. Cobalt is required by *Rhizobia* for the fixation of elemental nitrogen, and from the standpoint of the practical production of legumes it would have to be considered essential.

The essentiality of cobalt to *Rhizobia* is its part in the formation of vitamin B_{12} (cyanocobalamine), which in turn is essential to the formation of haemoglobin needed for nitrogen fixation.

There are numerous references to research carried out under controlled conditions with legumes adequately inoculated with *Rhizobia* that failed to grow in the absence of cobalt. Experimental work with this element is difficult because of the low concentrations on the order of 0.1 to 1.0 part per billion needed to supply the requirements of these organisms. In this connection a response by subterranean clover grown in the field was obtained from an application of 8 oz./A. of cobalt sulfate. This work was done in South Australia. A significant increase was observed in the yields of dry matter and nitrogen but none in the percentage content of nitrogen in the plant.

Russian workers recently reported responses by cotton, beans, and mustard to applications of cobalt. It was said that it improved growth, transpiration, and photosynthesis and that in beans and mustard the reducing activity and chlorophyll content of the leaves were raised. In cotton an increase in boll number and a decrease in the number of fallen squares were attributed to the addition of cobalt. It was further stated that water content and catalase activity in leaves were increased and that a decrease in the concentration of the cell sap resulted from cobalt treatments. An increase of 9 to 21 per cent in the yield of cotton resulted from the cobalt addition.

The functions of cobalt in plants, in addition to its relation to the synthesis of vitamin B_{12}, are not known. It has been reported to be one of several metals that activate the enzymes arginase, lecithinase, oxalacetic decarboxylase, and the malic enzyme, but the part it takes is certainly not specific.

Vitamin B_{12} is essential to nonruminant nutrition. If an inorganic salt of cobalt is fed to ruminants, they are able to synthesize this vitamin which they also require. Cobalt deficiencies of ruminants have been reported in the United States, Australia, and New Zealand. The Australians have developed a clever method for supplying cobalt to sheep. A small dense pellet, termed a *cobalt bullet,* which consists of cobalt oxide and some inert matrix material such as clay, is placed in the esophagus of the sheep. It lodges in the rumen where it yields a steady supply of the element to the rumen fluid.

Vanadium. Vanadium is definitely considered an essential nutrient for the green alga *Scenedesmus,* but proof of its essentiality for the growth of higher green plants has not been shown. As previously mentioned, vanadium may replace molybdenum to a certain extent in the nutrition of *Azotobacter.* Some workers have shown that it could replace molybdenum in the nutrition of *Rhizobia,* but here again this view is not universally accepted. Increases in higher plant growth attributable to vanadium have been reported for asparagus, rice, lettuce, barley, and corn.

The function of this element in plant nutrition is not known. It has been speculated, however, that it may function in biological oxidation-reduction reactions, but this supposition, of course, has not yet been proved.

Sodium. Sodium is absorbed by plants as the ion, Na^+. In a recent paper reporting on the results of studies carried out in the United Kingdom with sugar beets the claim is made that sodium is essential for this crop and that it is not simply a substitute for potassium. More recently sodium was shown by Australian workers to be an essential element for a group of plants exhibiting the so-called Hatch-Slack pathway of carbohydrate metabolism.

It has been known for years that the growth of some crops is stimulated by applications of sodium. In some cases this benefit has been observed when there was a deficiency of potassium, in other cases when the supply was adequate. Crops so responding to sodium have been classified accordingly and are listed in Table 3-8.

Sodium is reported to influence water relations in sugar beets and to increase their resistance to drought. In low sodium soils the beet leaves are dark green, thin, and dull in hue. The plants wilt more rapidly and may grow horizontally from the crown. There may also be an interveinal necrosis similar to that observed in a potassium deficiency of sugar beets.

Silicon. The essentiality of silicon for the growth of higher plants is a moot question. It has been found to be necessary to the growth of certain diatoms, and a recent rash of papers, largely from Japan, has suggested that silicon is required by rice and perhaps other crops, including cucumbers, gherkins, and barley. Silicon is reported to increase the top length, number of stems, and fresh and dry weight of rice plants grown in nutrient cultures. If silica were withheld during the reproductive period, there was a decrease in the number of spikelets per panicle and in the percentage of ripened grain.

TABLE 3-8. Crops Responding to Sodium Fertilization in the Absence and Presence of Adequate Potassium*

Degree of benefit with a deficiency of potassium		Degree of benefit with a sufficiency of potassium	
None to slight	Slight to medium	Slight to medium	Large
Buckwheat	Asparagus	Cabbage	Celery
Corn	Barley	Celeriac	Mangel
Lettuce	Broccoli	Horseradish	Sugar beet
Onion	Brussels sprouts	Kale	Swiss chard
Parsley	Caraway	Kohlrabi	Table beet
Parsnip	Carrot	Mustard	Turnip
Peppermint	Chickory	Radish	
Potato	Cotton	Rape	
Rye	Flax		
Soybean	Millet		
Spinach	Oat		
Squash	Pea		
Strawberry	Rutabaga		
Sunflower	Tomato		
White bean	Vetch		
	Wheat		

* Harmer et al., *Soil Sci.*, **60**:137 (1946). Reprinted with permission of The Williams & Wilkins Co., Baltimore.

Significant increases in the growth of sugar cane resulting from the application of calcium silicate slags and coral sands have been reported. These results were obtained on highly weathered tropical soils containing large amounts of iron oxides and small amounts of siliceous minerals.

The action of silicon in plant metabolic and physiological processes has not been determined.

Summary

1. All elements absorbed by plants are not necessarily essential to plant growth. The term *functional or metabolism nutrient* was introduced to include any element that functions in plant nutrition, regardless of whether its action is specific. It was suggested that this term might avoid the confusion that sometimes occurs in a definition of *essential* plant nutrients.
2. Twenty elements have been found to be essential to the growth of plants. Not all are required by all plants, but all are necessary to some plants.
3. The elements required by plants are carbon, hydrogen, oxygen, ni-

trogen, phosphorus, potassium, calcium, magnesium, sulfur, boron, iron, manganese, copper, zinc, molybdenum, chlorine, cobalt, vanadium, sodium, and silicon. The first three, with nitrogen, phosphorus, and sulfur, constitute their living matter or protoplasm. Elements other than carbon, hydrogen, and oxygen are termed *mineral nutrients* and are obtained by plants from the soil. The elements nitrogen, phosphorus, and potassium have been classed as major nutrients; calcium, magnesium, and sulfur, as secondary elements; and the remaining mineral nutrients, as microelements. These classifications are arbitrary and are probably based on the quantities required. As a very general rule, the major elements are needed in the largest amounts and the microelements, in the smallest.

4. Nitrogen is used largely in the synthesis of proteins, but structurally it is also a part of the chlorophyll molecule. Many proteins are enzymes, and the role of nitrogen can be considered as both structural and metabolic. The principal function of phosphorus is in the numerous energy transfer reactions that are effected by various phosphorylation and dephosphorylation reactions. They come about generally in the linking of phosphates to other compounds and in the breaking of the phosphate-compound bonds. The specific actions of potassium are numerous and it is necessary to many plant functions, especially those dealing with carbohydrate metabolism. It is also an activator of several enzymes.

5. Sulfur plays an important part in protein synthesis and is essential to the functioning of several enzyme systems. It is also believed to be necessary to oil formation, though this point has not been definitely established.

6. The remaining mineral elements are generally involved in the activation of various enzyme systems. Magnesium, in addition, is an essential component of the chlorophyll molecule.

7. The role of elements such as chlorine and silicon is not known.

Questions

1. Can you as a commercial grower do anything to supply plants with carbon, hydrogen, and oxygen? What, specifically?
2. In what ways does nitrogen function in plant growth?
3. Visually, how would you differentiate between nitrogen and potassium deficiencies of corn?
4. Is nitrogen a mobile element in plants? What visual proof is there?
5. Phosphorus is important in many plant functions. What, however, is probably its most important over-all function?
6. Nitrogen, phosphorus, and potassium are arbitrarily classed as macro-or major elements. In terms of their importance in plant nu-

trition, is this terminology justified? Why? What justification is there for such classification?

7. Fruit growers in a certain region of the Pacific Northwest decided to change their nitrogen fertilizer and use ammonium nitrate. A year or two later their trees turned a uniform light yellow-green. Tissue tests showed that the leaves were high in nitrogen. The trees were irrigated with water from a glacier-fed river into which no industrial wastes had been emptied. The area was far from industrial activities. Phosphorus, potassium, and magnesium levels were adequate, and soil pH was satisfactory. None of the microelements was in short supply. What element was most likely to be deficient? How would you correct this deficiency?

8. A deficiency of calcium is sometimes observed under very dry soil conditions. Can you explain this?

9. In what ways do the symptoms of magnesium and potassium deficiencies resemble each other? In what ways are they dissimilar?

10. What function of magnesium is unique?

11. Which of the three elements, nitrogen, phosphorus, and potassium, has recently been shown to be important to animal nutrition? In what ways is it thought to function in animal nutrition?

12. Which of the essential or metabolic elements are structurally a part of protoplasm?

13. What element is specifically involved in nitrate reduction in plants?

14. What element is required specifically by *Rhizobia* in the fixation of elemental nitrogen?

15. Which of the so-called microelements recently shown to be necessary for nitrogen fixation by legume bacteria is also essential to animal nutrition? Exclusive of ruminant animals, in what form is this element utilized by higher animals, including man?

16. Sulfur is an integral part of certain amino acids. Name the amino acids. Can ruminants synthesize these sulfur-containing amino acids from inorganic sulfur and nitrogen compounds?

17. What crops have responded to applications of sodium?

18. What precaution must be observed in applying elements such as copper, zinc, boron, cobalt, molybdenum, and magnesium to crops?

19. What is hypomagnesemia? It is often observed following application of large amounts of potassium- and ammonium-containing fertilizers. Does this suggest to you that the fertilization rate of these elements is too high or that the level of soil magnesium is too low? How can this condition be corrected permanently?

20. Name several elements, deficiencies of which are exhibited first in the apical region of the growing plant. What does this imply?

21. If you saw a field of dwarfed corn, reddish purple in color, and the corn plant tissue tested high in nitrate nitrogen, you might suspect that these plants were deficient in what element?

22. List the essential mineral elements required in plant nutrition and give the principal functions of each.

Selected References

Adams, S. N., "The effect of sodium and potassium fertilizer on the mineral composition of sugar beet," *J. Agri. Sci.*, **56**:383 [*Soils Fertilizers*, **24**:375 (1961)].

Ahmed, S., and H. J. Evans, "Cobalt: A micronutrient element for the growth of soybean plants under symbiotic conditions." *Soil Sc.*, **90**:205 (1960).

Albert, L. S., and C. M. Wilson, "Effect of boron on elongation of tomato root tips." *Plant Physiol.*, **36**:244 (1961).

Amberger, A., "The effect of boron nutrition on respiration intensity and quality of crops." *Landwirtsch. Forsch. Sonderh.*, **14**:107 [*Soils Fertilizers*, **23**:347 (1960).]

Arnon, D. I., and G. Wessel, Vanadium as an essential element for green plants. *Nature (London)*, **172**:1039 (1953).

Benson, N. R., E. S. Degman, I. C. Chmelir, *et al.*, "Sulfur deficiency in deciduous tree fruit." *Proc. Am. Soc. Hort. Sci.*, **83**:55 (1963).

Brownell, P. F., and C. J. Crossland, "The requirement for sodium as a micronutrient by species having the C_4 dicarboxylic photosynthetic pathway." *Plant Physiol.*, **49**:794 (1972).

Broyer, T. C., and P. R. Stout, "The macronutrient elements." *Ann. Rev. Plant Physiol.*, **10**:277 (1959).

Cairns, R. R., and R. B. Carson, "Effect of sulfur treatments on yield and nitrogen and sulfur content of alfalfa grown on sulfur-deficient and sulfur-sufficient grey wooded soils." *Can. J. Plant Sci.*, **41**:715 (1961).

Delwiche, C. C., C. M. Johnson, and H. M. Reisenauer, "Influence of cobalt on nitrogen fixation by Medicago." *Plant Physiol.*, **36**:73 (1961).

Evans, H. J., and G. J. Sorger, "Role of mineral elements with emphasis on the univalent cations." *Ann. Rev. Plant Physiol.*, **17**: 7 (1966).

Fox, R. L., J. A. Silva, O. R. Younge, D. L. Plucknett, and G. D. Sherman, "Soil and plant silicon and silicate response by sugar cane." *SSSA Proc.*, **31**:775 (1967).

Gauch, H. G., "Mineral nutrition of plants." *Ann. Rev. Plant Physiol.*, **8**:31 (1957).

Gilliam, J. W., "Hydrolysis and uptake of pyrophosphate by plant roots." *SSSA Proc.*, **34**:83 (1970).

Hewitt, E. J., and G. Bond, "Molybdenum and the fixation of nitrogen in *Casuarina* and *Alnus* root nodules." *Plant Soil*, **14**:159 (1961).

Hudson, J. P., "General effects of potash on the water economy of plants." *Potassium Symp.* (1958), p. 95.

Hunger Signs in Crops: A Symposium, 3rd ed., Howard B. Sprague, Ed. New York: David McKay, 1964.

Ishibashi, H., and M. Kawano, "The effect of silica on the growth of rice plants in water culture." *Bull. Fac. Agri. Yamaguti Univ.*, **8**:689 (1957).

Jackson, W. A., and H. J. Evans, "Effect of Ca supply on the development and composition of soybean seedlings." *Soil Sci.*, **94**:180 (1962).

Johnson, C. M., P. R. Stout, T. C. Broyer, and A. B. Carlton, "Comparative chlorine requirements of different plant species." *Plant Soil,* **8:**337 (1957).

Josephson, L. M., "Effects of potash on premature stalk drying and lodging of corn." *Agron. J.,* **54:**179 (1962).

Jung, G. A., and D. Smith, "Influence of soil potassium and phosphorus content on the cold resistance to alfalfa." *Agron. J.,* **51:**585 (1959).

Kandler, O., "Energy transfer through phosphorylation mechanisms in photosynthesis." *Ann. Rev. Plant Physiol.,* **11:**37 (1960).

Kilmer, V. J., S. E. Younts, and N. C. Brady, Eds., *The Role of Potassium in Agriculture.* Madison, Wisconsin: ASA, CSSA, SSSA (1968).

Levitt, J., C. Y. Sullivan, N. O. Johansson, and R. M. Pettit, "Sulfhydryls – a new factor in frost resistance. I. Changes in SH content during frost hardening." *Plant Physiol.,***6:**611 (1961).

Lewin, J., and B. E. F. Reimann, "Silicon and plant growth." *Ann. Rev. Plant Physiol.,* **20:**289 (1969).

Lowe, R. H., and H. J. Evans, "Cobalt requirement for the growth of Rhizobia." *J. Bact.,* **83:**210 (1962).

Marre, E., "Phosphorylation in higher plants." *Ann. Rev. Plant Physiol.,* **12:**195 (1961).

Mortvedt, J. J., P. M. Giordano, and W. L. Lindsay, Eds., *Micronutrients in Agriculture.* Madison, Wisconsin: SSSA, 1972.

Muth, O. H., and J. E. Oldfield, Eds., *Symposium: Sulfur in Nutrition.* Westport, Connecticut: The Avi Publishing Company, 1970.

Nicholas, D. J. D., "Minor mineral nutrients." *Ann. Rev. Plant Physiol.,* **12:**63 (1961).

Odelien, M., "The effect of sulfur supply on the quality of plant products." *Tidskr. Norske Land bruk.,* **70:**35 (1963). Translated by J. Platou, The Sulphur Institute.

Olson, R. A., T. J. Army, J. J. Hanway, and V. J. Kilmer, Eds., *Fertilizer Technology & Use.,* 2nd ed. Madison, Wisconsin: SSSA, 1971.

Plant Physiology – A Treatise, Vol. III. *Inorganic Nutrition of Plants,* F. C. Steward, Ed. New York and London: Academic, 1963.

Possingham, J. V., and D. Spencer, "Manganese as a functional component of chloroplasts." *Australian J. Biol. Sci.,* **15:**58 (1962).

Reisenauer, H. M., "Cobalt in nitrogen fixation by a legume." *Nature (London),* **186:**375 (1960).

Reisenauer, H. M., "The effect of sulfur on the absorption and utilization of molybdenum by peas." *SSSA Proc.,* **27:**553 (1963).

Spencer, K., "Growth and chemical composition of white clover as affected by sulfur supply." *Australian J. Agr. Res.,* **10:**500 (1959).

Teel, M. R., "Nitrogen-potassium relationships and biochemical intermediates in grass herbage." *Soil Sci.,* **93:**50 (1962).

Tisdale, S. L., R. L. Davis, A. F. Kingsley, and E. T. Mertz, "Methionine and cystine content of two strains of alfalfa as influenced by different concentrations of the sulfate." *Agron. J.,* **42:**221 (1950).

Underwood, E. J., *Trace Elements in Human and Animal Nutrition.* New York and London: Academic, 1962.

Venema, K. C. W., "Some notes regarding the function of the sulfate-anion in the metabolism of oil producing plants, especially oil palms." *Potash Trop. Agr.,* **5,** No. 3 (July 1962).

Venema, J. C. W., "Some notes regarding the function of the sulfate-anion in the matabolism of oil producing plants, especially oil palms." *Potash Trop. Agr.,* **5,** No. 4 (October 1962).

Wallace, A., *Regulation of the Micronutrient Status of Plants by Chelating Agents and Other Factors.* Los Angeles: Arthur Wallace, 1971.

Wallace, T., *The Diagnosis of Mineral Deficiencies in Plants by Visual Symptoms,* 2nd ed. New York: Chemical, 1961.

Whittington, W. J., "The role of boron in plant nutrition." Univ. of Nottingham, Rep. School Agric. 1959, p. 51–53.

Wong You Cheong, Y., and P. Halais, "Needs of sugar cane for silicon when growing in highly weathered latosols." *Expl. Agric.,* **6:**99 (1970).

4. BASIC SOIL-PLANT RELATIONSHIPS

The purpose of this chapter is to review briefly the phenomenon of ion exchange in soils and to consider some of the suggested mechanisms for the movement of ions in the soil solution and into the cells of the absorbing roots.

ION EXCHANGE IN SOILS

Ion exchange is simply the reversible process by which cations and anions are exchanged between solid and liquid phases. If two solid phases are in contact, exchange of ions may also take place between their surfaces.

Of the two phenomena, cation and anion exchange, the first is generally considered to be more important in soil. This may be attributed in part to a greater degree of study, for it is generally accepted that more is known about cation than about anion exchange.

Cation Exchange. Soils are composed of the three forms of matter—solids, liquids, and gases. The solid phase is made up of organic and inorganic materials, the organic fraction of which consists of the residues of plants and animals in all stages of decomposition, and the stable phase is usually termed humus.

The inorganic fraction of soil solids is composed of primary and secondary minerals, and in fact consists of particles of rock size or larger to sizes that are of colloidal dimensions.

The fractions of the soil that are the seats of ion exchange are the organic and the mineral fractions, with effective particle diameters of less than 20μ. This includes a portion of the silt and all of the clay fraction as well as colloidal organic matter.

Because cations are positively charged, they are attracted to surfaces which are negatively charged. In the organic fraction these charges arise from the -COOH and -OH groups and perhaps to some extent from the -NH$_2$ groups. The charge on the inorganic clay fraction generally arises from two sources. The first is isomorphous substitution and the second is caused by the ionization of hydroxyl groups attached to the silicon atoms at the broken edges of the tetrahedral planes.

The charge resulting from isomorphous substitution is fairly uniformly distributed over the plate-shaped clay particles. It arises from the substitution of a silicon or aluminum atom by an atom of similar geometry but of lower charge. This causes an excess negative charge to develop. The negative charges at the edge of the clay plates arise essentially by the reaction illustrated in the following equation:

$$SiOH + H_2O \rightleftharpoons SiO^- + H_3O^+$$

Mineral clays in soils are of two general classes — 2:1 and 1:1. The 2:1 clays are composed of layers, each of which consists of two silica sheets between which is a sheet of alumina. Examples of the 2:1 clays are montmorillonite, beidellite, and vermiculite; 1:1 clays are composed of layers, each of which consists of one silica sheet and one alumina sheet. The charge on the 2:1-type clays arises largely from isomorphous substitution, whereas that on the 1:1 type comes from the ionization of hydrogen from OH groups at the broken edges of the particles. The charge on the 2:1 clays is greater than that on the 1:1 clays.

The negative charge that develops on organic and mineral colloids is neutralized by cations attracted to the surfaces of these colloids. The quantity of cations expressed in milliequivalents per 100 g. of oven-dry soil is termed the *cation exchange capacity* (CEC) of the soil. It is one of the important chemical properties in soil and is usually closely related to soil fertility. A thorough understanding of cation exchange is necessary to an understanding of soil fertility and acidity. Therefore the following brief review is given of the way in which this quantity is determined. Procedures differ for measuring CEC of soils, but the following simplified description illustrates the basic features.

Cation exchange, as previously pointed out, means the exchange of one cation for another on the surface of a colloid. Soil colloids have adsorbed to their exchange sites numerous cations, including calcium, magnesium, potassium, sodium, ammonium, aluminum, iron, and hydrogen. These ions are held with varying degrees of tenacity, depending on their charges and their hydrated and unhydrated radii. As a rule, ions with a valence of 2 or 3 are held more tightly than monovalent cations. Also, the greater the degree to which the ion is hydrated, the less tightly it will be held.

When a soil containing this population of different ions is extracted with a fairly concentrated aqueous salt solution, such as one normal

($1.0N$) ammonium acetate, all of the adsorbed cations are replaced by the ammonium ions. The soil colloidal fraction is then saturated with ammonium. If this ammonium-saturated soil is extracted with a solution of a different salt, say $1.0N$ KCl, the potassium ions will replace the ammonium ions. If the soil-potassium chloride suspension is filtered, the filtrate will contain the ammonium ions adsorbed by the soil. The quantity of ammonium ions in the leachate is a measure of the CEC of the soil in question and can easily be determined.

To illustrate, suppose that 20 g. of oven-dry soil were extracted with 200 ml. of $1.0N$ NH_4Ac (ammonium acetate). The extraction is accomplished by intermittent shaking over a period of 30 minutes. The soil-ammonium acetate solution is filtered and the soil is washed with alcohol to remove the *excess* solution. The soil containing the adsorbed ammonium ions is next extracted with 200 ml. of a solution of $1.0N$ KCl. The soil-potassium chloride solution is filtered and the ammonium contained in the filtrate is determined. Suppose that 0.054 g. of NH_4^+ were found. This was, of course, retained by the 20 g. of soil extracted (0.054 g. is 3 meq. — i.e., $0.054/0.018 = 3$, as 0.018 g. is the milliequivalent weight of 1 meq.). Because 3 meq. were present in 20 g. of soil, the CEC of the soil is 15 meq./100 g.

The CEC of a soil will obviously be affected by the nature and amount of mineral and organic colloid present. As a rule, soils with large amounts of clay and organic matter will have higher exchange capacities than sandy soils low in organic matter. Also, soils with predominately 2:1 colloids will have higher exchange capacities than soils with predominately 1:1 mineral colloids.

Generally, 1:1 mineral colloids have CEC values of 10 to 20 meq./100 g.; 2:1 mineral colloids, about 40 to 80 meq./100 g., and organic colloids, 100 to 200 or more meq./100 g.

Effective CEC. The use of ammonium acetate for determining the CEC of soils has been and is still used by many laboratories in the USA. However, some workers believe that CEC can better be estimated by extraction with an unbuffered salt which would give a measure of the CEC at the soils normal pH. Use of neutral normal ammonium acetate will result in a high CEC value if the soil is acid simply because of the adsorption of NH_4^+ ions to the so-called pH-dependent exchange sites. Extraction with a neutral unbuffered salt will not.

Coleman, Kamprath, Thomas and other workers in North Carolina have defined exchangeable cations in acid soils as those cations extracted with a neutral unbuffered salt, with the sum of these cations being termed the effective cation exchange capacity. Such an unbuffered salt solution ($1.0N$ KCl in this case) will extract only the cations held at active exchange sites at the particular pH of the soil. The exchangeable acidity thus extracted is Al. To measure the effective CEC, one sample of soil is extracted with neutral normal NH_4Ac to determine the exchangeable basic cations such as K, Ca, Mg, and Na. Another sample of the same soil is extracted with $1.0N$ KCl to determine the exchangeable Al. The sum of the milliequivalents of basic cations plus Al is the effective CEC.

Base Saturation. One of the important properties of a soil is its degree of base saturation. It is defined as the percentage of total CEC occupied by such basic cations as calcium, magnesium, sodium, and potassium. To illustrate how this quantity is calculated, suppose that in the example given the following ion quantities were found in the ammonium acetate extract from the leaching of the 20 g. of soil:

Ca	0.02 g.
Mg	0.006 g.
Na	0.0115 g.
K	0.0195 g.

The milliequivalent weights of calcium, magnesium, sodium, and potassium are, respectively, 0.02, 0.012, 0.023, and 0.039. The milliequivalents of each of these ions present is

Ca	= 0.02/0.02	= 1 meq.
Mg	= 0.006/0.012	= 0.5 meq.
Na	= 0.0115/0.023	= 0.5 meq.
K	= 0.0195/0.039	= 0.5 meq.
Total		2.5 meq./20 g. of soil

(2.5 meq. of bases/20 g. of soil is 12.5 meq./100 g. of soil). The total CEC of this soil was 15 meq./100 g., and so the percentage of base saturation is $(12.5/15) \times 100$, or 83.3.

As a general rule, the degree of base saturation of normal uncultivated soils is higher for arid than for humid region soils. Though not always true, especially in humid regions, the degree of base saturation of soils formed from limestones or basic igneous rocks is greater than that of soils formed from sandstones or acid igneous rocks. Base saturation is, of course, related to soil pH and to the level of soil fertility. For a soil of any given organic and mineral composition, the pH and fertility level increase with an increase in the degree of base saturation.

The ease with which cations are absorbed by plants is related to the degree of base saturation. For any given soil the availability of the basic cations to plants increases with the degree of base saturation. For example, a soil with a base saturation of 80 per cent would provide cations to growing plants far more easily than the same soil with a base saturation of only 40 per cent. The relation between per cent base saturation and cation availability is modified by the nature of the soil colloid. As a rule, soils with large amounts of organic or 1 : 1 colloids can supply basic cations to plants at a much lower degree of base saturation than soils high in 2 : 1 colloids.

As will be seen in the following section, the CEC, hence the percentage of base saturation, can be rather arbitrary figures unless the method by which they are measured is clearly defined. As a general rule, however, the statements made concerning base saturation and plant availability of cations are true.

Nature of Charge and CEC. The foregoing discussion of cation exchange and degree of base saturation was deliberately oversimplified to set forth clearly the fundamental reaction in this basic soil phenomenon. Actually, studies have shown that the cation exchange capacity of a soil is not a fixed quantity but is dependent on the pH of the extracting solution used for its determination. The total negative charge on soil colloids which gives rise to cation exchange is caused by isomorphous substitution of ions in the lattice structure of clay minerals, the ionization of hydroxyl groups from hydrated iron and aluminum oxides, and organic matter. Numerous studies have shown that the cation exchange capacity of soils is a continuous function of pH, with this value being lowest in the acid range, pH 3–4, and increasing continuously as the pH increases up to the alkaline range, pH 8–9. This increase in CEC with increasing pH is caused by the ionization of the OH groups at the edges of the clay lattice and on the hydrous Al and Fe oxides and from the carboxyl and phenolic groups present in soil organic matter.

When the CEC of a soil is determined using an unbuffered neutral salt solution, the value obtained will be lower than would be the case if it were measured using a highly buffered solution at a pH of 7, 8, or 9. The effective CEC discussed in the preceding section therefore is probably a more meaningful value as far as plant growth, fertilizer additions, and liming are concerned than the CEC determined with the buffered solutions at high pH values. The CEC values found by using neutral normal ammonium acetate are somewhere between the values found by using the unbuffered salt solution and barium chloride, triethanol amine. For routine CEC determinations, the ammonium acetate method is rapid and convenient and is still used in many laboratories in the United States. If one keeps in mind that the value obtained will be greater than that obtained using the unbuffered salt solution and accordingly makes the necessary allowances, it is a perfectly satisfactory method for measuring this important soil property.

ANION EXCHANGE

It has been known for a long time that phosphates do not leach from soils but are retained in forms that may be removed only by extraction with various salt, acid, and alkaline solutions. A fraction of the phosphorus appears to be held in forms that are quite insoluble. The topic of phosphate retention is covered more fully in Chapter 6. More recently it has been found that much larger amounts of sulfate can be extracted from soils high in 1:1 clays and the hydrous oxides of iron and aluminum with a solution of potassium phosphate than can be extracted with water.

These findings have led to the realization that soils do indeed possess anion exchange properties, and subsequent studies have shown that anions such as chlorides and nitrates may be adsorbed, though not to the extent of phosphates and sulfates.

Contrary to cation exchange, the capacity for retaining anions increases with a decrease in soil pH. Further, anion exchange is much greater in soils high in 1:1 clays and those containing hydrous oxides of iron and aluminum than it is in soils with predominately 2:1 clays.

The mechanism of anion adsorption is not clearly understood. However, many workers believe it to be the result of OH ion displacement from the hydrated oxides of iron and aluminum. These oxides are found in large quantities in highly leached soils of the tropics and subtropics, and it is in such soils that anion exchange is greatest. The reaction can be illustrated as follows:

$$R-OH + HSO_4^- \rightleftharpoons R-OH_2^+SO_4^-$$

In strongly acid soils, where the concentration of OH^- ions is extremely low, other anions are assumed to be in competition with the hydroxyl ions found in the hydrous oxides present in the soil. Therefore, at low pH values and in soils with large amounts of hydrated Fe and Al oxides, the amounts of such anions as SO_4^{-2} that are retained may be quite high.

Anion exchange is the result of both a low soil pH and the presence of the hydrous oxides of iron and aluminum, and it is largely a pH-dependent phenomenon—the more acid the soil, the greater the extent of anion adsorption. Under most field conditions in the U. S., soils have pH values at which anion adsorption is at a minimum and as a result, anions, with the exception of phosphate and to a lesser degree sulfate, will be lost by leaching. The soils of the southeast and certain of the soils formed under the high rainfall conditions in Washington and Oregon are exceptions to this general statement. The subsurface horizons of such soils are frequently acid and high in their content of Fe and Al hydrous oxides, and anion adsorption, particularly sulfate, is observed.

The retention of chlorides and nitrates by anion adsorption, while possibly demonstrable under laboratory conditions, is generally not considered to be of any great practical significance in most agricultural soils in North America.

A more extensive treatment of phosphate and sulfate retention in soils will be found in Chapters 6 and 8.

CONTACT EXCHANGE

The discussion of ion exchange so far has dealt only with the exchange of ions between liquid and solid phases. It is believed by some that this exchange can take place between ions held on the surfaces of solid phase particles and that it does not have to occur via the liquid phase. The extension of this theory leads to the conclusion that ions attached to the surface of root hairs (such as H^+ ions) may exchange with those held on the surface of clays and organic matter in soils because of the intimate contact that exists between roots and soil particles.

The mechanism that permits such an exchange could be described in this way. Clays and plant roots both have CEC properties. Ions are believed to be held at certain spots or sites on both roots and colloidal soil surfaces. The ions held by electrostatic or van der Waals forces at these sites tend to oscillate within a certain volume. When the oscillation volumes of two ions overlap, the ions exchange places. In this way a calcium ion on a clay surface could then presumably be absorbed by the root and utilized by the plant.

Contact exchange is generally considered to be a real phenomenon, but it is not thought to contribute greatly to the total amount of nutrients absorbed by most plants.

ROOT CATION EXCHANGE CAPACITIES

Soil colloids are not the only component of the soil-plant system to exhibit cation exchange properties. It has been observed that plant roots themselves may also possess this property.

The cation exchange capacity of roots is usually measured by electrodialyzing a given mass of roots for some specified period of time and then titrating it with calcium hydroxide. By this method values ranging from 12 to 60 meq./100 g. of oven-dry roots have been reported. Here again there is no universality of opinion among soil scientists in regard to what is actually being measured when roots are electrodialyzed and titrated with a base. It is presumed, of course, that during the dialysis all of the cations adsorbed to the root surface migrate to the cathode cell and that the root exchange sites are left saturated with H^+. The neutralization of the adsorbed H^+ would then be a measure of the CEC of the root. However, during electrolysis and the resultant increase in acidity there may also be some removal of organic constituents as well as of inorganic ions. If organic acids were set free, the consumption of the base during titration would include the presence of these acids and the H^+ adsorbed by the root. Questions concerning the determination of root CEC and the magnitude of the values measured suggest that further work is required to establish the importance of this phenomenon.

Plants differ considerably in the magnitude of their measured root CEC's. Legumes generally have higher values than grasses. These observations have been correlated with others showing that legumes absorbed divalent cations preferentially over monovalent cations and that the converse is true with grasses. These observations have also been used to explain why, in grass-legume pastures in which the soil K^+ level is less than adequate, the grass survives but the legume disappears. The grasses are considered to be more effective absorbers of potassium than legumes.

MOVEMENT OF IONS FROM SOILS TO ROOTS

In order for ions to be absorbed by plant roots, they must come in contact with the root surface. There are generally three ways in which this contact is effected. These are: (a) contact exchange, which was discussed in a preceding section; (b) diffusion of ions in the soil solution; and (c) movement of ions by mass movement of water in the soil.

The importance of contact exchange as a mechanism for ion absorption is enhanced by the growth of new roots through the soil mass. As the root system develops and penetrates new soil regions, soil surfaces containing adsorbed ions are presented to the root mass and absorption of these ions by the contact exchange mechanism is effected. Its total contribution to the uptake of nutrients, however, is considered to be small, as was pointed out previously.

Movement of ions in the soil solution to the surfaces of roots is an important factor in ion uptake. This movement is accomplished largely by mass flow and diffusion. Mass flow refers to the movement of water together with dissolved electrolytes (ions) through the soil. This movement occurs as a result of rainfall or irrigation but more importantly as a result of the diffusion pressure gradient set up by water absorption by plant roots. The continued absorption of water causes the surrounding mass of soil moisture to move toward the plant root system bringing with it the dissolved nutrient ions such as NO_3^-, Ca^{2+}, and Mg^{2+}.

Work has shown that while mass flow and contact exchange contribute to the ion population presented to the root surface, ion diffusion is probably the most important process whereby $H_2PO_4^-$ and K^+ ions reach the root surface. As plant roots absorb nutrients from the surrounding soil solution, a diffusion gradient is set up. This gradient results in the continuous movement of additional ions to the root surface and their absorption by the plant. NO_3^-, Ca^{2+}, and Mg^{2+} move to the root surface mainly by mass flow.

Under some conditions the concentration of certain ions may build up at the root surface because the root is unable to absorb them at a sufficiently rapid rate. This results in a phenomenon known as "back diffusion" in which the concentration gradient, and hence the movement of certain ions, will be away from the root surface and back toward the soil solution. Normally such a condition will not occur, but as roots do not absorb all nutrient ions at the same rate, there may on occasion be a build-up of those ions that are less rapidly absorbed, particularly during periods when the plant is absorbing moisture rapidly.

The importance of both diffusion and mass flow in supplying the root surface with ions for absorption depends upon the ability of the solid phase of the soil to supply the liquid phase with these ions. Solution concentrations of ions will be influenced by the nature of the colloidal fraction of the soil and the degree to which these colloids are saturated with basic cations. The nature of the adsorbed cations is also important. Results of several studies have shown, for example, that the ease of replacement of calcium from colloids, either by dilute hydrochloric acid or by plant uptake, varies in this order: peat > kaolinite > illite > montmorillonite. Mehlich in North Carolina, for example, showed that an 80 per cent calcium-saturated beidellite clay (a 2 : 1 clay) gave the same percentage release of this ion as a 35 per cent calcium-saturated kaolinite or a 25 per cent calcium-saturated peat. Other workers have observed similar relationships.

There is another phenomenon affecting the nature of ions in the soil solution. It is known as the *complementary ion effect* and is defined as the influence of one adsorbed ion on the release of another from the surface of a colloid. High concentrations of potassium in the soil will reduce the uptake of calcium and magnesium by plants. This would appear to be an antagonistic effect of potassium on the uptake of the other two elements. However, such antagonism is generally not observed in straight solution cultures in which no soil or colloid is present in the rooting medium. Ions that are very similar *do* exhibit antagonism. For example, potassium will interfere with the uptake of rubidium, or vice versa, in solution cultures. A similar relation exists between calcium and strontium.

Potassium is not similar to calcium or magnesium. Therefore, the effects observed when plants are grown in soils must be caused by something other than simple ion antagonism. It is generally explained as the result of the 'complementary ion effect.' Soil colloids retain divalent cations more tightly than monovalent cations. If the exchangeable bases on a soil colloid consist of large amounts of potassium and NH_4^+ in proportion to the amounts of calcium and magnesium, these monovalent ions will be replaced more easily than the divalent ions. Hence the plant absorption of the former will be greater than the absorption of the latter. In extreme cases, in which the ratio of monovalent to divalent ions is very wide, plant deficiencies of the divalent ions may be induced. A good example is found in hypomagnesemia of ruminants, mentioned in Chapter 7. The addition of large amounts of K^+ (and perhaps also NH_4^+) results in decreased absorption of magnesium by plants. Ruminants grazing such forage may suffer from a magnesium deficiency, known as grass tetany or hypomagnesemia. The obvious correction is to increase the percentage of magnesium saturation of the soil with applications of a high magnesium dolomitic limestone or the addition of a magnesium-containing fertilizer.

The addition of fertilizers to the soil is of equal or greater importance than the soil colloids in maintaining a high concentration of nutrient ions in the soil solution. As the plant roots absorb the ions presented to their surface, the presence of added fertilizer will ensure the maintenance of the steep diffusion pressure gradient needed to keep the root surface supplied with these ions.

Two other factors play a role in modifying the delivery of ions to the plant root surface by diffusion and mass flow. These are soil texture and moisture content. The finer the texture of the soil, the less rapid will be the movement of soil moisture and the diffusion of ions through the water. A fine-textured soil offers greater resistance to moisture movement than does a coarse-textured sandy soil. Also, ions diffusing through soil moisture in clay soils are much more likely to be attracted to adsorption sites on the clay than would be the case in a sandy soil.

A reduction in soil moisture has a similar effect on both water movement and ion diffusion. As soil moisture is reduced (an increase in soil moisture tension), water movement slows down. Thus the movement of moisture to the root surface is slowed. Similarly, as the moisture content of the soil is lowered, the moisture films around the soil particles become thinner and the diffusion of ions through these films is restricted. Delivery of nutrients to the root surface is probably most rapid at a soil moisture content corresponding to field capacity. The various mechanisms discussed above which affect the movement of nutrient ions to root surfaces are illustrated in Figure 4-1.

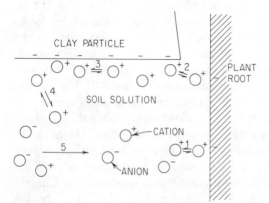

Figure 4–1. A diagrammatic sketch of the processes involved in the movement of cations to the root surface: (1) Solution diffusion; (2) diffusion from the particle to the root; (3) particle diffusion; (4) replenishment; and (5) mass flow. [Barber, *Soil Sci.,* **93**:39 (1962).]

ION ABSORPTION BY PLANTS

The last subject to be treated in this review of soil-plant relationships is that of the absorption of ions by plants. This again is a topic of great complexity and one that can be given only passing treatment in this book. The present consensus concerning the mechanics of ion absorption is that ions enter roots by exchange, by diffusion, and by the action of carriers or metabolic ion-binding compounds. These three mechanisms are associated with two components of the root system. One is termed *outer space,* or apparent free space, and the other is termed *inner space.* Absorption of ions into the outer space is believed to be governed by the processes of simple diffusion and exchange adsorption. Absorption of ions into the inner space is metabolic — that is, an expenditure of energy by the root cell is required for this type of absorption. In contrast to absorption in the outer space, it is largely irreversible.

The characteristics of the modes of entry of ions into the two types of root space have been summarized in Table 4-1.

The concept of an ion-binding compound or carrier is generally accepted, and active accumulation (accumulation against a concentration gradient) apparently involves a combination of the ions with a protoplasmic component, thus accounting for the selectivity of ions in this type of absorption. The various carrier theories all envision a metabolically produced substance that combines with free ions. This carrier-ion complex can then cross membranes and other barriers not permeable to free ions. After the transfer is accomplished, the ion-carrier complex is broken, the ion is released into the inner space of the cell, and the carrier is believed in some cases to be restored.

TABLE 4–1. Characteristics of Ion Movement into Inner and Outer Cell Spaces*

Outer	Inner
Diffusion and exchange adsorption	Ion-binding compounds or carriers
Nonlinear with time, equilibrium approached in short times	Linear with time
Ions stoichiometrically exchangeable for other ions	Ions essentially nonexchangeable and dialyzable
Not highly selective	Specific with regard to site and entry
Nonmetabolic	Dependent on aerobic metabolism
Ions in solution or adsorbed in outer space	Ions presumably in vacuoles

* Gauch, *Ann. Rev. Plant Physiol.,* **8**:31 (1957). Reprinted with permission of the author and the publisher, Annual Reviews, Inc., Palo Alto, California.

As indicated in Table 4-1, transfer into the inner space of cells is a highly selective process. Although potassium, rubidium, and cesium compete for the same carrier, they do not compete with elements such as calcium, strontium, and barium. The last three elements do, however, compete among themselves for another carrier. Selenium will compete with sulfate but not with phosphate nor with monovalent anions. Interestingly, $H_2PO_4^-$ and HPO_4^{2-} apparently have separate carriers and do not compete with one another for entry into the inner space.

The nature of the ion-binding compounds is not known but numerous suggestions concerning their nature have been advanced. Among these is one which proposes that adenosine triphosphate may react with certain metabolic intermediates (specifically of the so-called Krebs cycle) to form or destroy ion carriers. Another hypothesis suggests that the carriers are ribonucleoproteins in which nucleic acid binds the cations and the protein moiety binds the anions. Still other workers suggest that the carriers are phosphorylated nitrogen-containing intermediates in protein synthesis and that the carrier releases the ions on incorporation into the protein at the site of synthesis.

The cell mitochondria have also been suggested as ion carriers, and some work has indicated that cations and anions accumulate in these bodies. Considerable work has been done in Sweden in which the cytochromes are implicated in the active transport of ions. It has been pointed out by other workers, however, that the observed high correlation between cytochrome activity and ion uptake does not constitute proof that ion transport is achieved by the operation of the cytochrome system.

Considerable attention has recently been given to the passive absorption of ions by plant roots. This absorption takes place in the outer-space volume of the roots, and studies have shown that a considerable fraction of the total volume of plant cells is accessible for the entry of ions by diffusion or mass flow. Structurally, the free space is believed to include intercellular spaces, wet cell walls, and, for many solutes, the cytoplasm surrounding the tonoplast. The tonoplast itself is generally regarded as the boundary between outer and inner space.

The possible relation between outer space and translocation of ions to the tops of plants is of considerable interest. Some workers maintain that ions actively accumulated by roots are not exchangeable, hence are not free to move to the tops of plants. The ions in the outer space, however, apparently are transported quite freely to plant tops. Other workers have observed that the movement of ions from roots to shoots is determined by the rates of water absorption and transpiration, suggesting that mass flow may be important in the movement of ions. They also observed that the relative contributions of the active and passive components to the total accumulation of ions in shoots depended both on water absorption and the concentration of the medium. The passive component is more important at high than at low solute concentrations.

Because active absorption is believed to result in fixation in the tissue of the ions so absorbed, the concepts of free space and passive absorption are useful in explaining the movement of salts through root systems. Many workers believe that the cells of roots and plant tops may accumulate ions from the transpiration stream, for obviously any ions that enter the roots and are carried upward by mass flow come in contact with many adsorption sites on numerous actively accumulating cells. This in turn would lead to a considerable gradient in the concentration of ions in the transpiration stream and may support the observation that ions from dilute external solutions do not appear to reach plant tops through passive absorption.

Mass flow as a means of explaining ion transport is not without its flaws. Selectivity of ions in mass flow is most difficult to explain. An excellent example is found in sodium, which is virtually excluded from the tops of some plant species. Despite the incompleteness of the picture in regard to ion absorption, the present concepts of outer and inner space and their relation to active and passive absorption of ions are useful in explaining many of the observed facts. As in many other incompletely understood biological phenomena, future research will undoubtedly unveil more and more that is presently hidden from view.

For those readers interested in pursuing this subject, several excellent references are listed at the end of this chapter.

Summary

1. Ion exchange, which is defined as the reversible process by which cations and anions are exchanged between solid and liquid phases, was reviewed with emphasis on the phenomenon as it occurs in soils.
2. The determination of cation exchange capacity (CEC) in soils was reviewed, and the factors affecting this important property were discussed. The CEC of a soil is related to the nature and amounts of the mineral and organic colloids present.
3. Base saturation, which is the degree to which the exchange capacity of a soil is saturated with basic cations, was covered in relation to the nature of the charge on the exchange complex. The dependence of the CEC on the way in which this property is measured was discussed.
4. Anion exchange can take place in soils, but for all practical purposes it is confined to the phosphate and sulfate ions. Unlike cation exchange, anion exchange increases with a decrease in soil pH. The reason that other anions do not undergo adsorption in most agricultural soils was explained.
5. Contact exchange, which is the exchange of ions between the sur-

faces of two solids without movement through a liquid phase, was described. Its possible importance in soils was considered because plant roots themselves exhibit the property of cation exchange. This later phenomenon was also briefly discussed.

6. Plant nutrient ions are brought into contact with the absorbing surfaces of roots by: (a) contact exchange which is enhanced by the growth of roots through the soil mass; (b) diffusion of ions in the soil solution; and (c) mass flow of soil water which brings the nutrient ions into contact with the plant roots. Absorption of moisture by the roots is one of the principal causes of mass flow of soil water.

7. Some of the suggested mechanisms by which ions are actually absorbed by plants were considered. Ions are absorbed by both active and passive mechanisms. Active absorption is thought to take place by metabolically produced carriers which transfer the ions across otherwise impassable barriers. Passive absorption is governed largely by adsorption and diffusion phenomena. Actively absorbed ions are believed to be taken into what is termed *inner space* in roots, whereas ions that are passively absorbed move into the so-called *outer space* of roots. It was pointed out that much is still to be learned about the mechanics of ion absorption by roots.

Questions

1. Define ion exchange.
2. From what sources does the charge on soil colloids arise?
3. Why does anion adsorption appear to be of no importance in most agricultural soils?
4. Potassium acetate solution was used to determine the cation exchange capacity of a soil. 10 g. of oven-dry soil were extracted with 200 ml. of $1.0N$ KAc. It was found that 0.078 g. of potassium were retained by the soil. What is its CEC?
5. Can one determine by the method described in question 4 above the per cent base saturation of this soil? Why?
6. In the example listed on page 108, what assumption must be made if the percentage of base saturation measured is to be considered valid? Under what conditions would you assume that this assumption is valid?
7. Cation exchange in soils appears to increase as the pH _____.
   Anion exchange in soils appears to increase as the pH _____.
8. A soil was found to have a CEC of 24 meq./100 g. If the exchange capacity were saturated with sodium, to how many grams of sodium chloride would this be equivalent?
9. The CEC of a soil as measured using $1.0N$ neutral ammonium acetate gives lower values than those obtained when a solution of $BaCl_2$-triethanolamine buffered at pH 8.3 is used. Why?

10. What is the origin of the effective CEC in mineral soil colloids?
11. What is contact exchange?
12. A soil has a CEC of 25 meq./100 g. on an oven-dry basis. Suppose that this soil was 5 per cent saturated with potassium and you wished to increase the saturation to 9 per cent. Assuming that all of the added potassium would be adsorbed, how much 100 per cent potassium chloride would have to be added to an acre furrow slice of this soil to raise the potassium saturation to the desired level? Assume that an acre furrow slice of oven-dry soil weighs 2 million lb.
13. What mechanism is believed to be the cause of anion adsorption?
14. Under what soil conditions will the SO_4^{2-} ion be adsorbed to the greatest extent?
15. Based on methods presently used to measure the CEC of roots, what type of plant root appears to have the greater CEC – grasses or legumes? Has the CEC been related to the behavior of these plants growing in the field? Explain.
16. What is your opinion as to the relative importance of contact exchange, simple diffusion, and mass flow in bringing nutrient ions into contact with the absorbing surfaces of plant roots? Would the importance of these three mechanisms be altered by soil texture? By plant species? Why?
17. Assume that 50 lb. of sulfur as sulfate were lost per acre per year by leaching and run off. Assume that another 30 lb. were lost through crop removal. How many milliequivalents of cations per 100 g. of soil would have to accompany this sulfur to maintain electrical neutrality? If calcium were the only ion involved, to how many pounds per acre of calcium would this removal amount?
18. What is active and passive absorption of elements by plant root cells? In what way are these types of absorption related to the so-called inner and outer space in roots?
19. What, generally, is the mechanism that has been proposed to account for the active absorption of ions by roots?
20. What is the complementary ion effect? In what way does it influence plant uptake of ions? Is it of any practical consequence to commercial crop production? Explain. Give some specific examples to support your answer.
21. Why is cation exchange such an important item in a study of soil fertility and in commercial crop production as well, for that matter?
22. Assume that you are addressing a group of farmers and businessmen who are well versed in crop production but who are not so conversant with the technical aspects of plant nutrition and soil fertility. Your mission is to explain to this group the nature of cation exchange and why it is important to crop production. How would you proceed?

23. During the discussion after your speech, one member of the audience asks why chlorides and nitrates will leach from soils but phosphates, which also have a negative charge, will not. What is your answer?

Selected References

Adams, F., "Ionic concentrations and activities in soil solutions." *SSSA Proc.*, **35**:420 (1971).

Asher, C. J., and P. G. Ozanne, "The cation exchange capacity of plant roots and its relation to the uptake of insoluble nutrients." *Australian J. Agr. Res.*, **12**:755 (1961).

Barber, S. A., "A diffusion and mass flow concept of soil nutrient availability." *Soil Sci.*, **93**:39 (1962).

Black, C. A., *Soil-Plant Relationships*, 2nd Ed. New York: Wiley, 1968.

Bradfield, R., "A quarter century in soil fertility research and a glimpse into the future." *SSSA Proc.* **25**:439 (1961).

Chao, Tsun Tien, and M. E. Harward, "Nature of acid clays and relationships to ion activities and ion ratios in equilibrium solutions." *Soil Sci.*, **93**:246 (1962).

Coleman, T. R., E. J. Kamprath, and S. B. Weed, "Liming." *Advan. Agron.*, **10**:475 (1958).

Cooke, G. W., "Chemical aspects of soil fertility." *Soils Fertilizers*, **25**:417 (1962).

Crooke, W. M., and A. H. Knight, "An evaluation of published data on the mineral composition of plants in light of the cation exchange capacity of their roots." *Soil Sci.*, **93**:365 (1962).

Epstein, E., "Mineral nutrition of plants: Mechanisms of uptake and transport." *Ann. Rev. Plant Physiol.*, **7**:1 (1956).

Franklin, R. E., "Exchange and absorption of cations by excised roots." *SSSA Proc.*, **30**:177 (1966).

Fried, M., and H. Broeshart, *The Soil-Plant System*. New York and London: Academic, 1967.

Gauch, H. G., "Mineral nutrition of plants." *Ann. Rev. Plant Physiol.*, **8**:31 (1957).

Heintze, S. G., "Studies on cation-exchange capacities of roots." *Plant Soil*, **13**:365 (1961).

Huffaker, R. C., and A. Wallace, "Possible relationships of cation exchange capacity of plant roots to cation uptake." *SSSA Proc.*, **22**:392 (1958).

Huffaker, R. C., and A. Wallace, "Variation in root cation exchange capacity within plant species." *Agron. J.*, **51**:120 (1959).

Jackson, M. L., "Aluminum bonding in soils: A unifying principle in soil science." *SSSA Proc.*, **27**:1 (1963).

Khasawneh, F. E., "Solution ion activity and plant growth." *SSSA Proc.*, **35**:426 (1971).

Lagerwerff, J. V., "The contact exchange theory amended." *Plant Soil*, **13**:253 (1960).

Mouat, M. C. N., and T. W. Walker, "Competition for nutrients between grasses and white clover. I. Effect of grass species and nitrogen supply." *Plant Soil*, **11**:30 (1959).

Mouat, M. C. N., and T. W. Walker, "Competition for nutrients between grasses and white clover. II. Effect of root cation-exchange capacity and rate of emergence of associated species. *Plant Soil,* **11**:41 (1959).

Ohlrogge, A. J., "Some soil-root-plant relationships." *Soil Sci.,* **93**:30 (1962).

Oliver, S., and S. A. Barber, "An evaluation of the mechanisms governing the supply of Ca, Mg, K, and Na to soybean roots." *SSSA Proc.,* **30**:82 (1966).

Olsen, R. A., "The driving force on an ion in the absorption process." *SSSA Proc.,* **32**:660 (1968).

Pearson, R. W., and F. Adams, Eds., *Soil Acidity and Liming.* Agronomy **12**:1 (1967).

Rovira, A. D., "Plant-root excretions in relation to the rhizosphere effect. I. Nature of root exudate from oats and peas." *Plant Soil,* **7**:178 (1956).

Rovira, A. D., "Plant-root excretions in relation to the rhizosphere effect. II. A study of the properties of root exudate and its effect on the growth of micro-organisms isolated from the rhizosphere and control soil." *Plant Soil,* **7**:195 (1956).

Rovira, A. D., "Plant-root excretions in relation to the rhizosphere effect. III. The effect of root exudate on the numbers and activity of micro-organisms in soil." *Plant Soil,* **7**:209 (1956).

Rovira, A. D., "Plant-root exudates in relation to the rhizosphere microflora." *Soils Fertilizers,* **25**:167 (1962).

Schuffelen, A. C., "Growth substance and ion absorption." *Plant Soil,* **1**:121 (1949).

Sommerfeldt, T. D., "Effect of anions in the system on the amount of cations adsorbed by soil materials." *SSSA Proc.,* **26**:141 (1963).

Wiersum, L. K., "Utilization of the soil by the plant root system." *Plant Soil,* **15**:189 (1961).

Wiersum, L. K., and K. Bakema, "Competitive adaptation of the CEC of roots." *Plant Soil,* **11**:287 (1959).

Williams, D. E., "Anion-exchange properties of plant root surfaces." *Science,* **138**:153 (1962).

Williams, D. E., and N. T. Coleman, "Cation exchange properties of plant root surfaces." *Plant Soil,* **2**:243 (1950).

5. SOIL AND FERTILIZER NITROGEN

The ultimate source of the nitrogen used by plants is the inert gas N_2, which constitutes about 78 per cent of the earth's atmosphere. In its elemental form, however, it is useless to higher plants. The primary path-ways by which nitrogen is converted to forms usable by higher plants are these:

1. Fixation by *Rhizobia* and other microorganisms which live symbiotically on the roots of legumes and certain nonleguminous plants.
2. Fixation by free-living soil microorganisms and perhaps by organisms living on the leaves of tropical plants.
3. Fixation as one of the oxides of nitrogen by atmospheric electrical discharges.
4. Fixation as ammonia, NO_3^{2-}, or CN_2^{2-} by any of the various industrial processes for the manufacture of synthetic nitrogen fertilizers.

The supply of elemental nitrogen is for all intents and purposes inexhaustible. This inert nitrogen is in dynamic equilibrium with the various fixed forms. Even as nitrogen is fixed by the different processes just indicated, so is there a release of elemental nitrogen to the atmosphere from these fixed forms by microbiological and chemical processes which are discussed in this chapter. The so-called nitrogen cycle, which illustrates these transformations, is shown in Figure 5-1.

NITROGEN FIXATION BY
RHIZOBIA AND OTHER SYMBIOTIC BACTERIA

For centuries the use of legumes in crop rotations and the application of animal manures were the principal ways of supplying additional ni-

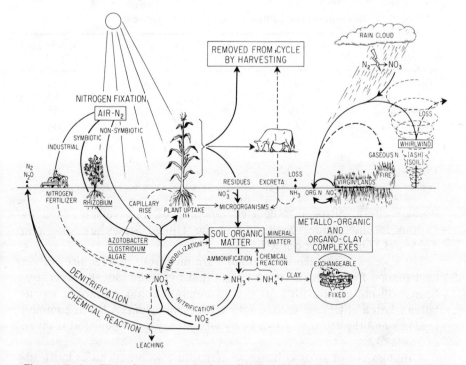

Figure 5–1. The nitrogen cycle in soil. (Sauchelli, *Fertilizer Nitrogen—Its Chemistry and Technology,* 1964. Reprinted with permission of Reinhold Publishing Corp., New York.)

trogen to nonleguminous crops. Although they are still major sources of fixed nitrogen for agriculture, the importance of legumes and manure is dwindling with each passing year because of the rapid increase in the production of low-cost synthetic nitrogen compounds. In 1973, 8 million tons of nitrogen (N) in the form of various synthetic fertilizers were applied to crops grown in the United States alone. It is estimated that by 1980 this figure will have risen to about 10 to 12 million tons. The growth in the consumption of synthetic nitrogen materials is largely the result of the efficiency with which the nitrogen industry is operated and the low cost at which these materials are offered to farmers.

The quantities of nitrogen fixed by *Rhizobia* differ with the *Rhizobial* strain, the host plant, and the environmental conditions under which the two develop. Amounts as high as 500 lb. of nitrogen fixed per acre under a crop of clover have been reported from New Zealand, where the climate is extremely favorable for growth the year round. Although this figure is considerably above the average generally reported, much nitrogen is fixed via the legume route in a large part of New Zealand, a country that presently depends almost entirely on this phenomenon for

TABLE 5-1. Average Fixation of Nitrogen by Legumes

Legume	Nitrogen fixed (lb./A.)	Legume	Nitrogen fixed (lb./A.)
Alfalfa	194	Lespedezas (annual)	85
Ladino clover	179	Vetch	80
Sweet clover	119	Peas	72
Red clover	114	Soybeans	100
Kudzu	107	Winter peas	50
White clover	103	Peanuts	42
Cowpeas	90	Beans	40

its agricultural nitrogen. Reported amounts of this element fixed by various legumes under conditions in the United States are shown in Table 5-1. These are too low for good yields.

The species of the genus *Rhizobium* are numerous and require certain host plants. For example, the bacteria that live symbiotically with soybeans will not do so with alfalfa. It is imperative that the farmer inoculating a batch of legume seed use the correct inoculum. A list of common legumes and the *Rhizobial* strains by which they are inoculated is given in Table 5-2.

Although not of great concern to agriculture as practiced in most advanced countries, fixation of nitrogen by leguminous trees is important to the ecology of tropical and subtropical forests. Numerous legume tree species, widely distributed throughout the tropical and temperate zones of the world, fix appreciable amounts of nitrogen. Two well-known examples in the United States are mimosa and acacia.

Some nonleguminous plants also fix nitrogen by a mechanism similar to that of the symbiotic relationship between legumes and *Rhizobia*. Such plants are widely distributed. Certain members of the following plant families are known to bear root nodules and to fix nitrogen: Betulaceae, Elaeagnaceae, Myricaceae, Coriariaceae, Rhamnaceae, and Casuarinaceae.

Legume fixation of nitrogen is at a maximum only when the level of available soil nitrogen is at a minimum. It is sometimes advisable to include a small amount of nitrogen in the fertilizer of agricultural legume crops at planting time to ensure that the young seedlings will have an adequate supply until the *Rhizobia* can become established on their roots. Large or continued applications of nitrogen, however, reduce the activity of the *Rhizobia* and therefore are generally uneconomic.

FREE-LIVING SOIL MICROORGANISMS

Nitrogen fixation in soils is also brought about by certain free-living organisms, which include numerous species of the blue-green algae and

TABLE 5–2. **Classification Scheme of *Rhizobium*-Legume Associations***

Rhizobium species	Cross-inoculation group	Host genera	Legumes included
R. meliloti	Alfalfa	*Medicago*	Alfalfa
		Melilotus	Sweet clover
		Trigonella	Fenugreek
R. trifolii	Clover	*Trifolium*	Clover
R. leguminosarum	Peas	*Pisum*	Peas
		Vicia	Vetch
		Lathyrus	Sweetpeas
		Lens	Lentils
R. phaseoli	Beans	*Phaseolus*	Beans
R. lupini	Lupine	*Lupinus*	Lupine
		Orithopus	Serradella
R. japonicum	Soybeans	*Glycine*	Soybeans
	Cowpeas†	*Vigna*	Cowpeas
		Lespedeza	Lespedeza
		Crotalaria	Crotalaria
		Pueraria	Kudzu
		Arachis	Peanuts
		Phaseolus	Lima beans

* Courtesy of Professor F. J. Stevenson, University of Illinois.
† This group has not attained species status.

certain free-living bacteria. The most important are *Rhodospirillum*, which is photosynthetic, *Clostridium*, which is an anaerobic saprophyte, and the aerobic saprophytes, *Azotobacter* and *Beijerinckia*.

The blue-green algae occur under a wide range of environmental conditions, including rock surfaces and barren wastelands. They are completely autotropic and require only light, water, free nitrogen (N_2), carbon dioxide (CO_2), and salts containing the essential mineral elements. Because they need light, they probably make only minor contributions to the nitrogen in upland agricultural soils. They are, however, believed to play a major role in the nitrogen economy of rice-paddy soils in tropical countries. The benefit derived from the blue-green algae probably results from the nitrogen they supply to other organisms during the early stages of soil formation.

The agricultural importance of nitrogen fixation by free-living bacteria is greater than that of the blue-green algae. These organisms, with the exception of *Rhodospirillum,* require a source of available energy, which is presented in the form of organic residues. Part of the energy from the oxidation of these residues is used to fix elemental nitrogen. There has been considerable speculation about the amounts of nitrogen actually fixed by these free-living organisms. Some estimates have been as high as 20 to 45 lb./A. annually, but a more generally accepted figure based on recent work is about 6 lb./A.

Considerable attention has been given to the rhizosphere of plant roots, the soil area immediately adjacent to the roots, high in energy-rich materials because of their exudation of organic compounds and their sloughing off of tissue. It has been suggested that this zone is the site of nitrogen fixation by *Azotobacter* and *Clostridium.* Russian agriculturalists have claimed that inoculation of seed with these organisms brought about increased plant growth. Workers in the United States Department of Agriculture have been unable to confirm these claims. Other workers in California found that the addition of soluble organic substrates (glucose, sucrose) to soils increased the fixation of atmospheric nitrogen. There was a small increase in fixation from the inversion of a disc of growing sod but none from the addition of grass cuttings, straw, or alfalfa meal to the soil. A portion of their data is shown in Table 5-3. The inoculation of these soils with a heavy population of *Azotobacter* did not increase nitrogen fixation.

The organism *Beijerinckia* inhabits the leaf surfaces of many tropical

TABLE 5-3. The Nonsymbiotic Fixation of Nitrogen in a Yolo Sandy Loam Treated in Different Ways.[*]

| Treatment | Soil | Ammonia (meq./g. soil) | | Nitrate (meq./g. soil) | | Total nitrogen fixed | |
		Final	Fixed	Final	Fixed	(meq./g. soil)	(lb./A. 6 in. deep)
10 g. Yolo soil + 780 mg. sucrose	Davis	0.66	0.055	0.75	0.028	12.3	344
Growing lawn, photosynthesizing	Davis	1.24	–	0.88	0.0004	0.0004	0.011
	Berkeley	1.78	0.004	1.39	0.001	0.005	0.140
Inverted lawn, decaying	Davis	1.49	0.18	5.44	0.26	0.44	12
	Berkeley	3.79	0.087	7.46	0.047	0.134	3.75
Soil with grass removed	Davis	1.07	0.001	9.99	0.007	0.008	0.22
	Berkeley	0.67	0.001	0.46	0.002	0.003	0.085

[*] Delwiche et al., *Plant Soil,* 7:113 (1956). Reprinted with permission of Martinus Nijhoff, The Hague.

plants and is thought by some to engage in its nitrogen-fixing operations on these leaves rather than in the soil. *Beijerinckia* is found almost exclusively in the tropics, and it has been suggested that it is a leaf inhabitant rather than a true soil bacterium.

ATMOSPHERIC FIXATION THROUGH ELECTRICAL DISCHARGE

Nitrogen compounds are present in the atmosphere and are returned to the earth in rainfall. The nitrogen is in the form of ammonia, NO_3^-, NO_2^-, and nitrous oxide and in organic combinations. The ammonia comes largely from industrial sites which manufacture or use ammonia. Some undoubtedly is present in the ammonia that escapes from the soil surface because of the reactions taking place. The organic nitrogen can probably be accounted for by the finely divided organic residues which are swept into the atmosphere from the earth's surface.

The soil has a pronounced capacity for adsorbing ammonia gas from the atmosphere. Laboratory studies carried out in New Jersey with six soil types and with atmospheres to which known amounts of ammonia gas had been added indicated that from 50 to 67 pounds per acre per year of NH_3 could be adsorbed by these soils. Sorption was positively related to NH_3 concentration and to temperature. In localized areas where atmospheric NH_3 concentrations are above normal, significant quantities of this gas may be adsorbed by soils. This, of course, is independent of that which may be added in rainfall.

Because of the small amount of NO_2^- present in the atmosphere, it is usually lumped in with the figures reported for NO_3^-. The presence of NO_3^- has been attributed to its formation during atmospheric electrical discharges, but recent studies suggest that only about 10 to 20 percent of the NO_3^- in rainfall and atmosphere arises from fixation by electrical discharges. The remainder is thought to come from industrial waste gases or possibly from the soil. Atmospheric nitrogen compounds are continually being returned to the soil in rainfall. The total amount of fixed nitrogen thus brought down has been variously estimated to range between 1 and 50 lb./A. annually, depending on location. These figures are generally higher around areas of intense industrial activity and as a rule are greater in tropical than in polar or temperate zones.

INDUSTRIAL FIXATION OF NITROGEN

From the standpoint of commercial agriculture, as practiced in the United States, Europe, and other advanced countries of the world, the industrial fixation of nitrogen is by far the most important source of this element as a plant nutrient. Because of the scope of this topic, it is treated in Chapter 9, which deals with the fundamentals of fertilizer manufacture.

FORMS OF SOIL NITROGEN

The nitrogen found in the soil can generally be classified as inorganic or organic. By far the greater total amount occurs as a part of the soil organic matter complex.

Inorganic Nitrogen Compounds. The inorganic forms of soil nitrogen include NH_4^+, NO_3^-, NO_2^-, N_2O, NO, and elemental nitrogen, which is, of course, inert except for its utilization by *Rhizobia*. It is also thought that hydroxylamine (NH_2OH) exists, but because it is believed to be an intermediate in the formation of NO_2^- from ammonium it is unstable and does not persist. Some recent work based on the calculated free energy changes associated with inorganic nitrogen oxidations has led to the suggestion that hydroxylamine is not an intermediate in the conversion of NH_4^+ to NO_2^-. Further work is necessary before this point can be resolved.

From the standpoint of soil fertility, the NH_4^+, NO_2^-, and NO_3^- forms are of greatest importance; nitrous oxide and nitric oxide are also important in a negative way, for they represent forms of nitrogen which are lost to crop use through denitrification. The ammonium, nitrite, and nitrate forms arise either from the normal aerobic decomposition of soil organic matter or from the additions to the soil of various commercial fertilizers.

Organic Nitrogen Compounds. The organic forms of soil nitrogen occur as consolidated amino acids or proteins, free amino acids, amino sugars, and other complex, generally unidentified compounds. This last group is believed to include materials which result from (1) the reaction of ammonium with lignin, (2) polymerization of quinones and nitrogen compounds, and (3) the condensation of sugars and amines.

The group consisting of consolidated amino acids or proteins is usually found in strong combination with clays, lignin, and perhaps other materials. This has been suggested as one of the reasons for their resistance to decomposition. The existence of these proteins is deduced from the presence of amino acids found in acid soil hydrolyzates. It is assumed that because proteins are formed by a combination of amino acids the presence of these amino acids in the hydrolyzates is proof of the existence of proteins in soils.

Recent developments in analytical techniques have made possible the isolation of free amino acids from soils which are not in peptide linkages or in combination with high-molecular-weight organic polymers, clays, or lignin. The suitability of these substrates for biological oxidation would suggest that they will not build up in large quantities in soils. The ease with which they are decomposed also suggests that they may be a more important source of NH_4^+, the substrate for the nitrifying bacteria, than the nitrogen in the more insoluble consolidated amino acids, the amino sugars, and the so-called lignin and humic complexes.

NITROGEN TRANSFORMATIONS IN SOILS

Plants absorb most of their nitrogen in the NH_4^+ and NO_3^- forms. The quantity of these two ions presented to the roots of agricultural plants depends largely on the amounts supplied as commercial nitrogen fertilizers and released from the reserves of organically bound soil nitrogen. The amount released from these organic reserves (and, to a certain extent, those remaining as such in the soil after the addition of ammonium or nitrate fertilizers) depends on the balance that exists among the factors affecting nitrogen mineralization, immobilization, and losses from the soil. By way of definition, nitrogen mineralization is simply the conversion of organic nitrogen to a mineral (NH_4^+, NO_2^-, NO_3^-) form. Nitrogen immobilization is the conversion of inorganic or mineral nitrogen to the organic form. Reactions associated with these phenomena, as well as nitrogen losses from the soil, are discussed in the following sections.

Organic-Mineral Nitrogen Balance in Soil. Soil organic matter is an ill-defined term used to cover organic materials in all stages of decomposition. Broadly speaking, soil organic matter can be placed in two categories. The first is a relatively stable material, termed *humus*, which is somewhat resistant to further rapid decomposition. The second includes those organic materials that are subject to fairly rapid decomposition, materials that range from fresh crop residues to those that, by a chain of decomposition reactions, are approaching a degree of stability.

Nitrogen in some form is needed for the decomposition of organic matter by heterotrophic soil microorganisms. If the decomposing organic material has a small amount of nitrogen in relation to the carbon present (wheat straw, mature corn stalks), the microorganisms will utilize any NH_4^+ or NO_3^- present in the soil to further the decomposition. This nitrogen is needed to permit rapid growth of the microbial population which accompanies the addition to the soil of a large supply of carbonaceous material.

If, on the other hand, the material added contains much nitrogen in proportion to the carbon present (alfalfa or clover turned under), there will normally be no decrease in the level of mineral nitrogen in the soil. There may even be a fairly rapid increase in this fraction of soil nitrogen, caused by its release from the decomposing organic material.

The ratio of the percentage of carbon to that of nitrogen is termed the carbon:nitrogen ratio, or simply the C:N ratio, which defines the relative quantities of these two elements in fresh organic materials, humus, or in the whole soil body. The C:N ratio of stable soil organic matter is about 10:1.

As a general rule, when organic materials with a C:N ratio of greater than 30 are added to soils, there is immobilization of soil nitrogen during the initial decomposition process. For ratios between 20 and 30 there may be neither immobilization nor release of mineral nitrogen. If the organic materials have a C:N ratio of less than 20, there is usually a release of mineral nitrogen early in the decomposition process. These are general rules of thumb *only,* for many factors other than the C:N ratio influence the decomposition of organic materials and the release or immobilization of nitrogen.

The pattern just discussed is illustrated diagrammatically in Figure 5-2. During the initial stages of the decomposition of fresh organic material there is a rapid increase in the numbers of heterotrophic organisms, accompanied by the evolution of large amounts of carbon dioxide. If the C:N ratio of the fresh material is wide, there will be a net immobilization of nitrogen, as shown in the shaded area under the top curve. As decay proceeds, the C:N ratio narrows and the energy supply (carbon) diminishes. Some of the microbial population dies because of the decreased food supply, and ultimately a new equilibrium is reached. The attainment of this new equilibrium is accompanied by the release of mineral nitrogen (indicated by the crosshatched area under the top curve). The result is that the final soil level of this form of nitrogen may be

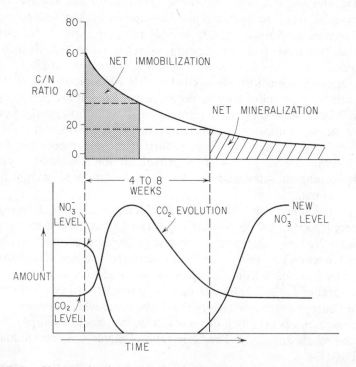

Figure 5-2. Changes in nitrate levels of soil during the decomposition of low-nitrogen crop residues. (Courtesy of Professor B. R. Sabey, University of Illinois.)

higher than the original level. There may also be an increase in the level of stable organic matter or humus, depending on the quantity and type of fresh organic material originally added. The time required for this decomposition cycle to run its course depends on the quantity of organic matter added, the supply of utilizable nitrogen, the resistance of the material to microbial attack (a function of the amount of lignins, waxes, and fats present), temperature, and moisture levels in the soil.

In undisturbed (uncultivated) soil the humus content tends to reach a level that is determined by soil texture, topography, and climatic conditions. As a rule, the humus level is higher in cooler than in warmer climates. Further, for any given level of mean annual temperature and type of vegetation the content of stable soil organic matter rises with an increase in effective precipitation. In general, humus contents are greater in fine-textured than in coarse-textured soils. Organic matter contents are higher under grassland vegetation than under forest cover. These relations are generally true for well-drained soil conditions. Under conditions of poor drainage or waterlogging aerobic decomposition is impeded and organic residues build up to high levels, regardless of temperature or soil texture.

As a rule, the $C:N$ ratio of the undisturbed *topsoil* in equilibrium with its environment is about 10 or 12 to 1. It narrows in the subsoil in many cases, partly because of the higher content of NH_4^+ nitrogen and the generally lower amounts of carbon. Under equilibrium conditions the soil microbial population remains about the same, a consistent amount of organic residues is returned to the soil, depending on the vegetative cover, and there is a fairly fixed and usually low rate of nitrogen mineralization. If the soil is disturbed, as in plowing, there is an immediate and rapid increase in mineralization. Continued cultivation without the return of adequate crop residues with sufficient nitrogen will ultimately lead to a decline in the humus content of soils. Continued cultivation with adequate use of commercial fertilizers, coupled with the return of crop residues, not only can maintain the level of soil organic matter but may actually increase it. This is well illustrated by the graph in Figure 5-3.

The importance of organic matter is certainly not to be underestimated. It is necessary to maintain good soil structure, especially in fine-textured soils. It increases the cation exchange capacity, thereby reducing leaching losses of elements such as potassium, calcium, and magnesium. It serves as a reservoir for soil nitrogen. It improves water relations, and its mineralization provides a continuous, though limited, supply of nitrogen, phosphorus, and sulfur to the crop. The idea that the maintenance of a high level of soil organic matter should be an end in itself in the farming enterprise is wrong. The ultimate objective of any farming enterprise is *sustained* maximum economical production. The

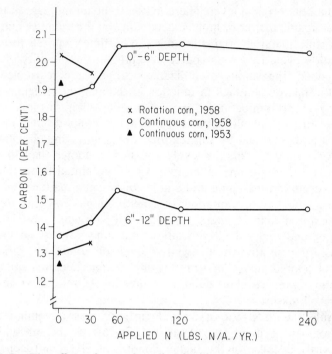

Figure 5–3. The effect of rate of applied fertilizer nitrogen on the carbon content of soil. [Sutherland et al., *Agron. J.*, **53**:339 (1961).]

judicious use of lime, fertilizers, and sound management and cultural practices will lead to this objective and, incidentally, will maintain the level of soil organic matter.

The mineralization and immobilization of soil nitrogen and the turnover of organic materials in the soil is effected by the heterotrophic soil organisms, including bacteria and fungi. Their requirement for energy is met by their oxidation of the carbonaceous material in the soil. This decomposition increases with a rise in temperature. It is further enhanced by adequate, though not excessive, soil moisture and a good supply of oxygen. Decomposition proceeds under waterlogged conditions, though at a slower rate, and is incomplete. Aerobic, and to a lesser extent anaerobic, respiration releases the contained nitrogen in the form of NH_4^+. This is the first step in the mineralization of nitrogen, the subject to be next considered.

Mineralization of Nitrogen Compounds. The mineralization of organic nitrogen compounds takes place in essentially three step-by-step reactions: aminization, ammonification, and nitrification. The first two are effected through the medium of heterotrophic microorganisms and the third is brought about largely by autotrophic soil bacteria. The heterotrophs require organic carbon compounds for their source of energy. Autotrophic organisms obtain their energy from the oxidation of inorganic salts and their carbon from the carbon dioxide of the surrounding atmosphere.

AMINIZATION. The population of heterotrophic soil microorganisms is composed of numerous groups of bacteria and fungi, each of which is responsible for one or more steps in the numerous reactions in organic-matter decomposition. The end products of the activities of one group furnish the substrate for the next, and so on down the line until the material is decomposed. One of the final stages in the decomposition of nitrogenous materials is the hydrolytic decomposition of proteins and the release of amines and amino acids. This step is termed *aminization* and is a function of some of the heterotrophic organisms. It is represented schematically by the following:

$$\text{proteins} \longrightarrow \text{R-NH}_2 + CO_2 + \text{energy} + \text{other products}$$

AMMONIFICATION. The amines and amino acids so released are further utilized by still other groups of heterotrophs with the release of ammoniacal compounds. This step is termed ammonification and is represented as follows:

$$\text{R-NH}_2 + \text{HOH} \longrightarrow NH_3 + \text{R-OH} + \text{energy}$$

The ammonia so released is subject to several fates in the soil:

1. It may be converted to nitrites and nitrates by the process of nitrification.
2. It may be absorbed directly by higher plants.
3. It may be utilized by heterotrophic organisms in further decomposing organic carbon residues.
4. It may be fixed in a biologically unavailable form in the lattice of certain expanding type clay minerals.

NITRIFICATION. Some of the NH_4^+ released by the processes of ammonification is converted to nitrate nitrogen. This biological oxidation of ammonia to nitrate is known as nitrification. It is a two-step process in which the ammonia is first converted to nitrite (NO_2^-) and thence to nitrate (NO_3^-). Conversion to nitrite is brought about largely by a group of obligate autotrophic bacteria known as *Nitrosomonas* by a reaction that can be represented by the following equation:

$$2NH_4^- + 3O_2 \longrightarrow 2NO_2^- + 2H_2O + 4H^+$$

It has also been shown that numerous heterotrophic organisms can convert reduced nitrogen compounds to nitrite (NO_2^-). The organisms include bacteria, actinomycetes, and fungi. The substrates from which the nitrite is produced include not only NH_4^+ but also amines, amides, hydroxylamines, oximes, and a number of other reduced nitrogen compounds. *Nitrosomonas*, however, is considered to be the most important of the soil organisms bringing about the conversion of NH_4^+ to NO_2^-. The conversion from nitrite to nitrate is effected largely by a second group of obligate autotrophic bacteria termed *Nitrobacter*. The equation representing this reaction may be written as follows:

$$2NO_2^- + O_2 \longrightarrow 2NO_3^-$$

Though *Nitrobacter* is probably by far the most important organism bringing about the conversion of NO_2^- to NO_3^-, some few heterotrophs, mostly fungi, will also produce nitrates, though a few bacterial strains will also effect this conversion. *Nitrosomonas* and *Nitrobacter* are usually referred to collectively as the *Nitrobacteria*.

Three important and very practical points are brought out by these nitrification equations, an understanding of which will make clearer the reactions taking place when commercial nitrogen fertilizers of the organic or ammoniacal form are applied to the soil. In the first place the reaction requires *molecular* oxygen. This means simply that it will take place most readily in well-aerated soils. A second point is that the reaction releases hydrogen ions (H^+). It is the release of these ions that results in the acidification of the soil when ammoniacal and most organic nitrogen fertilizers are converted to nitrates. Continued use of such forms of nitrogen will lower the soil *p*H. The use of lime in a farming program, however, will prevent this acid condition from developing. A third point of importance is that because microbial activity is involved, the rapidity and extent of the transformation will be greatly influenced by soil environmental conditions such as moisture supply and temperature. This point is considered in a subsequent section of this chapter.

In well-drained neutral to slightly acid soils the rate of oxidation of NO_2^- to NO_3^- is normally higher than that of NH_4^+ to NO_2^-. The rate of NO_2^- formation is equal to or greater than the rate of formation of NH_4^+. As a consequence, nitrate is the form that tends to accumulate in soils or, if plants are growing thereon, will be the form most used by them.

Factors Affecting Nitrification. Factors influencing the activity of the nitrifying bacteria have a pronounced effect on the amount of nitrates produced and consequently on the utilization of nitrogen by plants. As a general rule of thumb, the environmental factors favoring the growth of most upland agricultural plants are those that also favor the activity of the nitrifying bacteria.

Factors affecting the nitrification pattern in soils are (1) supply of the ammonium ion, (2) population of nitrifying organisms, (3) soil reaction, (4) soil aeration, (5) soil moisture, and (6) temperature.

ABUNDANCE OF THE AMMONIUM ION. Because the substrate for the nitrifying bacteria is the ammonium ion, a supply of this ion is the first requirement for nitrification. If conditions do not favor the release of ammonia from organic matter (or if ammonium-containing fertilizers are not added to the soils), there will be no nitrification. Temperature and moisture levels favorable to nitrification are also favorable to ammonification. If, however, the ratio of carbon to nitrogen in the soil is too wide, any ammonia released from organic matter will be appropriated by the heterotrophic population that is decomposing the organic material. As shown in Figure 5-2, only when the C:N ratio has dropped to 20 or 25 will there be a net release of mineral nitrogen.

This phenomenon is of practical agricultural importance. If large amounts of small grain straw, mature dry corn stalks, or similar materials are plowed into soils with only limited quantities of nitrogen, this nitrogen will be used by the microorganisms in the decomposition of the carbonaceous residues. If crops are planted on such areas immediately after the plowing in of these residues, they may suffer from a shortage of nitrogen. This shortage can be prevented by the addition of sufficient fertilizer nitrogen at the time of disking in the material to supply the needs of the microorganisms as well as those of the growing crop. Such nitrogen deficiencies induced by organic matter are not common but localized examples have been observed in the field.

POPULATION OF NITRIFYING ORGANISMS. Soils differ in their ability to nitrify added ammonium compounds, even under similar conditions of temperature, moisture, and level of added ammonium. One factor that may be responsible is the variation in the numbers of nitrifying organisms present in the different soils. The impact that this could have on soil nitrification patterns was investigated several years ago by workers at Iowa State University. A portion of their results is illustrated in Figure 5-4.

The presence of different-sized populations of nitrifiers under field conditions would probably result in differences in the lag time between the addition of the ammonium source and the build-up of nitrate nitrogen in the soil. Because of the tendency of microbial populations to multiply rapidly in the presence of an adequate supply of substrate, the total amount of nitrification taking place in soils would likely not be affected by the number of organisms initially present, provided temperature and moisture conditions were favorable for sustained nitrification.

Others have suggested that differences in the nitrification patterns of soils may be attributed in part to volatile losses of nitrogen resulting from the accumulation of nitrite and its subsequent decomposition.

SOIL REACTION. The range of reaction over which nitrification takes

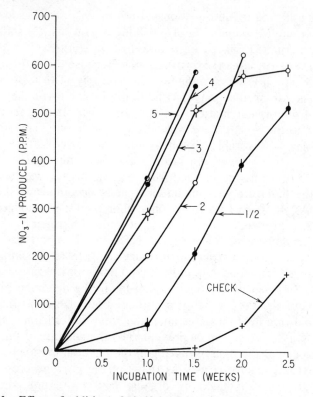

Figure 5–4. Effect of addition of nitrifying bacteria on the nitrification curve at 25°C. in a Hamburg silt loam subsoil with an initial *p*H of 8.3; 1/2, 2, 3, 4, and 5 refer to number of milliliters of concentrated inoculum of nitrifying organisms added. [Sabey et al., *SSSA Proc.*, **23**:465 (1959).]

place has generally been given as *p*H 5.5 to about 10.0, with the optimum around 8.5. It is known that nitrates are produced in some soils at *p*H values of 4.5, and nitrification has been reported in a pasture soil with a *p*H value of 3.8.

The nitrifying bacteria need an adequate supply of calcium and phosphorus, and a proper balance of the elements iron, copper, manganese, and perhaps others. The exact requirement for these mineral elements has not been determined. The influence of both soil *p*H and available calcium on the activity of the nitrifying organisms suggests the importance of liming in the farming enterprise. Enhancing nitrification during the growing season of the crop is one means of ensuring higher yields.

SOIL AERATION. The nitrobacteria, as previously indicated, are obligate autotrophic aerobes. They will not produce nitrates in the absence of molecular oxygen. The relation between oxygen level and nitrification is shown in Figure 5-5. In this study air with the indicated concentration

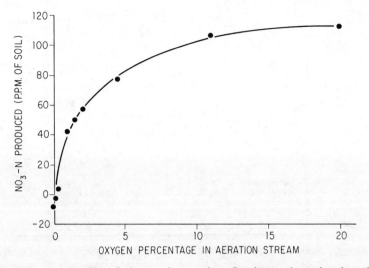

Figure 5-5. Production of nitrate nitrogen in a Carrington loam incubated with added ammonium sulfate and aerated with air-nitrogen mixtures with varying oxygen percentages. (Black, *Soil-Plant Relationships*, 1957. Reprinted with permission of John Wiley & Sons, Inc., New York.)

of oxygen was passed through soil to which ammonium sulfate had been added. The soil was then incubated under conditions of adequate moisture and temperature. Maximum nitrification occurred when the percentage of oxygen reached 20, which is about the same as the concentration of this gas in the aboveground atmosphere.

This example illustrates the importance of maintaining conditions that permit rapid diffusion of gases into and out of the soil. Soils that are coarse-textured or possess good structure (by virtue of an adequate supply of humus) facilitate this rapid exchange of gases and ensure an adequate supply of oxygen for the nitrobacteria.

SOIL MOISTURE.   The nitrobacteria are more sensitive to excess moisture than they are to dry soil conditions. This is well illustrated by the data graphed in Figures 5-6 and 5-7. In Figure 5-6 are shown the amounts of nitrates produced in a Grundy silt loam incubated in the laboratory at different soil moisture tensions, at two different temperatures, and for two time periods. It will be recalled that a tension of 100 cm. of water corresponds to the moisture content of a soil considerably higher than that at field capacity. Nitrification was severely curtailed at low moisture tensions (high moisture contents). It was much less affected by the higher tension of 275 cm. of water, which is still wetter than a soil at field capacity.

The data in Figure 5-7 illustrate the effect of high moisture tension (low moisture content) on nitrification taking place in a Millville soil in-

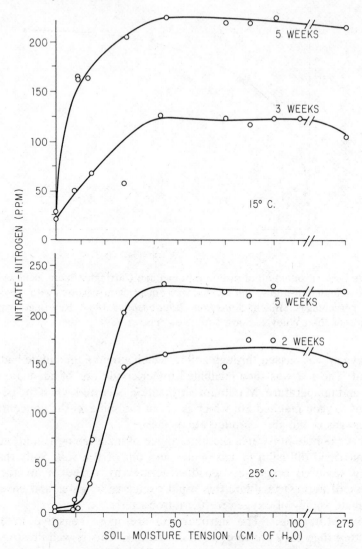

Figure 5–6. Nitrate-nitrogen produced at various moisture tensions in a Grundy silt loam to which 100 ppm. of NH_4–N had been added. [Parker et al., *SSSA Proc.*, **26**:238 (1962).]

cubated for various periods at two moisture tensions. As indicated in the graph, 150 ppm. of nitrogen as ammonium sulfate were applied to the soil in this study. A <u>bar</u> is roughly equivalent to an <u>atmosphere</u>. It will be seen from the graph that even at the approximate wilting percentage more than half the ammonium was nitrified at the end of twenty-eight days. At the lower moisture tension (7 bars, which is comparatively dry)

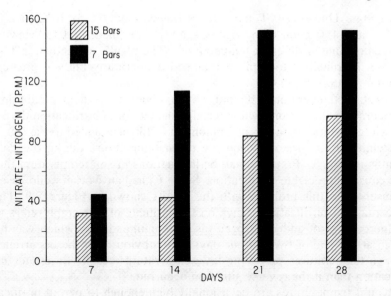

Figure 5–7. The effect of moisture levels near the wilting point on the nitrification of 150 ppm. of nitrogen applied as ammonium sulfate to a Millville loam and incubated at 25°C. [Justice et al., *SSSA Proc.*, **26**:246 (1962).]

100 per cent of the ammonium was nitrified at the end of twenty-one days. It appears from these data that the nitrobacteria function well even in reasonably dry soils.

TEMPERATURE. The relationship between nitrification and soil temperature has been under study for a long, long time. However, there has been a renewed interest in this topic because of larger supplies of ammoniacal fertilizers and the accompanying change in the pattern of farm application of these materials. In areas in which soil temperatures are low during the winter, off-season application of these fertilizers can mean a saving to the grower of both time and money. It is important, however, that winter temperatures be low enough to prevent nitrification of the added ammonium nitrogen or it may be lost by leaching before it can be used by the crop the following spring.

Recent studies in Georgia have shown that some nitrification of added ammonia compounds takes place at 37°F. At 42°F. nitrification was appreciable at the end of both three- and six-week incubation periods. At 52°F., and with 50 ppm. of nitrogen as NH_4^+, nitrification was essentially complete at the end of nine weeks. When the same amount of nitrogen was applied as aqua ammonia (NH_3), nitrification was not complete until the end of twelve weeks. The effect of free ammonia on soil microorganisms is discussed in a subsequent section of this chapter.

Another study on temperature-nitrification relationships was made at

Iowa State University. Temperatures ranged from 16 to 30°C. (61 to 86°F.), and 150 ppm. of nitrogen as NH_4^+ was incubated for varying periods of time at different temperatures. The results are shown in Figure 5-8. Nitrification took place at all soil temperatures but was greatest at a temperature of 30°C.

Constant temperatures do not obtain under most field conditions. Temperature fluctuations will determine the extent of nitrification during the winter months. Thus, if an ammonium fertilizer is added to the soil in the winter in an area in which the mean temperature during the cold months is, say, 37°F., there may be fluctuations in soil temperature that will cause appreciable nitrification. Some Canadian workers addressed themselves to this problem with the results shown in Figure 5-9. The percentage of nitrification shown on the ordinate of this graph refers to the percentage of added nitrogen (as ammonium sulfate) which was nitrified at the end of twenty-four days. It is obvious that the occurrence of high temperatures preceding low temperatures results in greater nitrification than if the reverse situation occurred.

Even if temperatures are occasionally high enough to permit nitrification of fall-applied ammoniacal fertilizers, this in itself is not detrimental if leaching does not occur. In many areas of the east and west North Central states, moisture movement through the soil profile during the winter months is insufficient to remove any nitrates that may accumulate because of temperature fluctuations. In other areas of the U.S., water

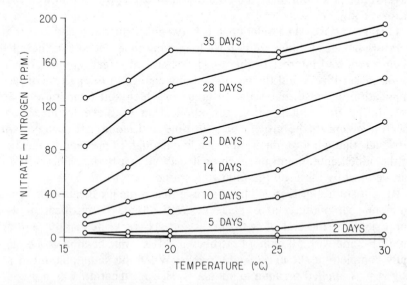

Figure 5–8. The effect of time and temperature on the nitrification of 150 ppm. of NH_4–N added to a Grundy silt loam and incubated for 35 days. [Parker et al., *SSSA Proc.*, **26**:238 (1962).]

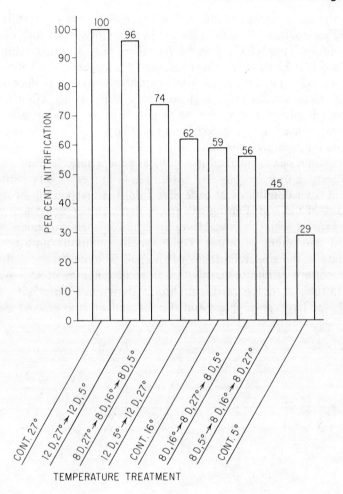

Figure 5-9. Nitrification as affected by time-temperature relationships. [Chandra, *Can. J. Soil Sci.,* **42**:314 (1962).]

movement through the soil profile is excessive, and NO_3^- losses will occur. Whether or not ammoniacal fertilizers can be fall-applied without a great risk of nitrate loss depends on local soil and weather conditions. Information concerning these patterns is available at the offices of local State and Federal agricultural agencies.

RETENTION OF IONIC NITROGEN IN SOIL

The cationic nature of NH_4^+ permits its adsorption and retention by soil colloidal material. It is generally not so subject to removal by leaching waters as the nitrate form. Ammonium nitrogen may be retained in soils for long periods of time if conditions for nitrification are unfavorable. As already indicated, it is possible to apply fertilizers containing ammonium nitrogen in the fall in cool climates to soils of fine texture without appreciable loss by leaching, provided that temperatures remain below 37 to 40°F. The presence of nitrogen in the cationic form, however, does not ensure its loss against leaching. It is necessary that the soil have a sufficiently high exchange capacity to retain the added ammonium nitrogen or it will be removed in percolating water. Sandy soils with low exchange capacities permit appreciable movement of ammonium nitrogen into the subsoil.

Once ammonia is nitrified it is subject to leaching. Nitrate nitrogen is completely mobile in soils and within limits moves largely with the soil water. Under conditions of excessive rain it is leached out of the upper horizons of the soil. During extremely dry weather and when capillary movement of water is possible there is an upward movement with the upward movement of water. Under such conditions nitrates will accumulate in the upper horizons of the soil or even on the soil surface.

The pattern of nitrate distribution in some columns of soils which differed in their particle-size distribution is shown in Figure 5-10. The percentage of large pore space and the amount of coarse sand decreased

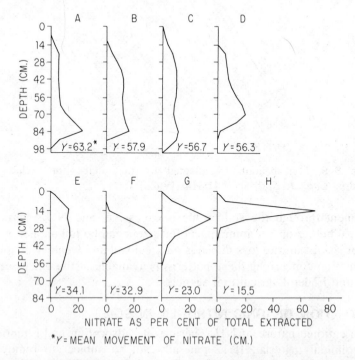

Figure 5–10. Distribution of nitrates through columns of coarse-textured soils after adding 3.29 cm. of water. [Bates et al., *SSSA Proc.*, **21**:525 (1957).]

from Sample A to Sample H, although the total pore space remained relatively constant. These data show that the coarser the texture and the greater the large pore space, the greater the mean downward movement of nitrates under the influence of a given quantity of added water.

Both ammonium nitrogen and nitrate nitrogen can be immobilized by soil microflora, as already pointed out in this chapter. In this immobilized organic state the nitrogen is not lost by leaching. The rate at which the immobilization of added nitrate nitrogen can take place in the presence of a large amount of rapidly decomposing organic material (wheat straw) is illustrated in Figure 5-11. Nitrogen (as nitrate) to the extent of 2 per cent of the weight of the straw was added to the soil and straw, and the mixture was incubated for seventy-five days. The amount of nitrogen released during this time was measured. It is apparent that during the period of active decomposition the added nitrogen was rapidly immobilized. As the rate of microbial activity subsided (as indicated by a decrease in the rate of carbon dioxide evolution), there was gradual release of the immobilized nitrogen.

AMMONIUM FIXATION

One of the possible fates of NH_4^+ nitrogen in soils is its fixation by clays with an expanding lattice. The mechanism of NH_4^+ fixation is simi-

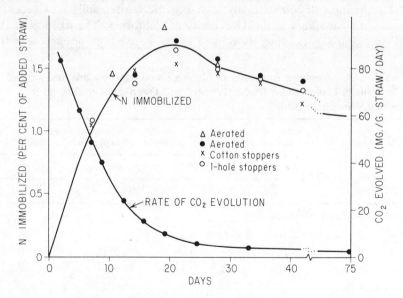

Figure 5–11. The immobilization and release of nitrogen and the rate of carbon dioxide formation in a soil receiving wheat straw and nitrate-nitrogen. [Allison et al., *Soil Sci.*, **93**:383 (1962). Reprinted with permission of The Williams & Wilkins Co., Baltimore.]

lar to that of K^+ fixation. It comes about by a replacement of NH_4^+ for interlayer cations in the expanded lattices of clay minerals. The fixed ammonium can be replaced by cations which expand the lattice (Ca^{2+}, Mg^{2+}, Na^+, H^+) but not by those that contract it (K^+, Rb^+, Cs^+).

The clay minerals largely responsible for ammonium fixation are montmorillonite, illite, and vermiculite. As a rule, fixation occurs to a much greater extent in subsoils than in topsoils.

It appears that there are small but significant amounts of native fixed ammonium in subsoils. The moisture content and temperature of the soil will affect the fixation of added ammonium compounds. Some of these effects are illustrated in Table 5-4. The data indicate that, at least in the soil types included in this study, appreciable quantities of native fixed ammonium were present and that freezing and drying increased the fixation.

Ammonium fixation is increased by raising the levels of soil potassium, shown by the data in Figure 5-12. The nitrification pattern of ammonium sulfate and ammonium-saturated Wyoming bentonite to which had been added different amounts of K^+ are shown. It is obvious that (1) not all of the bentonite-fixed ammonium was nitrified even at the end of the twenty-eight-day experimental period and (2) that the presence of K^+ blocked the release of the fixed ammonium. The degree of blocking was positively related to the amount of K^+ added.

The fixation of ammonia by high organic matter soils was recently reported by workers at the University of California. The fixation of the added ammonia was linearly related to the percentage of carbon in the

TABLE 5–4. Average Amounts of Native Fixed Ammonium and Added Ammonium Fixed Under Moist, Frozen, and Oven-dry Conditions in Several Wisconsin Soils*

Horizon groupings	Average native fixed ammonium	Average fixation of applied ammonium under three conditions		
		Moist	Frozen	Oven-dried
		(meq./100 g.)		
Gray-brown podzolic soils				
$A_p + A_1$	0.54	0.08	0.14	0.68
$A_2 + A_3$	0.41	0.06	0.06	0.35
$B_1 + B_2$	0.60	0.15	0.25	0.82
Brunizem soils				
$A_p + A_1$	0.64	0.07	0.10	0.56
A_3	0.65	0.07	0.11	0.72
$B_1 + B_2$	0.60	0.15	0.16	0.67

* Walsh et al., *Soil Sci.*, **89**:183 (1960). Reprinted with permission of The Williams & Wilkins Co., Baltimore.

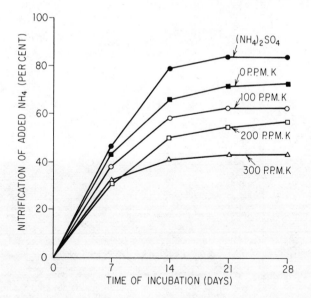

Figure 5–12. Nitrification of the ammonia in oven-dried, ammonium-saturated Wyoming bentonite as affected by added potassium. [Welch et al., *Soil Sci.,* **90**:79 (1960). Reprinted with permission of The Williams & Wilkins Co., Baltimore.]

organic matter, illustrated in Figure 5-13. The effect of oxygen and clay on fixation of the added ammonia is also illustrated. The mechanism of this fixation reaction is not completely understood, although it is suggested that hydroxyl groups present in the organic matter may be the site of the reaction with the added ammonia.

The agricultural significance of ammonium fixation is not generally considered to be great. However, it does occur, and under certain farm practices, such as the deep placement of ammonia or ammonium fertilizers in subsoils containing large amounts of expanding lattice clays, it could be a significant factor in modifying the efficiency of the added fertilizer.

GASEOUS LOSSES OF NITROGEN

Numerous studies for many years have shown that there are losses of nitrogen from soil in ways other than leaching and crop removal. These losses occur when nitrogen gas, nitrous oxide, nitric oxide, and ammonia are released because of certain biological and chemical reactions taking place in the soil. Three mechanisms have been suggested as causes of these losses.

1. Denitrification, which is the biochemical reduction of nitrates under anaerobic conditions.

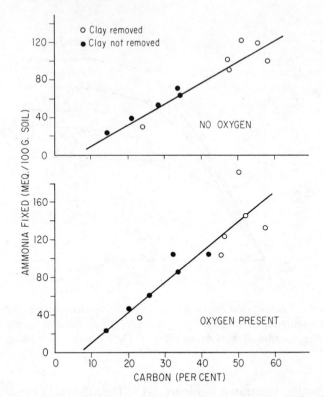

Figure 5–13. Relation between the carbon content and ammonia fixation of soil in the presence and absence of oxygen. [Burge et al., *SSSA Proc.*, **25**:199 (1961).]

2. Chemical reactions involving nitrites under aerobic conditions.
3. Volatile losses of ammonia gas (NH_3) from the surface of alkaline soils.

DENITRIFICATION

When soils become waterlogged, oxygen is excluded and anaerobic decomposition takes place. Some anaerobic organisms have the ability to obtain their oxygen from nitrates and nitrites with the accompanying release of nitrogen and nitrous oxide. The probable pathways whereby these losses come about are indicated in the equation:

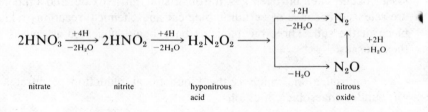

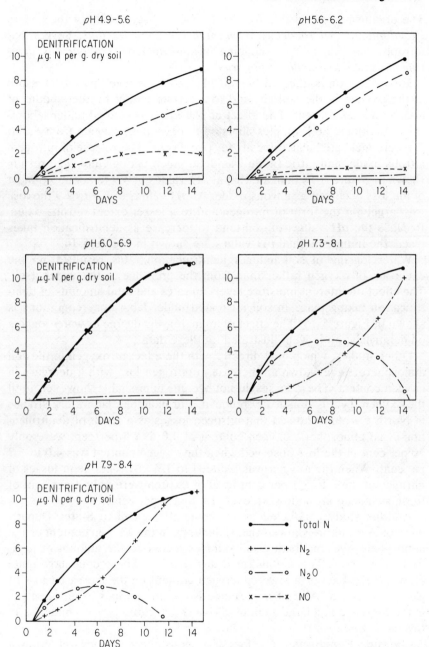

Figure 5–14. The effect of initial soil pH on the denitrification of added nitrogen as alfalfa meal in a Yolo silty clay loam. [Wijler et al., *Plant Soil,* **5**:155 (1954). Reprinted with permission of Martinus Nijhoff, The Hague.]

The organisms responsible for denitrification are species of the genera *Pseudomonas, Micrococcus, Achromobacter,* and *Bacillus.* Several autotrophs also capable of reducing nitrates include *Thiobacillus denitrificans* and *Thiobacillus thioparus.*

Denitrification is affected by soil pH, moisture level, partial pressure of the oxygen in the soil air, and to a certain extent by the amount of organic matter present. The effect of soil pH was studied under laboratory conditions. Soil samples adjusted to pH values ranging from 4.9 to 7.9 were incubated anaerobically for two weeks. The release of nitrogen, nitrous oxide, and nitric oxide was determined periodically. The results shown in Figure 5-14 indicate that at pH values of 4.9 to about 5.6 most of the loss occurred as nitrous oxide. At pH values of 7.3 to 7.9 the loss was largely in the form of nitrogen and to a lesser extent nitrous oxide. Because the pH values of soil tend to increase as denitrification takes place, the initial and final pH values are shown in Figure 5-14.

Water logging of soil induces denitrification. This results from the exclusion of oxygen rather than from the presence of the water itself. The effect of different moisture levels on rates and total amounts of denitrification taking place in soil incubated under laboratory conditions is shown in Figure 5-15. The effect of an increasing degree of water logging on denitrification is well illustrated by these data.

Other studies, which deal directly with the effect of oxygen in the soil atmosphere, have shown an increase in nitrogen loss with a decrease in oxygen content. These losses do not become appreciable, however, until the oxygen level is drastically reduced. For example, laboratory studies in North Carolina showed that nitrogen losses as a result of denitrification at an atmospheric oxygen content of 7.0 to 8.5 per cent were only 20 per cent of the loss observed when the oxygen content was 4.0 to 5.7 per cent. When the oxygen was reduced to 1.0 to 1.6 per cent, losses of nitrogen at the 7 to 8.5 per cent level of oxygen were only 4 per cent of those sustained at the lowest level (1.8 to 1.6 per cent).

Another study conducted by scientists at the United States Department of Agriculture showed that reductions in the oxygen content of the atmosphere over incubating soil samples increased the amount of denitrification losses. This is illustrated in Table 5-5. These data show quite clearly the impact of a lowered oxygen content on the increased loss of gaseous nitrogen. Also of importance is the greater loss of nitrogen gas in the presence of a large amount of oxidizable carbonaceous material, in this case, glucose.

Chemical Reactions.   The loss of gaseous nitrogen from well-drained acid soil has been suggested from time to time. How much nitrogen is removed under such conditions is generally not known. At least three mechanisms have been suggested, all of which relate to the decomposition of NO_2^-. The first is the decomposition of ammonium nitrite,

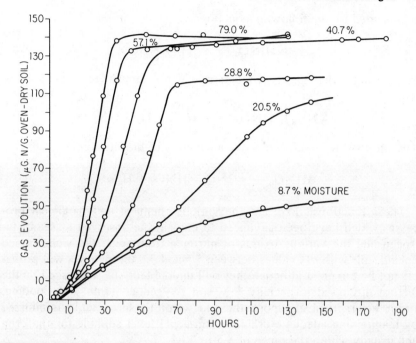

Figure 5–15. The effect of soil water content on denitrification. [McGarity, *Plant Soil,* **14:**1 (1961). Reprinted with permission of Martinus Nijhoff, The Hague.]

TABLE 5–5. Nitrogen Loss from Soils Aerated with 0.46 and 2.27 Per Cent Oxygen in Nitrogen Gas (all results/100 g. soil)*

Incubation period (days)	No glucose				With 0.5 per cent glucose			
	Nitrate-N		Total-N		Nitrate-N		Total-N	
	Found (mg.)	Decrease (%)	Found (mg.)	Lost (mg.)	Found (mg.)	Decrease (%)	Found (mg.)	Lost (mg.)
0.46% oxygen								
0	78.3		154.7		78.3		154.7	
5	72.5	7.3	147.9	6.8	23.2	70.4	150.8	3.9
10	63.5	18.8	143.4	11.3	1.4	98.2	114.8	39.9
2.27% oxygen								
0	78.3		154.7		78.3		154.7	
5	79.0	1.0	149.0	5.7	55.5	29.1	150.5	4.2
10	76.5	2.2	152.4	2.3	58.5	25.2	143.6	11.1

* Allison et al., *SSSA Proc.,* **24:**283 (1960).

represented by the following equation:

$$NH_4NO_2 \longrightarrow 2H_2O + N_2$$

The second is the van Slyke reaction, which is familiar to organic chemists:

$$RNH_2 + HNO_2 \longrightarrow ROH + H_2O + N_2$$

The third is the spontaneous decomposition of nitrous acid:

$$3HNO_2 \longrightarrow 2NO + HNO_3 + H_2O$$

These reactions, then, are the various chemical and biological processes which have been advanced to explain the observed losses of nitrogen and the various oxides of nitrogen from soils. On well-drained upland soils with pH values between 5.5 and 7.0 these losses will probably not be too great although they will undoubtedly be influenced by the factors discussed, especially low soil oxygen contents and flooding. From the practical standpoint these losses on the farm can be minimized by maintaining adequate drainage and a pH level suitable for the crop and usually within the range of 5.5 to 7.0

Recently there has been a renewed interest in denitrification. Because of the reported increase in the NO_3^- concentration of stream, lake, and ground waters in some areas which have been associated correctly or incorrectly with the increased use of nitrogen fertilizers, the role of denitrification as a means of reducing the NO_3^- concentration of irrigation waters has been investigated. In an experimental area in California, waters leaving an irrigated area are treated prior to return to normal stream flow. The treatment consists basically of passing the water through tanks to which methanol is added. The water and alcohol in these tanks are inoculated with denitrifying organisms which, using the methanol as a source of carbon, denitrify the NO_3^-, releasing the nitrogen as N_2. The essentially NO_3^- free water is then returned to normal stream flow channels or it can be reused in the irrigation system.

Whether such treatment plants for irrigation waters will be developed on a large scale will depend upon the cost and the degree to which the nitrate content of waters is considered to be a hazard to the public.

VOLATILIZATION OF AMMONIA

Ammonium salts in an alkaline aqueous medium react as follows:

$$NH_4^+ + H_2O + OH^- \longrightarrow NH_3 + 2H_2O$$

Free ammonia gas escapes. If fertilizer salts containing ammonium nitrogen are placed on the surface of alkaline soils, free ammonia may be

TABLE 5–6. Gaseous Losses of Nitrogen from Aerated Branchville Sandy Loam as Influenced by Liming and Rates of Added Ammonium Sulfate*

Addition per 100 g. soil[†]	Loss of total N (mg.)	Loss of added N[15]-N[‡] (%)	Soil reaction	
			At the beginning pH	After 6 weeks pH
Without $CaCO_3$				
25 mg. N	2.1	2.6	5.2	4.5
50 mg. N	4.4	5.8	5.0	4.7
75 mg. N	4.7	6.1	4.9	4.9
100 mg. N	6.2	6.2	4.9	4.9
0.25% $CaCO_3$				
25 mg. N	1.6	3.5	6.7	6.1
50 mg. N	9.0	13.1	6.7	6.1
75 mg. N	25.4	36.1	6.6	6.6
100 mg. N	32.1	36.9	6.6	6.6

* Carter et al., *SSSA Proc.*, 25:484 (1961).
† Nitrogen applied as tagged $(NH_4)_2SO_4$.
‡ LSD 5% = 4.8; 1% = 6.7.

lost because of this reaction. This is illustrated by the data shown in Table 5-6. This study was carried out in the laboratory and the nitrogen was added as $(NH_4)_2SO_4$. The ammonium sulfate was mixed in 50 g. of soil. The moisture tension was kept at $\frac{1}{3}$ atm. and the soil incubated at 30°C. At the end of the experiment the released ammonia was determined. The loss was negligible in the soil to which no calcium carbonate had been added but appreciable when it was present. Losses also increased with rising rates of applied nitrogen.

Normally, ammonia losses resulting from surface volatilization can be prevented by placing the nitrogen materials several inches under the soil surface or working them in with several inches of the top soil. These losses are aggravated by high soil temperatures and rapid evaporation of water.

Losses of ammonia precipitated by surface applications of urea $[CO(NH_2)_2]$, a common fertilizer material, take place regardless of soil pH. They also occur when anhydrous ammonia or ammonia solutions are improperly applied. These factors are discussed in the section dealing with nitrogen fertilizers and their behavior in soils.

Gaseous Loss of NO_3^-. It has also been suggested that there may actually be volatile losses of NO_3^- as nitric acid in soils containing appreciable amounts of KCl-exchangeable Al^{3+} plus H^+. Under laboratory conditions greater losses of NO_3^- were observed on those soils with the greater amount of exchangeable acidity, and the losses increased with increases in temperature. The extent of such losses under field conditions is not known.

FERTILIZER NITROGEN MATERIALS

During the last two decades a phenomenal increase in the consumption of commercial fertilizers has been noted in the United States and in other countries as well. The increase in the use of fertilizer nitrogen has been particularly impressive and, as already stated, is due in no small part to the efficiency of the nitrogen producers and the low cost with which these materials have been offered to growers. Nitrogen fertilizer compounds are subject to the immobilization and mineralization reactions covered in the first part of this chapter. Some knowledge of the different properties of the different materials in relation to the reactions of nitrogen fertilizers in the soil should result in greater efficiency in their on-farm use.

TYPES OF NITROGEN FERTILIZER

Nitrogen fertilizers may be classified broadly as either *natural organic* or *chemical*. The natural organic materials are of plant or animal origin; the chemical sources are neither plant nor animal.

TABLE 5-7. Average Composition of Some Common Natural Organic Materials

Source	Per cent						
	N	P_2O_5	K_2O	CaO	MgO	S	Cl
Activated sewage sludge	6.0	2.2	—	2.5	1.5	0.4	0.5
Blood, dried	13.0	—	—	0.5	—	—	0.6
Bone meal (raw)	3.5	22.0	—	31.5	1.0	0.2	0.2
Bone meal (steamed)	2.0	28.0	—	33.0	0.5	0.2	—
Castor pomace	6.0	1.5	0.5	0.5	0.5	—	0.3
Cocoa meal	4.0	1.5	2.5	0.5	1.0	—	—
Cocoa shell meal	2.5	1.0	3.0	1.5	0.5	—	—
Cocoa tankage	2.5	1.5	1.2	17.0	—	—	—
Cottonseed meal	6.6	2.5	1.5	0.5	1.5	0.2	—
Fish scrap (acidulated)	5.7	3.0	—	8.5	0.5	1.8	0.5
Fish scrap (dried)	9.5	6.0	—	8.5	0.5	0.2	1.5
Garbage tankage	2.5	1.5	1.0	4.5	0.5	0.4	1.3
Peanut meal	7.2	1.5	1.2	0.5	0.5	0.6	0.1
Peanut hull meal	1.2	0.5	0.8	—	—	—	—
Peat	2.7	—	—	1.0	0.5	1.0	0.1
Peruvian guano	13.0	12.5	2.5	11.0	1.0	1.4	1.9
Process tankage	8.2	—	—	0.5	—	0.4	—
Soybean meal	7.0	1.2	1.5	0.5	0.5	0.2	—
Tankage, animal	7.0	10.0	—	15.5	0.5	0.4	0.7
Tobacco stems	1.5	0.5	5.0	5.0	0.5	0.4	1.2
Whale guano	8.5	6.0	—	9.0	0.5	—	—

Organic Forms. Before 1850 virtually all of the fertilizer nitrogen consumed in the United States was in the form of natural organic materials. In 1973 these materials accounted for less than 0.5 per cent of the total sold in this country. For all intents and purposes these materials are of historical interest only, although small amounts still find their way into so-called specialty fertilizers for lawns, gardens, and shrubs and in some fertilizers applied to flue-cured tobacco. The average plant nutrient content of some natural organics is shown in Table 5-7.

Natural organic materials at one time were thought to release their nitrogen slowly, thereby supplying the crop with its nitrogen requirement as needed and also reducing the losses of this element by leaching. This was shown not to be the case, however, as most of the nitrogen that became available did so within the first three weeks. This is illustrated by the data graphed in Figure 5-16.

Under conditions optimum for nitrification at best only about half of the total nitrogen was converted to a form available to plants at the end of fifteen weeks. In addition, of the nitrogen mineralized during the fifteen-week period, 80 per cent had been converted to NO_3^- at the end of the first three weeks. It is obvious that under warm, moist conditions

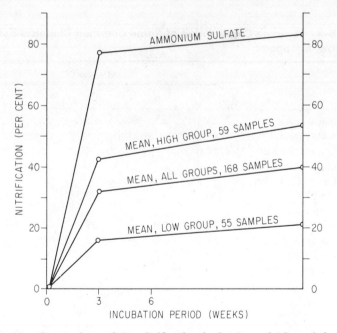

Figure 5–16. Comparison of the nitrification in 3-, 6-, and 15-week incubation periods of the nitrogen in ammonium sulfate and the water-insoluble nitrogen in groups of certain mixed fertilizers marketed in five Southeastern states. [Clark et al., *Agron. J.*, **43**:57 (1951).]

slow release of nitrogen from these materials is not effected and that the amount becoming available to the crop is but a fraction of the total amount they contain.

Chemical Sources of Nitrogen.   In terms of tonnage consumed, chemical sources of nitrogen are by far the most important of the fertilizer nitrogen compounds. Most are ammonia derivatives — that is, they are derived from the compound ammonia. Most of the ammonia in the United States is produced synthetically by reacting nitrogen and hydrogen gas, though some is still recovered as a by-product of the coking of coal. However, recovered ammonia makes up a significant fraction of the total produced in continental Europe. The fundamentals of the production of nitrogen materials and other fertilizers are covered in Chapter 9.

From the basic compound, NH_3, many different fertilizer nitrogen compounds are manufactured. A few materials do not originate from synthetic ammonia, but they constitute only a small percentage of the nitrogen fertilizer tonnage used in the United States. For convenience the various nitrogen compounds are grouped into four categories: ammoniacal, nitrate, slowly available, and other. The composition of some common chemical sources of nitrogen is shown in Table 5-8.

TABLE 5-8. Average Composition of Some Common Chemical Sources of Fertilizer Nitrogen

Source	Per cent						
	N	P_2O_5	K_2O	CaO	MgO	S	Cl
Ammonium sulfate	20.5	—	—	—	—	23.4	—
Anhydrous ammonia	82.2	—	—	—	—	—	—
Ammonium chloride	28.0	—	—	—	—	—	—
Ammonium nitrate	32.5	—	—	—	—	—	—
Ammonium nitrate with lime (ANL)	20.5	—	—	10.0	7.0	0.6	—
Ammoniated ordinary superphosphate	4.0	16.0	—	23.0	0.5	10.0	0.3
Monoammonium phosphate	11.0	48.0	—	2.0	0.5	2.6	—
Diammonium phosphate	20.0	54.0	—	—	—	—	—
Ammonium phosphate-sulfate	16.5	20.0	—	—	—	15.0	—
Calcium nitrate	15.5	—	—	27.0	2.5	—	0.2
Calcium cyanamide	22.0	—	—	54.0	—	0.2	—
Potassium nitrate	13.4	—	44.2	0.5	0.5	0.2	1.2
Sodium nitrate	16.0	—	—	—	—	—	0.6
Urea	46.0	—	—	—	—	—	—
Urea-sulfur	40.0	—	—	—	—	10.0	—

AMMONIACAL SOURCES. The following ammoniacal compounds are used as sources of fertilizer nitrogen: anhydrous ammonia, NH_3 (82% N); anhydrous ammonia-sulfur (74% N, 10% S); aqua ammonia and solutions of ammonia and other nitrogen salts (24–49% N); ammonium nitrate (32–33% N); ammonium nitrate with lime, ANL (20.5% N); ammonium nitrate-sulfate (30% N, 5% S); ammonium sulfate (20.5% N, 23.4% S); monoammonium phosphate, MAP (11% N, 21% P); diammonium phosphate, DAP (16–21% N, 21–23% P); ammonium phosphate-sulfate (16% N, 9% P, 15% S); ammonium chloride (26% N); urea (45% N); urea-sulfur (40% N, 10% S); and urea-phosphates. The properties and behavior in the soil of these materials are covered in the following sections.

Anhydrous Ammonia (NH_3). Anhydrous ammonia contains the highest percentage of nitrogen of any nitrogenous fertilizer currently on the market. It is stored under pressure as a liquid, and its application in the field requires the use of high-pressure tanks and metering devices. Bulk storage at atmospheric pressure requires refrigerating equipment to liquify the ammonia gas volatilized but does not depend on the construction of high-pressure tanks. Much of this material is custom applied so that the farmer does not have to purchase his own tanks and equipment for on-the-farm handling and storage. Anhydrous ammonia may be applied directly to the soil by injection through tubes running down the rear of a blade-type applicator. In some areas, notably the irrigated areas of the West, ammonia is applied by metering into irrigation water (but not, however, in overhead systems).

Because it is a gas at atmospheric pressure, some anhydrous ammonia may be lost to the aboveground atmosphere during and after application. Factors associated with this loss are the physical condition of the soil during application, soil texture and moisture content, and depth and spacing of placement. If the soil is hard or full of clods during application, the slit behind the applicator blade will not close or fill, and some of the released ammonia will escape to the atmosphere.

Soil moisture content and spacing of the applicator blades affect ammonia retention in the soil, as illustrated in Figure 5-17. An increase in the moisture content had a pronounced effect on lowering the gaseous loss of ammonia. A point of practical importance is the reduced loss that accompanied the narrower applicator spacing when the same rate of ammonia was applied. Greater losses at the wider spacing undoubtedly resulted from the greater localized concentration at the point of injections.

Data from field experiments conducted by scientists at Cornell University are shown in Table 5-9. Losses of ammonia under the conditions of these tests were negligible, even when the application was made to an established alfalfa sod. As a general rule of thumb, if no ammonia vapor can be seen during the actual application and if the odor of ammonia is not unduly strong, it may be reasonably assumed that the material is being properly applied and that minimum losses are being incurred.

Free ammonia is toxic to living organisms. As such, it might be expected to delay the nitrification process. Workers in Florida studied this problem in some laboratory experiments on three sandy soils. Results of these tests showed that anhydrous ammonia ultimately stimulated nitrification. However, an initial and marked reduction in the numbers of soil microorganisms was noted. On recovery, the numbers of bacteria increased more rapidly than the numbers of fungi following the ammonia application. The experimenters attributed this stimulation to the more favorable pH that resulted from the reaction of the ammonia with the soil and favored the activity of the nitrifying bacteria. The reaction of ammonia with soil moisture would, of course, produce ammonium hydroxide which would increase soil pH.

The ultimate effect of applying anhydrous ammonia to soil is to lower the pH. It will be recalled from the preceding discussion on nitrification that the conversion of ammonia to nitrate is accompanied by the production of acid. Thus, even though the initial effect of ammonia to soil is to

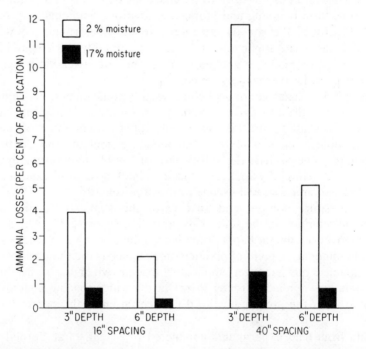

Figure 5–17. Ammonia losses from a Putnam silt loam as influenced by soil moisture and depth and width of spacing. [McDowell et al., *SSSA Proc.,* **22:**38 (1958).]

TABLE 5–9. Field Losses of Anhydrous Ammonia Immediately Behind the Applicator Blade*

	Condition of soil			Application of NH_3		
Soil type	Surface	pH	Moisture content (%)	Depth (in.)	Rate[†] (lb. N/A.)	Ammonia lost[‡] (%)
Sassafras f. s. l.	plowed	5.9	18	4	257	0.001
Lima silt loam	plowed	7.2	16	4	224	0.008**
		7.2	16	4	261	0.05
Lima silt loam	alfalfa sod	6.9	13	5	261	0.01

* Baker et al., *Agron. J.*, **51**:361 (1959).
† On basis of 14-in. spacing.
‡ Ammonia lost expressed as percent of the amount applied.
** This test covered a distance of 200 ft.; the other tests covered a distance of 100 ft.

raise the pH, the ultimate effect is a lowering of this soil property. The extent of this reduction will depend on the buffering capacity of the soil (amount of clay and organic matter present), its initial pH, and the amount of ammonia added. This lowering of soil pH from the application of ammoniacal fertilizers is common to all ammonium salts, ammonia and ammonia solutions, and urea. The acidifying effect of various nitrogen carriers is discussed in a subsequent section of this chapter.

The distribution and nitrification of anhydrous ammonia was investigated under field conditions by scientists at Iowa State University. A portion of their results is shown in Table 5-10. The soil type used in this study was a Nicollet sandy clay loam which is fairly common in the prairie area of Iowa. The rate of applied nitrogen was 120 lb./A. injected to a depth of 4 to 6 in. The data show that under the conditions of this experiment, which was carried out in August and September, nitrification was taking place at a rapid rate twenty-eight days after application. Initially, nitrification is more rapid at the periphery of the zone of retained ammonia. It progresses inward as the toxic effect of the gas subsides after its reaction with soil water, clay, and organic matter.

Work in Oregon has shown that anhydrous ammonia may be fixed in a nonexchangeable or difficultly exchangeable form by the eighteen Pacific Northwest soils included in the study. Of the ammonia retained by these soils, 1 to 8 per cent in the surface and 2 to 31 per cent in the subsurface horizons was fixed by the mineral fraction. It was also observed that mineral fixation of ammonia by air-dry soils exceeded wet fixation of nitrogen from aqua ammonia by several-fold. The full practical import of these observations is not known, but, as already pointed out, arable agricultural soils under field conditions are seldom air dry below 2 or 3 in. for any long period of time. It is unlikely that fixation in a totally

TABLE 5-10. Ammonia Retention Around the Point of Application in the Field Following the Period of Rapid Nitrification*

Depth (in.)	Analysis	Inches from applicator row								
		4	3	2	1	0	1	2	3	4
0	pH			6.3	6.0	6.3	6.3	6.3	6.2	
	NH$_4$-N, ppm.			4	13	30	31	18	31	
	NO$_3$-N, ppm.			91	198	164	123	151	185	
2	pH	6.1	6.1	6.1	5.7	5.3	4.5	5.4		5.4
	NH$_4$-N, ppm.	0	4	98	117	84	133	104		28
	NO$_3$-N, ppm.	29	123	207	174	103	224	215		170
3	pH			4.5	4.7	4.8	5.0	4.4	4.5	
	NH$_4$-N, ppm.			7	113	138	188	159	156	
	NO$_3$-N, ppm.			117	226	192	174	219	280	
4	pH	6.6	5.8	5.7	6.3	6.3	6.3	5.7		6.1
	NH$_4$-N, ppm.	0	15	104	226	295	281	383		18
	NO$_3$-N, ppm.	29	110	174	231	274	301	306		194
5	pH	6.5	6.0	5.9	5.5	5.4	5.3	5.3		6.7
	NH$_4$-N, ppm.	0	0	17	135	104	138	97		0
	NO$_3$-N, ppm.	24	83	119	209	261	273	242		76
6	pH	lost	6.3	6.4	5.8	5.4	5.6	5.7		6.7
	NH$_4$-N, ppm		2	0	10	17	24	18		0
	NO$_3$-N, ppm.		36	54	89	102	123	95		56
7	pH	6.1	6.9	7.0	6.9	6.6	6.4	6.6		6.8
	NH$_4$-N, ppm.	0	0	0	0	0	0	28		0
	NO$_3$-N, ppm.	33	30	44	54	65	71	67		55
8										

* McIntosh et al., *SSSA Proc.*, **22**:402 (1958).
120 lb. N/A., applied Aug. 27 and sampled Sept. 24, 1956.

unavailable form will account for large losses of applied anhydrous ammonia.

Perhaps the most important practical criterion of the value of anhydrous ammonia is its effect on plant growth. This problem was studied by workers at the University of Illinois who measured the effect of rate, depth, and time of application of ammonia before planting on the germination and early growth of corn. Corn was planted directly over the ammonia band which in turn had been applied at different depths (4, 7, and 10 inches) and at different rates of N (100, 200, 400, and 600 pounds per acre). The corn was then planted at 0, 1, and 2 weeks after the ammonia application. A portion of the results, shown graphically in Figure 5-18, illustrates the importance of the proper placement of ammonia.

Though not particularly germane to the subject of soil fertility, it is interesting to note that anhydrous ammonia is used to defoliate crops such as cotton. The extent to which this practice will expand will depend on its cost in relation to the cost of other defoliants.

Anhydrous Ammonia-Sulfur ($NH_3 + S$). Elemental sulfur is soluble in anhydrous ammonia to the extent shown in Table 5-11. Use has been made of this solvent action of ammonia to develop an ammonia-sulfur fertilizer material for direct soil application. Limited quantities are being produced and sold in some of the Intermountain states.

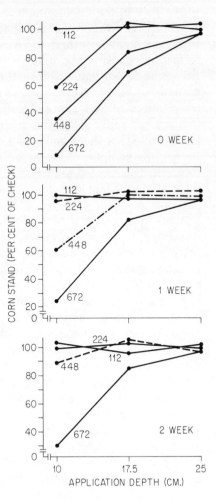

Figure 5-18. Effect of time, depth, and rate of NH_3 application on stand 27 days after planting (numbers by lines are kg./ha. of N; Bayes lsd$_{(.05)}$ = 6 per cent). [Colliver and Welch, *Agron. J.*, **62**:341 (1970).]

TABLE 5-11.   The Solubility of Elemental Sulfur in Anhydrous Ammonia.

Temperature (°C.)	Sulfur in solution (%)
−78.0	38.6
−20.5	38.1
0.0	32.3
+16.4	25.6
30.0	21.0
40.0	18.5

The sulfur in the NH_3-S is deposited in a very finely divided condition in the soil as the liquid ammonia is converted to a gas and it is subsequently oxidized to the sulfate form by sulfur bacteria. It does not appear that anything other than small quantities of S-N compounds are formed, for most of the sulfur is recovered as the elemental form when the ammonia is volatilized from the NH_3-S solution. Results of field tests carried out in the Northwest and other areas have shown its suitability as a source of nitrogen and sulfur for small grains and other crops.

Ammonia-sulfur solution is corrosive to certain parts of the applicator equipment. This shortcoming can be easily corrected by minor modifications in the equipment. However, the failure of many operators to make these changes has led to the predicted difficulties with the result that this otherwise excellent nitrogen-sulfur fertilizer has met with only limited acceptance.

Aqua Ammonia and Nitrogen Solutions.   Aqua ammonia is made by dissolving ammonia gas in water. Nitrogen solutions are made by dissolving nitrogen salts such as ammonium nitrate and urea either in water or aqua ammonia. All nitrogen solutions are classed as either pressure or nonpressure. The pressure solutions are those that have an appreciable vapor pressure because of the free ammonia. The nonpressure solutions contain no free ammonia.

Obviously, the nonpressure solutions can be handled and stored without the use of high-pressure tanks and equipment. The pressure solutions, however, may require modified tanks and equipment, especially if their vapor pressure is appreciable at operating temperatures.

Nitrogen solutions containing dissolved salts exhibit a phenomenon known as salting out, which is simply the precipitation of the dissolved salts when the storage temperature reaches a certain degree. The salting-out temperature determines the extent to which outside storage may be practiced in the winter and the time of the year at which these solutions may be field-applied. The composition and properties of pressure and nonpressure solutions are listed in Tables 5-12 and 5-13. By way of explanation of these tables, a system for characterizing nitrogen solutions has generally been accepted by the fertilizer industry. In this system the percentage of total nitrogen, with the decimal point omitted, is followed in parentheses by the percentage composition of ammonia, ammonium nitrate, urea, and a fourth figure which represents any other significant source of nitrogen, if present. All of these figures are rounded to the nearest whole number and stated in the order given. The company name, a trade-mark, or a trade name will often precede the total percentage of nitrogen. For example, a solution labeled 206(25-0-0) would be an ammonia solution containing 20.6 per cent total nitrogen and 25 per cent ammonia. A 414(19-66-6) solution would contain 41.4 per cent total nitrogen and 19, 66, and 6 per cent, respectively, ammonia, ammonium nitrate, and urea.

Low-pressure or no-pressure solutions are easy to use. Nitrogen solutions reduce labor because the pump and the tank replace the sack and the back. The equipment for applying the no-pressure solutions consists of a pump, barrel, hose, gage, bypass valve, and boom. A mechanically adept farmer can assemble this equipment himself though more and more, custom application by the dealer is being employed. With a boom 30 or 40 ft. long, more than 100 A./day can be covered. Planes are also used to spread these solutions at a rate of 100 A./hr.

Solutions with no pressure can be dribbled on or applied in a small stream directly to the foliage of grasses and small grains with little

TABLE 5-12. Composition and Properties of Pressure Nitrogen Solutions Used as Direct Application Materials[*]

	Forms of nitrogen				Vapor pressure (gage) at					Weight of solution per gallon at 60°F. (lb.)	Weight of nitrogen per gallon of solution at 60°F. (lb.)
	Ammonia										
Solution nomenclature	Neutralizing (% N)	Combined (% N)	Nitrate (% N)	Urea (% N)	90°F. (psi)	104°F. (psi)	120°F. (psi)	Salting-out temperature (°F.)	Specific gravity		
201 (24-0-0-)	20.1	—	—	—	0	0	7	−62	0.913	7.61	1.53
247 (30-0-0)	24.7	—	—	—	4	11	20	−112	0.896	7.47	1.85
370 (17-67-0)	13.7	11.7	11.7	—	−3	1	7	50	1.184	9.87	3.65
410 (19-58-11)	15.6	10.2	10.2	5.1	4	10	18	7	1.162	9.69	3.97
410 (22-65-0)	18.3	11.4	11.4	—	5	10	19	21	1.138	9.49	3.89
410 (26-56-0)	21.6	9.7	9.7	—	—	16	—	−23	1.077	8.98	3.68
411 (50-0-0)	41.1	—	—	—	55	74	103	Below −100	0.832	6.93	2.85

[*] Sauchelli, *Fertilizer Nitrogen—Its Chemistry and Technology*, 1964. Reprinted with permission of the Reinhold Publishing Corp., New York.

Each product represents one or more similar commercial products; slight variations occur in the published data for similar solutions.

TABLE 5–13.   The Composition of Nonpressure Nitrogen Solutions*

Solution nomenclature[a]	Combined ammonia (% N)	Nitrate (% N)	Urea (% N)	Salting-out temperature (°F.)	Specific gravity at 60°F.	Weight of solution per gallon at 60°F. (lb.)	Weight of nitrogen per gallon of solution at 60°F. (lb.)
160 (0-46-0)	8.0	8.0	–	11	1.207	10.07	1.61
170 (0-31-0-36)[b]	5.4	11.6	–	32	1.499[c]	12.50[d]	2.13[d]
170 (0-49-0)	8.5	8.5	–	16	1.222	10.19	1.73
175 (0-50-0)	8.8	8.8	–	22	1.231	10.27	1.81
180 (0-51-0)	9.0	9.0	–	23	1.235	10.30	1.85
190 (0-54-0)	9.5	9.5	–	33	1.253	10.45	1.99
200 (0-57-0)	10.0	10.0	–	41	1.264	10.54	2.11
200 (0-0-44)[e]	–	–	20.3	41	1.124	9.36	1.87
210 (0-60-0)	10.5	10.5	–	49	1.286	10.73	2.25
228 (0-65-0)	11.4	11.4	–	65	1.307	10.90	2.49
230 (0-0-50)[e]	–	–	28.0	63	1.152	9.60	2.21
245 (0-70-0)	12.3	12.3	–	86	1.346[f]	11.23[f]	2.75[f]
280 (0-39-31)	6.9	6.9	14.3	−1	1.279	10.67	3.00
280 (0-40-30)	7.0	7.0	14.0	−1	1.283	10.70	3.00
280 (0-80-0)	14.0	14.0	–	136	1.360[g]	11.34[h]	3.18[h]
290 (0-83-0)	14.5	14.5	–	154	1.380[g]	11.51[h]	3.34[h]
300 (0-42-33)	7.4	7.4	15.3	15	1.301	10.85	3.27
315 (0-0-68)[i]	–	–	31.5	135	1.170[j]	9.76[j]	3.07[j]
320 (0-44-35)	7.8	7.8	16.5	32	1.327	11.07	3.55
320 (0-45-35)	7.9	7.9	16.3	32	1.325	11.05	3.55
338 (0-0-72)[i]	–	–	33.8	155	1.18[k]	9.84[k]	3.33[k]
376 (0-0-80)[i]	–	–	37.6	190	1.19[l]	9.92[l]	3.73[l]

* Sauchelli, *Fertilizer Nitrogen — Its Chemistry and Technology*, 1964.   Reprinted with permission of the Reinhold Publishing Corp., New York.
[a] Each product represents one or more similar commercial products; slight variations occur in the published data for similar solutions.
[b] Contains 36.2 per cent calcium nitrate.
[c] At 68°F/60°F.
[d] At 68°F.
[e] Represents two solutions; one standard and the other with a low biuret content.
[f] At 70°F.
[g] At 68°F/60°F.
[h] At 158°F.
[i] Contains 3.0 per cent biuret.
[j] At 140°F.
[k] At 160°F.
[l] At 194°F.

danger of burning. If they are sprayed on, however, considerable damage to the foliage may result. This method has been used to kill weeds in corn and cotton, and the same solutions can be added to overhead irrigation water. The application of nitrogen solutions in the first part of the irrigation period should be followed by clear water to wash the nitrogen salts off the plants.

Numerous agronomic tests have shown that there is no real difference between solids and solutions as sources of nitrogen for crop production, provided the materials are properly handled. The factor determining the product used will undoubtedly be the on-farm cost of application. The behavior of the nitrogen solutions in soils will be the same as that of their solid and gaseous (NH_3) components.

Ammonium Nitrate (NH_4NO_3) *and Ammonium Nitrate with Lime.* These two products contain 32 to 33.5 per cent and 20.5 per cent nitrogen, respectively. They differ only in nitrogen content and in the fact that one is coated with finely ground dolomite. Theoretically, the dolomite neutralizes the acidity produced by the nitrification of the ammonium ion, which is initially adsorbed by the clay and held until nitrified. Some may be appropriated by heterotrophic microorganisms or fixed by certain clay minerals. If applied to the surface of alkaline or calcareous soils, some ammonia gas may be volatilized.

Both materials have excellent handling qualities. However, because ammonium nitrate is hygroscopic, it should not be left in open sacks for long periods of time in humid climates. Ammonium nitrate has on occasion picked up a bad reputation because of improper storage and handling. When in intimate contact with oxidizable carbonaceous materials, it forms an explosive mixture. A mixture of oil and ammonium nitrate can be detonated and is often used in blasting work. This property makes it imperative that the product be handled and stored with care. If due caution is observed, agricultural grade ammonium nitrate does not constitute an explosion hazard.

Both ammonium nitrate and ammonium nitrate with lime are excellent sources of fertilizer nitrogen and are widely used throughout the United States, particularly on crops which require side-dress applications of nitrogen.

Ammonium Nitrate-Sulfate. Ammonium nitrate-sulfate is a double salt of ammonium nitrate and ammonium sulfate [$NH_4NO_3 \cdot (NH_4)_2SO_4$] made by neutralizing a mixture of nitric and sulfuric acids with ammonia. Its behavior in soils is the same as that of ammonium nitrate. It is currently being produced by the Tennesses Valley Authority (TVA) as an experimental fertilizer and was developed largely as a means of introducing sulfur as a plant nutrient into this high-analysis solid nitrogen material. It has excellent handling and storage qualities and should prove to be useful on crops in which both nitrogen and sulfur are deficient.

Ammonium Sulfate [$(NH_4)_2SO_4$]. Ammonium sulfate contains 20.5 per cent nitrogen and 23.4 per cent sulfur. It is one of the oldest chemical sources of ammoniacal nitrogen, having been manufactured early in this country as a by-product of the coking of coal. It has good handling and storage qualities and is also a good source of sulfur on soils deficient in this element.

When added to the soil, the ammonium ion is temporarily retained by the colloidal fraction of the soil until it is nitrified. Because of the accompanying sulfate anion, this form of nitrogen tends to be somewhat more acid in its reaction in the soil than sources such as ammonium nitrate. Long-time field studies have shown that the continued use of ammonium sulfate without the addition of lime will reduce the soil pH to a level unsuitable for the economical production of crops. However, when the soil was limed periodically to maintain a suitable pH level, ammonium sulfate produced yields no different from those obtained from the less acid types of nitrogenous fertilizers.

Ammonium Phosphates. Mono-$(NH_4H_2PO_4)$ and diammonium phosphate $[(NH_4)_2HPO_4]$ and ammonium phosphate-sulfate are generally considered to be more important as sources of phosphorus than of nitrogen. Therefore their properties and reactions in the soil are covered in Chapter 6, "Soil and Fertilizer Phosphorus."

Ammonium Chloride. Ammonium chloride (NH_4Cl) contains about 26 per cent nitrogen. It is not employed to any great extent in the United States, though it is used in the Orient as a nitrogen fertilizer for rice. The economical production of ammonium chloride depends on a cheap source of hydrochloric acid or chlorine.

Ammonium chloride has been claimed by some to be a better source of nitrogen for rice than ammonium sulfate. Paddy rice is grown on flooded soils in which extreme reducing conditions obtain. When ammonium sulfate is applied to these soils, the sulfate is reduced to hydrogen sulfide (H_2S), which is toxic. Normally, this H_2S will be rapidly precipitated as the sulfides of such heavy metals as iron and manganese. However, years of continuous rice cultivation have in some areas left soils low in these metals, with the result that the free hydrogen sulfide is not rapidly precipitated. It is believed to come in contact with the rice roots, thereby inducing *Akiochi*, or late summer disease. There is no universal agreement among scientists working on rice fertilization that *Akiochi* is caused by hydrogen sulfide toxicity. Nonetheless, it has been partly responsible for an increase in the use of ammonium chloride under rice, for no hydrogen sulfide is released from that source.

Ammonium chloride has been compared with ammonium sulfate in this country as a source of nitrogen for grass. At lower rates of application it produced yields equal to those obtained from ammonium sulfate. At higher rates the ammonium chloride gave poorer results than the sulfate, in part because of the greater injury to the grass when in contact with the leaves. The acceptance of this material by farmers in the United States will undoubtedly depend on any economic advantage it might offer over existing sources of nitrogen.

Urea. Urea $[CO(NH_2)_2]$ is produced by reacting ammonia with carbon dioxide under pressure and at an elevated temperature. It contains the highest percentage of nitrogen of any solid material currently available (45%).

Though not an ammonium fertilizer in the form in which it is marketed, it hydrolyzes to ammonium carbonate very quickly when added to the soil, as shown by the following equation:

$$CO(NH_2)_2 + 2H_2O \longrightarrow (NH_4)_2CO_3$$

Ammonium carbonate is an unstable compound and decomposes to ammonia and carbon dioxide. The NH_3 or NH_4^+ so released is adsorbed by the colloidal fraction of the soil and subsequently nitrified. Because of the ammonium ion produced by the hydrolysis of this material, it is somewhat acid in its ultimate reaction with the soil.

The hydrolysis of this material is greatly increased in the presence of the enzyme urease, which is found to varying degrees in soils. In most soils it is present in sufficient concentrations to bring about the rapid conversion of urea to NH_4^+. Once in the NH_4^+ form, it behaves like any other ammoniacal source of nitrogen.

Urea is an excellent fertilizer material. However, it possesses several properties which should be understood in order that the greatest benefit may be derived from its use. The first is related to its rapid hydrolysis. If urea is applied to a bare soil surface or to soil in a sod cover, significant quantities of ammonia may be lost by volatilization because of its rapid hydrolysis to ammonium carbonate. Losses of ammonia from urea ap-

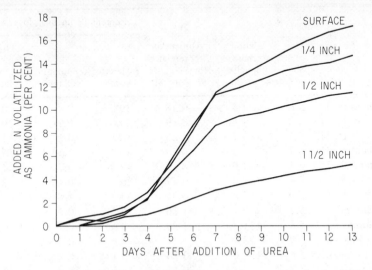

Figure 5–19. Cumulative loss of added nitrogen from urea mixed with surface soil layers of different thickness; 100 lb. urea-N/A applied to pH 6.5 Dickson silt loam and aerated at 75°F. [Ernst et al., *SSSA Proc.*, **24**:87 (1960).]

plied to the soil surface and when mixed with the surface soil to different depths are shown in Figure 5-19. These data were obtained from a laboratory study carried out at 75°F. with a Dickson silt loam to which urea at the rate of 100 lb. N/A. had been applied. The losses were appreciable, enhanced no doubt by the fact that the air was moving through the incubation flasks at a steady rate. However, the effect of mixing the urea with the soil on the reduction of ammonia losses is apparent.

Field studies conducted at the University of Florida have shown that losses of ammonia occur from the bare surface of acid soils. The results of some of this research are given in Table 5-14. It is obvious that losses were greater from some soils than from others and that the losses were greater from pelleted urea than from urea applied in solution. The volatilization of ammonia from ammonium sulfate applied to the Lakeland fine sand IV (*p*H 6.7) and the Perrine marl (*p*H 7.8) illustrate the losses from ammoniacal sources other than urea which can be expected when such materials are applied to neutral or alkaline soils.

When urea is applied to sod, losses of ammonia can be significant. This is illustrated by the curves in Figure 5-20. The losses were appreciable and tended to level off about six to seven days after application.

Burning as a practical means of controlling ammonia losses from surface applications of urea was suggested by research workers at the Tifton, Georgia, Experiment Station. The results of some of their studies

TABLE 5–14.   Gaseous Losses of Ammonia During Seven Days Following Surface Application of 100 lb. of Nitrogen per Acre to Bare, Moist Soils*

				Percentage nitrogen loss from		
Soil type	Soil *p*H	Cation exchange capacity (meq./100 g.)	Absorption potential, (mg. NH₃-N /cc. of soil)	Pelleted urea	Ammonium sulfate	Urea-ammonium nitrate solution†
Lakeland fs. I	5.6	1.5	0.38	39.8	0.4	0.6
Lakeland fs. II	6.3	1.6	0.19	59.0	0.9	29.4
Lakeland fs. III	5.4	4.7	1.23	16.8	0.2	1.6
Lakeland fs. IV	6.7	3.5	0.38	48.9	5.5	15.2
Lakeland fs. V	6.3	1.9	0.39	39.4	0.7	4.4
Leon fs. I	4.4	2.8	0.47	26.9	0.1	3.2
Leon fs. II	5.9	5.8	0.75	35.8	1.6	7.5
Red Bay fsl.	5.3	7.2	1.16	19.5	0.2	2.3
Arredondo fsl.	5.8	11.5	1.59	8.6	0.2	0.4
Fellowship fsl.	5.9	23.4	1.93	7.6	0.5	0.1
Brighton peat	5.6	120.0	3.74	3.1	0.2	–
Perrine marl	7.8	7.4	0.31	14.6	36.9	14.4

* Volk, *Agron. J.*, **51**:746 (1959).
† Contained 32% nitrogen—16.5% from urea and 15.5% from ammonium nitrate.

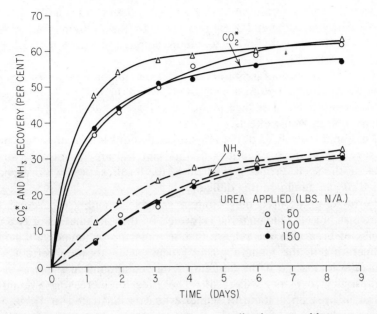

Figure 5–20. The influence of rates of urea applications to a bluegrass sod on the amounts of CO_2 and NH_3 lost to the atmosphere. [Simpson et al., *SSSA Proc.*, **26**:186 (1962).]

with Coastal Bermuda grass are listed in Table 5-15. Burning the sod before urea application significantly increased hay yield which was attributed to a reduction in ammonia losses from urea and a more efficient crop utilization of the applied nitrogen. This, in turn, was doubtless the result of the destruction of the enzyme urease in the plant tissue.

The hydrolysis of urea can be altered by the use of several compounds called urease inhibitors. These inhibitors inactivate the enzyme and thereby prevent the rapid hydrolysis of the urea when it is added to the soil. Some laboratory work done in Iowa showed that gaseous losses of

TABLE 5–15. The Influence of Nitrogen Source and Sod Treatment on Coastal Bermuda Grass Hay Production in 1958–1960[*]

Source of nitrogen (200 lb./A.)	Sod treatment	Relative yield (three-year average)
NH_4NO_3	Burned	100
Urea	Burned	97
Urea	None	86

[*] Jackson et al., *Agron. J.*, **54**:47 (1962).

N as NH_3 from a urea-treated sandy soil was reduced from 61.1 per cent to 0.3 per cent at the end of fourteen days by the addition of an inhibitor known as 2,5-dimethyl-*p*-benzoquinone applied at the rate of 2.3 parts per 100 parts of urea.

The effectiveness of inhibitors in reducing ammonia losses is generally greatest on coarse-textured soils. Also, while there are several types of compounds that will reduce urease activity, the substituted *p* quinones seem to be the most effective.

The rapid hydrolysis of urea in soils is responsible for ammonia injury to seedlings if large quantities of this material are placed with or too close to the seed. Proper placement of the fertilizer urea with respect to the seed can eliminate this difficulty.

Fertilizer-grade urea may contain variable amounts of a compound known as biuret. This material is formed by the combination of two molecules of urea with the release of one molecule or ammonia when the temperature in the manufacturing process goes above a certain level. Biuret is toxic, and if the content of this compound in the urea is too high, injury to plants will result. The effect of increasing amounts of biuret in urea on a stand of winter wheat is illustrated in Table 5-16, which includes a portion of the data from research carried out in South Dakota. The toxic effect of high quantities of biuret is obvious from the data given.

The foregoing discussion is not intended to discourage the use of urea as a fertilizer. Urea is, indeed, an excellent source of nitrogen, and most of the drawbacks mentioned can be overcome by its proper placement and by making certain that the biuret content is less than 1.5 to 2.0 per cent. Urea manufacturers are well aware of the toxicity of biuret, and high concentrations of this material are seldom found in commercial ureas.

Urea-Sulfur. Urea-sulfur is a relatively new compound only recently introduced as a fertilizer material. It contains 40 per cent nitrogen and

TABLE 5–16. The Effect of Biuret Content of Urea on a Stand of Winter Wheat*

Fertilizer source	Biuret (%)	Stand, per cent of check at indicated rates of nitrogen		
		10	20	40
Ammonium nitrate	—	95.6	101.3	88.0
Ammonium nitrate	2.5	94.2	81.9	83.5
Urea	2.5	80.4	62.3	26.7
Urea	5.0	60.7	42.0	19.9
Urea	10.0	41.5	28.3	9.1

* Brage et al., *SSSA Proc.*, **24**:294 (1960).

10 per cent sulfur. The melting points of urea and sulfur are within a few degrees of each other, and in the molten state the two materials are miscible. They can be prilled to form a product of excellent physical properties.

When urea-sulfur is applied to the soil, the urea dissolves, leaving a thin honeycombed skeleton of elemental sulfur, which is ultimately converted to the sulfate form by the sulfur-oxidizing bacteria in the soil. The urea behaves as previously described.

Urea-Phosphate. To supply a material based on phosphoric acid and containing a higher ratio of nitrogen to phosphorus than is found in the ammonium phosphates, urea phosphates have been developed by adding a slurry of urea to a slurry of diammonium phosphate before or duing granulation. A product containing 29 per cent nitrogen and 12.7 per cent phosphorus (29.0% P_2O_5) can be produced in this way. Urea-phosphate may also be produced by reacting urea with phosphoric acid. The product formed by this reaction contains 17.7 per cent N and 19.6 per cent P (44.9% P_2O_5).

NITRATE SOURCES. Several nitrate salts are sources of fertilizer nitrogen: sodium nitrate ($NaNO_3$, 16% N), potassium nitrate (KNO_3, 13.8% N, 36.5% K), calcium nitrate [$Ca(NO_3)_2$, 15.5% N, 19.5% Ca], and the ammonium nitrate materials covered in the preceding section. In terms of tonnages consumed, the ammonium nitrates are the most important of the nitrate carriers, but the other forms are consumed locally.

Nitrate fertilizers are quite soluble, and when added to the soil the nitrate ion is readily absorbed by plant roots. Nitrates are also susceptible to losses by leaching under conditions of high rainfall and in coarse-textured soils. They may also be appropriated by soil microorganisms in the decomposition of organic residues.

The nitrates of sodium, potassium, and calcium are not acid forming, as are ammonium nitrate and the ammonium fertilizers in general, because no acid is produced, as in ammonium salts. In addition, the nitrate is rapidly absorbed, generally more so, in fact, than the accompanying cation. This leaves an excess of basic cations and, of course, a somewhat more alkaline condition. Prolonged use of sodium nitrate has shown that the original soil pH may be maintained or even increased.

Sodium Nitrate. Sodium nitrate is sold in two forms. One is a natural product which is recovered from vast deposits in Chile and which contains trace amounts of microelements such as boron. At one time this source of nitrogen constituted almost the sole supply of chemical nitrogen used in the United States.

Though its importance has somewhat decreased, the absolute tonnages of this material consumed annually have remained relatively constant. It is used as a top dressing for row crops, largely in the southeastern United States.

Synthetic sodium nitrate, also containing 16 per cent nitrogen, is produced in the United States. It is manufactured by reacting nitric acid with either sodium chloride or sodium carbonate. It is a pure salt, as marketed, and does not contain the quantities of trace elements found in the natural product.

Potassium Nitrate. Potassium nitrate (KNO_3), production of which was begun only recently in the United States, can be made by reacting sodium nitrate with potassium chloride and separating the potassium nitrate by crystallization and filtration. It is more commonly produced, however, by reacting nitric acid with potassium chloride.

Like sodium nitrate, potassium nitrate is physiologically alkaline in its effect on soil *p*H. It finds its greatest use as a side-dress material for tobacco, citrus, and specialty crops which respond to delayed applications of both potassium and nitrogen.

Calcium Nitrate. Calcium nitrate is not an important fertilizer in the United States. It is produced in Europe, where it finds its largest use, by reacting nitric acid with calcium carbonate. It is also recovered as a by-product of one of the processes for manufacturing nitric phosphate fertilizers.

SLOWLY AVAILABLE NITROGEN COMPOUNDS. Crop recoveries of nitrogen seldom exceed 60 to 70 per cent of that added in the fertilizer. Although there are some losses from volatilization of nitrogen and ammonia and from the fixation of ammonia by clay mineral entrapment and immobilization by bacteria, the principal loss results from the leaching of nitrates. These losses have prompted a search for fertilizer materials that release their nitrogen over a long period of time so that the nitrates will be absorbed by the expanding root system during the entire growing period of the plant.

During the last few years several slowly available forms of nitrogen have been developed and a few have been used commercially on sod and certain specialty crops. Among the products developed are sulfur-coated urea, urea-formaldehyde compounds, metal ammonium phosphates, ox-amide, crotonylidene diurea, isobutylidene diurea, and a few others of purely academic interest, such as thiourea, urea-pyrolyzate, and dicyandiamide. The control of nitrification by certain inhibitors of bacterial activity has also received attention. The properties and reaction of some of these compounds and the action of nitrification inhibitors are briefly considered.

Sulfur-coated urea (SCU). Sulfur-coated urea is produced by spraying molten sulfur onto granular urea while the latter is passing through a rotary drum. Wax, a microbiocide, and a conditioner are then added to the sulfur-coated product. The final product will contain 7 to 12 per cent S, 2 per cent wax, 2 per cent conditioner, and a fraction of per cent of the microbiocide. A typical material produced by the Tennessee Valley Authority, which developed this material in the U.S., contains 39 per cent N and 10 per cent S.

The results of numerous agronomic tests have shown that sulfur-coated urea is an excellent slow-release nitrogen fertilizer, particularly for long-season crops. The seasonal utilization of the applied nitrogen is improved and leaching losses are reduced. SCU has been shown to be a particularly effective source of nitrogen on Bermuda grass and on rice growing under delayed and intermittent flooding.

Urea-Formaldehyde Compounds. These materials are produced commercially by reacting urea with formaldehyde. A whole series of compounds, ranging from quite soluble to completely insoluble, is possible, depending on the ratio of urea to formaldehyde in the final product. To be classed as fertilizer materials, however, they must contain at least 35 per cent nitrogen, largely in insoluble but slowly available form. The water-insoluble nitrogen in these products must test not less than 40 per cent active by the nitrogen activity index for urea-formaldehyde compounds, defined by the Association of Official Agricultural Chemists. The activity index (AI) for urea-formaldehyde compounds is defined as

$$AI = \frac{\% \text{ CWIN} - \% \text{ HWIN}}{\% \text{ CWIN}} \times 100$$

where AI is the activity index, CWIN is the % N insoluble in cold water (25°C), and HWIN is the % N insoluble in hot water (98–100°).

The suitability of these compounds as fertilizers is dependent on the following:

1. The *quantity* of cold-water insoluble nitrogen, which is the source of the slowly available nitrogen.
2. The *quality* of the cold-water insoluble nitrogen determined by its activity index, which reflects the rate at which the cold-water insoluble nitrogen will become available.

The relation between the AI of urea-formaldehyde products and the rate and extent to which they are nitrified is shown in Figure 5-21.

Urea-formaldehyde compounds can be used as a nitrogen fertilizer for direct application to crops or they can be included in complete N-P-K fertilizers. A popular urea-formaldehyde product commonly marketed in the United States contains 38 per cent nitrogen; 28 per cent of the nitrogen in this material is insoluble in water and has an activity index of 50.

Metal Ammonium Phosphates. Several metal ammonium phosphates are known, but the one that has been commercially produced as a fertilizer is magnesium ammonium phosphate. Its formula is $MgNH_4PO_4 \cdot H_2O$, and contains about 8.96 per cent nitrogen. It is very slightly soluble in water. When added to the soil, it releases its ammonia only slowly.

Greenhouse and field tests with this material have shown it to be a

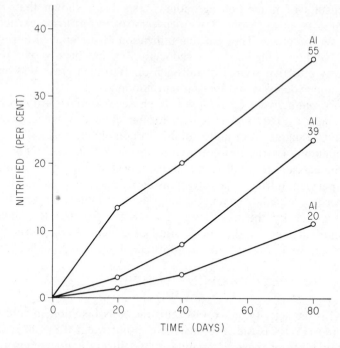

Figure 5–21. Relationship between the activity index (AI) and the nitrification of water-insoluble fractions of urea-formaldehyde products. (Sauchelli, *Fertilizer Nitrogen — Its Chemistry and Technology,* 1964. Reprinted with permission of Reinhold Publishing Corp., New York.)

suitable slowly available form of nitrogen (as well as of magnesium and phosphorus) for turf, fruit trees, and specialty crops. It can be applied to crops, even directly on sod, with little or no injury to the plants.

Oxamide. Oxamide is largely a laboratory curiosity but it has some desirable properties which may encourage its commercial development as a source of nitrogen. Oxamide is a water-insoluble compound containing 31.8 per cent nitrogen and having the formula

$$NH_2COCONH_2$$

It hydrolyzes in soils according to reactions represented by the following formulas:

$$NH_2COCONH_2 + H_2O \longrightarrow NH_2COCOOH + NH_4OH$$

oxamic
acid

$$NH_2COCOOH + H_2O \longrightarrow (COOH)_2 + NH_4OH$$

oxalic
acid

The released ammonia is nitrified. As predicted by its initial insolubility and its tendency to hydrolyze, the conversion of oxamide to nitrate nitrogen in soils is significantly influenced by the granule size of the material. Its rate of hydrolysis and the conversion of the released ammonium to nitrate are closely related to the size of the individual particles and that its availability pattern in the soil is determined by the distribution of particle sizes in the product.

Crotonylidene Diurea (CDU). Crotonylidene diurea was developed in Germany and has recently received some attention as a source of slowly available nitrogen. It is prepared by reacting crotonaldehyde with urea in a 1:2 mole ratio in an aqueous medium. It contains 28 per cent nitrogen, of which about 10 per cent is present as nitrate.

Results of tests with this material conducted in Germany have shown that it provided a long-lasting and steady supply of nitrogen to ryegrass and wheat (grown in pots), that only insignificant leaching losses occurred, and that it was better tolerated by wheat and sunflowers at high rates of application. It was reported that field studies with a wheat-corn-oats rotation confirmed its controlled release of nitrogen and its resistance to leaching losses.

Little else is known about this material, and its potential for development must await further agronomic tests and its economical commercial development.

Isobutylidene diurea (IBDU). IBDU is a condensation product of urea and isobutyraldehyde. It contains 31 per cent N, and the material is nonhygroscopic. Its conversion to a plant-available form of nitrogen depends upon the granules with the smaller particles dissolving more rapidly. Its mineralization in soil increases with pH, temperature, and moisture within the limits which also favor plant growth.

Miscellaneous Slowly Available Nitrogen Compounds. Dicyandiamide is one of the products formed during the hydrolysis of calcium cyanamide in the soil. Its formula is $NH_2C\text{=}NHNHCN$, and it contains 42.0 per cent nitrogen. It is generally considered to be toxic, and little is currently known about its value as a slowly available form of fertilizer nitrogen. The behavior of calcium cyanamide in the soil is covered in a subsequent section of this chapter.

Thiourea is a urealike compound in which sulfur is substituted for oxygen in the molecule. Its formula is $CS(NH_2)_2$, and it contains 36.8 per cent nitrogen. Thiourea is known to be toxic to some organisms and has been used as a weed killer. It is nitrified slowly in the soil, but its suitability as a source of fertilizer nitrogen has not yet been adequately evaluated.

Nitrification Inhibitors. Certain materials are toxic to the nitrifying bacteria and will, when added to the soil, temporarily inhibit nitrification. Work conducted in North Carolina has shown that when certain compounds were added to control nematode infestation in tobacco soils there was a reduction in the nitrification of added ammonium fertilizer. The materials used for nematode control were ethylene dibromide, a mixture of dichloropropene and dichloropropane, and methylbromide. These studies were not concerned primarily with nitrification. They were made because of deleterious effects on tobacco quality resulting from the use of the nematicides. These ill effects were later found to decrease with increasing amounts of nitrate applied to the crop.

There is on the market a compound, 2-chloro-6-(trichloromethyl) pyridine, offered specifically for the inhibition of nitrification and the delayed release of NO_3^- nitrogen. Results of research have

TABLE 5–17. Nitrification of Aqua-Ammonia Treated with Various Rates of 2-Chloro-6-(trichloromethyl)pyridine and Band-Applied to Various Soils Under Nonleaching Conditions[*]

Soil number and type	Concentration of chemical[†] (%)	Recovery of ammonium nitrogen applied (%)			
		4 weeks	9 weeks	13 weeks	18 weeks
8. Loamy sand	0	8	0	0	0
	0.5	20	3	8	3
	2	47	7	14	3
13. Clay	0	63	1	0	0
	0.5	81	16	10	0
	2	95	42	28	0
15. Loam	0	16	0	0	0
	0.5	30	10	1	5
	2	42	29	8	6
16. Loam	0	22	8	0	0
	0.5	27	12	8	4
	2	33	26	10	4
17. Sandy loam	0	35	3	0	0
	0.5	63	32	17	4
	2	88	53	31	7
19. Loam	0	34	6	5	0
	0.5	90	20	21	1
	2	99	52	37	26

[*] Goring, *Soil Sci.*, **93**:431 (1962). Reprinted with the permission of The Williams & Wilkins Co., Baltimore.
[†] Expressed as per cent of fertilizer nitrogen.

shown that the minimum concentration required in the soil to delay the conversion of NH_4^+ to NO_3^- for at least six weeks ranges from a low of 0.05 ppm. to as high as 20. Higher concentrations delay the conversion for longer periods of time. After the concentration of the chemical in the soil reaches a nontoxic level, the interval required for the recovery of the nitrifying organisms decreases with increasing pH, organic matter level, rate of reinfestation, and temperature.

The effect of soil type and rate of application of the inhibiting chemical on the recovery of applied ammonium nitrogen over different periods of time is shown in Table 5-17. The effect of rate of application on inhibiting nitrification is immediately apparent from a study of these data. The effect of texture is less obvious, and evidently soil factors other than texture influenced the ability of this material to delay nitrification.

The use of nitrification inhibitors is a novel approach to the problem of reducing nitrate losses and increasing the efficiency with which nitrogen fertilizers can be utilized by crops. The extent to which slowly available sources of nitrogen and nitrification inhibitors can be applied in agriculture will depend on their cost and their ability to increase crop yields. Their advantages have not been widely demonstrated with most farm crops. With turf, specialty, or high-value crops, the use of some of the slowly available nitrogen sources has been sufficiently successful to justify their commercial production.

Other Sources of Nitrogen. The only important source of fertilizer nitrogen that has not been touched on is calcium cyanamide. This material can be manufactured by reacting calcium carbide (CaC_2) with elemental nitrogen. Its formula is $CaCN_2$, and it contains 21 to 22 per cent nitrogen. It leaves a basic residue and is hydrolyzed to urea when added to the soil.

The conversion to urea is not a straightforward process and it is influenced by several factors. The following reactions are believed to take place in part when cyanamide is added to the soil:

$$2CaCN_2 + 2H_2O \longrightarrow Ca(HCN_2)_2 + Ca(OH)_2 \qquad (1)$$

$$2Ca(HCN_2)_2 + 2H_2O \xrightarrow[\text{soil}]{\text{moderately acid}} (CaOH)_2CN_2 + 3H_2CN_2 \quad (2)$$

$$Ca(HCN_2)_2 + 2H_2O \xrightarrow[\text{soil}]{\text{moderately alkaline}} (H_2CN_2)_2 + Ca(OH)_2 \qquad (3)$$

$$H_2CN_2 + H_2O \longrightarrow CO(NH_2)_2 \qquad (4)$$

Other side reactions are postulated, but these equations will suffice to indicate the nature of the changes taking place.

The intermediate products of the hydrolysis of cyanamide are toxic to plants. When used as a fertilizer, it should be applied at least two to three weeks before planting and should be well mixed with the soil. Thorough mixing permits the toxic intermediates to be completely and quickly converted to urea. If the cyanamide is not properly incorporated into the soil or if the soil is extremely dry, a surface coating forms on the granules, which traps the nitrogen and reduces its availability to plants. Not only does this coating reduce the availability of the contained nitrogen, but inadequate moisture results in the formation of the stable and somewhat toxic dicyandiamide $(H_2CN_2)_2$. In addition to dry weather, improper mixing can also result in the formation of this compound by producing localized zones with a high soil pH. As shown by Equation 3, an alkaline medium also favors the formation of this stable toxic material. When the precautions indicated are observed, however, calcium cyanamide is an excellent source of fertilizer nitrogen.

The toxicity of calcium cyanamide has value in the control of weeds in plant beds. It is applied in the fall at the rate of 1 to 2 lb./yd.², and the crop seed is planted in the late winter or early spring. In powdered form it is used as a defoliant.

Cyanamide has also been added to mixed fertilizers, for it is an excellent conditioning agent. Its total consumption, however, has steadily declined over the years, for its cost per unit of nitrogen is rather high in relation to the cost of other nitrogen materials.

THE ACIDITY AND BASICITY OF NITROGEN FERTILIZERS

Some fertilizer materials leave an acid residue in the soil, others a basic residue, and still others seemingly have no influence on the soil pH. Results of numerous experiments have shown that among the plant nutrients nitrogen, phosphorus, and potassium, the carriers of phosphorus and potassium have little or no influence on soil acidity. The carriers of nitrogen, however, have a considerable effect on both the soil pH and the loss of cations by leaching. The development of acidity is illustrated by the nitrification equation given earlier in this chapter (see p. 133).

A method for determining the acidity or basicity of fertilizers was developed by Pierre in 1933. His method is based on the assumption that (1) the sulfur, chlorine, one third of the phosphorus, and half of the contained nitrogen reduce the lime content, hence the pH, of the soil and (2) that the calcium, magnesium, potassium, and sodium increase its lime status. He further assumed that half of the nitrogen added to the soil was absorbed by the plant as nitric acid and the other half as a salt, such as calcium nitrate. He accordingly calculated that 1.8 lb. of CP calcium carbonate would be required to neutralize the acidity resulting from the addition of each pound of fertilizer nitrogen. Sources of nitrogen such as sodium nitrate or calcium nitrate would then leave a basic residue because of the nature of the accompanying ion. This method, modified for estimating the equivalent acidity or basicity of complete fertilizers, was adopted by the Association of Official Agricultural Chemists and is recognized presently as the official procedure.

Pierre's method has been criticized by Andrews on the basis that it gives lime equivalent values that are lower than those actually required to neutralize the acid formed by each of the various materials. Andrews maintains that each pound of fertilizer nitrogen as NH_3 will require 3.57 lb. of CP calcium carbonate to neutralize the acidity if converted to the nitrate form. Also every pound of nitrogen leached from the soil as the nitrate (NO_3) takes with it 3.57 lb. of $CaCO_3$ or its equivalent in basic cations. He has accordingly calculated the amounts of limestone he maintains will be required to neutralize the acidity formed by the various sources of nitrogen. These values, along with those determined by the Pierre method, are shown in Table 5-18. The figures currently listed for the acidity or basicity of mixed fertilizers and straight goods are determined by the official A.O.A.C. procedure, which is that of Pierre.

The removal of basic cations from soils is the reason that soils become acid. Crop removal and leaching account for the greatest losses of these metallic ions. *A priori* reasoning would lead to the conclusion that nitrate sources carrying a basic cation should be less acid forming than ammoniacal sources. Further, ammoniacal sources that carry an acidic anion such as SO_4^{2-}, which is not absorbed as rapidly as the NO_3^- ion, would be more acidic than a material such as ammonia or urea. Experimental evidence has shown this to be the case.

Pierre and his associates recently reviewed the factors which contribute to the acidity or basicity of fertilizer nitrogen materials. Greenhouse studies showed that in uncropped soil the acidity developed by ammonium nitrate was about equal to the theoretical value, provided that soluble salts were removed from the soil before the pH measurements were made.

It was also found that deviations in acidity from the theoretical values were affected by the type of crop which was grown. Their results showed that these deviations were quantitatively explained by the numbers of chemical equivalents of N and excess bases taken up by the crop, excess bases being defined as total cations (Ca^{+2}, Mg^{+2}, K, and Na^+) minus total anions (Cl^-, SO_4^{-2}, and $H_2PO_4^-$).

Regardless of which method, whether that of Pierre or Andrews, more closely estimates the acidifying effects of the different forms of fertilizer nitrogen, the fact remains that some materials are acid forming and others are not. The importance of this property fades into insignificance in a well-run farming enterprise. In such an operation the maintenance of proper soil pH in an adequate liming program is mandatory, and in such an operation the factors determining the choice of fertilizer nitrogen are its applied cost, market availability, and ease of application.

TABLE 5–18. Equivalent Acidity and Basicity of Nitrogenous Fertilizer Materials According to Andrews and Pierre*

Material	Per cent nitrogen	Pure lime necessary to make lime salts†			Pounds of pure lime; official method for neutralizing fertilizers		
		Per lb. of nitrogen	Per 20 lb. of nitrogen	Per 100 lb. of material	Per lb. of nitrogen	Per 20 lb. of nitrogen	Per 100 lb. of material
Inorganic sources of nitrogen							
Sulfate of ammonia	20.5	7.14	143	146	5.35	107	110
Ammo-phos A	11.0	6.77	135	74	5.00	100	55
Anhydrous ammonia	82.2	3.57	72	293	1.80	36	148
Calcium nitrate	15.0	0.42	8	6	1.35B	27B	20B
Calnitro	16.0	0.66	13	11	1.31B	26	21
Calnitro	20.5	1.77	35	36	0	0	0
Crude nitrogen solution	44.4	2.98	60	132	1.20	24	53
Nitrate of soda	16.0	0.00	0	0	1.80B	36B	29B
Potassium nitrate	13.0	0.00	0.00	0.00	2.00B	40B	26B
Manufactured organic nitrogen							
Cyanamid	22.0	1.18B	24B	26B	2.85B	57B	63B
Urea	46.6	3.57	71	166	1.80	36	84
Urea-ammonia liquor	45.5	3.57	71	162	1.80	36	82
Natural organic nitrogen							
Cocoa shell meal	2.7	2.37	47	6	0.60B	12B	2B
Castor pomace	4.8	2.67	53	13	0.90	18	4
Cottonseed meal	6.7	3.17	63	21	1.40	28	9
Dried blood	13.0	3.52	70	46	1.75	35	23
Fish scrap	9.2	2.67	53	25	0.90	18	8
Fish scrap	8.9	1.78	36	16	0.01	2	0
Guano, Peruvian	13.8	2.72	54	38	0.95	19	13
Guano, white	9.7	2.22	44	21	0.45	9	4
Milorganite	7.0	3.47	69	24	1.70	34	12
Tankage, animal	9.1	1.92	38	17	0.15	3	1
Tankage, garbage	2.5	0.93B	19B	2B	2.70B	54B	7B
Tankage, high grade	8.4	2.52	50	21	0.75	15	6
Tankage, low grade	4.3	5.43B	109B	23B	7.20B	144B	31B
Tankage, packing house	6.0	0.12	2	1	1.65B	33B	10B
Tankage, process	7.4	3.32	66	25	1.55	31	12
Tobacco stems	1.4	16.03B	321B	22B	17.80B	356B	25B
Tobacco stems	2.8	2.53B	51B	7B	4.30B	86B	12B

TABLE 5–18. (Continued)

Material	Pure lime necessary to make lime salts†				Pounds of pure lime; official method for neutralizing fertilizers		
	Per cent nitrogen	Per lb. of nitrogen	Per 20 lb. of nitrogen	Per 100 lb. of material	Per lb. of nitrogen	Per 20 lb. of nitrogen	Per 100 lb. of material
Sources of potash							
Manure salts	0	0	0	0	0	0	0
Muriate of potash	0	0	0	0	0	0	0
Potassium nitrate	13.0	0	0	0	2.00B	40B	26B
Sulfate of potash	0	0	0	0	0	0	0
Sulfate of potash — magnesia	0	0	0	0	0	0	0
Sources of phosphorus							
Ammo-phos A	11.0	6.77	135	74	5.00	100	55
Precipitated bone	0	0	0	0	0	0	29B
Superphosphate	0	0	0	0	0	0	0
Triple super-phosphate	0	0	0	0	0	0	0

* Andrews, *The Response of Crops and Soils to Fertilizers and Manures,* 2nd Ed. Copyright 1954 by W. B. Andrews.
† Data to make lime salts from organic sources of nitrogen were obtained by adding 1.77 lb./lb. of nitrogen to data for neutral fertilizers.
B = lime in excess of that required to make neutral salts or neutral fertilizers.

CROP RESPONSES TO VARIOUS SOURCES OF FERTILIZER NITROGEN

Much research has been carried out in the United States to evaluate the effectiveness of various sources of fertilizer nitrogen. By far the greater proportion of these studies has been done on the acid soils of the humid region east of the Mississippi River. It is not the purpose of this book to review individually the results of these various experiments. However, a summary of these findings, as well as of the principles involved in nitrogen fertilization, is given.

1. The continued use of acid-forming fertilizer materials will lead to a decrease in pH with an accompanying decrease in crop yields unless lime sufficient to neutralize the acidity formed is applied to the soil.
2. When applied in amounts generally recommended, the use of basic fertilizers may maintain the soil pH at its original level, but in general on humid-region soils they will do little to increase the

degree of base saturation and, conversely, the soil pH. When applied in large quantities over a long period of time, these fertilizers may cause some increase in the soil pH.

3. The nitrification pattern of both ammoniacal and natural organic materials provides little justification for the belief that these forms in warm, well-aerated, and moist soils release their nitrogen slowly, thus reducing excessive losses by leaching. In cold, fine-textured soils the water-insoluble forms may be expected to lose less of their nitrogen by leaching than the soluble forms because of reduced mineralization under cold conditions. When early crops are planted on such soils, this difference may be reflected in yields if all the fertilizer nitrogen is applied before planting.

4. The principles of ion exchange generally influence the effectiveness of chemical sources of nitrogen. The ammoniacal form is retained briefly against leaching because of its adsorption by soil colloids. The nitrate form is not so retained. This difference will be greatest in fine-textured and least in coarse-textured soils. The downward movement is a factor to consider when top dressing with various sources of solid nitrogen fertilizers on soils of different texture. Differences in the leaching losses of these two forms are reduced when conditions favor rapid nitrification.

5. When conditions favor nitrification, the superiority of one form of nitrogen over the other may be related to the accompanying ion or to some element contained as an impurity. This is illustrated by the use of a material such as ammonium sulfate. If applied on a sulfur-deficient soil, it would give an apparently better response than a nonsulfur-containing nitrogen carrier if sulfur were not included in some other component of the fertilizer. In such cases the way in which the limiting element should be supplied will be dictated by economic considerations.

6. Calcium cyanamide must be applied well before seeding and must be incorporated thoroughly into the soil. The occasional observations of the inferiority of this material can almost always be traced to improper application in the field.

7. Some nitrogen sources such as anhydrous ammonia, nitrogen solutions, urea, and other ammoniacal materials may lose ammonia by volatilization as a result of improper placement, surface application to alkaline soils, or, in the case of urea, surface application to soil or sod. In addition, if placed too close to seed or plants, injury may result from ammonia toxicity. These difficulties can most generally be corrected by proper placement and by adjusting the time of application.

8. Some crops, such as flue-cured tobacco, may be adversely affected by prolonged concentrations of the ammonium ion in the

soil which may be caused by the use of soil fumigants that inhibit nitrification. A significant portion of the applied fertilizer nitrogen should be in the nitrate form.

9. When differences in acid-forming properties, secondary or trace element content, and method and time of application and placement are recognized and handled accordingly, one source of fertilizer nitrogen is often as effective as any other in increasing crop yields. The determining factor in the selection of a source of nitrogen is then governed by economic considerations. The material purchased should be that from which maximum return on the fertilizer-nitrogen dollar can be expected.

Summary

1. Atmospheric nitrogen is fixed in soils by various free-living and symbiotic bacteria. The amounts fixed by these organisms are generally inadequate for the sustained high yields of crops required in commercial farming.
2. The various forms of soil nitrogen and their turnover in soils were discussed. Important to the immobilization and release of nitrogen are such factors as the supply of carbon, phosphorus, and sulfur, soil aeration, and temperature.
3. Considerable attention was given to the reactions of inorganic nitrogen compounds in soils, especially those of ammonification and nitrification. The retention of the various forms of inorganic nitrogen in soils as well as gaseous losses of this element were covered.
4. The four general classes of nitrogen fertilizer were discussed: ammoniacal, nitrate, slowly available forms, and miscellaneous materials. Most of the ammoniacal forms are acid forming and their continued use will lower soil pH values. The nitrate form is subject to loss by leaching. In coarse-textured soils, under high rainfall, such losses can be serious.
5. Crop responses to the various forms of nitrogen were covered. It is generally concluded that when the nitrogen alone is considered the results from one form are as good as another. However, method of application, accompanying elements in the carrier, and placement in the soil may cause differences in crop responses to the various carriers. The cost per unit of nitrogen applied to the land is an important item in determining the selection of the nitrogen fertilizer.

Questions

1. What are the ways, exclusive of synthetic nitrogen fixation, by which atmospheric nitrogen is made usable to higher plants?
2. What are the various microorganisms responsible for nitrogen fixation?

3. Define ammonification and nitrification. What are the factors affecting these reactions in soils?

4. If leaching losses of nitrogen are to be minimized after the fall application of ammoniacal nitrogen, soil temperatures during these winter months should not rise above what point?

5. As a general rule, is the fall application of nitrate fertilizers a sound practice? Why?

6. Nitrification, specifically, is defined how? It is a two-step reaction. What are the two steps and what organisms are responsible for each?

7. Describe the environmental and soil conditions under which you would expect to get significantly lower leaching losses of NH_4^+ nitrogen as contrasted with NO_3^- nitrogen. Under what soil and environmental conditions would you expect not to get these differences?

8. You have disked in a large amount of barley straw just about a week before planting fall wheat. At planting time you applied fertilizer which supplied 20 lb. of nitrogen, 20 lb. of phosphorus, and 40 lb. of potassium. The wheat germinates and shortly thereafter turns yellow. Tests show no NO_3^- nitrogen in the tissue. What is wrong with the wheat and why? The farmer on whose field this is observed asks you, as County Farm Adviser, what to do. What is your answer?

9. Why is nitrification important? Would you consider this phenomenon a mixed blessing? Why? Be precise.

10. What is ammonia fixation? What are the soil conditions under which it occurs? Discuss the role that potassium and ammonia play, each in the fixation or release of the other. How important do you consider this factor to be in the over-all nitrogen fertilization picture?

11. In what forms may nitrogen as a gas be lost from soil? Discuss the conditions under which each form is lost and indicate the reactions that are thought to take place.

12. How would you prevent or minimize the various gaseous losses of nitrogen?

13. Classify the various forms of nitrogen fertilizers.

14. What is the single most important *original* source of fertilizer nitrogen today?

15. What is the commercial nitrogen fertilizer with the highest percentage of nitrogen?

16. What solid nitrogenous material has the highest percentage of nitrogen?

17. Why is it sometimes unwise to apply urea to the surface of the soil?

18. In a properly managed farming enterprise is the acidity formed by the various sources of ammoniacal nitrogen a serious problem? Why?

19. What is perhaps the most important factor governing the selection of the source of fertilizer nitrogen?
20. Urea (45% N) at a price of $150.00 per ton is more or less expensive per pound of nitrogen than ammonium nitrate (33.5% N) at $106.30 per ton?
21. What precaution should be observed in applying ammoniacal nitrogen fertilizers to calcareous soils?
22. Intensive cultivation of land leads to a rapid decomposition of the organic matter and a more rapid rate of nitrification. Why?
23. What is the difference between nitrogen fixation and nitrification?
24. Why, specifically, have ammonia forms of nitrogen an acidifying effect on the soil?
25. In general, what types of nitrogen solutions can be dribbled onto the soil surface without losses of nitrogen to the atmosphere?
26. How much CP calcium carbonate would be necessary to neutralize the acidity formed by the application of 500 lb./A. of ammonium sulfate, according to the calculations of Pierre and of Andrews?

Selected References

Adriano, D. C, P. F. Pratt, and S. E. Bishop, "Nitrate and salt in soils and ground waters from land disposal of dairy manure." *SSSA Proc.*, **35**:759 (1971).

Abruna, F., R. W. Pearson, and C. B. Elkins, "Quantitative evaluation of soil reaction and base status changes resulting from field application of residually acid forming nitrogen fertilizers." *SSSA Proc.*, **22**:539 (1958).

Allison, F. E., J. N. Carter, and L. D. Sterling, "The effect of partial pressure of O_2 on denitrification in soil." *SSSA Proc.*, **24**:283 (1960).

Allison, F. E., and C. J. Klein, "Rates of immobilization and release of nitrogen following additions of carbonaceous material and nitrogen to soil." *Soil Sci.*, **93**:383 (1962).

Anderson, O. E., "The effect of low temperatures on nitrification of ammonia in Cecil sandy loam." *SSSA Proc.*, **24**:286 (1960).

Anderson, O. E., L. S. Jones, and F. C. Boswell, "Soil temperature and source of nitrogen in relation to nitrification in sodded and cultivated soils." *Agron. J.*, **62**:206 (1970).

Andrews, W. B., *The Response of Crops and Soils to Fertilizers and Manures*, 2nd ed. State College, Mississippi: W. B. Andrews, 1954.

Baker, J. H., M. Peech, and R. B. Musgrave, "Determination of application losses of anhydrous ammonia." *Agron. J.*, **51**:361 (1959).

Bartholomew, W. V., "Soil nitrogen, supply processes and crop requirements." *N. C. State University Technical Bulletin No. 6*, 1972.

Bates, T. E., and S. L. Tisdale, "The movement of nitrate nitrogen through columns of coarse-textured soil materials." *SSSA Proc.*, **21**:525 (1957).

Beaton, J. D., "Fertilizers and their use in the decade ahead." *Proc. 20th Ann. Meeting, Agric. Res. Inst.*, Oct. 1971.

Bennett, A. C., and F. Adams, "Concentration of NH_3(aq) required for incipient NH_3 toxicity to seedlings." *SSSA Proc.*, **34**:259 (1970).

Black, C. A., *Soil-Plant Relationships*, 2nd ed. New York: Wiley, 1968.

Blue, W. G., and C. F. Eno, "Distribution and retention of anhydrous ammonia in sandy soils." *SSSA Proc.*, **18**:420 (1954).

Bremner, J. M., and L. A. Douglas, "Effects of some urease inhibitors on urea hydrolysis in soils." *SSSA Proc.*, **37**:225 (1973).

Broadbent, F. E., G. N. Hill, and K. B. Tyler, "Transformations and movement of urea in soils." *SSSA Proc.*, **22**:303 (1958).

Broadbent, F. E., and T. Nakashima, "Plant uptake and residual value of six tagged nitrogen fertilizers," *SSSA Proc.*, **32**:388 (1968).

Brown, J. R., and G. E. Smith, "Soil fertilization and nitrate accumulation in vegetables." *Agron. J.*, **58**:209 (1966).

Burge, W. D., and F. E. Broadbent, "Fixation of ammonia by organic soils." *SSSA Proc.*, **25**:199 (1961).

Carter, J. N., O. L. Bennett, and R. W. Pearson, "Recovery of fertilizer nitrogen under field conditions using nitrogen-15." *SSSA Proc.*, **31**:50 (1967).

Chandra, P., "The effect of shifting temperatures on nitrification in a loam soil." *Can. J. Soil Sci.*, **42**:314 (1962).

Clark, F. E., W. E. Beard, and D. H. Smith, "Dissimilar nitrifying capacities of soils in relation to losses of applied N." *SSSA Proc.*, **24**:50 (1963).

Clark, K. G., and T. G. Lamont, "Correlation between nitrification and activity indexes of urea-formaldehyde in mixed fertilizer." *J. Assoc. Offic. Agr. Chemists*, **43**:504 (1960).

Colliver, G. W., and L. F. Welch, "Toxicity of preplant anhydrous ammonia to germination and early growth of corn: I. Field Studies." *Agron. J.*, **62**:341 (1970).

Colliver, G. W., and L. F. Welch, "Toxicity of preplant anhydrous ammonia to germination and early growth of corn: II. Laboratory studies." *Agron. J.*, **62**:346 (1970).

Cooke, I. J., "Damage to plant roots caused by urea and anhydrous ammonia." *Nature (London)*, **194**:1262 (1962).

Court, M. N., R. C. Stephen, and J. S. Waid, "Nitrate toxicity arising from the use of urea as a fertilizer." *Nature (London)*, **194**:1263 (1962).

Diamond, R. B., and F. J. Myers, "Crop responses and related benefits from SCU." *Sulphur Inst. J.*, **8**:9 (1972).

Dommengues, Y., "Nitrogen mineralization at low moisture contents." *Trans. 7th Intern. Congr. Soil Sci.*, **2**:672 (1960).

Doxtader, K. G., and M. Alexander, "Nitrification by heterotrophic soil microorganisms." *SSSA Proc.*, **30**:351 (1966).

Eno, C. F., and W. G. Blue, "The effect of anhydrous ammonia on nitrification and the microbial population in sandy soils." *SSSA Proc.*, **18**:178 (1954).

Eno, C. F., and W. G. Blue, "The comparative rate of nitrification of anhydrous ammonia, urea, and ammonium sulfate in sandy soils." *SSSA Proc.*, **21**:392 (1957).

Erdman, L. W., "Legume inoculation; what it is—what it does." *USDA Farmers' Bull. 2003* (1959).

Ernst, J. W., and H. F. Massey, "The effects of several factors on volatilization of ammonia formed from urea in the soil." *SSSA Proc.*, **24:**87 (1960).

Giordano, P. M., and J. J. Mortvedt, "Release of nitrogen from sulfur-coated urea in flooded soil." *Agron. J.*, **62:**612 (1970).

Fisher, W. B., Jr., and W. L. Parks, "Influence of soil temperature on urea hydrolysis and subsequent nitrification." *SSSA Proc.*, **22:**247 (1958).

Goring, C. A. I., and R. T. Martin, "Diffusion and sorption of aqua ammonia injected into soils." *Soil Sci.*, **88:**338 (1959).

Goring, C. A. I., "Control of nitrification by 2-chloro-6-(trichloro-methyl)pyridine." *Soil Sci.*, **93:**211 (1962).

Hanawalt, R. B., "Environmental factors influencing the sorption of atmospheric ammonia by soils." *SSSA Proc.*, **33:**231 (1969).

Harding, R. B., T. W. Embleton, W. W. Jones, and T. M. Tyan, "Leaching and gaseous losses of nitrogen from some non-tilled California soils." *Agron. J.*, **55:**515 (1963).

Harmsen, G. W., and D. A. van Schreven, "Mineralization of organic nitrogen in soil." *Advan. Agron.*, **7:**300 (1955).

Jackson, J. E., and G. W. Burton, "Influence of sod treatment and nitrogen placement on the utilization of urea nitrogen by Coastal Bermuda grass." *Agron. J.*, **54:**47 (1962).

Jacob, K. D., Ed. "Fertilizer Technology and Resources in the U. S." Vol. III, *Agronomy: A Series of Monographs.* New York: Academic, 1953.

Jones, M. B., C. M. McKell, and S. S. Winans, "Effect of soil temperature and nitrogen fertilization on the growth of soft chess at two elevations." *Agron. J.*, **55:**44 (1963).

Justice, J. K., and R. L. Smith, "Nitrification of ammonium sulfate in a calcareous soil as influenced by combinations of moisture, temperature, and levels of added nitrogen" *SSSA Proc.*, **26:**246 (1962).

Koeke, H., "The effects of fumigants on nitrate production in soils." *SSSA Proc.*, **25:**204 (1961).

Kresge, C. B., "Ammonia volatilization losses from nitrogen fertilizers when applied to soils." *Dissertation Abstr.*, **20:**448 (1959).

Kresge, C. B., and D. P. Satchell, "Gaseous losses of ammonia from nitrogen fertilizers applied to soils." *Agron. J.*, **52:**104 (1960).

Kresge, C. B., and S. E. Younts, "Effect of nitrogen source on yield and nitrogen content of blue grass forage." *Agron. J.*, **54:**149 (1962).

Kuntz, L. T., L. D. Owens, and R. D. Hauk, "Influence of moisture on the effectiveness of winter applied nitrogen fertilizers." *SSSA Proc.*, **25:**40 (1961).

Larsen, S., and D. Gunary, "Ammonia loss from ammoniacal fertilizers applied to calcareous soils." *J. Sci. Food Agr.*, **13:**566 (1962).

Leo, W. M., T. E. Odland, and R. S. Bell, "Effect on soils and crops of long continued use of sulfate of ammonia and nitrate of soda with and without lime." *Rhode Island Agr. Exp. Sta. Bull. 344* (1959).

Long, F. L., and G. M. Volk, "Availability of nitrogen from condensation products of urea and formaldehyde." *Agron. J.*, **55:**155 (1963).

McCants, C. B., E. O. Skogley, and W. G. Woltz, "Influence of certain soil fumigation treatments on the response of tobacco to ammonium and nitrate forms of nitrogen." *SSSA Proc.*, **23:**466 (1959).

McClure, G. W., Jr., and A. S. Hunter, "Investigations of ammonium chloride as a nitrogen fertilizer for forage and corn." *Agron. J.,* **54**:443 (1962).

McDowell, L. L., and G. E. Smith, "The retention and reactions of anhydrous ammonia on different soil types." *SSSA Proc.,* **22**:38 (1958).

McIntosh, T. H., and L. R. Frederick, "Distribution and nitrification of anhydrous ammonia in a Nicollet sandy clay loam." *SSSA Proc.,* **22**:402 (1958).

Meek, B. D., L. B. Grass, and A. J. MacKenzie, "Applied nitrogen losses in relation to oxygen status of soils." *SSSA Proc.,* **33**:575 (1969).

Meyer, R. D., R. A. Olson, and H. F. Rhoades, "Ammonia losses from fertilized Nebraska soils." *Agron, J.,* **53**:241 (1961).

Moberg, E. L., D. V. Waddington, and J. M. Duich, "Evaluation of slow-release nitrogen sources on Merion Kentucky bluegrass." *SSSA Proc.,* **34**:335 (1970).

Moe, P. G., "Nitrogen losses from urea as affected by altering soil urease activity." *SSSA Proc.,* **31**:380 (1967).

Morris, H. D., and J. Giddens, "Response of several crops to NH_4 and NO_3 forms of nitrogen as influenced by soil fumigation and liming." *Agron. J.,* **55**:372 (1963).

Mortland, M. M., "Reactions of ammonia in soils." *Advan. Agron.,* **10**:325 (1958).

Olsen, R. J., R. F. Hensler, O. J. Attoe, S. A. Witzel, and L. A. Peterson, "Fertilizer nitrogen and crop rotation in relation to movement of nitrate nitrogen through soil profiles." *SSSA Proc.,* **34**:448 (1970).

Olson, R. A., Ed., *Fertilizer Technology & Use,* 2nd ed. Madison, Wisconsin: Soil Sci. Soc. Amer., Inc., 1971.

Overrein, L. N., and P. G. Moe, "Factors affecting urea hydrolysis and ammonia volatilization in soil." *SSSA Proc.,* **31**:57 (1967).

Owens, L. D., "Nitrogen movement and transformations in soils as evaluated by a lysimeter study utilizing isotopic N." *SSSA Proc.,* **24**:372 (1960).

Parker, D. T., and W. E. Larson, "Nitrification as affected by temperature and moisture content of mulched soils." *SSSA Proc.,* **26**:238 (1963).

Parr, J. F., and R. I. Papendick, "Greenhouse evaluation of the agronomic efficiency of anhydrous ammonia." *Agron. J.,* **58**:215 (1966).

Pearson, R. W., H. V. Jordan, and O. L. Bennett, "Residual effects of fall and spring applied nitrogen fertilizers on crop yields in the Southeastern United States." *USDA Tech. Bull. 1254* (1961).

Pierre, W. H., J. Meisinger, and J. R. Birchett, "Cation-anion balance in crops as a factor in determining the effect of nitrogen fertilizers on soil acidity." *Agron. J.,* **62**:106 (1970).

Platou, J., "SCU—A progress report." *Sulphur Inst. J.,* **8**:9 (1972).

Power, J. F., "The effect of moisture on fertilizer nitrogen immobilization in grasslands." *SSSA Proc.,* **31**:223 (1967).

Reichman, G. A., D. L. Grunes, and F. G. Viets, Jr., "Effect of soil moisture on ammonification and nitrification in two Northern Plains soils." *SSSA Proc.,* **30**:363 (1966).

Russell, E. W., *Soil Conditions and Plant Growth,* 9th ed. London: Longmans, Green, 1961.

Sabey, B. R., "The influence of nitrification suppressants on the rate of ammonium oxidation in Midwestern USA field soils." *SSSA Proc.,* **32**:675 (1968).

Sabey, B. R., L. R. Frederick, and W. V. Bartholomew, "The formation of nitrate from ammonium nitrogen in soils: III. Influence of temperature and initial population of denitrifying organisms on the maximum rate and delay period." *SSSA Proc.*, **23**:462 (1959).

Sauchelli, V., Ed., *Fertilizer Nitrogen: Its Chemistry and Technology, ACS Monograph 161*. New York: Reinhold, 1964.

Scarsbrook, C. E., "Urea-formaldehyde fertilizer as a source of nitrogen for cotton and corn." *SSSA Proc.*, **22**:442 (1958).

Schwartzbeck, R. A., J. M. MacGregor, and E. L. Schmidt, "Gaseous nitrogen losses from nitrogen fertilized soils measured with infrared spectroscopy." *SSSA Proc.*, **25**:186 (1961).

Simpson, D. M. H., and S. W. Melsted, "Gaseous ammonia losses from urea solutions applied as a foliar spray to various grass sods." *SSSA Proc.*, **26**:186 (1962).

Simpson, D. M. H., and S. W. Melsted, "Urea hydrolysis and transformations in some Illinois soils." *SSSA Proc.*, **27**:48 (1963).

Smith, D. H., and F. E. Clarke, "Volatile losses of nitrogen from acid or neutral soils or solutions containing nitrite and ammonium ions." *Soil Sci.*, **90**:86 (1960).

Stevenson, F. J., and A. P. S. Dhariwal, "Distribution of fixed ammonium in soils." *SSSA Proc.*, **23**:121 (1959).

Stevenson, F. J., "The nitrogen cycle in soils." Lecture notes, ICA Short Course, University of Illinois, 1963.

Sutherland, W. N., W. D. Shrader, and J. T. Pesek, "Efficiency of legume residue nitrogen and inorganic nitrogen in corn production." *Agron. J.*, **53**:339 (1961).

Temme, J., "Transformation of calcium cyanamide in the soil." *Plant Soil*, **1**:145 (1949).

Thomas, G. W., and D. E. Kissel, "Nitrate volatilization from soils." *SSSA Proc.*, **34**:828 (1970).

Tisdale, S. L., *et al.*, "Sources of nitrogen in crop production." *North Carolina Agr. Exp. Sta. Tech. Bull. 96* (1952).

Tuller, W. N., Ed., *The Sulphur Data Book: Freeport Sulphur Company*. New York: McGraw-Hill, 1954.

Tyler, K. B., and F. E. Broadbent, "Nitrite transformations in California soils." *SSSA Proc.*, **24**:279 (1960).

Tyler, K. B., F. E. Broadbent, and G. N. Hill, "Low-temperature effects on nitrification in four California soils." *Soil Sci.*, **87**:123 (1959).

Vines, H. M., and R. T. Wedding, "Some effects of ammonia on plant metabolism and a possible mechanism for ammonia toxicity." *Plant Physiol.*, **35**:820 (1960).

Volk, G. M., "Volatile loss of ammonia following surface application of urea to turf or bare soils." *Agron. J.*, **51**:746 (1959).

Volk, G. M., "Ammonia losses from soils, gaseous losses of ammonia from surface applied nitrogenous fertilizers." *J. Agr. Food Chem.*, **9**:280 (1961).

Volk, G. M., "Efficiency of fertilizer urea as affected by method of application, soil moisture, and lime." *Agron. J.*, **58**:249 (1966).

Volk, G. M., "Gaseous loss of ammonia from prilled urea applied to slash pine." *SSSA Proc.*, **34**:513 (1970).

Walsh, L. M., and J. T. Murdock, "Native fixed ammonium and fixation of applied ammonium in several Wisconsin soils." *Soil Sci.*, **89**:183 (1960).

Welch, L. F., and A. D. Scott, "Nitrification of fixed ammonium in clay minerals as affected by added potassium." *Soil Sci.*, **90**:79 (1960).

Wijler, J., and C. C. Delwiche, "Investigations on the denitrifying process in soil." *Plant Soil*, **5**:155 (1954).

Wilkinson, S. R., and A. J. Ohlrogge, "Influence of biuret and urea fertilizers containing biuret on corn plant growth and development." *Agron. J.*, **52**:560 (1960).

Wing-To Chan, and A. F. MacKenzie, "Effects of shading and nitrogen on growth of grass-alfalfa pastures." *Agron. J.*, **63**:667 (1971).

Young, J. L., and R. A. Cattani, "Mineral fixation of anhydrous NH_3 by air-dry soils." *SSSA Proc.*, **26**:147 (1963).

6. SOIL AND FERTILIZER PHOSPHORUS

The importance of phosphorus in plant nutrition was discussed in Chapter 3. This element is present in plant tissues and in soils in smaller amounts than are nitrogen and potassium. The generally small quantities of phosphorus in soils and its tendency to react with soil components to form compounds, relatively insoluble, hence unavailable to plants, make it a topic of major importance in the realm of soil fertility.

PHOSPHORUS CONTENT OF SOILS

The total phosphorus content varies from soil to soil, but it is generally higher in young virgin soils in areas in which rainfall is not excessive. The average phosphorus content of virgin soils to a depth of 1 ft. is shown in Figure 6-1. The low phosphorus level of the uncultivated soils of the humid regions of the Southeast is apparent, as is the high level of native phosphorus in the soils of the prairie and western states.

The available phosphorus level in fertilized cultivated soils is quite another story. Because little of this element is lost in percolating water from most soils and crop removals are generally small, it tends to accumulate in the surface horizon of cultivated soils. This is well illustrated by the data in Table 6-1. Available phosphorus refers to the amount of this element that can be absorbed from the soil by crop plants. It is estimated in various ways, usually involving extraction of the soil with a dilute acid, alkali, or salt solution. These techniques of soil analysis are discussed in detail in Chapter 12, which deals with soil fertility evaluation. It is enough to mention at this point that the total quantities of soil phosphorus are much greater than those of the available phosphorus, but the latter is of greater importance to plant growth. The data in Table 6-1 point out the extent to which levels of available soil phosphorus may be increased by long-continued additions of fertilizers containing this element.

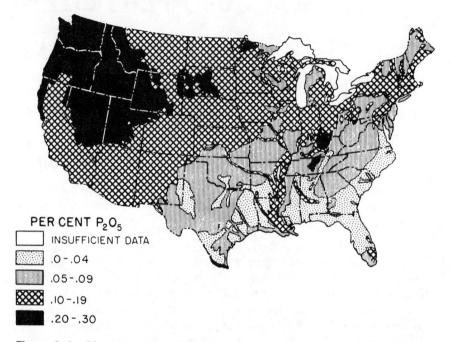

Figure 6–1. Phosphate content in the surface foot of soils in the United States. [Pierre and Norman, Eds., *Agronomy,* **4:**401 (1953). Reprinted with permission of Academic Press, Inc., New York.]

TABLE 6–1.   The Available Phosphorus Content of Virgin and Cultivated Soils*

State	Soil series	Major crop	Available phosphorus as P_2O_5 in	
			Virgin soil (lb./A.)	Cultivated soil (lb./A.)
Alabama	Norfolk-Ruston	Potato	12	300
Maine	Caribou	Potato	34	219
Virginia	Sassafras	Potato	45	808
Florida	Norfolk	Citrus	63	543
New York	Sassafras	Potato	15	778

* Pierre and Norman, Eds., *Agronomy,* **4:**401 (1953). Reprinted with permission of Academic Press, Inc., New York.

FORMS OF SOIL PHOSPHORUS

The phosphorus in soil can be classed generally as organic or inorganic, depending on the nature of the compounds in which it occurs. The organic fraction is found in humus and other organic materials which may or may not be associated with it. The inorganic fraction occurs in numerous combinations with iron, aluminum, calcium, fluorine, and other elements. These compounds are usually only very slightly soluble in water. Phosphates also react with clays to form generally insoluble clay-phosphate complexes.

The content of inorganic phosphorus in soils is almost always higher than that of organic phosphorus. An exception to this rule would, of course, be the phosphorus contained in predominantly organic soils. In addition, the organic phosphorus content of mineral soils is usually higher in the surface horizon than it is in the subsoil because of the accumulation of organic matter in the upper reaches of the soil profile.

Soil Solution Phosphorus. As indicated in Chapter 3, phosphorus is absorbed by plants largely as the primary and secondary orthophosphate ions ($H_2PO_4^-$ and HPO_4^{2-}), which are present in the soil solution. Some very small quantities of soluble organic phosphate may also be absorbed, but they are generally considered to be of minor importance. The concentration of these ions in the soil solution and the maintenance of this concentration are of the greatest importance to plant growth.

Plants absorb phosphorus from solutions in proportion to the concentration of phosphate ions in the solution. Further, if other factors are not limiting, growth will be proportional to the amounts of phosphorus absorbed by the plant, as described by the growth equations in Chapter 2. For plants growing in soils it follows, then, that the maintenance of a suitable phosphorus concentration in the soil solution is essential to the production of agricultural crops. The rate at which the phosphorus concentration in the soil solution is renewed therefore becomes *the* principal item, particularly so because plant roots do not absorb phosphorus uniformly from the entire soil mass.

The importance of the renewal of the phosphorus content of the soil solution can be illustrated by the following. If a soil with a bulk density of 1.2 contains 25 per cent moisture on the dry-weight basis and 0.03 ppm. of P dissolved in the water, the total amount of phosphorus in solution in one acre of this soil to a depth of two feet is 0.04 pounds. If absorption and renewal of P in the soil solution were to take place alternately with complete removal and renewal in each cycle, the soil solution would have to be replenished 249 times to supply a crop with the approximately 10 pounds of P it would need for its growth.

The actively absorbing surface of plant roots is near the root tips. As the roots move through the soil, new areas are contacted from which phosphorus absorption did not previously occur. Replenishment of soil solution is therefore probably taking place rapidly in small areas near the root tips but more slowly in larger areas around the older portions of the root where absorption is slower.

The maintenance of a suitable concentration of phosphorus in the soil solution depends among other things on the relative rate of organic matter formation and decomposition, and on the ability of the soil's inorganic fraction to react with, or fix, soluble orthophosphates in an insoluble or slightly soluble form. These equilibria may be schematically illustrated.

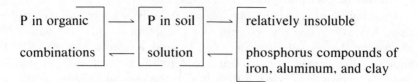

They can also be temporarily disturbed by the addition of soluble phosphate fertilizers, by immobilization of soluble phosphorus by microorganisms, and by rapid mineralization of the organic matter that would occur in plowing and cultivation.

The native phosphorus in soils originated largely from the disintegration and decomposition of rocks containing the mineral apatite, $Ca_{10}(PO_4)_6(F, Cl, OH)_2$. Phosphorus is found in the soil as finely divided fluorapatite, hydroxyapatite, or chlorapatite, as iron or aluminum phosphates, as some of the compounds shown on page 198, or in combination with the clay fraction. It also occurs in combination with humus and other organic fractions.

The nature and reactions of phosphorus with the organic and inorganic soil components and the effect of added fertilizer phosphorus on the phosphate equilibria in soils are covered in the following sections of this chapter.

Organic Soil Phosphorus. The nature and reactions of organic soil phorphorus are not so well understood as are those of inorganic soil phosphorus. Nonetheless, some understanding of currently held views on this topic will give a better understanding of the over-all importance of phosphorus in soil fertility. It is in this light that the following brief summary is offered.

FORMS OF ORGANIC SOIL PHOSPHORUS. The organic phosphorus which has been specifically identified in soils occurs in three principal forms: phospholipids, nucleic acids, and inositol phosphates. Evidence of the existence of phospholipids in soils is of two kinds. In the first place, choline, a product of the hydrolysis of lecithin, one of the phospholipids, was isolated from soil extracts many years ago. The second line of evidence stems from the finding of organic phosphorus in alcohol and ether extracts. Solubility in alcohol and ether is a property characteristic of phospholipids. It should be emphasized that the highest reported value for phospholipids is only 34 ppm. of phosphorus in the form of this compound. Many workers report that the amounts present seldom exceed 3 ppm. of phosphorus as phospholipids. It appears that these compounds do not form a large proportion of the total organic phosphorus in soils. However, final judgment must await further research, for the analytical techniques used in earlier work may not have been adequate.

The presence of nucleic acid in soil has also been deduced from indirect evidence. The first such evidence was the finding of constituent parts of the nucleic acid molecule in the hydrolyzate of soil extracts. The second is the susceptibility of nucleic acid and other organic phosphates to dephosphorylation. The third line of evidence is the differential solubility of nucleic acid and other organic phosphorus compounds.

In studies conducted over the years many compounds, most of which are hydrolytic products of nucleic acid, have been isolated from soil hydrolyzates. These compounds include phosphoric acid, pentose sugar, cytosine, adenine, guanine, uracil, hypoxanthine, and xanthine. The last two are not primary hydrolytic products of nucleic acid but are derived from guanine and adenine. These findings suggest that constituents of nucleic acids are present in soil, but they do not necessarily prove the existence of intact nucleic acid.

Evidence for the existence of nucleic acid comes from research showing that phytin and a mixture of some of the lower phosphoric acid esters of inositol were precipitated quantitatively by calcium hydroxide. Yeast nucleic acid and guanidine, on the other hand, remained soluble and were not precipitated. Organic phosphorus compounds in several soil extracts were subjected to appropriate chemical and enzyme tests to determine their susceptibility to dephosphorylation. The results of these experiments showed that both phytin and nucleic acid were present in the soil extracts.

Several types of inositol phosphate have been identified in soils. Inositol is a saturated, cyclic, six-carbon alcohol with an alcoholic group on each carbon atom. Evidence of the presence of these compounds is perhaps stronger than that of the presence of phospholipids and nucleic acid.

Soil inositol phosphates have been investigated for several years. Some workers have shown that inositol is present in soil hydrolyzates and have deduced the presence of inositol phosphate. Other evidence is based on studies of the resistance of soil preparations which contain added amounts of inositol hexaphosphate to dephosphorylation by chemical and enzymatic attack.

Still other work has shown that the ratio of inositol to phosphorus from soil extracts agreed with the theoretical ratio for that in inositol hexaphosphoric acid. Soil extracts have been analyzed by chromatographic methods and the products verified by the ratio of inositol to phosphorus in the compounds separated.

In studies carried out in Iowa samples of forty-nine soils were analyzed for inositol hexaphosphate. The presence of this material accounted for an average of 17 per cent of the total organic phosphorus with a range of 3 to 52 per cent. The percentage of this phosphorus component was higher in soils developed under forest vegetation (av. 24%) than in soils developed under grassland (14%). It decreased with an increase in soil pH in both types of soil.

In this same work data were presented to show that at least two types of inositol hexaphosphate have their origin in the decomposition of added organic material. Organic and inorganic nutrients were incubated for several months with mixtures of parent materials, sand and clay. The two forms of inositol phosphate were found in the mixtures after incubation but not before. Other forms of inositol phosphate are believed to be synthesized by higher plants and are thus added to the soil when these plant residues are plowed under.

Work in Australia also confirms the presence of inositol phosphates in soils and suggests that certain forms of this compound are of microbiological origin.

All of the organic phosphorus in soils may not be intimately associated with the humus fraction of the organic matter. Workers in Australia, reviewing this complex topic of organic phosphorus, point out that, unlike carbon or nitrogen, all of the organic phosphorus can be fairly easily removed from soil by alkaline extracting reagents. They further indicate that phosphorus does not appear to occur in the soil in any characteristic proportion in relation to nitrogen. Although about 40 to 50 per cent of the organic soil phosphorus can be identified as nucleic acids, inositol phosphate, and perhaps phospholipids, the remainder is largely unidentified.

ORGANIC PHOSPHORUS TURNOVER IN SOILS. In soils in or near equilibrium with their environment the relationship between the amount of nitrogen and the amount of carbon present is fairly close. Workers in Australia have shown that a similar relationship exists between nitrogen and organic sulfur. This clear-cut relation, however, does not obtain between nitrogen and organic phosphorus.

Because soil organic matter is decomposed with the liberation of ammonium and nitrate nitrogen, it would be expected that this would be accompanied by the release of inorganic phosphate. It undoubtedly does take place, but the evidence is not so clear-cut as that of the mineralization of nitrogen in which the released ammonium or nitrate is easily determined over short periods of time.

Evidence for the mineralization of organic phosphate is of two kinds. The first is based on the lowering of the organic phosphorus level in soils as a result of long-continued cultivation. It has been observed that when virgin soils are brought under cultivation the content of organic matter decreases. With this decrease in organic matter there is an initial increase in the citric acid soluble phosphorus, a measure of the level of inorganic phosphate in soils, but within a few years this also decreases.

The second type of evidence is based on the results of short laboratory experiments in which decreases in the organic phosphorus content of soils are related to increases in the dilute acid extractable inorganic phosphate. Numerous studies have shown this relationship to be a fairly close one.

Several studies show the behavior of organic phosphorus to be not entirely analogous to that of organic carbon and nitrogen. Experimental work carried out in Iowa indicated that mineralization of organic phosphorus increased with rises in soil pH but that mineralization of organic carbon and nitrogen did not. These same pH effects on mineralization in the field are indicated by the fact that the ratios of total organic carbon and total nitrogen to total organic phosphorus increased with soil pH in the same samples.

The mineralization of organic phosphorus has been studied in relation to the ratio of C:N:P in the soil. A C:N:P ratio of 100:10:1 for soil organic matter has been suggested, but values ranging from 229:10:0.39 to 71:10:3.05 have been found. It is obvious that no one set of figures will describe the ratio for all soils. It has been suggested that if the carbon:inorganic P ratio is 200:1 or less, mineralization of phosphorus will occur and that if the ratio is 300:1 immobilization will occur. Some Australian workers believe the N:P ratio to be closely tied in with mineralization and immobilization of phosphorus and suggest that the decreased supply of one results in the increased mineralization of the other. Thus, if nitrogen were limiting, inorganic phosphate might accumulate in the soil and the formation of soil organic matter would be inhibited. The addition of fertilizer nitrogen under such conditions could result in the immobilization not only of some of the accumulated inorganic phosphorus but also of some of the added fertilizer nitrogen.

Much has yet to be learned about the immobilization and mineralization of phosphorus in soils and its relation to the supplies of carbon, nitrogen, and sulfur present. These reactions could have an important bearing on fertilizer practices but the extent of this influence is not known. It is probably safe to assume the following:

1. If adequate amounts of nitrogen, phosphorus, and sulfur are added to soils to which crop residues are returned, some of the added elements may be immobilized in fairly stable organic combination with carbon compounds.

2. Continued cropping of soils without the addition of supplemental nitrogen, phosphorus, and sulfur will result in the mineralization of these elements and their subsequent depletion in such soils.

3. If nitrogen, phosphorus, or sulfur is present in insufficient

amounts, the synthesis of soil organic matter may be curtailed. All of these reactions presuppose the presence of adequate carbon and conditions conducive to the synthesis and breakdown of soil organic matter.

Inorganic Soil Phosphorus. Plants absorb phosphorus largely as the orthophosphate ions, $H_2PO_4^-$ and HPO_4^{2-}. The concentration of these ions in the soil solution at any one time is small, generally never more than a few parts per million and frequently less than 1 ppm. As crop removal of phosphorus is usually between 4 and 20 lb./A., the phosphorus in the soil solution must be continuously replaced or the crop would not have sufficient phosphorus to grow to maturity. The mainte-

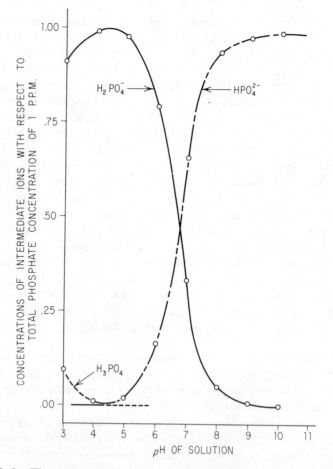

Figure 6–2. The concentration of various orthophosphate ions in solution as related to *p*H. [Buehrer, *Arizona Exp. Sta. Tech. Bull. 42* (1932).]

nance of this low concentration is of paramount importance to the growth of crops.

The concentration of the various phosphate ions in solutions is intimately related to the pH of the medium. The $H_2PO_4^-$ ion is favored in more acid media whereas the HPO_4^{2-} ion is favored above pH 7.0. These relationships are shown graphically in Figure 6-2. The curves presuppose that no ions such as iron, aluminum, calcium, and magnesium are present. When they are, the insoluble phosphates of iron and aluminum will be precipitated in acid soils and the insoluble phosphates of calcium and magnesium at pH values greater than 7.0. A series of phosphorus compounds of varying solubilities is formed under different soil conditions. As a general rule of thumb, maximum availability of phosphorus to most agricultural crops occurs within the soil pH range of 5.5 to 7.0.

INORGANIC PHOSPHORUS COMPOUNDS IN SOILS. Inorganic phosphorus exists in soils in many forms. Some of these have been prepared by scientists at TVA and other locations and have also been evaluated in terms of their availability to crop plants. Some of these data are shown in Table 6-2. In the soil they are formed in part by the reaction of fertilizer salts with certain soil components, and many are termed *phosphate reaction products*. They are discussed more fully in a subsequent section of this chapter. The data showing the availability of these materials to plants were obtained by comparing the uptake of phosphorus by plants receiving this element as mono- or dicalcium phosphate to the uptake of phosphorus by plants receiving the same amount in the form of these various compounds. Phosphorus uptake by plants receiving the mono- or dicalcium phosphate is assigned a value of 100. Those materials giving values greater than 100 were as available to plants, or more so, than the mono- or dicalcium phosphate. Those with values less than 100 were not so available. Other phosphorus compounds have been identified in soils, but those listed here are probably the most important.

PHOSPHORUS IN ACID SOILS. When a solution containing a soluble phosphate salt such as $Ca(H_2PO_4)_2$ is drawn through a column of soil, the effluent solution contains only a fraction, if any at all, of the phosphate originally present. In addition, repeated extractions with water or even salt and weak acid solutions remove only a small part of the retained phosphorus. This reduction in the solubility of added phosphates has come to be known as *phosphate retention or fixation*. Although these two terms are frequently used interchangeably, in the strict sense retention refers to that portion of the phosphorus which is loosely held by the soil and which can generally be extracted with dilute acids. This phosphorus is considered to be largely available to plants. *Fixed* phosphorus refers to that portion which is not extractable in dilute acids

TABLE 6-2. **The Comparative Value of Various Phosphate Compounds as Sources of Phosphorus for Plants**[*]

Phosphate compound	Particle size mesh (US-NBS)	Test conditions Application rate (mg.P/3 kg. soil)	Crop and number of harvests	Relative P uptake from neutral soils (mono-calcium phosphate = 100)
Products from acid water-soluble orthophosphate fertilizers				
$CaHPO_4 \cdot 2H_2O$	−100	60	Maize 3	123[†]
dicalcium phosphate	−40 + 60	60	Maize 3	102[†]
dihydrate	−20 + 40	60	Maize 3	66[†]
$CaHPO_4$	−100	26	Ryegrass 1, Sudangrass 1	110
Anhydrous dicalcium	−60 + 100	60	Wheat 1, oats 1	65
phosphate	−10 + 14	60	Wheat 1, oats 1	27
$Ca_4H(PO_4)_3 \cdot 3H_2O$	−100	60	Ryegrass 1, Sudangrass 1	77
octocalcium phosphate				
$Ca_{10}(PO_4)_6(OH)_2$	−100	60	Ryegrass 1, Sudangrass 1	9
hydroxyapatite				
$Ca_{10}(PO_4)_6F_2$	−325	200	Maize 3	2
fluorapatite				
$CaFe_2(HPO_4)_4 \cdot 5H_2O$	−200	200	Maize 3	83
calcium ferric phosphate				
$K_3Al_5H_6(PO_4)_8 \cdot 18H_2O$	−200	200	Maize 3	79
potassium taranakite	−200	150	Maize 1	40
$(NH_4)_3Al_5H_6(PO_4)_8 \cdot 18H_2O$	−200	150	Maize 1	10
ammonium taranakite				
$KFe_3H_8(PO_4)_6 \cdot 6H_2O$	−200	200	Maize 3	6
acid potassium ferric phosphate				
$(NH_4)Fe_8H_8(PO_4)_6 \cdot 6H_2O$	−325	200	Maize 3	38
acid ammonium ferric phosphate				
Aluminum phosphate (colloidal)	−200	200	Maize 3	74
Ferric phosphate	−200	200	Maize 3	71
(colloidal)	−325	200	Maize 3	92
$FePO_4 \cdot 2H_2O$ strengite	−325	200	Maize 3	3
Products from alkaline water-soluble orthophosphate fertilizers				
$MgNH_4PO_4 \cdot 6H_2O$	−35 + 60	60	Oats 1	138[†]
Struvite	−20 + 35	60	Oats 1	124[†]
$K_3CaH(PO_4)_2$	−35 + 60	60	Oats 1	123[†]
	−20 + 35	60	Oats 1	136[†]
$KAl_2(PO_4)_2OH \cdot 2H_2O$	−200	200	Maize 3	3
basic potassium aluminum phosphate				
$KFe_2(PO_4)_2OH \cdot 2H_2O$	−200	200	Maize 3	0
basic potassium ferric phosphate				
Products from soluble pyrophosphate fertilizers				
$Ca(NH_4)_3P_3O_7 \cdot H_2O$	−60 + 100	60	Wheat 1, oats 1	108
	−10 + 14	60	Wheat 1, oats 1	54
$Ca_3(NH_4)_3P_3O_7)_2 \cdot 6H_2O$	−60 + 100	60	Wheat 1, oats 1	54
	−10 + 14	60	Wheat 1, oats 1	25
$Ca_2P_3O_7 \cdot 2H_3O$	−60 + 100	60	Wheat 1, oats 1	15
calcium pyrophosphate	−10 + 14	60	Wheat 1, oats 1	7

[*] Lindsay et al., *Trans. 7th Intern. Congr. Soil Sci.*, **3**:580 (1960).
[†] Data relative to −100 mesh anhydrous dicalcium phosphate = 100.

and is generally not considered to be readily available to plants. The mechanism by which the phosphorus is rendered inactive under various soil conditions is somewhat better understood than it was a few years ago.

Phosphate retention by the mineral fraction of acid soils is generally believed to result from the reaction of orthophosphate ions with iron and aluminum and possibly with silicate clays. The mechanism of these reactions is briefly considered.

Reaction with Iron and Aluminum. In acid mineral soils the exchange complex contains appreciable quantities of adsorbed aluminum and smaller but significant amounts of iron and manganese. These ions combine with phosphates to form insoluble compounds of aluminum, iron, and perhaps manganese. The extent to which phosphates may react with exchangeable aluminum is shown in Figure 6-3. The resulting compounds may be precipitated from solution or adsorbed on the surface of iron and aluminum oxides or on clay particles. These freshly deposited compounds generally become less soluble with the passage of time.

As clays become more acid, they tend to contain more adsorbed aluminum and iron. Hence in acid soils the products of phosphorus fixation are largely complex phosphates of iron and aluminum. The more acid

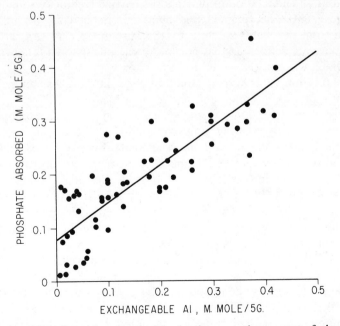

Figure 6–3. The effect of exchangeable aluminum on the amount of phosphorus adsorbed by suspended clay. [Coleman et al., *Soil Sci.*, **90**:1 (1960). Reprinted with the permission of The Williams & Wilkins Company, Baltimore.]

the soil, the greater the amount of phosphorus fixed in this manner and the greater the precautions which have to be taken to ensure reasonable crop utilization of fertilizer phosphates added to such soils.

Reaction with Silicate Clays. Another mechanism has been suggested to explain phosphorus fixation. It is the reaction of phosphates with silicate clays. Soil clays are composed of layers of silica and alumina combined to form silica-alumina sheets. The two principal types of clay material, of course, are 2:1 and 1:1. It has been suggested that phosphate ions may combine directly with these clays by (1) replacing a hydroxyl group from an aluminum atom or (2) forming a clay-Ca-phosphate linkage. It is known that clays with a low $SiO_2:R_2O_3$ ratio will fix larger quantities of phosphate than will clays with a high ratio. This is in keeping with point (1) and is associated with a greater number of hydroxyl groups exposed in the 1:1 than the 2:1 type clay.

Several studies have shown that a calcium-saturated clay will adsorb greater quantities of phosphate than will clays saturated with sodium. Since the exchange capacity of the 2:1 clays is generally greater than that of the 1:1 type, it might be expected that the latter linkage would account for a greater amount of phosphate fixation in soils containing appreciable amounts of calcium-saturated 2:1 clays.

The effect of increasing the calcium saturation of the clay separate on the retention and fixation of phosphorus by two Kentucky soils is shown in Figure 6-4. The retained phosphorus is defined as the amount removed from the water solution by the suspended clay separate. Fixed phosphorus is defined as that portion of the phosphorus retained on the clay which was not recovered by extraction with $0.005N$ H_2SO_4. At calcium saturation values greater than 70 per cent there was an increase in both phosphorus retention and phosphorus fixation.

Phosphorus retention by clays suggests that anion exchange may occur in soils. By treating soils containing adsorbed phosphorus with solutions containing such anions as arsenate, hydroxyl, and oxalate the phosphorus may be replaced. Generally, the replacement is not stoichiometric, as is cation exchange. Phosphate or anion replacement is greatest for the 1:1 clay, and the capacity to adsorb phosphates can be increased by grinding. The reaction is thought to be one of the phosphate ions in solution exchanging for hydroxyls of the Al-OH groups of the gibbsite layer of the clays.

The comments in this section may be summarized by stating that phosphorus fixation in acid soils is caused primarily by the formation of iron and aluminum compounds which have the general formula $M(H_2O)_3(OH)_2H_2PO_4$, where M is iron or aluminum. The iron and aluminum-containing soil minerals, including the clay minerals, are the sources of the iron and aluminum ions. The formation of these compounds is governed by the principles of the solubility product, the

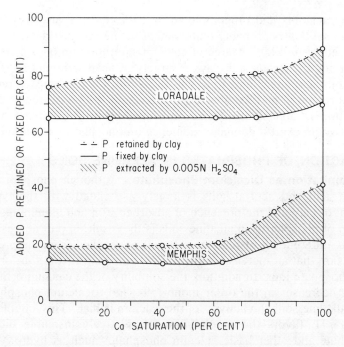

Figure 6–4. The effect of soil type and degree of calcium saturation on the retention and fixation of phosphorus by clay separates in a water suspension. [Ragland et al., *SSSA Proc.*, **21:**261 (1957).]

common ion effect, and the salt effect. Under certain conditions the compounds form a precipitate, whereas under other conditions they are adsorbed to the surfaces of soil clays and minerals. Regardless of whether they are adsorbed or precipitated, the compounds formed and the mechanisms of reaction appear to be the same. In a generalized manner the fixation of phosphorus in acid soils by iron and aluminum can be visualized. The symbol M stands for iron or aluminum and the symbol A represents oxide or hydroxide.

$$[\text{M}(\text{H}_2\text{O})_x\text{A}_y]^{3+} \text{ to }^{1+} + \text{A} + \text{H}_2\text{PO}_4^- \rightleftharpoons [\text{M}(\text{H}_2\text{O})_x\text{A}_2\text{H}_2\text{PO}_4]\downarrow$$

$$\uparrow \quad \uparrow$$
$$\text{M}_2\text{O}_3, \ \text{M}(\text{OH})_3$$

precipitated
or adsorbed

clay minerals
and exchange
sites

When fertilizer phosphates containing potassium, calcium, and ammonium are added to the soil, compounds such as ammonium and potassium taranakites and others shown in Table 6-2 are also formed by a mechanism that is probably quite similar to the one just described.

The effect of a high degree of calcium saturation on phosphate retention in noncalcareous soil clays is largely of academic interest. In truly acid soils aluminum and iron will have the greatest impact on phosphorus retention and fixation. In alkaline soils a high degree of calcium saturation is almost always accompanied by free calcium carbonate which would exert a dominant influence on phosphorus retention.

REACTION OF PHOSPHATES IN ALKALINE SOILS

Precipitation as Dicalcium Phosphate. Although phosphorus fixation in acid and neutral soils is largely associated with the type and amount of clay and the presence of adsorbed iron and aluminum, the retention of phosphorus in alkaline soils is the result, in part at least, of an entirely different set of reactions. It will be recalled from the preceding discussion that an increase in the pH of a medium favors the formation of diphosphate ions. In addition, the solubility of the calcium orthophosphates decreases in the order mono-, di-, and tricalcium phosphate. In most alkaline soils the activity of the calcium is high. This, coupled with a high pH, favors the precipitation of relatively insoluble dicalcium phosphate and other basic calcium phosphates such as hydroxyapatite and carbonatoapatite.

Surface Precipitation on Solid Phase Calcium Carbonate. In alkaline soils that contain free calcium carbonate a second mechanism is responsible for decreasing the activity of phosphorus. Phosphate ions coming in contact with solid phase calcium carbonate are precipitated on the surface of these particles. The amount of precipitation taking place is influenced by the amount of surface exposed by the calcium carbonate and by the concentration of the phosphate in the surrounding solution. At least the initial stage of this fixation process is considered to be a surface phenomenon. Subsequent deposition may be of the mass action type, the rate of precipitation being governed by the concentration of the reactants in the soil solution. Regardless of the nature of the reaction, the end product seems to be a relatively insoluble salt of calcium, phosphorus, and perhaps CO_3^{2-} or OH^-.

A third mechanism generally considered to be responsible for phosphorus fixation in alkaline soils is the retention of phosphate by clays saturated with calcium. As already indicated, clays saturated with this ion can retain greater amounts of phosphorus than those saturated with sodium or other monovalent ions. A linkage such as clay-Ca-H_2PO_4 has been suggested. Such reactions might occur at pH values slightly less than 6.5, but in soils more basic than this dicalcium phosphate would probably be directly precipitated from solution.

It follows from this discussion that the concentration (or, more precisely, the activity) of phosphorus in the soil solution in alkaline or calcareous soils will be largely governed by three factors:

1. Ca^{2+} activity.
2. The amount and particle size of free calcium carbonate in the soil.
3. The amount of clay present.

The activity of the phosphorus will be lower in those soils that have a high Ca^{2+} activity, a large amount of finely divided calcium carbonate, and a large amount of calcium-saturated clay. Conversely, in order to maintain a given level of phosphate activity in the soil solution, it is necessary to add larger quantities of phosphate fertilizers to such soils. Workers at Colorado have studied phosphate equilibria in calcareous soils intensively. The results of one study, which illustrates the effect of texture and amount of added concentrated superphosphate (CSP) on phosphate activity, are shown in Figure 6-5. The phosphorus activity is expressed in terms of the mean activity of dicalcium phosphate in the soil solution. Two important points are illustrated in this figure. First, in any one soil the phosphate activity in the soil solution increased with additions of monocalcium phosphate. Second, larger additions of superphosphate were required to reach a given level of phosphate activity in

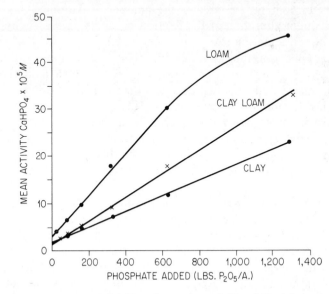

Figure 6–5. Phosphorus solubility (the mean activity of dicalcium phosphate in solution) as a function of the amounts of CSP added to three calcareous soils of different texture. [Cole et al., *SSSA Proc.*, **23**:119 (1959).]

fine-textured soils than were needed to reach the same activity in coarse-textured soils. Because a high level of phosphate activity is associated with a high level of phosphate availability to plants, the points illustrated in Figure 6-5 are of obvious practical concern.

FACTORS INFLUENCING PHOSPHORUS RETENTION IN SOILS

Several of the factors influencing phosphorus retention in soils are apparent from the preceding discussion on the mechanisms involved in these reactions. Others are not. Because of the importance of retention and fixation in modifying the effectiveness of applied fertilizer phosphorus, these factors, and the extent to which they influence fixation, are briefly considered.

Type of Clay. Phosphorus is retained to a greater extent by 1:1 than by 2:1 clays. Soils high in kaolinitic clays, such as those found in areas of high rainfall and high temperatures, will fix or retain larger quantities of added phosphorus than those containing the 2:1 type. The presence of hydrous oxides of iron and aluminum also contributes to the retention of added phosphorus. These compounds are found in soils containing large amounts of 1:1 clays. Soils containing large amounts of clay will fix more phosphorus than those containing small amounts. In other words, the more surface area exposed with a given type of clay, the greater the amount of fixation taking place.

Time of Reaction. The greater the time the soil and added phosphorus are in contact, the greater the amount of fixation. This results from subsequent alteration of the fixation products such as dehydration and crystal reorientation. An important practical consequence is the time after application during which the plant is best able to utilize the added fertilizer phosphorus. On some soils with a high fixing capacity this period may be short, whereas with other soils the period of utilization may last for months or even years. This time period will determine whether the fertilizer phosphorus should be applied at one time in the rotation or in smaller, more frequent applications. Also important is the placement of phosphorus in the soil. Band placement and broadcast applications of phosphates are discussed in a subsequent section of this chapter.

Soil Reaction. Soil pH is one of the factors affecting phosphorus utilization which the farmer can easily alter. In most soils phosphorus availability is at a maximum in the pH range 5.5 to 7.0, decreasing as the pH drops below 5.5 and decreasing as this value goes above 7.0. At low pH values the retention results largely from the reaction with iron and aluminum and their hydrous oxides. As the pH increases, the activity of these reactants is decreased until, within the pH range just given, the activity of phosphorus is at a maximum. Above pH 7.0 the ions of calcium and magnesium, as well as the presence of the carbonates of these metals in the soil, cause precipitation of the added phosphorus, and its availability again decreases.

The pattern of phosphorus availability discussed is generally the one observed. If, however, a soil were alkaline because of the presence of cations such as sodium rather than calcium, a decrease in phosphorus availability would not necessarily be observed with a continuing increase in the soil pH. The presence of calcium or magnesium ions must accompany high pH values if there is to be a continued decrease in the solubility of soil phosphorus.

Temperature. Although the speed of chemical reactions generally increases with a rise in temperature, the extent to which this factor influences the fixation of soil phosphorus under field conditions is not well understood. However, the soils of the warmer climates are generally much greater fixers of phosphorus than the soils of more temperate regions. These warmer climates also give rise to soils with higher contents of the hydrous oxides of iron and aluminum.

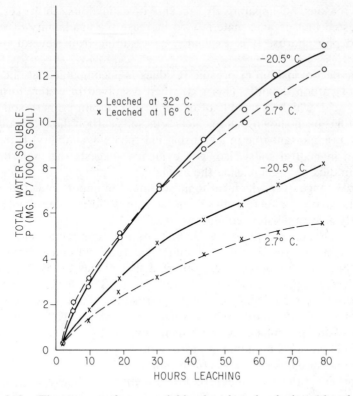

Figure 6–6. The amount of water-soluble phosphate leached at 16 and 32°C. from a Bedford soil preconditioned at −20.5 and 2.7°C. [Mack et al., *SSSA Proc.*, **24**:381 (1960).]

Some work along this line was carried out by scientists at Purdue University. Samples of a Bedford silt loam were incubated for several months at -20.5 and $2.7°C$. The soils were then leached with water at two temperature levels, 32 and $16°C$. The results are shown in Figure 6-6. Even though the temperature of the leaching water affected the amount of phosphorus removed, the lower amount was removed in both cases from the soil incubated at the higher temperature. Similar findings were reported by workers from the Northern Plains Field Station at Mandan, North Dakota. Their studies showed that both the water- and $NaHCO_3$-soluble phosphate extracted from soils to which phosphate fertilizer had been added decreased when the soil incubation temperatures were about $59°F$. However, on soils with no added phosphate the amounts of P extracted were not affected by soil incubation temperatures ranging from 45 to $80°F$.

Organic Matter. It is generally agreed that the turning under of stable or green manures results in a better utilization of phosphorus by subsequent crops. This has been difficult to show experimentally for several reasons. Comparatively recent investigations, however, have suggested that organic materials do increase the availability of soil and added phosphorus. It is explained as resulting from several different causes.

The decomposition of organic residues is accompanied by the evolution of carbon dioxide. This gas, when dissolved in water, forms carbonic acid, which is capable of decomposing certain primary soil minerals. It has been shown that in calcareous soils carbon dioxide production plays an important role in increasing phosphate availability. It has been shown in neutral soils, also, and evidence suggests the importance of carbon dioxide in increasing the availability of phosphorus in acid soils. The pH range over which carbonic acid may be important is rather wide but its greatest effect in dissolving soil phosphorus is probably under slightly acid to alkaline conditions.

The effect on phosphorus availability of other compounds arising from the decomposition of organic residues has received considerable attention. Numerous workers have reported that humus extracts from soils have increased the solubility of phosphorus. This has been variously described as resulting from (1) the formation of phosphohumic complexes which are more easily assimilable by plants, (2) anion replacement of the phosphate by the humate ion, and (3) the coating of sesquioxide particles by humus to form a protective cover and thus reduce the phosphate fixing capacity of the soil.

Workers at the Massachusetts Experiment Station have suggested that certain organic anions arising from the decomposition of organic matter will form stable complexes with iron and aluminum, thus preventing their reaction with phosphorus. It was further stated that these complex

TABLE 6-3.   The Leaching of Fertilizer P from an Acid Organic (Muck) Soil[*]

P added mg./column	pH	Exch. Al meq/100 g.	P leached	Fertilizer P absorbed
			mg./column	
Surface Soil				
0	4.6	0.02	0.48	—
2.5	4.6	0.02	2.95	0.03
10.0	4.6	0.02	8.67	1.81
Subsurface Soil				
0	3.3	—	0.54	—
10.0	3.3	—	10.22	0.32

[*] Fox and Kamprath, *SSSA Proc.*, **35:**154 (1971).

ions release phosphorus previously fixed by iron and aluminum by the same mechanism. The anions that are most effective in replacing phosphates are citrate, oxalate, tartrate, malate, and malonate, some of which may also occur as degradation products during organic matter decay.

On the basis of the available evidence, it seems clear that the addition of organic materials to mineral soils may increase the availability of soil phosphorus. Further work is necessary before the mechanisms involved can be definitely established.

Soils that consist largely of quartz sand and soils which are primarily muck and peat are subject to leaching losses of added phosphate fertilizer. This was demonstrated by laboratory work in North Carolina in which monocalcium phosphate was added to columns of two soils, one a muck and the other a soil containing 90 per cent coarse sand and 10 per cent organic matter. A portion of the results of this study is shown in Table 6-3.

It is apparent from these data that there was very little retention of added phosphate by either the surface or subsurface horizon of this soil. This is due to the absence of Al and Fe compounds which are largely responsible for the retention of mineral phosphates. In fact, the North Carolina workers found that by adding $AlCl_3$ to the two soils the leaching of added phosphorus in these studies was almost completely stopped.

Phosphorus Status of the Soil.   Of considerable importance in the fixation of added fertilizer phosphorus is the degree of phosphorus saturation of the soil or the amount of this element previously fixed by the soil. It has been found in North Carolina, for example, that there is a pronounced relationship between the amount of fixation of added fertil-

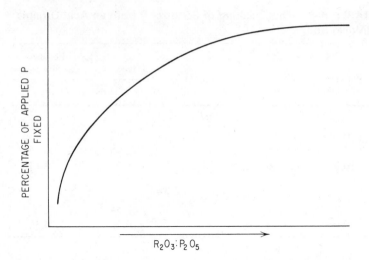

Figure 6–7. Phosphorus fixation as related to the $R_2O_3 : P_2O_5$ ratio of the soil. (Courtesy of Professor Adolf Mehlich, North Carolina Agricultural Experiment Station, Raleigh.)

izer phosphorus and the $R_2O_3 : P_2O_5$ ratio of the soil. This relationship is shown graphically in Figure 6-7. The ratio $R_2O_3 : P_2O_5$ is a measure of the amount of phosphorus present in relation to the iron and aluminum oxide content of the soil. A wide ratio indicates a small amount of phosphorus present or a low phosphorus saturation value. Under such conditions larger amounts of added phosphorus are fixed than when the ratio is narrow.

The practical implications of this relationship are quite important. On soils that have been heavily phosphated for several years it should be possible (1) to reduce the amount of phosphate currently applied in the fertilizer, (2) to utilize to a greater extent the phosphorus in the soil, or (3) to effect a combination of (1) and (2).

PLACEMENT OF ADDED FERTILIZER

The placement of fertilizer phosphorus in the soil is treated at greater length in a subsequent chapter, but because of the relationship of this topic to the subject matter covered here it will be considered briefly at this point. The finer the soil texture, the greater the retention of added fertilizer phosphorus. This would be predicted from a knowledge of the relation between the speed of a chemical reaction and the amount of the surface exposed by the reactants. If a finely divided soluble fertilizer phosphate is added to a soil by applying it broadcast and disking it in, the phosphate is exposed to a greater amount of surface; hence more fixation takes place than if the same amount of fertilizer had been applied in bands. Band placement reduces the surface of contact between the soil and fertilizer with a consequent reduction in the amount of fixation. Although this is not the only factor to consider in the placement of a phosphorus fertilizer, it is one of considerable importance when a crop is to be grown on a low phosphate soil with a high fixing capacity and when maximum return from the dollars spent on phosphorus is desired. Granulation of soluble phosphate fertilizers tends to have a similar effect, particularly in the case of larger granules such as those of 4 to 10 mesh.

Band placement generally increases the plant utilization of the water-soluble phosphates such as the superphosphates. There are, however, other phosphatic fertilizers, which are classed as water-insoluble, whose plant utilization seems to be greatest when mixed with the soil rather than when applied in bands.

PHOSPHATE FERTILIZERS

The development of the modern phosphate fertilizer industry began with the demonstration by Liebig in 1840 that the fertilizing value of bones could be increased by treatment with sulfuric acid. Shortly thereafter, in 1842, John B. Lawes patented a process by which phosphate rock was acidulated with this acid. He began the commercial production of this material in England in 1843. The first recorded sale of what is now commonly known as normal superphosphate manufactured in the United States did not take place until 1852.

From such beginnings has sprung the multimillion dollar fertilizer phosphate industry which uses essentially the same principle for the manufacture of soluble phosphates employed by John Lawes well over a hundred years ago. Ordinary superphosphate is a good fertilizer which is easily and inexpensively manufactured. In recent years, however, new materials have been developed which, because of their higher content of phosphorus, their handling and storage properties, and the economics of manufacture, are replacing ordinary superphosphate. Much progress has been made during the last several years because of the cooperative efforts of scientists and engineers in both private and public agencies to improve the quality of fertilizers. The properties and behavior of some of these phosphates are discussed in the following sections of this chapter. The principles of the manufacture of phosphate fertilizers are considered in Chapter 9.

Calculation of Phosphorus Content of Fertilizers. Historically, the phosphorus content of fertilizers has been expressed in terms of its P_2O_5 equivalent. More recently, attempts have been made to change this procedure and to have the phosphorus content expressed in terms of its phosphorus equivalent, $\%P$ rather than $\%P_2O_5$. It is likely that in the long run the $\%P$ proponents will have their way. The same groups are attempting to have potassium expressed as $\%K$ rather than as $\%K_2O$, as has historically been the case. It must be admitted that percentage figures expressed on the elemental basis are easier to calculate – and to discuss – than on the oxide basis. As a matter of interest, it was not too many years ago that nitrogen was guaranteed as $\%NH_3$ rather than as $\%N$, as is now done.

The conversion of $\%P$ to $\%P_2O_5$ and vice versa is simple; it is illustrated here with the salt, dicalcium phosphate ($CaHPO_4$). The percentage of phosphorus in this salt is

$$31/136 \times 100 = 23\%$$

To convert $\%P$ to $\%P_2O_5$ and vice versa the following expressions are used:

$$\%P = \%P_2O_5 \times 0.43$$
$$\%P_2O_5 = \%P \times 2.29$$

So the $\%P_2O_5$ in the example of dicalcium phosphate is

$$\%P_2O_5 = 23 \times 2.29 = 53$$

Throughout the remainder of this book the phosphorus and potassium content of fertilizers will, wherever possible, be expressed in terms of $\%P$ or $\%K$. However, because most fertilizer grades are listed, sold and discussed by the trade in terms of their P_2O_5 and K_2O content and, as such, grades are currently widely understood on this basis, so are they referred to here. In other cases the grade, or guarantee, is followed in parentheses by another figure in which it is expressed in terms of nitrogen, phosphorus, and/or potassium.

Phosphate Fertilizer Terminology. The solubility of the phosphorus in the different phosphate carriers is variable. The water-solubility of fertilizer phosphates is not always the best criterion of the availability of this element to plants, though it is a good one. Perhaps the most accurate measure of the plant availability of any nutrient element is the extent to which it is absorbed by plants under conditions favorable to growth. Such determinations are not easily made when the availability of fertilizer elements has to be determined quickly on large numbers of samples, as in fertilizer control work. Chemical methods have been developed which permit a fairly rapid estimate of the water-soluble, available, and total phosphorus content of phosphate fertilizers. The terminology currently accepted in the United States to describe the availability of phosphate materials is frequently employed by persons dealing with fertilizers. A brief description of the analytical procedure as well as a definition of the terms, follows.

WATER-SOLUBLE PHOSPHORUS. The terms frequently encountered in describing the phosphate contained in fertilizers are *water-soluble, citrate-soluble, citrate-insoluble, available,* and *total phosphorus* (currently as P_2O_5). A small sample of the material to be analyzed is first extracted with water for a prescribed period of time. The slurry is then filtered, and the amount of phosphorus contained in the filtrate is determined. Expressed as a percentage by weight of the sample, it represents the fraction of the sample that is *water-soluble.*

CITRATE-SOLUBLE PHOSPHORUS. The residue from the leaching process is added to a solution of neutral $1N$ ammonium citrate. It is extracted for a prescribed period of time by shaking, and the suspension is filtered. The phosphorus content of the filtrate is determined, and the amount present, expressed as a percentage of the total weight of the sample, is termed the *citrate-soluble* phosphorus.

CITRATE-INSOLUBLE PHOSPHORUS. The residue remaining from the water and citrate extractions is analyzed. The amount of phosphorus found is termed *citrate-insoluble*

AVAILABLE PHOSPHORUS. The sum of the water-soluble and citrate-soluble phosphorus represents an estimate of the fraction available to plants and is termed *available* phosphorus.

TOTAL PHOSPHORUS. The sum of available and citrate-insoluble phosphorus represents the total amount present. The total phosphorus can, of course, be determined directly without resorting to the step-by-step process described.

SUITABILITY OF PRESENT ANALYTICAL METHOD

Many states require that the guarantee of the phosphorus contained in fertilizers include both the percentage of available phosphorus and the total amount present. Crop responses are often correlated with the amount of available phosphorus in fertilizer materials, as determined by the method outlined. However, research on this subject has cast doubt on the value of the neutral ammonium citrate method as a means of determining available phosphorus in fertilizers containing large amounts of basic calcium phosphate. This work was carried out by scientists at the Mississippi Experiment Station. Forty-nine field experiments were conducted, in which corn, cotton, and wheat were grown at four locations over a five-year period. Investigated were the effects of five factors on the agronomic effectiveness of ammoniated ordinary superphosphate: (1) degree of ammoniation, (2) size of the fertilizer granules, (3) source of nonammoniating nitrogen, (4) ratio of nitrogen to phosphorus in the fertilizer, and (5) method of preparation of the fertilizer, that is, dry mixing versus granulation. In all of these tests the phosphates were applied in a band, and adequate nitrogen and potassium were added to ensure that these two factors did not limit plant growth.

As a result of these studies, it was concluded that ". . . the degree of ammoniation was the most important (factor) in influencing the effectiveness of the phosphorus in ordinary superphosphate" and that "excessive ammoniation drastically reduced the effectiveness of the phosphate."

These workers arrived at the following conclusions, which are excerpted from this publication, *Mississippi Agr. Exp. Sta. Tech. Bull. No. 52*, 1963:

> Comparisons of the agronomic effectiveness, as revealed by these experiments with chemical analyses used for control purposes in the United States and other countries, showed that the present method used in the United States does not reflect the true agronomic value of ammoniated ordinary superphosphate in mixed fertilizers and does not provide a suitable basis for quality control. On the other hand, there was a close relationship between alkaline citrate "available" phosphorus and agronomic effectiveness. Solubility of the phosphorus in water according to the A.O.A.C. procedure was also closely correlated with agronomic effectiveness. A study of the phosphorus soluble in neutral ammonium citrate and in alkaline ammonium citrate revealed that the heavily ammoniated fertilizers contained relatively large quantities of phosphorus compounds more basic than dicalcium phosphate.

This work, and other evidence that is being accumulated, may cause a revision in the methods presently employed to evaluate the availability of phosphate fertilizers in the United States. The alkaline ammonium citrate method, for example, has been used in Europe and other countries for many years. The effect of ammoniation on the reduction of the water-soluble content of superphosphates is given further coverage in this chapter in the section dealing with ammoniated superphosphates.

CLASS OF PHOSPHATE FERTILIZERS

The original source of phosphorus in the early manufacture of phosphatic fertilizer was bones, but the supply was soon exhausted. Today, and for many years past, the only important source of fertilizer and industrial phosphorus is rock phosphate. Deposits of phosphate rock occur in several areas of the world. Their location and the extent of the deposits are discussed at greater length in Chapter 9. The phosphate compound in these deposits is apatite, which has the general formula $Ca_{10}(PO_4,CO_3)_6(F, Cl, OH)_2$. This mineral occurs in several forms, such as the carbonato-, fluoro-, chloro-, or hydroxyapatite. The rock is either heat-treated or acid-treated to break the apatite bond and to render the contained phosphate more soluble.

Acid-Treated Phosphates. Commercially important fertilizer phosphates are classed as either acid-treated or thermal-processed; the former constitute by far the more important group. Acid-treated phosphates are classed in the indicated groups, each of which contains several materials which will be discussed in the following sections: phosphoric acids; superphosphates; ammonium phosphates; nitric phosphates; and miscellaneous and new phosphate materials.

PHOSPHORIC ACID. Phosphoric acid (H_3PO_4) is manufactured by treating rock phosphate with sulfuric acid or by burning elemental phosphorus to phosphorus pentoxide and reacting it with water. That produced from sulfuric acid is known as *green* or *wet process acid* and is used largely in the fertilizer industry. Elemental phosphorus is termed *white* or *furnace acid* and is used largely by the nonfertilizer segment of the chemical industry. White acid has a much higher degree of purity than the green and is therefore more expensive. During periods of peak supply, however, some white acid finds its way into the fertilizer industry, for the manufacturers make it available at a lower price to eliminate expensive storage costs. Improved methods of producing wet acid, especially wet process superphosphoric acid, have led to the increased use of wet acid for processes formerly requiring white acid.

Agricultural grade phosphoric acid, which contains 24 per cent phosphorus (55% P_2O_5), is used to acidulate phosphate rock to make triple superphosphate and is neutralized with ammonia in the manufacture of ammonium phosphates and liquid fertilizers. It can also be applied directly to the soil, particularly in alkaline and calcareous areas, by injection, but this method requires special equipment, so the acid is more frequently added to the irrigation water.

Phosphoric acid reacts with the soil components in essentially the same way as any other orthophosphate. In calcareous soils dicalcium phosphate and the hydroxy- and carbonatoapatites will ultimately be formed. In acid soils complex phosphates of iron and aluminum will result. Phosphoric acid may, when applied in a band, bring temporarily into solution certain otherwise insoluble trace elements such as zinc, iron, and manganese. In addition, because it is in the liquid form, it may sometimes prove superior to solid sources of phosphorus when it is applied well after planting, a practice not generally recommended. This superiority was, in fact demonstrated in Colorado. Band-placed phosphoric acid and triple superphosphate were applied to sugar beets as a late side dressing. The phosphoric acid was a better source of phosphorus than the triple superphosphate. In many other tests, when rate of applied phosphorus, method of placement, and time of application are comparable, phosphoric acid has been found to be equal to triple superphosphate as a source of phosphorus for most crops studied.

SUPERPHOSPHORIC ACID. Superphosphoric acid, another phosphoric acid, has recently been developed, largely through the efforts of the Tennessee Valley Authority. Originally, it was produced from white acid only, but it is now produced from the less expensive green acid, a development which can be very favorable to the fertilizer industry and the consumers of this acid.

214 Soil Fertility and Fertilizers

Superphosphoric acid made from white acid contains 33 to 37 per cent phosphorus (76–85% P_2O_5) in contrast to the 24 per cent phosphorus for the unconcentrated acid. It is made essentially by dehydrating phosphoric acid, in addition to which the final product contains about 35 to 50 per cent of such condensed phosphate radicals as tetra-, pyro-, and tripolyphosphates. Wet process super acid can be made by evaporating the water to concentrate orthophosphoric acid beyond the equivalent of 100 per cent of H_3PO_4. Because of the impurities in wet process acid, however, the maximum concentration of P_2O_5 in the super acid so made is only about 72 per cent (31% P). Superphosphoric acid is used for the manufacture of ammonium and calcium polyphosphates and liquid fertilizers, which are discussed in a subsequent section of this chapter.

CALCIUM ORTHOPHOSPHATES. The most important phosphate fertilizers, consisting of calcium orthophosphate, are ordinary superphosphate, 7 to 9 per cent phosphorus (16–22% P_2O_5), triple or concentrated superphosphate, 19 to 23 per cent phosphorus (44–52% P_2O_5), enriched superphosphates, 11 to 13 per cent phosphorus (25–30% P_2O_5), and ordinary and triple superphosphates which have been reacted with different amounts of ammonia. The last group goes by the term *ammoniated superphosphates.*

The superphosphates are *neutral fertilizers* in that they have no appreciable effect on soil pH, as have phosphoric acid and the ammonium-containing fertilizers. The ammoniated superphosphates have a slightly acid reaction, depending, of course, on the extent to which they have been ammoniated.

Ordinary superphosphate (OSP) is manufactured by reacting sulfuric acid with rock phosphate. This product, which is essentially a mixture of monocalcium phosphate and gypsum, contains 7 to 9.5 per cent phosphorus (16 to 22% P_2O_5), of which about 90 per cent is water soluble and essentially all is classed as *available.* In addition, it contains about 8 to 10 per cent sulfur as calcium sulfate. In areas in which the soils are deficient in that element the gypsum content has been an important contributor to satisfactory crop responses to this phosphate fertilizer. In fact, in numerous experiments conducted in the United States, in which various fertilizers were compared as sources of phosphorus for crops, the superiority of ordinary superphosphate to some of the other sources has frequently been associated with its sulfur content.

The phosphorus component of superphosphate reacts with the soil components, as does any water-soluble orthophosphate, in keeping with the reactions already discussed in this chapter. It is an excellent source of fertilizer phosphorus, but its low content of this element has resulted in its replacement in many areas by higher analysis materials. When consumed within two hundred miles of the point of manufacture, however, it is usually competitive with the higher analysis products. Too, with the increasing incidence of sulfur deficiency in many soils throughout North America, there have been localized increases in the demand for this material.

Ordinary superphosphate is used largely in the production of mixed fertilizers in which other dry, powdered, or finely granular materials are blended to effect a product that contains nitrogen, phosphorus, and potassium. It is also used for direct application.

Triple or concentrated superphosphate (TSP or CSP) contains 19 to 23 per cent phosphorus (44–52% P_2O_5), 95 to 98 per cent of which is water-soluble and nearly all of which is classed as available. It is essentially monocalcium phosphate and is manufactured by treating rock phosphate with *phosphoric* acid. CSP contains varying amounts of sulfur (usually less than 3%), depending on the manufacturing process. This low content of sulfur is insufficient to supply crop requirements on sulfur-deficient soils.

CSP is an excellent source of fertilizer phosphorus. Its high phosphorus content makes it particularly attractive when transportation, storage, and handling charges make up a large fraction of the total fertilizer cost. CSP is manufactured in pulverant and granular forms and is used in mixing and blending with other materials and in direct soil application.

Enriched superphosphates (ESP) are produced by treating rock phosphate with a mixture of phosphoric and sulfuric acids. Although any percentage of phosphorus between 8.5 and 19 per cent can be obtained, the final product usually contains 11 to 13 per cent (25–30% P_2O_5), of which 90 to 95 per cent is water-soluble. Nearly all of the contained phosphorus is available. The enriched superphosphates are mixtures of monocalcium phosphate and calcium sulfate. ESP obviously contains sulfur, the amount of which depends on the amount of sulfuric acid used in acidulating the rock.

ESP historically has never enjoyed the popularity in the United States that it has had in Europe. The phosphate component is as satisfactory a source of phosphorus as that in OSP and CSP. Further, the use of ESP is a simple way of upgrading the phosphorus component of the product and at the same time providing plant-nutrient sulfur on soils deficient in this element.

Ammoniated superphosphates are prepared by reacting anhydrous or aqua ammonia with ordinary or triple superphosphate. The total phosphorus content of the end product is decreased in proportion to the weight of the ammonia added. Ammoniation of superphosphates offers the advantage of inexpensive nitrogen but decreases the amount of water-soluble phosphorus in the product. The reduction in water-solubility is greater in OSP than in TSP, as shown in Figure 6-8.

The effect of excessive ammoniation of OSP on reducing crop yields

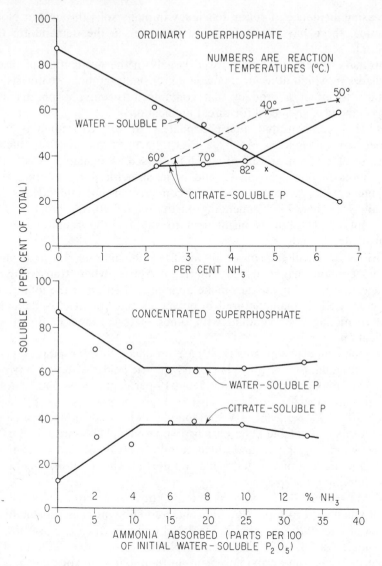

Figure 6–8. Influence of ammoniation on the solubility of phosphorus in OSP and CSP. [Hill, *USDA Res. Rept. 251* (1952).]

has been studied in Mississippi (see page 211). In the study referred to, the workers expressed the results in terms of superphosphate equivalents, which in effect is the extent, on a percentage basis, of the effectiveness of the variously ammoniated superphosphate samples compared to the nonammoniated material. A summary of their results, which include five different crops grown at four locations over a five-year

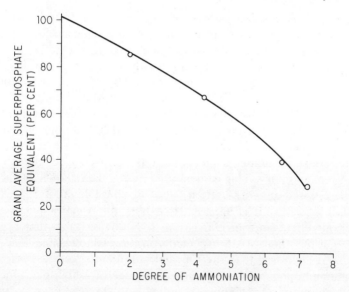

Figure 6-9. Average relationship between percentage of superphosphate equivalent and degree of ammoniation. [Wright et al., *Mississippi Agr. Exp. Sta. Tech. Bull. 52* (1963).]

period, is represented graphically in Figure 6-9. For crops responding to a high degree of water-soluble phosphorus (as these crops obviously did), it is apparent that a high degree of ammoniation of OSP will have a depressing effect on the plant availability of the contained phosphorus.

AMMONIUM PHOSPHATES. Ammonium phosphates are produced by reacting ammonia with phosphoric acid or a mixture of phosphoric and sulfuric acids. Some of the more widely used ammonium phosphate fertilizers are monoammonium phosphate (MAP), diammonium phosphate (DAP), and ammonium phosphate-sulfate (16-20-0) (8.6% P). Pure MAP contains 12 per cent nitrogen and 61 per cent phosphorus pentoxide (26% P). It has the grade 12-61-0. Another common fertilizer-grade MAP is 11-48-0 (21% P). Pure DAP has the grade 21-53-0 (23% P). Although small amounts of this relatively pure material are produced, other fertilizer grades of DAP are 16-48-0 (21% P) and 18-46-0 (20% P). Ammonium phosphate-sulfate is prepared by reacting ammonia with a mixture of phosphoric and sulfuric acids. The product contains 16 per cent nitrogen and 20 per cent phosphorus pentoxide (8.6% P), as indicated by its grade, 16-20-0.

The ammonium phosphates are completely water-soluble. Ultimately, they have an acid effect on soils because of the ammonia they contain, even though the initial reaction of DAP is alkaline, a point to be considered in more detail in a subsequent section of this chapter. The ammonium phosphates are usually offered in the granular form, though some crystalline materials are produced. They are used for formulating solid fertilizers, either by conventional mixing methods or in bulk blended goods (see Chapter 10) and (with the exception of 16-20-0) in manufacturing liquid and slurry fertilizers. They are also used for direct application as starter fertilizers, particularly in Midwestern and western United States and in the central and western provinces of Canada.

When DAP is applied at seeding, care must be taken to place it properly with respect to the seed. This is especially true when it is used on alkaline soils, for under such conditions released free ammonia may cause seedling injury. This is illustrated by the data shown in Table 6-4, which are based on work carried out at Purdue University by Ohlrogge and his associates. It is apparent that both the MAP and DAP had a deleterious effect on root elongation. It is obvious also that DAP was more injurious than MAP, especially on the Oaktown fine sand, which was coarser textured than the Fincastle silt loam.

Ammonium phosphates have the advantage of a high plant-food content which minimizes shipping, handling, and storage costs. As with the superphosphates, they have good handling properties, all of which combine to make them the fertilizer phosphates most rapidly increasing in popularity in the United States and abroad.

A relative newcomer to the field of the ammonium phosphates is ammonium polyphosphate. Developed by TVA, it is produced either by ammoniating superphosphoric acid directly or by a two-stage ammoniation technique employing orthophosphoric acid. In the latter process, the heat of neutralization is used to bring about the molecular dehydration of the orthophosphate that is required to form the polyphosphate. The

TABLE 6-4. Primary Seminal Root Lengths of Corn Germinated Near Mono- and Diammonium Phosphate Bands[*]

	Median lengths (cm.)					
	In Oaktown soil			In Fincastle soil		
Seed position	Band of $(NH_4)_2HPO_4$	Band of $NH_4H_2PO_4$	No band	Band of $(NH_4)_2HPO_4$	Band of $NH_4H_2PO_4$	No band
2 in. above and 2 in. to side	10	29	(23)	33	30	(42)
2 in. above	4	30	(23)	31	33	(42)
¼ in. above	0	0.5	(27)	0	0	(44)
1 in. to side	3	18	(27)	23	30	(44)
2 in. to side	23	18	(27)	33	30	(44)

[*] Allred et al., *Agron. J.*, **56**:309 (1964).

product made from electric furnace acid has the analysis 27-34-0 (15% P) and that made from wet acid somewhat lower. Ammonium polyphosphates can be used in the solid form for direct application or for bulk blending and they are also used in the manufacture of liquid fertilizers.

NITRIC PHOSPHATES. Nitric phosphates are manufactured by reacting nitric acid with rock phosphate. One of the reaction products, calcium nitrate, is objectionable because of its hygroscopicity. When some sulfuric or phosphoric acid or a sulfate salt is added, most of the calcium nitrate is converted to calcium sulfate or phosphate. Another modification removes the excess calcium by introducing carbon dioxide to precipitate the calcium as calcium carbonate, whereas still another employs refrigeration and centrifugation. The acidified slurry is ammoniated, so that the end products contain a complex assortment of salts such as ammonium phosphates, dicalcium phosphates, ammonium nitrate, calcium sulfate, and still others if potassium salts are added to make a complete fertilizer, as they frequently are.

The $N:P_2O_5$ ratio of materials produced by nitric phosphate methods ranges from 1:1 to 1:3. This is usually considered to be one of the principal objections to the production of nitric phosphates, for the possible number of grades that can be made is limited. In addition, most nitric phosphates generally have a lower degree of water-soluble phosphorus than have materials based on the neutralization of phosphoric acid with ammonia. Nitric phosphates are always produced in the granular form.

Nitric phosphates are not produced extensively in the United States but are used to a greater degree in certain European countries, especially France, Italy, and the Netherlands. They have been studied extensively here in the United States, principally by the TVA, which in fact pioneered product development and agronomic testing of the materials in this country. The nitric phosphate process originated in Europe a number of years ago.

The results of numerous agronomic tests have shown that these materials are generally satisfactory sources of fertilizer phosphorus. When used on crops responding to water-soluble phosphorus, nitric phosphates may be inferior to those materials containing a high degree of water-soluble phosphorus. The water-soluble fraction of the total phosphorus in nitric phosphates will range from zero to perhaps 80 per cent. The desired degree of water solubility may be obtained by replacing part of the nitric acid with phosphoric or sulfuric acids. The higher the degree of water solubility desired, the greater will be the amounts of phosphoric or sulfuric acids required. A high degree of water solubility may also be obtained by adding nitric acid in amounts greater than that which is theoretically required to react completely with the phosphate rock. The excess acid is subsequently neutralized with ammonia.

Nitric phosphates, in general, will give the best results on acid soils and under crops with a relatively long growing season, such as turf and sod crops. It must be re-emphasized, however, that when the degree of water-soluble phosphorus in these materials is kept high (60% or greater) they are usually just as effective as sources of phosphorus for most crops as the super- and ammonium phosphates.

Nitric phosphates are potentially important fertilizers in the United States because of the increasing supply of ammonia, hence nitric acid. The economics of producing these materials and the ability to increase their nitrogen content, in addition to the degree of water-soluble phosphorus, are the two factors that will probably determine the future of nitric phosphate fertilizers in this country.

MISCELLANEOUS PHOSPHATES. Several miscellaneous materials which have been used as phosphate fertilizers for several years and a few comparatively new compounds which show promise of becoming significant sources of fertilizer phosphorus include raw rock phosphate, potassium phosphate, dicalcium phosphate, ammonium phosphate-nitrate, magnesium ammonium phosphate, and ammonium polyphosphate.

Untreated rock phosphate is stable and quite insoluble in water. Commercially important sources contain between 11.5 per cent and 17.5 per cent total phosphorus (27–41% P_2O_5). The citrate solubility varies from 5 to about 17 per cent of the total phosphorus content and none of the contained phosphorus is water-soluble.

In its natural state it is of only limited value to plants, unless applied to the soil in an extremely finely ground condition and in quantities supplying three to five times the phosphorus equivalent normally applied in the form of more soluble phosphate carriers. Even in these large amounts it is of little or no use on alkaline soils. On slightly acid soils only certain crops are capable of direct utilization of the untreated rock.

The effect of raw rock phosphate on crop yields has been studied in literally thousands of experiments conducted throughout the United States. It is not possible within the scope of this text to consider individually the data from these experiments, but a few summary statements concerning the effectiveness of raw rock as a source of fertilizer phosphorus, based on the results of these experiments, will be of interest. In these studies the rock was compared in effectiveness to that of OSP and CSP.

On acid soils, low in phosphorus, raw rock phosphate may be considered profitable, but the treated phosphates are generally more economical. In some tests rock phosphate produced greater yields of the test crop than did OSP, but *only* when the rock was supplied in quantities that furnished two to three times more phosphorus. Rock phosphate has been reported to give better residual effects than superphosphate, but whenever this was so it was found that the rates of applied rock phosphate were considerably in excess of those of superphosphate.

Rock phosphate should never be applied directly under any short-season row crop with the idea that phosphorus will be supplied. Its availability is low, and only when used in large amounts and under rotation, including red and sweet clover which are strong feeders on rock phosphate, should it be considered. Such a program has been used in some states, notably in Illinois, and has proved to be successful on a long-term basis. If this system is initiated on a soil low in phosphorus, soluble phosphates, in addition to the rock phosphate, will be needed for the first few years to provide the plants with sufficient available phosphorus. When rock phosphate can be used, its choice should be based on the cost per unit of phosphorus compared with its cost per unit in the more available forms of fertilizer phosphate.

Workers in Australia have developed a granular material containing raw rock phosphate and finely ground elemental sulfur. The product, called Biosuper, is inoculated with the sulfur-oxidizing bacteria *Thiobacillus thiooxidans* to ensure the conversion of the sulfur to sulfuric acid. The acid in turn reacts with the phosphate rock, making the contained phosphorus more available to plants. Extensive field trials have shown that while the material is not as effective as superphosphate, it is a suitable product for use on pasture and range land. Its greatest use will probably be found in developing nations that do not have a sophisticated fertilizer industry.

Potassium phosphate is represented by two salts, KH_2PO_4 and K_2HPO_4, which have the grades 0-52-35 (22% P, 29% K) and 0-41-54 (18% P, 45% K), respectively. They are completely water-soluble and find their greatest market in soluble fertilizers sold in small packets for home and garden use. Their high content of phosphorus and potassium makes them attractive possibilities for commercial application on a farm scale. Developments in the economics of producing these salts will determine whether they can be manufactured on a large scale for use as commercial fertilizers.

Little has been published or is known about the behavior of this compound, but work conducted in Canada* has shown that when potassium phosphate is applied to a soil in pellet form the concentration of water-soluble phosphate in the reaction zone around the pellet two days after application is 18 ppm. of phosphorus. The next highest concentration of phosphorus, only about 12 ppm., was obtained with a pellet of ammonium phosphate, which suggests that the availability of these materials may be greater for a longer period of time than that from conventional phosphate sources. If this turns out to be true, the value of these materials would be even further enhanced.

Dicalcium phosphate is a component of many phosphate fertilizers

* Pers. Comm., Dr. J. D. Beaton, Cominco, Ltd.

that contain calcium. It arises from the ammoniation of solids or liquids in which there is monocalcium phosphate. The pure material has the formula $CaHPO_4$ and contains 23 per cent phosphorus (53% P_2O_5). Although this material could be used as a fertilizer, it cannot compete with the less expensive phosphates presently available. Dicalcium phosphate is used as a mineral supplement for animals and pregnant women.

Dicalcium phosphate can be manufactured in several ways. In one method the sulfuric acid extract of rock phosphate is neutralized with limestone. Another development involves acidulation with nitric acid, ammoniating, and carbonating the slurry with carbon dioxide. Because of the increasing availability of lower-cost nitric acid, this process holds promise for future commercial exploitation.

Ammonium phosphate-nitrate is essentially a mixture of ammonium phosphate and ammonium nitrate. These materials are completely water soluble and can be manufactured to contain different amounts of nitrogen and phosphorus. Sample grades are (on an N-P_2O_5-K_2O basis) 8-16-32, 15-15-15, and 30-10-0, the last of which has become quite popular and is now being produced on a commercial scale for use in direct soil application and in the bulk blending of fertilizers. The original research on these products was carried out by TVA.

Magnesium ammonium phosphate, covered in Chapter 5, is a slowly available source of nitrogen and phosphorus and is used as a specialty fertilizer for turf, fruit crops, and ornamentals.

Heat-Treated Phosphates. Heat-treated or thermal phosphate fertilizers are manufactured by heating phosphate rock to varying temperatures with or without additives, such as silica. In the United States thermal-process phosphates do not account for a large segment of the total fertilizer phosphates manufactured. Some of them are locally important in various parts of the world. The chief drawbacks to thermal-process phosphates are the following:

1. They are generally more expensive to produce than acid-derived phosphates.
2. They contain no water-soluble phosphorus and the available phosphorus is frequently considerably less than 100 per cent of the total present.
3. They do not lend themselves to ammoniation, hence are of no value in the manufacture of N-P-K fertilizers.

Heat-treated rock phosphate is frequently referred to as defluorinated phosphate. Defluorination may be effected by calcination or fusion. By calcination is meant the heating of phosphate rock, usually in the presence of steam and silica, to temperatures *below* the melting point of the mixture. Fusion is the term applied when similar mixtures are heated *above* the melting point so that the charge fuses or runs together, thus forming a glassy product. The calcined material is sintered and porous in appearance.

Several thermal process phosphates are of sufficient importance to warrant some discussion. These materials, with their total and citrate-soluble contents, are defluorinated phosphate rock, 9 per cent total phosphorus (21% P_2O_5), 8 per cent citrate-soluble phosphorus (18% P_2O_5); phosphate rock-magnesium silicate glass, 10 per cent total (22.5% P_2O_5), 8 per cent citrate-soluble (19% P_2O_5); Rhenania phosphate, 12 per cent total, 11.8 per cent citrate-soluble (28% and 27.5% P_2O_5, respectively); calcium metaphosphate, 27.5 per cent total, 27 per cent citrate-soluble (64% and 63% P_2O_5, respectively); and basic slag, 1.0 to 7.8 per cent total (2.3–18% P_2O_5), of which 60 to 80 per cent is citrate-soluble.

Defluorinated phosphate rock, commonly referred to as Coronet phosphate, is made by combining finely ground phosphate rock, high silica tailings from the phosphate rock mining process, and enough water to form a slurry. This slurry is passed into an oil-fired rotary kiln in which the temperatures in the burning zone are between 1480 and 1590°C. Heating at these temperatures for about 30 minutes produces a porous mass that is quenched in water. The product is ground so that 60 per cent will pass through a 200-mesh screen. Most of this material goes into animal feeds. However, it is also a satisfactory source of fertilizer phosphorus except on alkaline or calcareous soils.

Phosphate rock-magnesium silicate glass is formed by fusing rock phosphate and either olivine or serpentine in a furnace at 1550°C. It finds its greatest use in areas in which the supplies of sulfuric or other acids are inadequate. When finely ground, this material is a satisfactory source of plant-nutrient phosphorus on acid soils. It is not suitable for use on alkaline or calcareous soils.

Rhenania phosphate was first developed in Germany in 1917. It is prepared by calcining a mixture of soda ash, silica, and phosphate rock at 1100 to 1200°C. It is quenched with water and ground to pass a 180-mesh screen. Rhenania phosphate is not used to any extent in this country, though it is popular in some areas of Europe. It is a very satisfactory source of phosphorus when used on acid soils, but is generally unsatisfactory when applied to soils that are alkaline or calcareous.

Calcium metaphosphate was developed in Germany in 1929. Work on this compound in the United States has been done almost entirely by TVA. Its successful manufacture requires a relatively cheap source of electric power, which is perhaps the chief factor limiting its commercial production in this country.

Calcium metaphosphate is prepared by burning elemental phosphorus in the presence of finely divided phosphate rock. The melt is not quenched but poured onto a water-cooled drum flaker and then ground to the desired degree of fineness.

Subsequent studies by TVA have shown that calcium metaphosphate can be partly hydrolyzed by treatment with sulfuric acid. It can then be ammoniated and used as an orthophosphate in the manufacture of complete N-P-K fertilizers.

Calcium metaphosphate is a suitable source of phosphorus for most crops grown on acid soils but variable results have been obtained on alkaline or calcareous soils.

Basic slag, or Thomas slag, is a byproduct of the basic open-hearth method of making steel from pig iron. In the United States it is produced only in the area around Birmingham, Alabama, and its use is confined to a radius of several hundred miles around that city. However, it is a very popular phosphate fertilizer in Europe. Both in the United States and on the European continent its production is dependent entirely on the production of steel by the open-hearth process and supplies vary accordingly.

Basic slag in the United States runs around 3 per cent P_2O_5, whereas that in Europe is usually much higher, in a range of 14 to 18 per cent, because of the higher phosphate content of its iron ore. In addition to its phosphate content, basic slag has a neutralizing value of 60 to 80 per cent, which makes it particularly valuable on acid soils, for it will act like lime to neutralize soil acidity.

Basic slag is a good source of phosphorus on acid soils. It has not been used in the United States on alkaline soils, for, like other thermal phosphates, its application to such soils could not be recommended. In Europe, however, where the soil phosphate levels are frequently quite high because of centuries of intensive fertilization, basic slag is said by some to be a satisfactory source of phosphorus on neutral to slightly alkaline as well as on acid soils.

All of the thermal process phosphates must be finely ground to be effective sources of fertilizer phosphate. Although the fluoroapatite structure is broken by the heat treatment, the phosphate remains in a completely water-insoluble form. Therefore fine grinding is essential to ensure that a large total surface is exposed for contact by the plant's root system. The availability of these materials to plants is related directly to their specific surface, which is, of course, inversely related to the particle size of the material. These fertilizers should not be banded but applied broadcast and disked in for the same reason that fine grinding is necessary.

BACTERIAL PHOSPHATE FERTILIZATION

In the Soviet Union and several eastern European countries soils are inoculated with bacteria which apparently increase the plant availability of native and applied soil phosphorus. Several species of microorganisms are effective in this respect, but the one principally employed is *Bacillus megatherium* var. *phosphaticum.* Cultures of these organisms, which the Russians term *phosphobacterins,* are prepared commercially and distributed to the farmers. In 1958 it was estimated that about ten million hectares (about twenty-five million acres) were treated with phosphobacterin. Azotobacterin, discussed in Chapter 5, is also popular and is used in Russian agriculture.

The increase in available soil phosphorus results primarily from the decomposition of organic phosphorus compounds. The treatment is most effective on neutral to somewhat alkaline soils and on those high in organic matter. Some organisms, which produce considerable acidity as a result of their metabolic activity, will also dissolve some of the plant-unavailable soil mineral phosphates, such as hydroxyapatite. Some work has been reported in which the availability of rock phosphate and other water-insoluble phosphates inoculated with phosphobacterin was increased.

Yield increases from the use of phosphobacterin are reported to range from 0 to 70 per cent. The average, however, is about 10 per cent. Crops responding to this treatment are sugar beets, potatoes, cereals, vegetables, some grasses, and legumes.

The effectiveness of phosphobacterin has been investigated by workers at the United States Department of Agriculture, who showed that the culture readily decomposes glycerophosphates. It was also found that the yield of tomatoes grown in greenhouse tests was increased 7.5 per cent. There was no increase in the yield of wheat. However, neither phosphorus concentration nor total phosphate uptake was favorably influenced by phosphobacterin.

The use of phosphobacterin is undoubtedly of value in increasing crop yields in Russia, where use of chemical fertilizers is still rather limited. How popular it will remain when the supply of mineral fertilizers increases is a matter of conjecture. It is fairly certain that the inoculation of soil or fertilizers with phosphobacterin is not likely to be practiced on a large scale in the United States in the foreseeable future.

A summary of the properties and plant-nutrient contents of the most important phosphate fertilizers is given in Table 6-5.

BEHAVIOR OF PHOSPHATE FERTILIZERS IN SOILS

The reactions taking place in the soil and the properties of the various phosphate fertilizer materials combine to determine the effectiveness of any source of phosphorus under any set of soil and cropping conditions. To improve the effectiveness of phosphate fertilizers, different fertilizer particle sizes and different methods of placing the fertilizer in the soil with respect to the seed are employed. Granulation and placement are

TABLE 6–5. Composition of Phosphatic Fertilizer Materials

Material	Total Nitrogen (%)	Total Potassium (%)	Total Sulfur (%)	Total Calcium (%)	Total Magnesium (%)	Phosphorus Total (%)	Phosphorus Available* (% of total)
Ammonium phosphates							
21-53-0†	21	—	—	—	—	23	100
21-61-0	21	—	—	—	—	27	100
11-48-0	11	—	0–2	—	—	21	100
16-48-0	16	—	0–2	—	—	21	100
18-46-0	18	—	0–2	—	—	20	100
16-20-0	16	—	15	—	—	8.7	100
Ammoniated OSP	2–5	—	10–72	17–21	—	6.1–8.7	96–98
Ammoniated CSP	4–6	—	0–1	12–14	—	19–21	96–99
Ordinary super-phosphate	—	—	11–12	18–21	—	7–9.5	97–100
Conc. (triple) super-phosphate	—	—	0–1	12–14	—	19–23	96–99
Enriched super-phosphate	—	—	7–9	16–18	—	11–13	96–99
Dicalcium phosphate	—	—	—	29	—	23	98

Material							
Superphosphoric acid	–	–	–	–	–	34	100
Phosphoric acid	–	–	0–2	–	–	23	100
Potassium phosphate	29–45	–	–	–	–	18–22	100
Ammonium phosphate nitrate	–	30	–	–	–	4	100
Ammonium polyphosphate	–	15	–	–	–	25	–
Magnesium ammonium phosphate	–	8	–	–	14	17	–
Raw rock phosphate	–	–	–	33–36	–	11–17	14–65
Basic slag	–	–	0.2	32	3	3.5–8	62–94
Defluorinated phosphate rock	–	–	–	20	–	9	85
Phosphate rock-magnesium silicate glass	–	–	–	20	8.4	10	85
Rhenania phosphate	–	–	–	30	0.3	12	97
Calcium metaphosphate	–	–	–	19	–	27	99
Potassium metaphosphate	29–32	–	–	–	–	24–25	–

* By neutral 1.0N ammonium citrate procedure.
† Ammonium phosphate grades expressed as % N, % P_2O_5, % K_2O.

covered more thoroughly in Chapter 13, but the basic principles regarding the effect of these and other factors on the behavior of phosphate fertilizers are discussed here.

During the last few years much information has been gained on the behavior of fertilizer phosphates in different soils. It has been found that phosphates react with soil components to form specific and identifiable compounds, many of which are listed in Table 6-2. These compounds are referred to as *soil-fertilizer reaction products*. It is increasingly apparent that the nature of these materials determines the amount of phosphorus that will be available to the growing crop.

Monocalcium Phosphate.   When materials containing a high percentage of monocalcium phosphate (normal superphosphate or triple superphosphate) are applied to the soil, water vapor moves rapidly into each granule. The solution formed is saturated with monocalcium phosphate and dicalcium phosphate dihydrate, and a residue of dicalcium phosphate remains at the site of the original particle. The saturated solution that moves out of the granule is concentrated with respect to phosphorus and calcium and is extremely acid (pH 1.8).

This concentrated acid solution moves out in a front, reacting as it goes with different soil constituents. In acid soils the reaction is largely with compounds of iron, aluminum, and manganese, which are dissolved in the solution. In time the phosphates of these three ions are precipitated. In calcareous soils the moving phosphate front is precipitated as

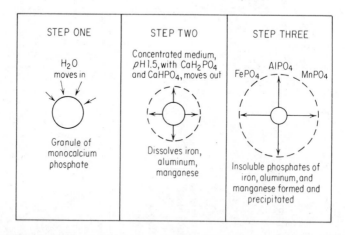

STEP ONE	STEP TWO	STEP THREE

Figure 6–10.   Diagrammatic representation of the reactions of a granule of CSP with soil. (As water enters a granule containing water-soluble monocalcium phosphate, a solution is formed which is saturated with monocalcium phosphate and dicalcium phosphate dihydrate. This solution moves out in a front to react with different soil constituents. In acid soils the reaction is largely with compounds of iron, aluminum, and manganese.)

dicalcium phosphate on the surface of particles of calcium carbonate. Possible further reversion to hydroxyapatite may occur. This phenomenon is illustrated diagrammatically in Figure 6-10. Phosphate ions also may react directly with calcium ions in the soil solution to form dicalcium phosphate and, under some circumstances, small amounts of hydroxyapatite.

Ammonium Phosphate. Ammonium phosphates move out from the granule as the monocalcium phosphates do. However, no residue of dicalcium phosphate remains at the granule site because of the absence of calcium in the fertilizer. Undoubtedly some calcium phosphate will be formed in the zone of the moving solution front, for one of the reaction products, especially in soils containing a large amount of exchangeable calcium, is dicalcium phosphate.

The *p*H of a saturated solution of DAP is about 9.0, whereas that of MAP is about 4.0. This difference in the *p*H of the saturated solution of the two salts affects the type of reaction product formed in different soils by the two compounds. In calcareous soils a greater proportion of insoluble reaction products will be formed than would be with MAP.

Ammonium Polyphosphate. Ammonium polyphosphate is a relatively new addition to the list of phosphatic fertilizers that are available through commercial channels. Because of this, its reactions and behavior in the soil have not been studied as intensively as some of the other phosphatic fertilizers. However, sufficient work has been done to establish that polyphosphates are as effective as orthophosphates as sources of P for crops. Plants can absorb and utilize the polyphosphates (which are primarily pyrophosphates with the formula $(NH_4)_3HP_2O_7 \cdot H_2O$) directly. Hydrolysis to the ortho form takes place rather rapidly in most soils so a major portion of the phosphate is probably absorbed as the ortho ion.

Because polyphosphates have the ability to form complex ions with some metals, it has been suggested that they may be effective in mobilizing Zn in soils in which deficiencies of this element have been induced by high *p*H or high phosphate levels. However, evidence from field experiments has been inconclusive on this point.

Dicalcium Phosphate. When a fertilizer granule contains a high proportion of water-insoluble material, such as dicalcium phosphate, no soil solution is formed, and dispersion of the reaction products into the soil surrounding the original fertilizer particle does not occur. Plant availability is therefore reduced, particularly in calcareous soils.

Effect of Adding Nitrogen and Phosphorus. When ammonium sulfate is mixed with a water-soluble phosphatic fertilizer and applied in a band there is great proliferation of roots in the band and a greatly increased uptake of phosphorus by the plant. That this effect does not occur when the nitrogen is in the nitrate form is illustrated by the curves

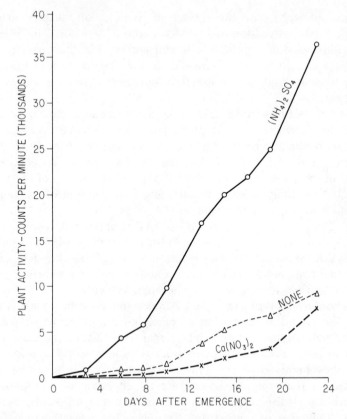

Figure 6–11. The effect of source of nitrogen on the plant uptake of phosphorus from a fertilizer band containing nitrogen of the source indicated and monocalcium phosphate. (Courtesy of Professor A. J. Ohlrogge, Purdue University.)

in Figure 6-11. The cause of this phenomenon is not explained. It could result from the decreased activity of calcium in the fertilizer zone, brought about by the presence of the SO_4^{2-} ion which could in turn produce precipitation of gypsum. It could possibly result also from the acidifying effect that accompanies the bacterial oxidation of the ammonium sulfate. This acidity would enhance the formation of the more soluble $H_2PO_4^-$ ion and allow greater uptake of phosphorus by the plant.

Effect of Size of Granule. The size and type of fertilizer granule, and the manner in which it is placed in the soil, have a marked influence on the relative effectiveness of phosphatic fertilizers.

As a general rule, best results are achieved with water-insoluble or slightly soluble phosphates on both acid and calcareous soils when they are applied in powdered form or in very fine granules (less than 35 mesh) and mixed thoroughly with the soil of the root zone. This is more important, however, on calcareous than on acid soils.

Water-soluble phosphates give best results on acid soils when they are granulated and applied in a band. Granule size is generally satisfactory in the range of 12 to 50 mesh, but the present trend is toward 8 to 12 mesh. Granular forms of water-soluble materials give good results on calcareous soils, although experimental evidence indicates that the best results (as in water-insoluble materials) will probably be obtained from pulverized materials thoroughly mixed with the soil.

Granular materials are preferred to the powdered or pulverized forms because of their greater ease of handling and spreading. Nitric phosphates are generally manufactured in a granular form and are coated and shipped in special moistureproof bags because of their hygroscopicity. As a result, granular nitric phosphates with their low water solubility (less than 50 per cent) are not suitable for use on calcareous soils, whereas granulated materials of high water solubility, such as monoammonium phosphate and triple superphosphate, are quite satisfactory.

Soil Moisture. Moisture content of the soil has a decided effect on the effectiveness and rate of availability of applied phosphorus in various forms. Experimental work has shown that when the soil water content is at field capacity, 50 to 80 per cent of the water-soluble phosphorus can be expected to move out of the fertilizer granule within a twenty-four-hour period. Even in soils with only 2 to 4 per cent moisture, 20 to 50 per cent of the water-soluble phosphorus will move out of the granule within the same time.

Under wet conditions response to granular phosphates of high water solubility is superior to that from powdered materials. Under dry conditions, however, powdered materials are likely to give better results.

Granule Distribution. Distribution of fertilizer granules in the soil, granule diameter, and phosphorus content all affect crop response to applied phosphates. Poor distribution may be caused by inadequate mixing with the soil or by relatively light applications, so that the few granules applied are widely spaced in the soil. Poor distribution, particularly unfavorable in materials of low water solubility, is explained simply by the decreased probability of a plant root reaching a fertilizer granule.

The effects of poor distribution may be partly overcome by the use of highly water-soluble granular materials because of their ability to disperse into soil zones surrounding the granule and to react quickly with the soil constituents.

Rate of Application. At low rates of applied phosphorus water solubility may be much more important than at high rates. This is due to the same factors that influence crop response under conditions of poor distribution and suggests that when optimum application rates cannot be expected for some time it is important, in order to get full benefit from the limited amounts of fertilizer applied, that materials of high water solubility be used. The effect of the degree of water-soluble phosphorus on corn yields in Iowa is shown in Figure 6-12.

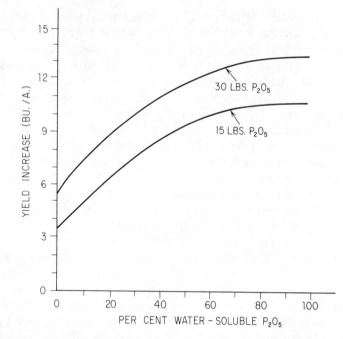

Figure 6–12. The effect of rate and water-solubility of applied phosphate fertilizer on the yield increase of corn. [Webb et al., *SSSA Proc.*, **22**:533 (1958).]

COMPARATIVE FERTILIZER VALUE OF VARIOUS PHOSPHATE MATERIALS

The suitability of various materials as sources of fertilizer phosphorus has been evaluated in thousands of greenhouse and field tests throughout North America and Europe. It is not the purpose of this book to include a discussion of the results of even a fraction of these experiments, for references which themselves include extensive bibliographies of the literature on this subject are cited at the end of this chapter.

Since OSP was almost the sole source of fertilizer phosphorus for many decades, it was only natural that it serve as a yardstick for the evaluation of more recently developed materials. OSP, however, contains sulfur, and many of the newer phosphate materials do not. Under conditions of sulfur deficiency OSP will give superior results, but this is not a valid test of the materials as carriers of *phosphorus*. For that reason CSP has become the preferred standard. If sulfur deficiency is suspected in the test area, it can be uniformly corrected by the addition of a suitable sulfur carrier.

On the basis of published experimental data, the following general conclusions may be drawn concerning the effectiveness of various phosphates under different circumstances:

1. Both water-soluble and citrate-soluble phosphates may be said to be *available* to plants. However, there may be a considerable difference in crop response under different circumstances.

2. For maximum yields short-season fast-growing crops and those with restricted root systems generally require a fertilizer containing a high proportion of water-soluble phosphorus. Such crops frequently give only limited response to applications of citrate-soluble materials.

3. A high degree of water solubility (greater than 60%) is less important on long-season crops and perennials with extensive root systems such as permanent pastures and meadows.

4. A high degree of water solubility may be desirable for early growth and stand establishment in crops such as small grains to be used for grazing and corn (maize).

5. When the amount of phosphate to be applied is limited, the greatest crop response will almost always be obtained when a large proportion of the fertilizer phosphorus is in water-soluble form and when the fertilizer is applied in a band slightly to the side and below the seed or transplant. This is particularly true on soils that are depleted or naturally low in phosphorus.

6. On acid to neutral soils granular fertilizers with a high degree of water solubility are more effective than powdered fertilizers containing the same proportion of water-soluble phosphorus when the fertilizer is to be mixed with the soil. Within limits, the larger the fertilizer granule, the greater its effectiveness under these conditions.

7. On acid to neutral soils band application of powdered fertilizers with a high degree of water solubility will give better results than mixing the fertilizer with the soil.

8. On calcareous soils granular forms of highly water-soluble phosphates will generally give good results, although experimental evidence indicates that better results are sometimes obtained with pulverized materials thoroughly mixed with the soil. Granular nitric phosphates of low water solubility (less than 50%) are not suitable for use on calcareous soils.

9. Best results may be achieved with materials of low water solubility when they are applied in powdered form or in very small granules and mixed thoroughly with the soil. This is somewhat more important, however, on calcareous than on acid soils.

10. Monoammonium phosphate will generally give better results than diammonium phosphate on calcareous soils, although both are water soluble.

11. With phosphates of low water solubility, effectiveness decreases with an increase in particle size.

12. The thermal-process phosphates, when finely ground, are satisfactory sources of phosphorus for most crops on acid soils, but they have failed generally to give favorable results on neutral and alkaline soils.

13. Maximum response will not be obtained from an applied phosphatic fertilizer, whether water soluble or water insoluble, unless adequate quantities of the other plant nutrients, including the secondary and micronutrient elements, are present. Experimental evidence indicates that phosphorus utilization by plants may be improved by the presence of sulfate and ammonium ions in the fertilizer material.

In general, it may be concluded that when the granule size and the degree of water solubility of the phosphorus in fertilizers are comparable there is likely to be little if any difference in the effectiveness of materials based on sulfuric, phosphoric, or nitric acid acidulation. However, highly water-soluble materials, such as ammonium phosphate and the superphosphates, will give satisfactory results under all circumstances in which crop response to an applied phosphate is possible. Although water-insoluble phosphates and those of low water solubility may give equally good results under some conditions, they are not so suitable as the water-soluble materials.

Summary

1. Phosphorus occurs in soils in both inorganic and organic forms. The concentration of the inorganic forms ($H_2PO_4^-$, HPO_4^{2-}) in the soil solution is the most important single factor governing the availability of this element to plants.

2. The concentration of phosphate ion in the soil solution is influenced by the rate and extent to which this element is immobilized by biological factors and by reaction with the mineral fraction of soils. Soils high in clays (especially those of the 1 : 1 type and the hydrous oxides of iron and aluminum) react with orthophosphates to fix them in a form that is largely unavailable to growing plants. The factors concerned with phosphorus turnover and its reaction with the mineral fraction of soils were covered.

3. Soils that are calcareous also decrease the availability of phosphates. In such soils the phosphate ions are adsorbed to the surface of finely divided calcium carbonate and subsequently converted to insoluble apatites or they are precipitated as insoluble calcium phosphates directly from the soil solution.

4. The availability of added water-soluble fertilizer phosphates can be considerably extended by placing them in a band in the soil. Similar results can be obtained by granulating the phosphate materials.

5. The terminology peculiar to phosphate fertilizers was discussed and the terms *water-soluble, citrate-soluble, available,* and *total phosphorus* were defined. The adequacy of the current method for assaying the availability of phosphate fertilizers was covered.

6. Phosphate fertilizers are classed generally on the basis of their manufacture as heat- or acid-treated phosphates. Heat-treated phosphates are either calcined or fused. Calcining is heat treatment below the melting point of the furnace charge, and fusing results from heating the charge above its melting point. The contained phosphorus is water insoluble. Acid-treated phosphates are those in which the phosphate rock is treated with a strong acid such as sulfuric, phosphoric, or nitric. The phosphorus in unammoniated acid-treated phosphates is largely water soluble.

7. The phosphobacterins, so widely publicized by Russian scientists, were discussed. Although these bacterial cultures may be required to increase native soil phosphates under conditions of Russian agriculture, they do not appear to be helpful in areas in which large amounts of inorganic phosphate fertilizers are used.

8. The behavior and properties of the various phosphate fertilizers in the soil were discussed. The reaction of soluble phosphate fertilizers with various soil components gives rise to what is termed *fertilizer-soil reaction products* and it is the solubility of these compounds that largely governs plant availability of added phosphate fertilizers.

9. The suitability of the various materials as sources of fertilizer phosphorus was covered. As a general rule, those materials that have a high percentage of the contained phosphate in the water-soluble form are more generally acceptable than are those with none or only a small amount. There are, however, certain crops and certain soil conditions with and on which the less water-soluble forms perform as well as those that are water soluble. As with nitrogen, the cost per unit of contained phosphorus should loom large as one of the determining factors in the selection of a phosphate fertilizer, but this decision should be tempered by the response of the crop to water-soluble forms.

Questions

1. What is the original source of soil phosphorus?
2. Is the soil phosphorus in organic combinations available to plants?
3. What are the factors affecting the retention of phosphorus in soils?
4. How is phosphorus availability influenced by soil *p*H?
5. What are soil phosphate reaction products?
6. What are probably the two most important factors that influence the uptake of phosphorus by plants?

7. What is phosphate fixation? Why is it important agriculturally? Is fixed phosphorus totally lost to plants?

8. What are the various mechanisms of phosphate retention in acid mineral soils?

9. What soil properties influence the retention or fixation of added fertilizer phosphorus?

10. What can be done to reduce the amount of fixation of fertilizer phosphorus?

11. What is the original source of most fertilizer phosphorus?

12. A fertilizer contains 46 per cent phosphorus pentoxide. To what per cent of phosphorus does this correspond?

13. Derive the conversion factor: $\%P = \dfrac{\%P_2O_5}{2.29}$.

14. What is meant by ammoniating superphosphates? What is the effect on the water- and citrate-soluble contents of ammoniating OSP and CSP? Why the difference?

15. What acids are commonly used to acidify phosphate rock? Why, specifically, does acid treatment of phosphate rock render the phosphorus more plant available?

16. Describe the soil conditions under which you might expect an appreciable downward movement of phosphorus through the soil profile.

17. What is the significance of a wide $R_2O_3 : P_2O_5$ ratio? A narrow ratio? Is this important to the grower? Why?

18. Under what soil conditions would the band placement of phosphorus result in its greatest utilization by the plant? If there were no such thing as phosphorus fixation, what method of fertilizer placement would probably result in the greatest utilization of this element by plants? Why?

19. Phosphates held in organic combination are generally considered to be of little value to plants during cold weather. Why?

20. A soil was reported to contain 20 ppm. of available phosphorus. To how many pounds per acre of ordinary superphosphate (8.5% P) does this correspond?

21. Chemically speaking, on what is the stability of rock phosphate based?

22. Under what types of soil and cropping conditions might the use of rock phosphate give satisfactory results? Explain.

23. With what type of crop would the use of a high-water-soluble phosphate be particularly recommended?

24. What types of phosphate fertilizer are not recommended for use on alkaline and calcareous soils?

25. On what basis should phosphate fertilizers be purchased—total or available phosphorus?

26. What advantages are offered by the high-analysis phosphates such as DAP, MAP, and CSP? What disadvantages?

Selected References

Adriano, D. C., G. M. Paulsen, and L. S. Murphy, "Phosphorus-iron and phosphorus-zinc relationships in corn (*Zea mays L.*) seedlings as affected by mineral nutrition." *Agron. J.* **63**:36 (1971).

Allred, S. E., and A. J. Ohlrogge, "Principles of nutrient uptake from fertilizer bands. VI. Germination and emergence of corn as affected by ammonia and ammonium phosphate." *Agron. J.* **56**:309 (1964).

Armiger, W. H., and M. Fried, "The plant availability of various sources of phosphate rock." *SSSA Proc.*, **21**:183 (1957).

Barrow, N. J., "Phosphorus in soil organic matter." *Soils Fertilizers*, **24**:169 (1961).

Beaton, J. D., "Fertilizers and their use in the decade ahead." *Proc. 20th Ann. Meet., Agric. Res. Inst.*, St. Louis, Mo. (1971).

Beaton, J. D., and N. A. Gough, "The influence of soil moisture regime and phosphorus source on the response of alfalfa to phosphorus." *SSSA Proc.*, **26**:265 (1962).

Beaton, J. D., and D. W. L. Read, "Phosphorus uptake from a calcareous Saskatchewan soil treated with mono-ammonium phosphate and its reaction products." *Soil Sci.*, **94**:404 (1962).

Beaton, J. D., and D. W. L. Read, "Effect of temperature and moisture on phosphate uptake from a calcareous Saskatchewan soil treated with several pelleted sources of phosphorus." *SSSA Proc.*, **27**:61 (1963).

Bixby, D. W., D. L. Rucker, and S. L. Tisdale, "Phosphatic fertilizers, properties and processes." *Tech. Bull. No. 8* (Rev.) Washington, D.C.: The Sulphur Institute, 1968.

Black, C. A., *Soil-Plant Relationships*, 2nd ed. New York: Wiley, 1968.

Bromfield, S. M., "Some factors affecting the solubility of phosphates during the microbial decomposition of plant material." *Australian J. Agr. Res.*, **11**:304 (1960).

Brown, A. L., and B. A. Krantz, "Source and placement of zinc and phosphorus for corn (*Zea mays L.*)." *SSSA Proc.*, **30**:86 (1966).

Bouldin, D. R., J. R. Lehr, and E. C. Sample, "The effect of associated salts on transformations of monocalcium phosphate monohydrate at the site of application." *SSSA Proc.*, **24**:464 (1960).

Bouldin, D. R., and E. C. Sample, "Laboratory and greenhouse studies with monocalcium, monoammonium, and diammonium phosphates." *SSSA Proc.*, **23**:338 (1959).

Bouma, D., "The effect of ammonium sulfate usage on the availability of soil phosphorus to citrus." *Australian J. Agr. Res.*, **11**:292 (1960).

Burns, G. R., D. R. Bouldin, C. A. Black, and W. L. Hill, "Estimation of particle size effects of water-soluble phosphate fertilizer in various soils." *SSSA Proc.*, **27**:556 (1963).

Caldwell, A. G., and C. A. Black, "Inositol hexaphosphate: II. Synthesis by soil microorganisms." *SSSA Proc.*, **22**:293 (1958).

Caldwell, A. G., and C. A. Black, "Inositol hexaphosphate: III. Content in soils." *SSSA Proc.*, **22**:296 (1958).

Calvert, D. V., H. F. Massey, and W. A. Seay, "The effect of exchangeable calcium on the retention of phosphorus by clay fractions of soils of the Memphis catena." *SSSA Proc.*, **24**:333 (1960).

Cole, C. V., and S. R. Olsen, "Phosphorus solubility in calcareous soils: II. Effects of exchangeable phosphorus and soil texture on phosphorus solubility." *SSSA Proc.*, **23**:119 (1959).

Coleman, N. T., J. T. Thorup, and W. A. Jackson, "Phosphate-sorption reactions that involve exchangeable Al." *Soil Sci.*, **90**:1 (1960).

Cook, R. L., K. Lawton, L. S. Robertson, et al., "Phosphorus solubility, particle size and placement as related to the uptake of fertilizer phosphorus and crop yields." *Com. Fertilizer*, **94**:41 (1957).

Cooper, R., "Bacterial fertilizers in the Soviet Union." *Soils Fertilizers*, **22**:327 (1959).

De Datta, S. K., R. L. Fox, and G. D. Sherman, "Availability of fertilizer phosphorus in Latosols of Hawaii." *Agron. J.* **55**:311 (1963).

Duncan, W. G., and A. J. Ohlrogge, "Principles of nutrient uptake from fertilizer bands: II. Root development in the band." *Agron. J.* **50**:605 (1958).

Duncan, W. G., and A. J. Ohlrogge, "Principles of nutrient uptake from fertilizer bands: III. Band volume, concentration, and nutrient composition." *Agron. J.*, **51**:103 (1959).

Englestad, O. P., and S. E. Allen, "Ammonium pyrophosphate and ammonium orthophosphate as phosphorus sources: Effects of soil temperature, placement, and incubation." *SSSA Proc.*, **35**:1002 (1971).

Ensminger, L. E., "Response of crops to various phosphate fertilizers." *Alabama Agri. Exp. Sta. Bull. No. 270* (1950).

Fox, R. L., and E. J. Kamprath, "Adsorption and leaching of P in acid organic soils and high organic matter sand." *SSSA Proc.*, **35**:154 (1971).

Franklin, W. T., and H. M. Reisenauer, "Chemical characteristics of soils as related to phosphorus fixation and availability." *Soil Sci.*, **90**:192 (1960).

Ganiron, R. B., D. C. Adriano, G. M. Paulsen, and L. S. Murphy, "Effect of phosphorus carriers and zinc sources on phosphorus-zinc interaction in corn." *SSSA Proc.*, **33**:306 (1969).

Gerretsen, F. C., "The influence of microorganisms on the phosphate intake by the plant." *Plant Soil*, **1**:51 (1949).

Gilliam, J. W., "Hydrolysis and uptake of pyrophosphate by plant roots." *SSSA Proc.*, **4**:83 (1970).

Gough, N. A., and J. D. Beaton, "Influence of phosphorus source and soil moisture on the solubility of phosphorus." *J. Sci. Food Agr.*, **14**:224 (1963).

Grunes, D. L., "Effect of nitrogen on the availability of soil and fertilizer phosphorus to plants." *Advan. Agron.*, **11**:369 (1959).

Hagin, J., and J. Berkovits, "Efficiency of phosphatic fertilizers of varying water solubility." *Can. J. Soil Sci.*, **41**:68 (1961).

Hashimoto, I., and J. R. Lehr, "Mobility of polyphosphates in soil." *SSSA Proc.*, **37**:36 (1973).

Hemwall, J. B., "The fixation of phosphorus in soils." *Advan. Agron.*, **9**:95 (1957).

Hignett, P. T., and J. A. Brabson, "Phosphate solubility, evaluation of water insoluble phosphorus in fertilizers by extraction with alkaline ammonium citrate solutions." *J Agr. Food Chem.*, **9**:272 (1961).

Hinman, W. C., J. D. Beaton, and D. W. L. Read, "Some effects of moisture and temperature on transformation of monocalcium phosphate in soil." *Can. J. Soil Sci.*, **42**:229 (1962).

Humphreys, F. R., and W. L. Pritchett, "Phosphorus adsorption and movement in some sandy forest soils." *SSSA Proc.*, **35**:495 (1971).

Jacob, K. D., Ed., "Fertilizer Technology and Resources in the United States." *Agronomy*, Vol. III. A series of monographs. New York: Academic, 1953.

Kamprath, E. J., "Residual effect of large applications of phosphorus on high phosphorus fixing soils." *Agron. J.*, **59**:25 (1967).

Larsen, J. E., R. Langston, and G. F. Warren, "Studies on the leaching of applied labeled phosphorus in organic soils." *SSSA Proc.*, **22**:558 (1958).

Larsen, J. E., G. F. Warren, and R. Langston, "Effect of iron, aluminum, and humic acid on phosphorus fixation by organic soils." *SSSA Proc.*, **23**:438 (1959).

Lathwell, D. J., J. T. Cope, Jr., and J. R. Webb, "Liquid fertilizers as sources of phosphorus for field crops." *Agron. J.*, **52**:251 (1960).

Lawton, K., C. Apostolakis, and R. L. Cook, "Influence of particle size, water solubility and placement of fertilizers on the nutrient value of phosphorus in mixed fertilizers." *Soil Sci.*, **82**:465 (1956).

Lawton, K., and J. A. Vomocil, "The dissolution and migration of P from granular superphosphate in some Michigan soils." *SSSA Proc.*, **18**:26 (1954).

Lindsay, W. L., A. W. Frazier, and H. F. Stephenson, "Identification of reaction products from phosphate fertilizers in soils." *SSSA Proc.*, **26**:446 (1962).

Lindsay, W. L., and E. C. Moreno, "Phosphate phase equilibria in soils." *SSSA Proc.*, **4**:177 (1960).

Lindsay, W. L., and H. F. Stephenson, "Nature of the reactions of monocalcium phosphate monohydrate in soils: I. The solution that reacts with the soil. *SSSA Proc.*, **23**:12 (1959).

Lindasay, W. L., and H. F. Stephenson, "Nature of the reactions of monocalcium phosphate monohydrate in soils: II. Dissolution and precipitation reactions involving iron, aluminum, manganese, and calcium." *SSSA Proc.*, **23**:18 (1959).

Lindsay, W. L., and H. F. Stephenson, "Nature of the reactions of monocalcium phosphate monohydrate in soils: IV. Repeated reactions with metastable triple-point solution." *SSSA Proc.*, **23**:440 (1959).

Lindsay, W. L., and A. W. Taylor, "Phosphate reaction products in soil and their availability to plants. *Trans. 7th Intern. Congr. Soil Sci.*, **3**:580 (1960).

Martin, W. E., J. Vlamis, and J. Quick, "Effect of ammoniation on availability of phosphorus in superphosphates as indicated by plant response." *Soil Sci.*, **75**:41 (1959).

Mattingley, G. E. G., "The agricultural value of some water and citrate soluble fertilizers: An account of recent work at Rothamsted and elsewhere." *Proc. Fertiliser Soc. (London)*, **75**:57 (1963).

McLean, E. O., and T. J. Logan, "Sources of phosphorus for plants grown in soils with differing phosphorus fixation tendencies." *SSSA Proc.*, **34**:907 (1970).

Miller, M. H., and A. J. Ohlrogge, "Principles of nutrient uptake from fertilizer bands: I. Effect of placement of nitrogen fertilizer on the uptake of band-placed phosphorus at different soil phosphorus levels." *Agron. J.,* **50:**95 (1958).

Miller, M. H., and V. N. Vij, "Some chemical and morphological effects of ammonium sulfate in a fertilizer phosphorus band for sugar beets." *Can. J. Soil Sci.,* **42:**87 (1962).

Miner, G., and E. Kamprath, "Reactions and availability of banded polyphosphate in field studies." *SSSA Proc.,* **35:**927 (1971).

Mishustin, E. N., and A. N. Naumova, "Bacterial fertilizers, their effectiveness and mechanism of action." *Mikrobiol.,* **31:**543 (1962). (*Soils Fertilizers,* **25:**382 (1962).)

Neller, J. R., "Effect on plant growth of particle size and degree of solubility of phosphorus labeled in 12-12-12 fertilizer." *Soil Sci.,* **94:**413 (1962).

Norland, M. A., R. W. Starostka, and W. L. Hill, "Crop response to phosphorus fertilizers as influenced by level of phosphorus solubility and by time of placement prior to planting." *SSSA Proc.,* **22:**529 (1958).

Nye, P. H., and W. N. M. Foster, "A study of the mechanism of soil phosphate uptake in relation to plant species." *Plant Soil,* **9:**338 (1958).

Olsen, S. R., F. S. Watanabe, and R. E. Danielson, "Phosphorus absorption by corn roots as affected by moisture and phosphorus concentration." *SSSA Proc.,* **25:**289 (1961).

Olson, R. A., A. C. Drier, and G. W. Lowrey, et al., "Availability of phosphoric acid to small grains and subsequent clover in relation to: I. Nature of soil and method of placement." *Agron. J.,* **48:**106 (1956).

Olson, R. A., T. J. Army, J. J. Hanway, and V. J. Kilmer, Eds., *Fertilizer Technology & Use,* 2nd ed. Madison, Wisconsin: SSSA, 1971.

Ozanne, P. G., D. J. Kirton, and T. C. Shaw, "The loss of phosphorus from sandy soils." *Australian J. Agr. Res.,* **12:**409 (1961).

Philen, O. D., Jr., and J. R. Lehr, "Reactions of ammonium polyphosphates with soil minerals." *SSSA Proc.,* **31:**196 (1967).

Phillips, A. B., R. D. Young, F. G. Heil, and M. M. Norton, "Fertilizer technology, high analysis superphosphate by the reaction of phosphate rock with superphosphoric acid." *J. Agr. Food Chem.,* **8:**310 (1960).

Pierre, W. H., and A. G. Norman, Eds., "Soil and Fertilizer Phosphorus in Crop Nutrition." *Agronomy,* Vol. IV: A series of monographs. New York: Academic, 1953.

Power, J. F., D. L. Grunes, G. A. Reichman, and W. O. Willis, "Soil temperature effects on phosphorus availability." *Agron. J.* **56:**545 (1964).

Power, J. F., D. L. Grunes, W. O. Willis, and G. A. Reichman, "Soil temperature and phosphorus effects upon barley growth." *Agron. J.,* **55:**389 (1963).

Power, J. F., W. O. Willis, D. L. Grunes, and G. A. Reichman, "Effect of soil temperature, phosphorus, and plant age on growth analysis of barley." *Agron. J.* **59:**231 (1967).

Ragland, J. L., and W. A. Seay, "The effects of exchangeable calcium on the retention and fixation of phosphorus by clay fractions of soil." *SSSA Proc.,* **21:**261 (1957).

Read, D. W. L., and R. Ashford, "Effect of varying levels of soil and fertilizer phosphorus and soil temperature on the growth and nutrient content of bromegrass and Reed canarygrass." *Agron. J.* **60:**680 (1968).

Robinson, R. R., V. G. Sprague, and C. F. Gross, "The relation of temperature and phosphate placement to growth of clover." *SSSA Proc.*, **23**:225 (1959).

Rogers, H. T., "Crop response to nitrophosphate fertilizers." *Agron. J.*, **43**:468 (1951).

Russell, E. W., *Soil Conditions and Plant Growth*, 9th ed. London: Longmans, Green, 1961.

Sauchelli, V., Ed., "Chemistry and Technology of Fertilizers." *Amer. Chem. Soc. Monograph No. 148.* New York: Reinhold, 1960.

Sauchelli, V., *Manual on Fertilizer Manufacture*, 3rd ed. Caldwell, New Jersey: Industry Publications, 1963.

Shapiro, R. E., W. H. Armiger, and M. Fried, "The effect of soil water movement vs. phosphate diffusion on growth and phosphorus content of corn and soybeans." *SSSA Proc.*, **24**:61 (1960).

Sheard, R. W., G. J. Bradshaw, and D. L. Massey, "Phosphorus placement for the establishment of alfalfa and bromegrass." *Agron. J.* **63**:922 (1971).

Simpson, K., "Factors influencing uptake of phosphorus by crops in Southeast Scotland." *Soil Sci.*, **92**:1 (1961).

Smith, J. H., and F. E. Allison, "Phosphobacterin as a soil inoculant. Laboratory, greenhouse, and field evaluation." *USDA Tech. Bull. 1263* (1962).

Smith, J. H., F. E. Allison, and D. A. Soulides, "Evaluation of phosphobacterin as a soil inoculant." *SSSA Proc.*, **25**:109 (1961).

Stanberry, C. O., W. H. Fuller, and N. R. Crawford, "Comparison of phosphate sources for alfalfa on a calcareous soil." *SSSA Proc.*, **24**:304 (1960).

Stanford, G., and D. R. Bouldin, "Biological and chemical availability of phosphate-soil reaction products." *Trans. 7th Intern. Congr. Soil Sci.*, **2**:388 (1960).

Striplin, M. M., Jr., D. McKnight, and G. H. Megar, "Fertilizer materials, phosphoric acid of high concentration." *J. Agr. Food Chem.*, **6**:298 (1958).

Swaby, R. J., and J. Sherber, "Phosphate dissolving microorganisms in the rhizosphere of legumes." *Proc. Univ. Nottingham 5th Easter School Agr. Sci. 289* (1958).

Taylor, A. W., W. L. Lindsay, and E. O. Huffman, "Potassium and ammonium taranakites, amorphous aluminum phosphate, and variscite as sources of phosphate for plants." *SSSA Proc.*, **27**:148 (1963).

Terman, G. L., D. R. Bouldin, and J. R. Lehr, "Calcium phosphate fertilizers: I. Availability to plants and solubility in soils varying in pH. *SSSA Proc.*, **22**:25 (1958).

Terman, G. L., D. R. Bouldin, and J. R. Webb, "Phosphorus availability, crop response to phosphorus in water-soluble phosphates varying in citrate solubility and granule size." *J. Agr. Food Chem.*, **9**:166 (1961).

Terman, G. L., J. D. DeMent, L. G. Clements, and J. A. Lutz, Jr., "Crop response to ammoniated superphosphates and DCP as affected by granule size, water solubility, and time of reaction with the soil." *J. Agr. Food Chem.*, **8**:13 (1960).

Terman, G. L., J. D. DeMent, and O. P. Engelstad, "Crop response to fertilizers varying in solubility of the phosphorus as affected by rate, placement and seasonal environment." *Agron. J.*, **53**:221 (1961).

Thien, S. J., and W. W. McFee, "Influence of nitrogen on phosphorus absorption and translocation in *Zea mays.*" *SSSA Proc.*, **34**:87 (1970).

Tisdale, S. L., and D. L. Rucker, "Crop response to various phosphates." *Tech. Bull. No. 9.* Washington, D.C.: The Sulphur Institute, 1964.

Tisdale, S. L., and E. Winters, "Crop response to calcium metaphosphate on alkaline soils." *Agron. J.* **45**:228 (1953).

Van Burg, P. F. J., "The agricultural evaluation of nitrophosphates with particular reference to direct and cumulative phosphate effects and to interaction between water solubility and granule size." *Proc. Fertiliser Soc. (London),* **75**:7 (1963).

Van Wazer, J. R., Ed., *Phosphorus and Its Compounds.* Vol. II. New York: Interscience, 1961.

Waggaman, W. A., *Phosphoric Acid, Phosphates, and Phosphatic Fertilizers,* 2nd ed. New York: Reinhold, 1952.

Walsh, L. M., and J. D. Beaton, Eds., *Soil Testing and Plant Analysis* (rev. ed.). Madison, Wisc.: Soil Sci. Soc. Amer., Inc., 1973.

Watanabe, F. S., S. R. Olsen, and R. E. Danielson, "Phosphorus availability as related to soil moisture." *Trans. 7th Intern. Congr. Soil Sci.,* **3**:450 (1960).

Webb, J. R. K. Eik, and J. T. Pesek, "An evaluation of phosphate fertilizers applied broadcast on calcareous soils for corn." *SSSA Proc.,* **25**:232 (1961).

Webb, J. R., and J. T. Pesek, "An evaluation of phosphorus fertilizers varying in water solubility: I. Hill applications for corn." *SSSA Proc.,* **22**:533 (1958).

Webb, J. R., and J. T. Pesek, "An evaluation of phosphorus fertilizers varying in water solubility: II. Broadcast applications for corn." *SSSA Proc.,* **23**:381 (1959).

Webb, J. R., J. T. Pesek, and K. Eik, "An evaluation of phosphorus fertilizers varying in water solubility: III. Oat fertilization. *SSSA Proc.,* **25**:22 (1961).

Welch, L. F., D. L. Mulvaney, L. V. Boone, G. E. McKibben, and J. W. Pendleton, "Relative efficiency of broadcast versus banded phosphorus for corn." *Agron. J.,* **58**:283 (1966).

Williams, C. H., and A. Steinbergs, "Sulphur and phosphorus in some Eastern Australian soils." *Australian J. Agr. Res.,* **9**:483 (1958).

Williams, C. H., E. G. Williams, and N. M. Scott, "Carbon, nitrogen, sulphur, and phosphorus in some Scottish soils." *J. Soil Sci.,* **11**:334 (1960).

Wright, B., J. D. Lancaster, and J. L. Anthony, "Availability of phosphorus in ammoniated ordinary superphosphate." *Mississippi Agr. Exp. Sta. Tech. Bull. 52* (1963).

7. SOIL AND FERTILIZER POTASSIUM, MAGNESIUM, CALCIUM, AND SODIUM

Potassium, calcium, magnesium, and sodium play an important role in soil-plant relations. Not only are these elements essential to the complex biochemistry of plant growth, but their presence in the soil in adequate amounts and in suitable proportions to one another and to the other exchangeable cations such as aluminum, hydrogen, and NH_4^+ is necessary if the soil is to be a suitable medium for plant-root development.

The principles of cation exchange and the importance of base saturation were covered in Chapter 4. The reactions of these cations in soils and the factors influencing their availability to plants are considered in this chapter. Ways of applying these elements in fertilizers are also covered.

POTASSIUM

Potassium is absorbed by plants in larger amounts than any other mineral element except nitrogen. Large deposits in the form of salts of chlorides and sulfates are found in several areas of the world. In some locations these salts occur in large deposits several hundred and sometimes several thousand feet below the earth's surface. In other locations they are found in the brines of dying lakes and seas.

Potassium Content of Soils. Unlike phosphorus, potassium is present in relatively large quantities in most soils. The phosphorus content of the earth's crust is only 0.11 per cent, whereas that of potassium is 2.40 per cent. Although these figures include all of the earth's crust (of which the soil is only a part), they do emphasize the relative abundance of the two elements.

The potassium content of soils is variable and may range from only a few hundred pounds per acre furrow slice in coarse-textured soils formed from sandstone or quartzite to 50,000 lb. or more in fine-textured soils formed from rocks high in the potassium-bearing minerals.

Soils of the southeastern and southern coastal plain areas of the United States are formed from marine sediments which have been highly leached and are generally low in their content of plant nutrients. Soils of the Midsouth are formed from igneous, sedimentary, and metamorphic rocks. Because of their age and the climate under which they were formed, these soils are low in potassium even though the parent rocks are frequently high in potassium-bearing minerals. The soils of the middle and far western states are formed from geologically young parent materials and under conditions of lower rainfall. The low content of potassium in the coastal soils of the Pacific Northwest is accounted for by high rainfall in that area.

In tropical soils the total content of potassium may be quite low because of their origin, high rainfall, and continued high temperatures. The last two factors have hastened the release and leaching of soil potassium over the years.

Origin of Soil Potassium. Exclusive of that added in fertilizers, the potassium contained in soils originates from the disintegration and decomposition of rocks containing potassium-bearing minerals. The minerals that are generally considered to be original sources of potassium are the potash feldspars $(KAlSi_3O_8)$, muscovite $[H_2KAl_3(SiO_4)_3]$, and biotite $[(H,K)_2(Mg,Fe)_2Al_2(SiO_4)_3]$. As far as plant response is concerned the availability of the potassium in these minerals, though slight, is of the order biotite > muscovite > potash feldspars.

Some experimental work has been carried out in Georgia in which finely ground granite was used as a source of plant-nutrient potassium for Coastal Bermuda grass. Granite contains feldspars and micas in the unweathered form. These studies showed that 8 lb. of potassium as granite meal had to be applied to get the same crop response as 1 lb. of potassium in muriate of potash.

Potassium is also found in the soil in the form of secondary or clay minerals: (1) illites or hydrous micas, (2) vermiculites, (3) chlorites, and (4) interstratified minerals in which two or more of the preceding types occur in more or less random arrangement in the same particle.

Potassium Equilibria in Soil. Of the total amount of potassium in most soils, only a fraction can be immediately utilized by plants. In fact, on many soils containing large amounts of total potassium, crops may respond to additions of a potassium fertilizer. Soil potassium exists in three forms termed (1) relatively unavailable, (2) slowly available, and (3) readily available. These forms are considered to be in equilibrium, as illustrated diagrammatically:

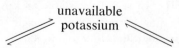

unavailable
potassium

slowly available potassium ⇌ readily available potassium

Unavailable potassium, or that which is not available to plants, occurs as a part of the crystal structure of unweathered or only slightly weathered primary and secondary micaceous and feldspathic minerals. Slowly available potassium is gradually taken up by plants through the reactions of such minerals as illite which appear alternately to release or fix it, depending on several factors. One is the concentration of available potassium in the soil solution and in the exchangeable form. Readily available potassium, the combination of water-soluble and exchangeable potassium, is present either in the soil solution or is held by the exchange fraction of the soil. It is the fraction of soil potassium that can be readily absorbed by growing plants. These divisions are arbitrary but they serve in a general way to define the types of potassium in the soil and the relative availability of each to plants. The boundaries are diffuse and not at all as sharp as these categories would imply.

Because of the continuous removal of potassium by crop uptake and leaching on some sandy soils, a static equilibrium probably never obtains. There is a continuous but slow transfer of potassium in the primary minerals to the exchangeable and slowly available forms. Under some soil conditions, including applications of large amounts of fertilizer potassium, some reversion to the slowly available form will occur. The unavailable form accounts for 90 to 98 per cent of the total soil potassium, the slowly available form, 1 to 10 per cent, and the readily available form, 0.1 to 2 per cent.

Several years ago Bray and DeTurk of Illinois suggested the importance of micaceous clays as a source of nonexchangeable potassium used by plants. They postulated that the micaceous clay represents a reservoir of unavailable potassium with which the exchangeable potassium slowly comes to equilibrium. Exchangeable potassium may be removed by extracting the soil for a short time with a solution of a neutral salt such as ammonium acetate.

To support this equilibrium hypothesis, these workers carried out several types of experiments. In the first they removed the exchangeable potassium from several Illinois soils and then incubated them at a suitable moisture content for six months. At the end of this period they again determined the quantity of exchangeable potassium that had accumulated. Their results are shown graphically in Figure 7-1. Although the amounts of potassium that became exchangeable after six months were not always as great as the amounts originally present, the correlation between the two values is obviously quite good.

The second type of experiment carried out by these workers involved the incubation for five years of samples of soil from which the potassium was not removed as in the preceding experiments. The samples were maintained in a suitably moist condition. At the end of the incubation period the exchangeable potassium was determined. It was found that this value had increased for some soils and decreased for others, but on

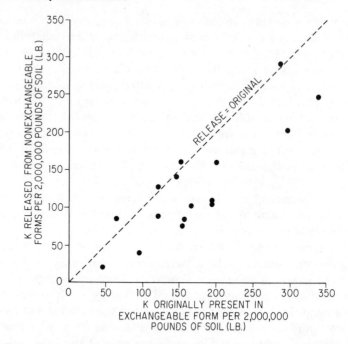

Figure 7-1. Potassium released from nonexchangeable to exchangeable forms in various soils after six months' incubation as related to the original level of exchangeable potassium before incubation. [Bray and DeTurk, *SSSA Proc.*, **3:**101 (1938).]

the average there had been no pronounced changes one way or the other.

The third type of experiment involved the addition of a soluble potassium salt to samples of different soils and subsequent incubation in the moist state for eight weeks. The exchangeable plus water-soluble potassium was determined at the end of that time. It was found that some of the added potassium had been converted to a nonexchangeable form.

Although the research of these Illinois workers was carried out in the middle and late 1930's, the concepts they developed concerning potassium equilibria in soils are generally accepted as principles today.

In today's intensive agriculture where the emphasis is on high yields, there is a large daily demand for potassium. On many soils considerable fertilizer potassium must be added to meet this need. Under such conditions the release of soil potassium from the slowly available forms will be much less as the concentration gradient for movement from the nonexchangeable to the exchangeable forms will be less.

UNAVAILABLE POTASSIUM. Unavailable potassium occurs in primary minerals such as micas and feldspars and in certain secondary minerals which entrap potassium in their lattice structure. These minerals, when exposed to the various weathering processes, will, over a period of time, gradually undergo decomposition, as illustrated by the following equation:

$$KAlSi_3O_8 + HOH \longrightarrow KOH + HAlSi_3O_8$$

With decomposition there is a release of K^+ ions which may be (1) lost in drainage waters or appropriated by living organisms, (2) held as an exchangeable ion on surrounding clay particles, or (3) converted to one of the slowly available forms of soil potassium.

As already suggested, the conversion of the unavailable forms to the available and slowly available forms of soil potassium is a reversible phenomenon. Such reversible reactions have been effected under laboratory conditions. For example, vermiculite was transformed to biotite and vice versa simply by alternate treatment with KCl and $MgCl_2$ solutions at 70°C. Vermiculite may be considered as an alteration product of biotite in which the K^+ ions in the biotite are replaced by Mg^{2+} ions. The transformation of biotite to vermiculite has also been effected experimentally by growing wheat in clay to remove the released potassium.

Several years ago the suggestion was made that in certain soils which had received large amounts of potash fertilizers the formation of mica-type minerals could be expected. This hypothesis was tested experimentally, and a strong indication of the presence of muscovite was found on an X-ray diffraction pattern of the clay fraction of a Hagerstown silt loam, which had received 50,000 lb. of muriate of potash over a 50-year period. Soil from adjacent plots which had not been treated with potash gave no positive test for the presence of this mineral. It has been said that its formation in soils is unlikely because of its hydrothermal origin. However, there may have been some conversion into a hydrous-mica compound as suggested by the results of these experiments.

SLOWLY AND READILY AVAILABLE POTASSIUM. *Readily available potassium* occurs in the soil solution and on the exchange complex and is readily absorbed by plants. *Slowly available* potassium is generally considered to be unextractable by procedures employed in the determination of the exchangeable form of this ion. It becomes available to plants slowly and over longer periods of time. However, it is much more available to plants than the potassium present in the primary minerals.

Consider the readily available potassium. The soil contains cations that are in equilibrium with similar cations adsorbed by the exchange fraction. This is represented diagrammatically as

$$\text{clay} \left.\begin{array}{l} K \\ Ca \\ Ca \\ H \end{array}\right] \begin{array}{l} \xrightarrow{} K^+ \\ Ca^{2+} \text{ (solution)} \\ \xleftarrow{} H^+ \end{array}$$

An increase in the concentration of the K^+ ions in solution will cause the reaction to proceed to the left, and some K^+ will be adsorbed by the clay and part may be converted to an unavailable form. This condition would result, for example, if a potassium fertilizer were added to a soil. The effect on plant growth of this reduction in soluble potassium because of its reaction with the soil components is illustrated in Figure 7-2. The spring-applied potassium was in contact with the soil for a shorter period than that applied in the fall. Less potassium was converted to the unavailable form and consequently more was available to the crop, hence the greater yield of barley on the soil treated with spring-applied potassium.

If the concentration of an ion in the soil solution is decreased, there will be a shift in the equilibrium to the right and some of that ion will enter the soil solution. This is brought about by the absorption of potassium ions from solution by plant roots or loss through removal in percolating waters. The potassium contained in the soil solution and that on the exchange fraction are in dynamic equilibrium and constitute that which is absorbed by plants during the growing season.

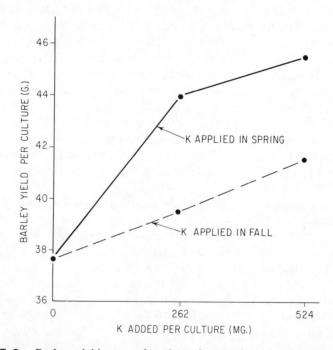

Figure 7-2. Barley yields as a function of potassium fertilizer applied just before planting or in the preceding fall. (After Chaminade, in Black, *Soil-Plant Relationships,* 1957. Reprinted with permission of John Wiley & Sons, Inc., New York.)

As already pointed out, readily available potassium, itself an equilibrium system, is in turn in equilibrium with the form termed slowly available potassium. This system may be expanded and indicated as

slowly available K $\rightleftharpoons$ exchangeable K^+ $\rightleftharpoons$ water-soluble K^+

It has been shown repeatedly that exchangeable potassium may be converted to forms no longer extractable with the reagents normally employed.

POTASSIUM FIXATION. This does not occur to the same extent in all soils or under all conditions. It reaches its maximum, however, in soils high in 2:1 clays and with large amounts of illite. This suggests that potassium fixation is the result of a re-entrapment of K^+ ions between the layers of the 2:1 minerals, especially those such as illite. Why, then, are not other cations similarly held? Presumably openings in the oxygen network of the silica sheets are of a size similar to that of the potassium ion. This permits its accommodation, and the potassium is then held very firmly by electrostatic forces. In this connection the NH_4^+ ion is also of almost the same ionic radius as the K^+ ion and is subject to fixation by 2:1 clays. The mechanism proposed for the fixation of the ammonium is similar to that for potassium. Cations such as Ca^{2+} and Na^+ have different ionic radii and are not subject to entrapment by expanding-type clays.

The interrelationship between the fixation of K^+ and NH_4^+ was touched on briefly in Chapter 5 in the discussion of subsoil retention of native and applied NH_4^+-nitrogen. Because NH_4^+ can be fixed by clays in a manner similar to the fixation of potassium, its presence will alter both the fixation of added potassium and the release of fixed potassium. Even as the presence of potassium can block the release of fixed NH_4^+, as pointed out in Chapter 5, so can the presence of NH_4^+ block the release of fixed potassium. This is illustrated by the data shown in Figure 7-3, which shows the reduction in the release of nonexchangeable potassium as affected by increasing amounts of added NH_4^+. The ammonium ions evidently are held in the openings in the oxygen network of the silica sheets, thus closing the adjacent sheets and further trapping the potassium already present. The greater the amount of ammonium present, the greater the amount of potassium entrapment. This phenomenon may be of some importance in fine-textured agricultural soils which have a high fixation capacity for both potassium and ammonium. However, it is not generally considered to be a serious factor in limiting crop response to either applied NH_4^+ or K^+.

Most 2:1 clays contract when dried and expand when wet. It might be supposed therefore that trapped potassium would be immediately released on rehydration of the clay. However, the bonding energy between

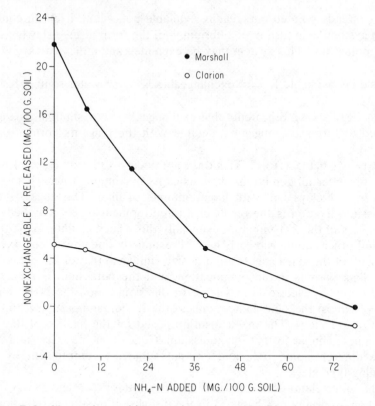

Figure 7–3. Nonexchangeable potassium released by Marshall and Clarion surface soils during a ten-day cropping period as influenced by the amount of added NH_4^+. [Welch and Scott, *SSSA Proc.*, **25**:102 (1961).]

the potassium ion and the entrapping silica sheets is great enough to prevent immediate re-expansion of all of the mineral layers. In those clays containing fixed potassium and ammonium subsequent hydrations will result in re-expansion of some of the layers but not of others. This pattern, repeated again and again with alternate wetting and drying, may aid in the slow release of the fixed potassium and ammonium ions. It has been suggested by workers in Wisconsin that drying causes the edges of some clay minerals to roll back, thus permitting exposure and release of potassium.

Soil Texture and Soil-Solution Potassium. The potassium concentration of the soil solution, which is the form absorbed to the greatest extent by plants, is related not only to the level of exchangeable K but also to the amount and type of clay mineral present. The effect of soil texture, which is a reflection of the mineralogical make-up of the soil on K concentration of the soil solution, is shown in Figure 7-4. The clay soil

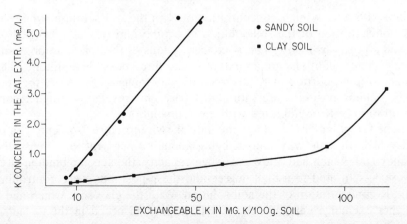

Figure 7–4. The K-concentration of the soil solution vs. the exchangeable K for two different textured soils. (Mengel, *Proc. Jap. Pot. Symp.,* Tokyo, 1971.)

shows a rather small increase in the K concentration of the soil solution with increasing amounts of exchangeable K, while the sandy soil shows a marked increase in solution K with increasing amounts of exchangeable K. This is due to the different binding intensities for K of 2:1 minerals such as illite, vermiculite, and chlorite. The binding intensity is related to the binding position on the clay mineral as illustrated in Figure 7-5. The planar position (p) is nonspecific for K but the edge (e) position and the inner (i) positions particularly have a high specificity or selectivity for K.

K-selectivity means that K ions bound by a *p*-position are in equilib-

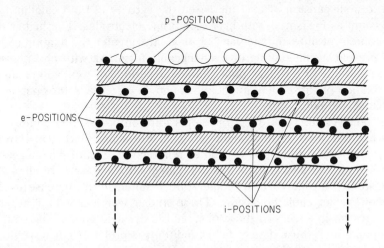

Figure 7–5. Cross section of a 2:1 clay micelle showing p-, e-, and i-positions of K-bonds. (Mengel, *Proc. Jap. Pot. Symp.,* Tokyo, 1971.)

rium with a fairly high concentration of K in the soil solution, while the K concentration of the soil solution in equilibrium with K adsorbed by *i*-and *e*-positions is low. The K concentrations of the soil solution found under field conditions are a mixture of the three possible equilibria. The higher the proportion of K adsorbed by clay minerals the more the specific binding positions are saturated by K. This results in a higher concentration of K equilibrated with the soil solution.

The foregoing also explains the data shown in Figure 7-4. The type of clay in both soils was similar, being about 80 per cent illite. With the sandy soil, which had a low exchange capacity, the specific binding sites were K-saturated to a high degree and the p-positions were controlling the K concentration of the soil solution. With the clay soil, which had a higher exchange capacity, the K was largely adsorbed in the *e*- and *i*-positions resulting in low concentrations of K in the soil solution. As the exchangeable K of the clay soil approached 100 mg./100 g. of soil, however, the more specific binding sites (*e*- and *i* positions) became saturated and the *p*-position then began to control the K concentration of the soil solution.

Factors Affecting Potassium Equilibria in Soils. Certain factors are known to influence the conversion of soil and added potassium to less available forms. Some of these are (1) type of colloid, (2) temperature, (3) wetting and drying, and (4) soil *p*H.

TYPE OF COLLOID. The 1:1 clays do not fix potassium ions in the manner described for the 2:1 type. Organic matter (humus), although possessed of a great capacity to retain K^+ and other cations in the exchangeable form, has no capacity whatever for the fixation of this element. In fact, some experimental work has shown that adding humic acid to a suspension of clay increased the activity of both calcium and potassium as measured with clay membrane electrodes. The higher the soil content of illite and similar 2:1 clays, the greater the fixation of K^+ ions present in the soluble or exchangeable form. As with phosphorus, large additions of fertilizer potassium over periods of time will result in less fixation of subsequent applications and an increase in the content of exchangeable potassium.

TEMPERATURE. The effect of temperature on potassium equilibria in soils has not been too thoroughly studied. Burns and Barber at Purdue University reported results of studies relating the level of exchangeable potassium in three soils to incubation temperature and time. In all of the soils studied an increase in temperature resulted in an increase in the level of exchangeable potassium. These studies were carried out at temperatures far in excess of those observed under field conditions, so that extrapolation of these data to field conditions is probably not warranted.

The freezing and thawing of moist soils may also be important in the release of fixed potassium and the fixation of the exchangeable form. Controlled laboratory studies have shown that, with alternate freezing and thawing, certain soils will release to the exchangeable form a fraction of the fixed potassium. In other soils, particularly those high in exchangeable potassium, no such release was observed, and, in fact, some of the exchangeable potassium was converted to the less readily available form. This phenomenon was observed in soils that contained appreciable quantities of illite. The importance of this reaction in terms of the amounts of potassium made available under various field conditions has not been determined. However, it seems probable that under alternate freezing and thawing this reaction may play a significant role in the potassium supply of certain soils. In fact, in some soils that contain only moderate amounts of exchangeable potassium the supply of this element made available by freezing and thawing may be sufficient to produce the first cutting of an alfalfa crop.

WETTING AND DRYING. When field-moist soils are dried, there is usually an increase in the exchangeable potassium. This is particularly true when the levels of soil potassium are medium to low. When the levels are high, however, the reverse situation may occur.

Soils under field conditions are alternately moist and dry. The effect of this alternating moisture level on potassium availability was studied in the laboratory. Undried samples of a Marshall subsoil, which contained 24 per cent moisture, were exposed in desiccators at 25°C. to relative humidities of 10 and 95 per cent. At the end of successive four-week intervals the moisture and exchangeable potassium were determined and the samples were switched from one desiccator to the other. The study was continued for twenty weeks. The results are illustrated in Figure 7-6. The soil moisture level shifted from 2 per cent in the dry desiccator to 8 per cent in the moist. The effect of the alternate wetting and dry-

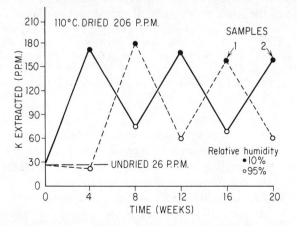

Figure 7–6. Exchangeable potassium in Marshall subsoil samples stored at 25°C. at relative humidities of 10 and 95 per cent and then switched after four-week intervals. [Scott et al., *Proc. 7th Intern. Congr. Soil Sci.,* **3**:72 (1960).]

ing of this soil on its content of exchangeable potassium is obvious. Although these effects are frequently observed, the reasons for their occurrence are not understood.

The effects of wetting and drying on the field availability of soil potassium are not known. It is important, however, in the testing of soils. Frequently, soil test procedures call for the air drying of samples before analysis. It is apparent from the foregoing discussion that this procedure could lead to high soil test values for potassium—and consequently to a low recommendation for potassium fertilization on some soils.

SOIL *p*H.   The effect of *p*H on the fixation and release of soil potassium has been a controversial subject among soil scientists for many years. Results of numerous experiments to determine the relation between soil acidity and potassium fixation have often been completely contradictory. This problem is an extremely important one because of the relation it bears to liming in a soil management program. A brief summary of some of its principles should contribute to an understanding of the problem.

Consider first the reaction

$$\text{clay} \begin{bmatrix} H \\ K \\ K \\ K \\ K \end{bmatrix} + CaSO_4 \longrightarrow \begin{bmatrix} H \\ Ca \\ K \\ K \end{bmatrix} \text{clay} + K_2SO_4$$

If a soil colloid is saturated with potassium and a neutral salt such as calcium sulfate is added, there will be a replacement of some of the adsorbed potassium ions by the calcium ions. The amount of replacement taking place will depend on the nature and amount of the added salt as well as that of the ionic population adsorbed on the clays.

Suppose that a soil condition is represented by the following equation:

$$\text{clay} \begin{bmatrix} Ca \\ H \\ Al \end{bmatrix} + KCl \longrightarrow \begin{bmatrix} \frac{1}{2}Ca \\ K \\ H \\ Al \end{bmatrix} \text{clay} + \frac{1}{2}CaCl_2$$

This soil clay contains adsorbed calcium, hydrogen, and aluminum ions to which potassium chloride has been added. Because the calcium is more easily replaced than the hydrogen, the added potassium will replace some of the calcium and will itself be adsorbed onto the surface of the clay. This reaction illustrates an important point—the greater the degree of calcium saturation, the greater the adsorption by clay of potassium from the soil solution. This is not inconsistent with the point previously illustrated in which the calcium from calcium sulfate replaced potassium from the colloid. Calcium, when added as a neutral salt, replaces hydrogen and aluminum only with great difficulty, and, if a soil clay contains, in the adsorbed form, ions such as K^+, Na^+, and NH_4^+ in addition to H^+ and Al^{3+}, these ions, rather than the hydrogen and aluminum, will be replaced. In such cases there will be a net loss of potassium to the soil solution.

Sandy soils with a high degree of base saturation lose less of their exchangeable potassium by leaching than soils with a low degree of base saturation. Liming is the common means by which the base saturation of soils is increased, and it must follow that liming decreases the loss of exchangeable potassium. Whether this exchangeable potassium in turn reverts to less available forms will depend on factors such as the nature and amount of clay and the degree of potassium saturation. Whether soil pH *per se* actually has an effect on the equilibrium

$$\text{exchangeable K} \rightleftharpoons \text{fixed K}$$

is not known.

Lime is capable of neutralizing an acid. How this is accomplished is illustrated by the following equation:

$$\text{clay} \begin{bmatrix} \text{Ca} \\ \text{H} \\ \text{H} \end{bmatrix} + \text{Ca(OH)}_2 \longrightarrow \text{Ca} \begin{bmatrix} \text{Ca} \\ \text{clay} \end{bmatrix} + 2\text{H}_2\text{O}$$

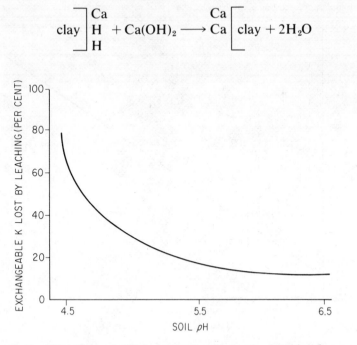

Figure 7–7. The effect of pH on potassium losses by leaching from a Creedmoor sandy loam. [Baver, *Soil Sci.*, **55**:121 (1943). Reprinted with permission of The Williams & Wilkins Co., Baltimore.]

Since water is thermodynamically a more stable compound than a hydrogen clay, the reaction proceeds to the right. This leaves a calcium-saturated clay. Results of a study conducted in North Carolina indicate that the fraction of exchangeable potassium lost by leaching is inversely related to the degree of base saturation (Figure 7-7).

Agricultural Significance of Potassium Equilibria. Throughout this chapter the terms exchangeable and available potassium have been used perhaps to imply that they are interchangeable. Actually they are not, for it is obvious that significant portions of potassium not removed during one or even several extractions with a neutral salt may actually be absorbed by plants. However, the amount of exchangeable potassium extracted with salts, such as neutral ammonium acetate, is an indication of the potassium fertility status of the soil at any point in time. This is well illustrated by the data presented in Figure 7-8. Plant uptake of potassium, of course, is the best criterion of the availability of this element, provided that other growth factors are not limiting. The agricultural significance of exchangeable potassium is considerable and is a property often used in soil testing procedures for making potassium fertilizer recommendations.

The retention of potassium in less available or fixed forms is likewise

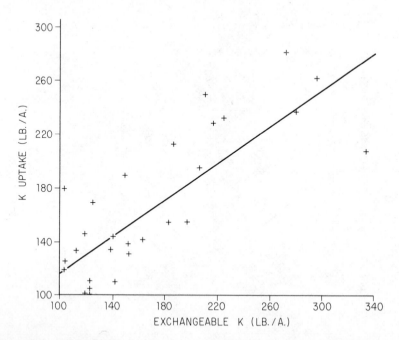

Figure 7-8. The relationship between potassium uptake and exchangeable soil potassium. [Jaworski et al., *Soil Sci.*, **87**:37 (1959). Reprinted with permission of The Williams & Wilkins Co., Baltimore.]

of considerable significance to the practical aspects of farming. As with phosphorus, the conversion of potassium to the slowly available or fixed forms reduces its immediate value as a plant nutrient. However, it must not be assumed that potassium fixation is completely unfavorable, for there are, in fact, certain beneficial aspects associated with this phenomenon.

Potassium fixation, in the first place, results in a conservation of this element which otherwise might be lost by leaching on sandy soils. In the second place, fixed potassium tends to become available over a long period of time and is thus not lost completely to plants, though crop plants do vary in their ability to utilize slowly available potassium.

Continued additions of potassium not only decrease the potassium-fixing power of soils, but they also produce increases in the exchangeable and soil solution potassium levels, which, in turn, are reflected in crop yields and, equally important, in crop responses to further

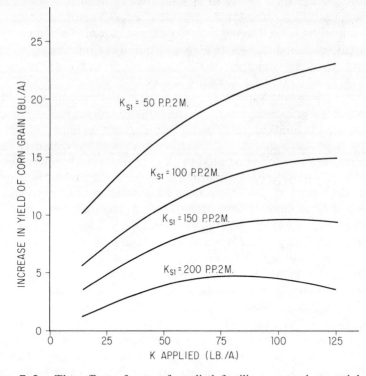

Figure 7–9. The effect of rate of applied fertilizer potassium and level of exchangeable soil potassium on increases of corn in the field. (The subscript S1 indicates that the sample was field moist and that it came from the 0-to-6-in. layer of soil.) [Hanway et al., North Central Regional Potassium Studies III. Field Studies with Corn, *North Central Regional Publication No. 135* (April 1962).]

additions of this element. As with phosphorus, or any other element for that matter, an increase in the level of available soil potassium decreases the crop responses obtained from future fertilizer applications of that element. This is very clearly illustrated by the data in Figure 7-9. It is not to be implied that fertilizer additions of some nutrient should be stopped because of its already high soil level. Maximum economic returns from a farming enterprise are almost always obtained only when high soil fertility is maintained, not only for potassium but for the other nutrients as well. What is implied, however, is that fertilizer applications of potassium in this case can be adjusted downward with increasing levels of available soil potassium. If the soil level of an element begins to decrease, fertilizer applications can again be increased. Soil testing will provide the farm operator with a measure of the fertility level of his soil and fertilizer applications of the various plant nutrients can be adjusted to maintain or increase crop production.

Plants differ in their ability to utilize native or fixed potassium. It has been proposed by workers at the Massachusetts Experiment Station that this condition is related to the cation exchange capacity of the plant root itself. It is suggested that those plants with roots of relatively low cation exchange capacity values are best able to utilize native soil potassium. This low cation exchange capacity seems to be associated with a low calcium requirement. Support for this contention stems from the observation that the grasses and cereals with roots of low exchange capacities respond least to applications of fertilizer potassium on many soils in contrast to plants such as clovers which have higher root cation exchange capacities. This is well illustrated by data from Australia shown in Figure 7-10. It is quite apparent that the grass was much better able to survive at low levels of potassium than was the clover, which responded tremendously to applications of fertilizer potassium. The graph also illustrates the need for high potassium fertilization of mixed grass-legume pastures and meadows and suggests one very important reason why legumes frequently play out in such systems.

Although the cation exchange capacity of roots may be important in determining the ability of plants to absorb the more slowly available forms of soil potassium, it is probably not the only factor involved. Important also are the extent of the root system, the rate of growth of the plant, competition from other species, and potash requirements.

LUXURY CONSUMPTION. The term *luxury consumption* has been grossly misused. It means that plants will continue to absorb an element in amounts in excess of that required for optimum growth. It results in an accumulation of the element in the plant without a corresponding increase in growth and suggests, in other words, inefficient and uneconomical use of that particular element. However, with the higher yields of crops that are being obtained today, a much higher concentration of

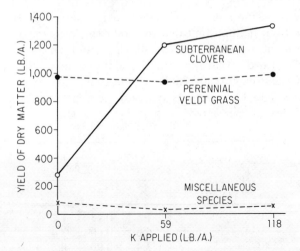

Figure 7–10. Yield of botanical components of a mixed meadow as influenced by rates of applied potassium fertilizer. (After Rossiter, in Black, *Soil-Plant Relationships,* 1957. Reprinted with permission of John Wiley & Sons, Inc., New York.)

potassium, and other nutrients, too, is required, suggesting that the term "luxury consumption" is less meaningful than it was a decade ago. As an example, 1.0 to 1.2 per cent K in alfalfa tissue was formerly thought to be adequate. Today, however, a K content of 2 to 3 per cent is considered necessary to maintain consistently high yields and good stands and to enable the plants to survive periods of stress with a minimum decrease in growth.

MOVEMENT TO PLANT ROOTS. Plant roots absorb nutrients as a result of root interception, mass flow of nutrients by movement of soil moisture, and diffusion of the nutrients to the root surface. The amount of potassium absorbed as a result of root interception is small for plant roots occupy only about 1 to 2 per cent of the soil volume.

Growing plants absorb soil moisture. This causes a mass movement of soil moisture and dissolved nutrients toward the roots. While this method also accounts for the presentation of nutrients to the root surface for absorption, only limited amounts of potassium are considered to be absorbed as a result of mass flow. The diffusion of ions through the soil moisture, however, is believed to be the mechanism responsible for the greatest amount of absorption of K and P, in particular. The rate of diffusion is dependent on the amount of moisture in the soil as well as the concentration gradient. If soil moisture levels are low (high pF) and moisture films are thin, diffusion is slow. Similarly, if the concentration gradient is low, i.e., the K or other nutrient level in the soil solution is low, diffusion is slow. The distances moved by the K ion by diffusion through the soil solution is only about ¼ in. and by phosphorus only about ⅛ in. So it is important that the soil contain adequate amounts of the needed plant nutrients so that diffusion can take place rapidly and continuously in order that the growth requirements of the crop can be met. These requirements are high as indicated below, showing the uptake of N, P, and K in pounds per acre per day of a corn crop that yielded 187 bushels per acre*:

Plant age (days)	Nutrients absorbed in pounds per acre per day		
	N	P	K
20–30	1.5	0.15	1.3
30–40 (knee-high)	6.0	0.60	7.4
40–50	7.4	0.90	8.6
50–60 (tasseling)	4.7	0.80	3.3

PLACEMENT OF FERTILIZER POTASSIUM. The factors determining the most effective placement of fertilizer potassium are those, generally, that influence the potassium equilibria in the soils covered in this chapter. If a soil is high in potassium-fixing clay minerals and limited quantities of potassium fertilizers are to be applied, best results will be obtained if the material is applied in bands rather than broadcast and mixed with the soil. This results from the reduced area of contact between the fertilizer potassium and the soil. In fact, banding of potassium fertilizers on high potassium-fixing river clay soils was 3.65 times more effective than broadcast application.

When soils have a high level of exchangeable potassium, usually one method of placement is as satisfactory as the other. In band-placing fertilizers containing soluble potassium salts care must be taken, of course, to place the band about 2 in. to the side and 1 in. below the seed or transplant crown. Failure to observe this precaution will frequently result in salt injury to the seedlings, reduced and uneven stands, and lowered yields. This point is treated at greater length in the chapter dealing with fertilizer application.

LOSS OF POTASSIUM BY LEACHING. The foregoing discussion of the various potassium equilibria in soils may suggest that this element will be lost to crop use by outgo in percolating waters. However, in most soils, except those that are quite sandy or subject to flooding, the losses of potassium in this manner are small. To illustrate these points, work

* Spires, C., Purdue Univ. 1973.

carried out in Illinois several years ago indicated that losses from the silt loam soil studied would amount to only about 1.4 lb. of potassium per acre per year. Other work carried out in Florida, however, showed that large amounts of potassium would be lost from coarse sandy soils with a low exchange capacity. In fact, in the study referred to, 126 lb. of potassium in a total of 135 lb. present (as native and added potassium) were removed in the leachate when 16 in. of water were passed through the soil, which was contained in a 5-gal. pot.

The wide differences illustrated by these two examples can be explained on the basis of the total cation exchange capacities of these soils and their relative degrees of potassium saturation. The silt loam soil had a much higher exchange capacity than the sandy soil. It was therefore able to hold a much larger amount of potassium. Additions of fertilizer potassium would result in adsorption of a much higher proportion of the added potassium than it would in the case of the sandy soil. The unadsorbed part, of course, remains in the soil solution and is removed in leaching waters, which explains why such a large quantity was lost from the sandy soil with the low CEC.

While organic soils, such a mucks, have high exchange capacities, the bonding strength for cations such as potassium is not great, and the exchangeable potassium level tends to vary somewhat with the intensity of rainfall. Thus emphasis should be placed on annual applications rather than on buildup of soil K. Since crops grown on organic soils characteristically have a high potassium need, it is important to monitor the fertility level with the help of soil tests.

Although losses of potassium from surface soils may be caused by leaching, this loss can be reduced by liming the soil to maintain it at a favorable pH level. In addition, potassium in percolating waters is usually readsorbed by clays in the subsoil, so that a buildup of this element in the lower reaches of the soil is frequently observed. In the final analysis leaching losses of potassium are of consequence only in coarse-textured or organic soils in areas of high rainfall.

CALCIUM

Form Utilized by Plants. Calcium is absorbed by plants as the ion, Ca^{2+}, which takes place from the soil solution and by contact exchange.

Source of Soil Calcium. The calcium present in soils, exclusive of that added as lime or in fertilizer materials, has its origin in the rocks and minerals from which the soil was formed. Calcium is contained in a number of minerals—dolomite, calcite, apatite, calcium feldspars, and amphiboles, to name but a few—and on their disintegration and decomposition calcium is released.

The fate of released calcium is less complex than that of potassium. Calcium ions set free in solution may (1) be lost in drainage waters, (2) absorbed by organisms, (3) adsorbed onto surrounding clay particles, or (4) reprecipitated as a secondary calcium compound, particularly in arid climates. As far as is known, there is no conversion in the soil of calcium to a form comparable to fixed or slowly available potassium.

As a general rule, coarse-textured, humid-region soils formed from rocks low in calcium-containing minerals are low in their content of this element. Soils that are fine-textured and formed from rocks high in the calcium-containing minerals are much higher in their content both of exchangeable and total calcium. However, in the humid region even soils formed from limestones are frequently acid in the surface layers because of the removal of calcium and other basic cations by excessive leaching. As water containing dissolved carbon dioxide percolates through the soil, the carbonic acid so formed displaces calcium (and other basic cations) in the exchange complex. If considerable percolation of such water through the soil profile takes place, soils gradually become acid.

The calcium content of soils of the arid regions is generally high, regardless of texture, as a result of low rainfall and little leaching. Many of the soils of the arid regions actually have within their profiles secondary deposits of calcium carbonate or calcium sulfate.

Behavior of Calcium in Soil. The calcium in acid, humid-region soils occurs largely in the exchangeable form and as undecomposed primary minerals. In most of these soils calcium, aluminum and hydrogen ions are present in the greatest quantity on the exchange complex. Like any other cation, the exchangeable and solution forms are in dynamic equilibrium. If the activity of calcium in the solution phase is decreased, as it might be by leaching or plant removal, there tends to be replacement from the adsorbed phase. Conversely, if the activity of calcium in the soil solution is suddenly increased, there tends to be a shift of equilibrium in the opposite direction, with subsequent adsorption of some of the calcium by the exchange complex.

Plants absorb calcium largely from the soil solution. In soils not containing excess calcium carbonate the amount of calcium in the soil solution is dependent on the amount of exchangeable calcium present. The soil factors believed to be of the greatest importance in determining the availability of calcium to plants are the following:

1. The amount of exchangeable calcium present.
2. The degree of saturation of the exchange complex.
3. The type of soil colloid.
4. The nature of the complementary ions adsorbed by the clay.

The absolute amount of exchangeable calcium present is frequently not so important to plant nutrition as the amount present in relation to the quantities and kinds of other cations held by the clay or the degree of calcium saturation. For example, a soil having only 2,000 lb. of exchangeable calcium per acre but with a low cation exchange capacity might well supply plants with more of this element than one containing 8,000 or 9,000 lb. of exchangeable calcium per acre but with a high cation exchange capacity. The degree of calcium saturation is of considerable importance in this respect, for as the amount of this element held in the exchangeable form by a clay decreases in proportion to the total exchange capacity of that clay the amount of calcium absorbed by plants decreases.

The type of clay influences the degree of calcium availability; 2:1 clays require a much higher degree of saturation for a given level of plant utilization than 1:1 clays. Montmorillonitic clays require a calcium saturation of 70 per cent or more before this element is released sufficiently rapidly to growing plants. Kaolinitic clays, on the other hand, are able to satisfy the Ca^{2+} requirements of most plants at saturation values of only 40 to 50 per cent.

Calcium is an extremely important mineral in plant nutrition. However, only few soils contain absolute amounts of calcium as such that are insufficient for plant growth. Many soils in humid, warm areas contain insufficient calcium to maintain a suitable degree of base saturation of the soil colloids. In such soils exchangeable aluminum dominates the exchange sites of the clay, contributing to excessive soil acidity and soluble aluminum, the latter of which is toxic to many plants. The most obvious way to correct this deficiency is by the application of calcitic or dolomitic lime. In the event that calcium is required without the increase in pH that would result from the use of lime, gypsum is also a satisfactory source of this element. The chemistry of the behavior in the soil of liming materials and gypsum is considered at greater length in a later chapter.

MAGNESIUM

Form Utilized by Plants. Magnesium is absorbed by plants as the ion Mg^{2+}. Like the calcium and potassium already discussed, this absorption takes place largely from the soil solution and by contact exchange.

Source of Soil Magnesium. Magnesium constitutes 1.93 per cent of the earth's crust. As with similar data for calcium and potassium, this figure represents the average of a wide range of values. The total magnesium content of soils is variable, ranging from only a fraction of 1 per cent in coarse, sandy soils in humid regions to perhaps several per cent in fine-textured, arid, or semiarid soils formed from high magnesium parent materials.

Magnesium in the soil originates from the decomposition of rocks containing minerals such as biotite, dolomite, chlorite, serpentine, and olivine. On decomposition of these minerals, the magnesium is set free into surrounding waters. It may then be (1) lost in these percolating waters, (2) absorbed by living organisms, (3) adsorbed by surrounding clay particles, or (4) reprecipitated as a secondary mineral. This last phenomenon would be expected to take place most readily in an arid climate.

Behavior of Magnesium in the Soil. The soil magnesium available to plants is in the exchangeable and/or water-soluble forms. The same general principles apply to its behavior as apply to calcium and potassium. The absorption of magnesium by plants depends on the amount present, the degree of magnesium saturation, the nature of the other exchangeable ions, and the type of clay. Like potassium, but to a lesser extent perhaps, magnesium may occur in soils in a somewhat slowly available form, in which it is in equilibrium with exchangeable magnesium. The formation of these relatively unavailable forms in acid soils would be favored by the presence of large quantities of soluble magnesium compounds and a 2:1 clay. Presumably there could be an entrapment of magnesium ions between the expanding and contracting sheets of the mineral. Such a reaction has been suggested, but the extent to which it takes place is not actually known.

The coarse-textured soils of the humid regions are those in which a deficiency of magnesium is generally manifested. These soils normally contain only small amounts of exchangeable magnesium, a condition that is aggravated by the addition of large quantities of fertilizer salts which contain little or none of this element. The magnesium in these soils is released by ion exchange when these fertilizers are added, and large quantities of chlorides and sulfates enhance its removal in percolating waters.

On many humid-region, coarse-textured soils the continued use of high calcic liming materials may result in an unfavorable calcium:magnesium balance and the consequent development of magnesium deficiency symptoms on certain crops.

Magnesium is a required component of fertilizers for certain crops grown under the soil conditions described. The fertilizer for flue-cured tobacco produced in several of the eastern seaboard States is a case in point and must contain a guaranteed percentage of magnesium oxide. Magnesium, of course, can be supplied in dolomitic limestone or as magnesium sulfate if no change in soil pH is required.

A problem of considerable importance in some areas is that of a low magnesium content of forage crops, particularly grass forages. Cattle consuming such forages may suffer from hypomagnesemia, more commonly known as grass tetany, which is an abnormally low level of blood Mg. High rates of applied nitrogen fertilizers, especially ammoniacal forms, or high levels of potassium may depress the magnesium level in the plant tissue. It has been shown, for example, that the magnesium level of young corn plants is markedly reduced when NH_4^+ rather than NO_3^- is the source of applied nitrogen (Fig. 12-11). As grass tetany often occurs in the spring, the nitrogen may still be in the NH_4^+ form, particularly if cool weather has prevailed.

Another factor of importance is that high protein content of ingested forages (and other feeds, too) will depress the absorption of magnesium by the animal.

Levels of soil magnesium may be increased through the use of dolomitic limestone, if liming is advisable, or through the use of magnesium-containing fertilizers, such as potassium-magnesium sulfate. Also, the inclusion of legumes in the forage program is advisable as these plants have a higher content of magnesium than do grasses. Cattle can also be fed a magnesium salt to help prevent grass tetany.

Although some workers maintain that hypomagnesemia is the result of excessive nitrogen or potassium fertilization, it seems more reasonable to class it as a magnesium deficiency and to treat it accordingly.

A deficiency of magnesium is less of a problem on finer-textured soils and soils of the arid regions. In some semiarid localities magnesium compounds may actually be precipitated in the soil profile. There may also be such large quantities present that this element may be toxic to plants. This condition is quite rare, but in certain western regions of the United States it has been reported on soils formed from rocks high in magnesium-containing minerals.

SODIUM

Although sodium constitutes an appreciable fraction of the earth's crust (2.63%), its presence in soils in any but very small amounts is restricted to those of arid and semiarid regions. In soils of the humid regions long-continued application of sodium nitrate will result in measurable quantities of this element in an exchangeable form. But sodium is one of the most loosely held of the metallic ions and is readily lost in leaching waters. Only if continued additions of sodium salts are made could it be expected that this element would constitute an appreciable part of the suite of exchangeable ions in humid-region soils. It is absorbed by plants as the Na^+ ion, largely from the soil solution.

Data from numerous sources of nitrogen fertilizer experiments conducted for many years, principally in the humid southeastern United States, have shown that continuous fertilization with sodium nitrate results in the maintenance of a suitable soil pH level. Ammoniacal forms of nitrogen used for similar periods lower the pH of the soil. The sodium substitutes for some of the calcium and magnesium that would otherwise accompany the anions lost in percolating waters. The result is the maintenance of a suitable degree of base saturation. Undoubtedly some sodium is adsorbed, but the low energy with which it is held by soil colloids suggests that it will be lost in leaching waters to a greater extent than potassium or the divalent cations.

The effect of sodium on the dispersal both of clay and organic matter is well known. For that reason its presence in large quantities in fine-textured soils is undesirable because of its effect on soil structure.

In semiarid soils this ion may accumulate. When it does, particularly on fine-textured soils, it almost prohibits plant growth. Such *alkali spots*, as they are termed, present quite a problem in certain western and Midwestern states. This condition may be alleviated by treatment with gypsum plus sufficient water to leach out the displaced sodium. The calcium ions replace the adsorbed sodium which in turn is removed in percolating waters as sodium sulfate. The improvement both in the physical and chemical conditions of the soil may eventually bring the land into profitable production.

SOME PRACTICAL CONSIDERATIONS

The foregoing discussion of the behavior of potassium, calcium, magnesium, and sodium in soils should have suggested several important points in regard to the use of fertilizers containing these ions.

Briefly, the application of lime to an acid soil in an amount required to reach a pH of 6.5–7.0 will result in a conservation of potassium, particularly when the lime is broadcast uniformly and worked into the soil. The application of gypsum or other neutral calcium salts will almost invariably raise the outgo of potassium in percolating waters.

The application of calcitic limestone to soils low in magnesium may tend to induce magnesium deficiency of plants growing on such soils. If soils are extremely low in potassium, the application of large amounts of lime or neutral calcium salts may also induce potassium deficiency of plants growing on these soils. This is a consequence that has been termed the *lime-potash law*, which states in effect that when the calcium concentration is high, plants may not be able to absorb potassium at a rate rapid enough to meet their needs. This law has not been found to be universally applicable, but it seems to hold fairly well for soils in which the absolute amount of potassium is quite low. Of course, if the lime increases plant growth, the removal of potassium will be greater.

Soils that are high in 2:1 clays will generally require a higher degree of calcium, potassium, and magnesium saturation for plant availability of these elements than soils that are high in 1:1 clays. Increasing the degree of base saturation in the majority of soils is probably best effected with liming materials containing both calcium and magnesium.

FERTILIZERS CONTAINING POTASSIUM, CALCIUM, MAGNESIUM, AND SODIUM

In the following sections is a description of some of the fertilizer materials which supply the elements potassium, calcium, magnesium, and sodium. Those containing potassium are the most numerous, and from the standpoint of production of large tonnages of fertilizer they are by far the most important. The basic processes by which these fertilizer materials are produced are covered in Chapter 9, which deals with the principles of fertilizer manufacture.

Potassium Fertilizers. Extensive deposits of soluble potassium salts are found in many areas of the world, most of them well beneath the surface of the earth but some in the brines of dying lakes and seas. The potassium salts in many of these deposits and brines are of a high degree of purity and therefore lend themselves to mining operations for the production of agricultural and industrial potassium salts, usually termed *potash salts* by the trade.

As with nitrogen and phosphorus fertilizers, there have been impressive increases in the consumption of potassium materials in recent years. As more knowledge of crop production is acquired, made urgent by the rise in world population and the reduction of acreage of arable land, it is axiomatic that the consumption not only of potassium but of all fertilizers will continue to increase and at a significant rate.

Like phosphorus, the potassium content of fertilizers is presently guaranteed in terms of its potassium oxide equivalent. This is determined analytically by measuring the amount of potassium salt that is soluble in an aqueous solution of ammonium oxalate. As mentioned in Chapter 6, there is some agitation for expressing the plant nutrient content of fertilizers in terms of the element instead of the oxide. Converting %K to %K_2O and the reverse can be accomplished by the two following expressions.

$$\%K = \%K_2O \times 0.83$$
$$\%K_2O = \%K + 1.2$$

Practically all of the potassium fertilizers are water-soluble. They consist essentially of potassium in combination with chloride, sulfate, nitrate, or polyphosphate. Some double salts exist, such as potassium-magnesium sulfate. The properties of the individual potassium carriers are covered in the following sections.

POTASSIUM CHLORIDE (KCl). This salt is sold under the commercial term *muriate of potash*. The term *muriate* is derived from muriatic acid, a common name for hydrochloric acid. Fertilizer grade muriate contains 50 to 52 per cent potassium (60–63% K_2O) and varies in color from pink or red to white, depending on the mining and recovery process used. There is no agronomic difference among the products.

Muriate of potash is generally marketed in three particle sizes: *standard, coarse,* and *granular.* Typical particle-size distributions for these three classes in sieve mesh openings are $-20 + 100$, $-10 + 35$, and $-6 + 14$, respectively. A steadily increasing demand for the coarse and granular types has been noted because of their suitability for blending with other products as well as for direct application. In addition, a soluble white muriate is produced for use by the rapidly growing fluid fertilizer market.

Muriate of potash is by far the most widely used potassium fertilizer. It is employed for direct application to the soil and for the manufacture of N-P-K fertilizers. When added to the soil, it dissolves in the soil moisture. The resulting K^+ and Cl^- ions behave in keeping with the pattern already discussed in this chapter.

POTASSIUM SULFATE (K_2SO_4). *Sulfate of potash* is the term usually applied to this salt by the fertilizer trade. It is a white material containing 42 to 44 per cent potassium (50 to 53% K_2O) and is produced by four different processes, two of which involve reactions of other salts with potassium chloride and two of which involve the reaction with sulfur or sulfuric acid.

Potassium sulfate finds its greatest use on potatoes and tobacco, which are sensitive to large applications of chlorides. Its behavior in the soil is essentially the same as that of muriate, but it has the advantage of supplying plant-nutrient sulfur, which is far more widely deficient in soils than chlorine.

POTASSIUM MAGNESIUM SULFATE (K_2SO_4, $MgSO_4$). This is a double salt of potassium chloride and potassium sulfate with small amounts of sodium chloride, which are largely removed in processing. The material contains 18 per cent potassium (22% K_2O), 11 per cent magnesium (18% MgO), and 22 per cent sulfur. It has the advantage of supplying both magnesium and sulfur and is frequently included in mixed fertilizers for that purpose on soils deficient in these two elements. Its behavior is not unique, for it reacts as would any other neutral salt when applied to the soil.

POTASSIUM NITRATE (KNO_3). This fertilizer material is also known as saltpeter or niter. It contains 13 per cent nitrogen and 37 per cent potassium (44% K_2O). Agronomically, it is an excellent source of fertilizer nitrogen and potassium. Its once high production cost restricted its use to crops of high acre value, but in 1963 commercial production was started in the United States. Before that time most of the fertilizer grade was imported, some of it from Chile, where it was manufactured in conjunction with the production of nitrate of soda.

Potassium nitrate is being marketed largely for use on fruit trees and on crops such as tobacco, cotton, and vegetables. If production costs can be lowered, it will undoubtedly compete with other sources of nitrogen and potassium for use on crops of a lower value per acre.

POTASSIUM POLYPHOSPHATE (KPO_3). A number of years ago scientists at TVA developed potassium metaphosphate which contains 33 per cent K (40% K_2O) and 27 per cent P (60% P_2O_5). It is water insoluble but hydrolyzes in water and the soil to the orthophosphate form. Production of K-meta was never commercialized because of the high cost. Numerous tests showed it to be an excellent fertilizer source of both P and K, and it has the unique advantage of having no salt effect.

Quite recently, however, a potassium polyphosphate has been developed which shows promise of successful commercial development. While not strictly a potassium metaphosphate, it does contain a mixture of polymerized phosphates which result from the removal of molecular water from the orthophosphate. The material presently being marketed is suitable for both liquid and solid fertilizers. A clear liquid product has the grade 0-25-20 (56% P_2O_5, 24% K_2O) and a solid material of the grade 0-50-40 (114% P_2O_5, 48% K_2O) can also be made.

POTASSIUM CARBONATE (K_2CO_3) AND POTASSIUM BICARBONATE ($KHCO_3$). These salts have been used only experimentally as sources of fertilizer potassium. Their high cost of manufacture has precluded their production as commercial fertilizers. Limited experimental work with potassium bicarbonate as a source of potassium for plant growth suggests that its use on acid soils will reduce loss of cations by leaching. It was also suggested that when applied with phosphate fertilizers the availability of the phosphate will be increased.

The important potassium fertilizers together with their plant nutrient contents are listed in Table 7-1.

Agronomic Value of Various Potassium-Containing Fertilizers. Potassium fertilizers have been compared in numerous field and green-

TABLE 7-1. The Plant Nutrient Content of Some Common Potassium Fertilizer Materials

Material	Nitrogen (%)	Phosphorus (%)	Potassium (%)	Sulfur (%)	Magnesium (%)
Muriate of potash	—	—	50–52	—	—
Sulfate of potash	—	—	44	18	—
Potassium magnesium sulfate	—	—	18	22	11
Potassium nitrate	13	—	37	—	—
Potassium polyphosphate	—	25–50	20–40	—	—
Potassium carbonate	—	—	56	—	—
Potassium bicarbonate	—	—	39	—	—

house trials, as have the sources of nitrogen and phosphorus. Here, again, the data are too voluminous to consider. However, the following points summarize the results of this experimental work:

1. In general, if the material is being used for its potassium content alone, one material is as good as any other. The selection should be based on the cost per pound of potassium applied to the soil.
2. The lower-grade potassium materials will have a greater content of soluble salts per pound of potassium supplied. With some crops and with some methods of fertilizer placement, the higher salt content may be injurious to plant growth.
3. Sources such as potassium nitrate and potassium polyphosphate, which contain the other major elements, are fully as effective as sources of potassium as potassium chloride and must be evaluated on the basis of economy of supplying potassium as well as nitrogen and phosphorus. Materials such as these may be completely absorbed by plants, and no anions such as chloride or sulfate will remain. Use of similar forms in combination with other fertilizer materials in greenhouse culture permits the maintenance of adequate nitrogen, phosphorus, and potassium without danger of accumulation of excess salts. The same principle would apply at least in part under some field conditions in which heavy fertilization is required.
4. Accompanying elements, such as sulfur, magnesium, and sodium, are agronomically important on many soils. The value of the accompanying element must be considered in choosing among the various sources. In addition, the economics of applying the elements in question in one material or in separate materials must be kept in mind.
5. Tobacco is a crop that is extremely sensitive to excessive amounts of chloride. Although available data indicate that as many as 20 lb./A. are beneficial, quantities in excess of 30 to 40 lb. will impair burning quality. In some potato, sweet potato, and citrus areas high quantities of chloride are avoided. Instead, sources such as potassium sulfate or potassium nitrate may furnish the major portion of the potassium.

Calcium and Magnesium Fertilizers.   Calcium and magnesium are often termed *secondary nutrient elements*, perhaps in part because large amounts occur incidentally in some fertilizers. Although these elements may be of secondary importance to the fertilizer manufacturer, whose purpose historically has been to supply the so-called primary nutrients, nitrogen, phosphorus, and potassium, they are as essential to plant growth as the major plant-nutrient elements.

SOURCES OF CALCIUM. Fertilizers are not manufactured as such simply to supply calcium. This element is more economically supplied through periodic applications of agricultural lime, a subject covered in detail in Chapter 11.

In the past most mixed fertilizers contained about 12 per cent calcium, for they were based largely on ordinary superphosphate. The increasing use of ammonium phosphates and phosphoric acid in fertilizer manufacture, however, is steadily reducing the calcium content of fertilizers. This fact emphasizes even more the importance of maintaining an adequate liming program.

Several fertilizer materials used today, which contain significant quantities of calcium, are shown in Table 7-2. Most of these materials, with the exception of gypsum, have already been discussed. Gypsum $(CaSO_4 \cdot 2H_2O)$ has been used as a fertilizer for a long time. It was applied in the early Greek and Roman times and was also used extensively in Europe in the eighteenth century. Deposits are found in several states in the United States and in many other areas of the world. Large amounts of by-product gypsum are on hand as a result of the manufacture of phosphoric acid.

Gypsum is a source of calcium for peanuts in the United States and is applied directly to the plant in early bloom. In several African countries, for example, Nigeria, Senegal, and Upper Volta, the gypsum contained in the superphosphate applied to this crop is as valuable for its sulfur content as it is for calcium. Large acreages of soil in these countries are severely sulfur-deficient. Gypsum has little effect on soil reaction, hence may have some value on crops that demand an acid soil, yet need considerable calcium. It is widely used on the alkali soils of the West. The calcium replaces sodium on the exchange complex, and the sodium sulfate is carried out in the drainage water. This replacement serves to flocculate the soil and make it more permeable to water.

TABLE 7-2. Average Calcium Content of Common Materials*

	Calcium (%)
Calcium nitrate	19.4
Ammonium nitrate-lime mixtures	8.2
Calcium cyanamide	38.5
Gypsum	22.3
Phosphate rock	33.1
Superphosphate, ordinary	19.6
Superphosphate, triple	14.3

* Mehring, *Soil Sci.*, **65**:9 (1948). Reprinted with permission of The Williams & Wilkins Co., Baltimore.

TABLE 7–3. Average Magnesium Content of Some Common Fertilizer Materials*

	Magnesium (%)
Basic slag, open hearth	3.4
Ammonium nitrate-lime mixtures	4.4
Epsom salt ($MgSO_4 \cdot 7H_2O$)	9.6
Kieserite, calcined ($MgSO_4 \cdot H_2O$)	18.3
Manure salts (20% K_2O)	3.5
Potassium magnesium sulfate (22.3% K_2O)	11.1
Magnesia	55.0

* Mehring, *Soil Sci.*, **66:**147 (1948). Reprinted with permission of The Williams & Wilkins Co., Baltimore.

SOURCES OF MAGNESIUM. An important source of magnesium is dolomitic limestone ($CaCO_3 \cdot MgCO_3$), a material used to supply calcium and magnesium as well as to correct soil acidity. As already stated, limestones are discussed in Chapter 11 and only sources of magnesium other than dolomite are covered in this section.

In contrast to calcium, few of the carriers of primary nutrients contain large amounts of magnesium. Potassium magnesium sulfate is a very important exception. When magnesium is required in mixed fertilizers, a material known principally for its magnesium content is added. Dolomitic limestone is used for this purpose, as are magnesium sulfate and potassium magnesium sulfate.

POTASSIUM MAGNESIUM SULFATE. This material was discussed in the section dealing with potash fertilizers. About the only other materials produced specifically for their magnesium content are magnesium sulfate and magnesia, which is manufactured by calcining magnesite or brucite to give magnesium oxide. It can also be made from well brines or sea water by treatment with calcium hydroxide and subsequent heating.

The magnesium content of fertilizer compounds containing this element are listed in Table 7-3.

Summary

1. The potassium content of soils is variable, but many soils contain large amounts of this element. The total amount present, however, is no criterion of the amount available to plants.
2. The availability of potassium to plants is governed by the equilibria in the soil system among the forms of potassium arbitrarily designated as unavailable, slowly available, and readily available. The unavailable forms are those occurring in the primary potassium-bearing

minerals. The slowly available forms are those resulting from the interaction of K^+ ions with certain clay minerals, whereas the readily available forms are made up of exchangeable and water-soluble potassium.

3. With today's intensive agriculture which demands the production of high-yielding crops, considerable potassium is required to fulfill the needs of these crops. Under such conditions, the potassium released from slowly available forms in the soil will be much less than that released under exhaustive cropping in pots or under low yield conditions in the field.

4. Plant availability of soil potassium is greatly influenced by the soil colloids present, temperature, wetting and drying of clays, and soil pH. In some cases applications of large amounts of potassium fertilizer may result in greater uptake of this element than the plant can efficiently utilize. This has been termed *luxury consumption,* but with modern, high yield agriculture it does not appear to be a serious problem.

5. Calcium and magnesium, which are somewhat similar in their behavior in soils, are adsorbed by soil colloids. Magnesium may become fixed in the lattice structure of certain clay minerals under some soil conditions, although calcium does not exhibit this behavior.

6. The principal fertilizer compounds containing potassium are potassium chloride, potassium sulfate, potassium magnesium sulfate, and potassium nitrate. The principal sources of magnesium are potassium magnesium sulfate, magnesium sulfate, magnesium oxide, and dolomitic limestone. Calcium is supplied by both dolomitic and calcitic limestone as well as gypsum. Some calcium is supplied incidentally in the calcium phosphate fertilizers.

7. The agronomic effectiveness of the principal potassium fertilizers is about equal. Obviously if potassium sulfate or potassium-magnesium sulfate were compared with potassium chloride on a sulfur-deficient soil, the first two materials would give superior results. However, if adequate sulfate were supplied with the potassium chloride, there would be no yield differences among the three sources. As with nitrogen and phosphorus, an important factor governing the selection of the potassium fertilizer is the *cost per unit of potassium* applied to the land.

Questions

1. Why are soils with much clay generally more fertile than sandy soils? Are they *always* more productive? Why?
2. Under what soil conditions is there most likely to be reversion of available or added potassium to less available forms?
3. A soil was found to contain 1.5 meq./100 g. of exchangeable potas-

sium. To how many pounds of muriate of potash per acre does this correspond?

4. What effect will the liming of an acid soil have on the retention of potassium?
5. Why does not the addition of gypsum to an acid soil result in an increased conservation of potassium?
6. What is the original source of soil potassium? Is finely ground granite dust a suitable source of potassium? Why?
7. What mechanisms are thought to account for potassium fixation in soil?
8. From your knowledge of cation exchange in soils, would you predict that the addition of sodium would tend to deplete or conserve the supply of soil potassium? Why?
9. By what processes is potassium transported to the plant root surface? What factors govern this movement?
10. What factors control the amount of potassium present in the soil solution?
11. Under what conditions (soil and environmental) would you expect to obtain the least and greatest crop response to surface-applied top dressings of a soluble potassium fertilizer?
12. By what common agricultural practice is magnesium most easily added to soils?
13. What is *luxury consumption* of potassium? Is it a serious problem under most soil and cropping conditions? How can it be minimized?
14. A fertilizer is guaranteed to contain 30 per cent potassium oxide. To what percentage of potassium does this correspond?
15. Under what soil conditions might you prefer to use potassium magnesium sulfate rather than muriate and dolomite or muriate alone?
16. Name the common sources of fertilizer potassium.
17. What are some incidental sources of plant-nutrient calcium?
18. How many pounds of potassium sulfate would be required to supply the same amount of potassium in 350 lb. of 60 per cent muriate of potash?
19. What are some common sources of fertilizer magnesium?
20. In what forms are potassium, magnesium, and calcium absorbed by plants?

Selected References

Adams, F., and J. B. Henderson, "Magnesium availability as affected by deficient and adequate levels of potassium and lime." *SSSA Proc.*, **26**:65 (1962).

Agarwal, R. R., "Potassium fixation in soils." *Soils Fertilizers*, **23**:375 (1960).

Attoe, O. J., "Potassium fixation and release in soils occurring under moist and drying conditions." *SSSA Proc.*, **11**:145 (1946).

Barber, S. A., "Relation of fertilizer placement to nutrient uptake and crop yield: II. Effects of row potassium, potassium soil level, and precipitation." *Agron. J.*, **51**:97 (1959).

Barber, T. E., and B. C. Matthews, "Release of non-exchangeable soil potassium by resin-equilibration and its significance for crop growth." *Can. J. Soil Sci.*, **42**:266 (1962).

Barshad, I., "Cation exchange in micaceous minerals: I. Replacement of interlayer cations of vermiculite with ammonium and potassium ions." *Soil Sci.*, **77**:463 (1954).

Barshad, I., "Cation exchange in micaceous minerals: II. Replaceability of ammonium and potassium from vermiculite, biotite, and montmorillonite." *Soil Sci.*, **78**:57 (1954).

Bartlett, R. J., and T. J. Simpson, "Interaction of ammonium and potassium in a potassium-fixing soil." *SSSA Proc.*, **31**:219 (1967).

Bates, T. E., and A. D. Scott, "Control of potassium release and reversion associated with changes in soil moisture." *SSSA Proc.*, **33**:566 (1969).

Black, C. A., *Soil-Plant Relationships*, 2nd ed. New York: Wiley, 1968.

Bolt, G. H., M. E. Summer, and A. Kamphorst, "A study of the equilibria between three categories of potassium in an illitic soil." *SSSA Proc.*, **27**:294 (1963).

Boswell, F. C., and O. E. Anderson, "Potassium movement in fallowed soils." *Agron. J.*, **60**:688 (1968).

Bray, R., "A nutrient mobility concept of soil-plant relationships." *Soil Sci.*, **78**:9 (1954).

Bray, R. H., and E. E. DeTurk, "The release of potassium from non-replaceable forms in Illinois soils." *SSSA Proc.*, **3**:101 (1938).

Buckman, H. O., and N. C. Brady, *The Nature and Properties of Soils*, 7th ed. New York: Macmillan, 1969.

Burns, A. F., and S. A. Barber, "The effect of temperature and moisture on exchangeable potassium." *SSSA Proc.*, **25**:349 (1961).

Claassen, M. E., and G. E. Wilcox, "Comparative reduction of calcium and magnesium composition of corn tissue by NH_4—N and K fertilization," *Agron. J.*, **66**:(in press) (1974).

Dennis, E. J., and R. Ellis, Jr., "Potassium in fixation equilibria and lattice changes in vermiculite." *SSSA Proc.*, **26**:230 (1962).

Dowdy, R. H., and T. B. Hutcheson, Jr., "Effect of exchangeable potassium level and drying on release and fixation of potassium by soils as related to clay mineralogy." *SSSA Proc.*, **27**:31 (1963).

Dowdy, R. H., and T. B. Hutcheson, Jr., "Effect of exchangeable potassium level and drying upon availability of potassium to plants." *SSSA Proc.*, **27**:521 (1963).

Drake, M., J. Vengris, and W. G. Colby, "Cation exchange capacity of plant roots." *Soil Sci.*, **72**:139 (1951).

Drechsel, E. K., "Potassium phosphates: The new generation of super phosphates." Paper presented to Amer. Chem. Soc., Div. Fert. and Soil Chem., August 28, 1973.

Fried, M., and H. Broeshart, *The Soil-Plant System*. New York and London: Academic Press, 1967.

Graham, E. R., and D. H. Kampbell, "Soil potassium availability and reserve as related to the isotopic pool and calcium exchange equilibria." *Soil Sci.,* **106:**101 (1968).

Grava, J., G. E. Spalding, and A. C. Caldwell, "Effect of drying upon the amounts of easily extractable potassium and phosphorus in Nicollet clay loam." *Agron. J.,* **53:**219 (1961).

Gray, B., M. Drake, and W. G. Colby, "Potassium competition in grass-legume associations as a function of root cation exchange capacity." *SSSA Proc.,* **17:**235 (1953).

Grimme, H., K. Nemeth, and L. C. v. Braunschweig, "Some factors controlling potassium availability in soils." *Proc. Int. Symp. Soil Fert. Evaln. New Delhi.* **1:**5 (1971).

Hanway, J. J., and A. D. Scott, "Soil potassium-moisture relations: II. Profile distribution of exchangeable K in Iowa soils as influenced by drying and rewetting." *SSSA Proc.,* **21:**501 (1957).

Hanway, J. J., and A. D. Scott, "Soil potassium-moisture relations: III. Determining the increase in exchangeable soil potassium on drying soils." *SSSA Proc.,* **23:**22 (1959).

Henderson, R., "The application of potassic fertilizers to pasture and the incidence of hypomagnesemia." *Potash Ltd. Tech. Ser.,* **1:**23 (1960).

Hood, J. T., N. C. Brady, and D. J. Lathwell, "The relationship of water-soluble and exchangeable potassium to yield and potassium uptake by Ladino clover." *SSSA Proc.,* **20:**228 (1956).

Jackson, J. E., and G. W. Burton, "An evaluation of granite meal as a source of potassium for Coastal Bermuda grass." *Agron. J.,* **50:**307 (1958).

Jacob, K. D., Ed., *Fertilizer Technology and Resources in the United States.* New York: Academic Press, 1953.

Kilmer, V. J., S. E. Younts, and N. C. Brady, Eds., *The Role of Potassium in Agriculture.* Madison, Wisconsin: American Society of Agronomy, 1968.

Matthews, B. C., and C. G. Sherrell, "Effect of drying on exchangeable potassium of Ontario soils and the relation of exchangeable K to crop yield." *Can. J. Soil Sci.,* **40:**35 (1960).

Mengel, K., "Potassium availability and its effect on crop production." *Proc. Jap. Pot. Symp. Tokyo, Japan* (1971).

Mortland, M. M., K. Lawton, and G. Uehara, "Fixation and release of potassium by some clay minerals." *SSSA Proc.,* **21:**381 (1957).

Mortland, M. M., K. Lawton, and G. Uehara, "Alteration of biotite to vermiculite by plant growth." *Soil Sci.,* **82:**477 (1958).

Nemeth, K., and H. Grimme, "Effect of soil pH on the relationship between K concentration in the saturation extract and K saturation of soils." *Soil Sci.,* **114:**349 (1972).

Nemeth, K., K. Mengel, and H. Grimme. "The concentration of K, Ca, and Mg in the saturation extract in relation to exchangeable K, Ca, and Mg." *Soil Sci.,* **109:**179 (1970).

Nuttall, W. F., B. P. Warkentin, and A. L. Carter, "'A' values of potassium related to other indexes of soil potassium availability." *SSSA Proc.,* **31:**344 (1967).

Olson, R. A., et al., Eds., *Fertilizer Technology & Use,* 2nd ed. Madison, Wisconsin: Soil Science Society of America, 1971.

Page, A. L., F. T. Bingham, T. J. Ganje, and M. J. Garber, "Availability and fixation of added potassium in two California soils when cropped to cotton." *SSSA Proc.*, **27:**323 (1963).

Page, A. L., W. D. Burge, T. J. Ganje, and M. J. Garber, "Potassium and ammonium fixation by vermiculitic soils." *SSSA Proc.*, **31:**337 (1967).

Pope, A., and H. B. Cheney, "The K-supplying power of several Western Oregon soils." *SSSA Proc.*, **21:**75 (1957).

Pratt, P. F., "Potassium removal from Iowa soils by greenhouse and laboratory procedures." *Soil Sci.*, **72:**107 (1951).

Richards, G. E., and E. O. McLean, "Release of fixed potassium from soils by plant uptake and chemical extraction techniques." *SSSA Proc.*, **25:**98 (1961).

Sauchelli, V., Ed., *The Chemistry and Technology of Fertilizers.* ACS Monograph 148. New York: Reinhold, 1960.

Sauchelli, V., *Manual on Fertilizer Manufacture*, 3rd ed. Caldwell, New Jersey: Industry Publications, 1963.

Scott, A. D., and T. E. Bates, "Effect of organic additions on the changes in exchangeable potassium observed in drying soils." *SSSA Proc.*, **26:**209 (1962).

Scott, T. W., and F. W. Smith, "Effect of drying upon availability of potassium in Parsons silt loam surface soil and subsoil." *Agron. J.*, **49:**377 (1957).

Spies, C., "Crops have big appetites." *Purdue Agron. Crops and Soils Notes No. 218.* July 1973.

Stevenson, F. J., and A. P. S. Dhariwal, "Distribution of fixed ammonium in soils." *SSSA Proc.*, **23:**121 (1959).

Thorne, D. W., and H. B. Peterson, *Irrigated Soils.* New York: Blakiston, 1954.

Thorup, R. M., and A. Mehlich, "Retention of potassium meta- and orthophosphates by soils and minerals." *Soil Sci.*, **91:**38 (1961).

Walsh, L. M., and J. D. Beaton, Eds. *Soil Testing and Plant Analysis* (revised ed.). Madison, Wisconsin: Soil Science Society of America, 1973.

Walsh, L. M., and J. T. Murdock, "Native fixed ammonium and fixation of applied ammonium in several Wisconsin soils." *Soil Sci.*, **89:**183 (1960).

Welch, L. F., "Availability of non-exchangeable potassium and ammonium to plants and micro-organisms." *Dissertation Abstr.*, **19:**1885 (1958).

Welch, L. F., and A. D. Scott, "Nitrification of fixed ammonium in clay minerals as affected by added potassium." *Soil Sci.*, **90:**79 (1960).

Welch, L. F., and A. D. Scott, "Availability of non-exchangeable soil potassium to plants as affected by added potassium and ammonium." *SSSA Proc.*, **25:**102 (1961).

Williams, D. E., "The absorption of potassium as influenced by its concentration in the nutrient medium." *Plant Soil*, **15:**387 (1961).

Wood, L. K., and E. E. DeTurk, "The absorption of potassium in soils in non-replaceable forms." *SSSA Proc.*, **5:**152 (1941).

Woodruff, C. M., "The energies of replacement of calcium by potassium in soils." *SSSA Proc.*, **19:**167 (1955).

York, E. T., Jr., R. Bradfield, and M. Peech, "Calcium-potassium interactions in soils and plants: I. Lime-induced potassium fixation in Mardin silt loam." *Soil Sci.*, **76:**379 (1953).

8. SULFUR AND MICROELEMENTS IN SOILS AND FERTILIZER

Sulfur is not a microelement, but it is considered with them in this chapter for the sake of convenience of presentation. Sulfur, as pointed out in Chapter 3, is required by plants in about the same quantities as is phosphorus, an element traditionally classed as a major plant nutrient. The microelements, on the other hand, are required by plants in very small quantities which range from only a few pounds per acre to amounts measured in ounces per acre or less. The microelements to be considered in this chapter are boron, copper, cobalt, iron, manganese, molybdenum, zinc, and chlorine.

SULFUR

Sources in Soils. The earth's crust contains about 0.06 per cent sulfur. It is present as sulfide, sulfates, and in organic combination with carbon and nitrogen. The original source of soil sulfur was doubtless the sulfides of metals contained in plutonic rocks. As these rocks were exposed to the weathering processes, the minerals decomposed and sulfide was oxidized and released as sulfate. These sulfates were then precipitated as soluble and insoluble sulfate salts in arid or semiarid climates, absorbed by living organisms, or reduced by other organisms to sulfides or elemental sulfur under anaerobic conditions. Some of the released sulfates, of course, found their way to the sea in drainage waters.

The sulfur in most arable land today is in the form of organic matter, soluble sulfates in the soil solution, or adsorbed on the soil complex. It will be recalled that sulfur is a component of proteins and that when these materials are returned to the soil and converted to humus a large fraction of the sulfur remains in organic combination. Much of the total sulfur found in the surface of humid-region soils is in the organic form. In arid soils, of course, the sulfates of calcium, magnesium, sodium, and potassium are frequently precipitated in large quantities in the soil profile. Appreciable amounts of exchangeable SO_4^{2-} may be present in subsoils that contain 1:1 clays and hydrous oxides of iron and aluminum.

Another source of soil sulfur is the atmosphere. Around centers of industrial activity, in which coal and other sulfur-containing products are burned, sulfur dioxide is released into the air, and much of this gas is later brought back to earth by the rain. Plants may also absorb sulfur dioxide by gaseous diffusion into the leaves, and this sulfur is utilized by the plant in its normal metabolic processes. If the concentration in the air is too great, however, there may be injury to plants.

In the United States about 32 million tons of SO_2 (16 million tons of sulfur) were estimated to have been emitted into the atmosphere in 1972. Most of this resulted from the combustion of fossil fuels, but industrial processes such as ore smelting, petroleum refining operations, and others contributed about 20 per cent of the total emitted. Much of this SO_2 is returned to the soil in rainfall and within a relatively small radius around the point of emission. Available data indicate that the amounts of sulfur brought down in rainfall in the USA range from about one pound per acre per year in rural areas to 100 pounds per acre per year near areas of industrial activity.

In addition to the sulfur brought down in the rain, there is movement into the soil of sulfur dioxide gas by diffusion. The extent to which this phenomenon takes place has not been accurately determined, but recent studies at Minnesota have indicated that SO_2 absorption by soils can take place to a significant degree.

Because of the growing concern over air pollution, legislation may ultimately be enacted to require the cleaning and scrubbing of all waste gases. Many industrial companies are presently required to clean effluent gases, and it is likely that this practice will spread and, in turn, will reduce the amounts of sulfur brought down by the rain and absorbed from the atmosphere directly by plants and soil. While the emission of sulfur compounds into the atmosphere by industrial activity has received a lot of unfavorable publicity, the fact remains that 70 per cent of the total content of sulfur compounds in the atmosphere is of nonmanmade origin. Volatile sulfur compounds are released in large quantities from volcanic activity, from tidal marshes, from decaying organic matter, and from other sources.

Behavior of Sulfur Compounds in Soils. The behavior of sulfur in the soil will be considered with reference to three types of compounds: (1) reduced forms of sulfur contained in organic combination, (2) sulfate sulfur, and (3) elemental sulfur and sulfides.

ORGANIC SULFUR. As described in Chapter 6, sulfur forms an important part of soil organic matter. Several Scottish soils had ratios of total C:N:nonsulfate sulfur of 113:10:1.3 for calcareous soils and

147:10:1.4 for the humid noncalcareous. The ratio of N:S in these studies was less variable than that for C:S.

Workers in Australia examined the relationship existing between total soil nitrogen and the nonsulfate sulfur content of 155 eastern Australian soils. A very close similarity exists between these two soil quantities, much closer, in fact, than that between soil nitrogen and soil organic phosphorus. The nitrogen:sulfur ratio of the Australian soils ranged from 10:1.21 on the acid soils to 10:1.52 on the calcareous. Studies in New Zealand have shown nitrogen:sulfur values of 10:1, whereas for some podzolic soils of Minnesota this figure was 10:1.3. For some chernozems and black prairie soils in the same state the nitrogen:sulfur ratio was 10:1.5. The fairly constant relation between these two elements in such a wide range of soils illustrates the importance of sulfur in the formation and decomposition of soil organic matter.

SULFUR TURNOVER IN SOILS. Sulfur turnover in soil has also been studied by workers in Australia, who found that the mineralization of sulfur (defined as for nitrogen in Chapter 5) from the decomposition of a wide range of organic materials depends on the sulfur content of the decomposing material in much the same way that the mineralization of nitrogen depends on the nitrogen content. This is illustrated by the data in Figure 8-1. It is apparent that smaller amounts of sulfate were liberated from the materials containing the smaller percentages of this element and that an analogous situation exists for the mineralization of nitrogen. In the samples containing less than about 0.15 per cent sulfur, there was, in fact, a reduction in the level of soil sulfate at the end of the incubation period, which may suggest immobilization of sulfur.

Sulfur may be immobilized in soils in which the ratio of either carbon

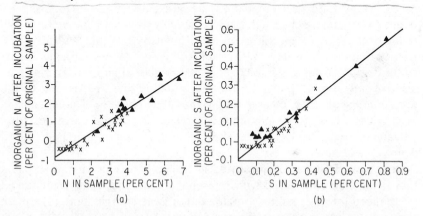

Figure 8-1. Relationships between: (a) the nitrogen mineralized and the nitrogen content of the original material; and (b) the inorganic sulfur present after incubation and the sulfur content of the original material. [Barrow, *Australian J. Agr. Res.*, **11**:960 (1960).]

or nitrogen to sulfur is too wide. Conversely, if the C:S or N:S ratio is too narrow and mineral nitrogen is added, some of the added nitrogen will be immobilized. The importance of maintaining an appropriate balance in the soil between nitrogen and sulfur is obvious.

The practical implication of the foregoing is well illustrated by some work done in Colorado. Wheat straw with a low sulfur content was incorporated into soils fertilized with N, P, and K or N, P, K, and S. Winter wheat was then planted on these soils. A portion of the results of this study are shown in Figure 8-2. Two points are clear from these data. The addition of wheat straw in increasing amounts to soil to which no fertilizer sulfur was added progressively decreased the growth of wheat plants. The addition of sulfur in the fertilizer overcame the limiting effect of the straw, and the growth of wheat was the same on the treatments receiving straw as it was on those receiving no straw.

Apparently the addition of straw with a low sulfur content to the soil used in this study tied up the available soil sulfur because of the immobilization by soil microorganisms during decomposition of the straw. This situation was aggravated by the addition of fertilizer N which further widened the N:S ratio of the soil, resulting in the immobilization of any available SO_4. As a result the wheat plants had insufficient sulfur available to them to permit proper growth. The addition of fertilizer sulfur to this soil, however, overcame this unfavorable situation.

In practical farming operations where large amounts of straw or stover are to be returned to the soil, the grower should take steps to ensure that adequate nitrogen and sulfur are available to promote rapid decomposition of the added straw. Otherwise, a temporary nitrogen or sulfur deficiency may be induced in the following crop.

EFFECT OF DRYING ON MINERALIZATION. The drying of soils has a pronounced effect on the mineralization of sulfur. Workers in Australia have shown that when soils are dried before incubation the increase in mineralization of sulfur is large. This is quite convincingly illustrated by

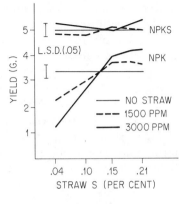

Figure 8-2. The growth of winter wheat as affected by incorporating different straws of varying sulfur contents into NPK fertilized soil, with and without fertilizer S. [Stewart et al., *SSSA Proc.*, **30**:355 (1966).]

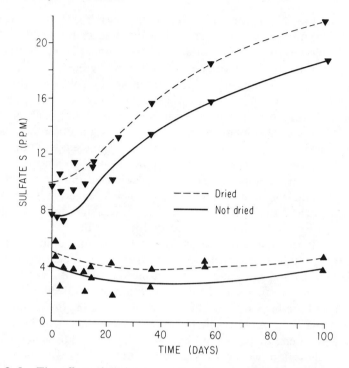

Figure 8–3. The effect of drying two soils before incubation on the mineralization of sulfur. [Barrow, *Australian J. Agr. Res.*, **12**:306 (1961).]

the curves in Figure 8-3. This effect of drying on the mineralization of sulfur would have an obvious influence on fertilizer recommendations based on soil-test results from sulfur analyses made on air-dry soils. The same would obtain if plant growth studies were made in the greenhouse on soils that had been dried before planting. The higher level of measured sulfur on one of these soils suggests that supplemental fertilizer applications of this element may not be needed, a conclusion that may have been entirely unwarranted in the undisturbed condition in the field.

The effect of drying on sulfur mineralization could have an impact on the growth of crops. In Australia it has been observed that protracted dry spells are followed by a flush of plant growth and this has been thought to be the result, at least in part, of the increased mineralization of organic sulfur. Australia and New Zealand are two of many countries in which there are extensive areas of sulfur-deficient soils.

ORGANIC SULFUR COMPOUNDS IN SOILS. A large proportion of the sulfur in soils, especially those of the humid regions, is in the organic form. The specific identity of such compounds is presently largely unknown, though recent work in Australia has thrown some light on this topic. It has been shown, for example, that much of the organic sulfur in soils may be in the form of organically bound sulfates. It has been suggested that sulfur-containing amino acids may also take part, perhaps as intermediates, because of their presence in most plant proteins.

Freney of Australia determined sulfur intermediates in soil by using a cysteine perfusion technique, which, very briefly, involves the perfusion of a solution of cysteine hydrochloride through soil until sulfate production reaches a maximum. The accumulated sulfate is then removed from the system by washing with water. As a result of this treatment, the soil has a population of microorganisms and a system of enzymes capable of oxidizing to sulfate cysteine or any of the intermediates in its transformation to sulfate. Such intermediates are quickly oxidized to sulfate, and those compounds that are not intermediates in the oxidation of cysteine to sulfate are converted at a much slower rate.

Results of a part of this work are shown in Figure 8-4. The enriched soil has been perfused with the cysteine solution, and the rapidity with which it was oxidized to sulfate identifies it as one of the intermediates

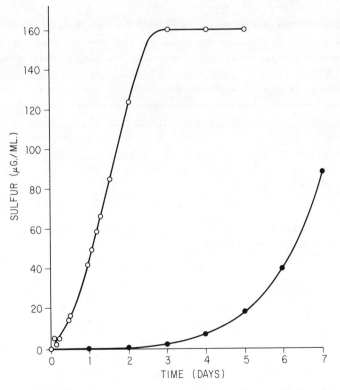

Figure 8-4. Sulfate production from a 0.01N cysteinesulphinic acid solution perfused through an enriched soil (∘) and a normal soil (•). [Freney, *Australian J. Biol. Sci.,* **13:**387 (1960).]

in the decomposition process. Several other intermediates identified were cysteic acid and three other amino acids not containing sulfur. The speed with which the cysteic acids were converted to sulfate suggests that they are not stable sulfur intermediates in soils. This perfusion technique is a useful tool for the identification of intermediate products of degradation, not only of sulfur compounds, but others as well. However, there is still considerable work needed to identify the stable organic sulfur compounds in soils.

RHIZOSPHERE EFFECT ON SULFUR MINERALIZATION. An interesting phenomenon concerning the mineralization of organic sulfur has been observed by Freney and Spencer in some Australian soils. It was found that growing plants increased the mineralization of organic soil sulfur. Plants were grown in pots containing soil to which sulfate sulfur had been added in amounts varying from 0 to 108 ppm. On a similarly treated set of pots no plants were grown. The results of this study are shown in Table 8-1. These data indicate that there was considerable immobilization of added sulfur in many of the soils on which no plants were grown. There was some immobilization of added sulfate in the pots that were planted and to which 108 ppm. of sulfate had been added. It is apparent, however, that in the presence of growing plants there was a net mineralization of sulfur on four of the five soils studied at levels of added sulfate of 0 to 36 ppm. On soil No. 4 there was a net mineralization only at the 0- and 4-ppm. levels.

This was explained as possibly being caused by the rhizosphere effect, for a greater concentration of soil microorganisms has been observed in the zone surrounding plant roots than in the remainder of the soil. The stimulating effect of these organisms and their secretions probably produced an increased breakdown of soil organic matter and a consequent release of sulfate sulfur.

The Australian workers further commented on the possible relation of

TABLE 8-1. Net Mineralization or Immobilization of Sulfate* in the Presence and Absence of Plants†

	Added sulfate sulfur (ppm.)									
	0		4		12		36		108	
Soil number	With‡	Without‡	With	Without	With	Without	With	Without	With	Without
1	1.2	0.3	2.6	−1.6	6.5	− 6.9	1.9	−17.7	−12.5	−55.2
2	1.1	1.1	0.9	−0.1	5.1	− 2.8	2.0	−12.9	−10.8	−19.6
3	6.9	3.8	6.5	0.4	6.4	− 3.2	6.1	−13.7	− 7.7	−41.7
4	2.2	−3.9	2.7	−6.9	0.6	−13.5	0.5	−31.2	−13.4	−70.6
5	2.1	1.1	4.3	−0.3	5.6	' 4.2	2.5	−10.2	− 4.5	−32.8

* All figures in parts per million of sulfate sulfur.
† Freney et al., *Australian J. Agr. Res.*, **11**:339 ,1960).
‡ With plants and without plants.

these findings to bare fallowing of agricultural lands, a practice followed not only in Australia but in the United States and other countries as well. Bare fallowing is the system of accumulating moisture and a supply of available nutrients for the next crop by the decomposition of the residues of the preceding crop. The results of this study suggest, however, that decomposition was greater wherever crops were grown. In addition, there was appreciable immobilization of added sulfate on the soil which had not been planted which suggests that prefallow applications of a sulfate fertilizer could be largely tied up in the organic form and not immediately available to the planted crop.

The various reactions discussed in the foregoing sections point to the considerable influence that organic sulfur may have on the level of plant-available sulfur in soils. In humid regions this fraction of the soil sulfur constitutes the principal reservoir of this element in the surface horizons and, together with hydrous oxides of iron and aluminum and 1 : 1 clays, it will determine the amounts of sulfur that will be retained by soils under leaching conditions.

Inorganic Sulfate Sulfur. Almost all of the inorganic sulfur in well-drained arable soils occurs as the sulfate ion in combination with cations such as calcium, magnesium, potassium, sodium, or NH_4^+ in the soil solution, precipitated as salts of these elements in arid climes, or adsorbed by 1 : 1 clays and the hydrous oxides of iron and aluminum. Some very small quantities of sulfides may occur, but such occurrences are rare and of no agricultural significance. Sulfides occur frequently, however, under water-logged conditions and are covered in a later section which deals with elemental sulfur and sulfides.

MOVEMENT OF SULFATE IN SOILS. Because of its anionic nature and the solubility of most of its common salts, leaching losses of sulfates are generally rather large. However, their tendency to disappear from soils varies widely. As an example, workers at the University of Georgia showed that cotton, grown for five years on two texturally different soils, responded differently to applications of sulfur at 0, 4, 8, 16, and 32 lb./A. per year. On the silt loam soil no responses to added sulfur were observed at the end of the five-year experiment. On the sandy loam soil, however, a sulfur deficiency developed during the fourth cropping year at the zero level of added sulfur.

Another piece of work conducted in California illustrates the rapidity with which sulfur can be lost from soils under conditions of heavy rainfall. Sulfur as gypsum was applied to a stand of clover growing on a field plot of sandy loam soil. Rates of applied gypsum were 100, 200, and 300 lb./A. and rainfall contributed about 21 lb./A. Measurements showed that in the growing season following fertilization 77 per cent of the sulfur applied at the 100-lb. rate of gypsum and 78 per cent of that applied at the 300-lb. rate were accounted for in the percolating water collected from these plots. There was a significant yield response to the added sulfur, and about 31 and 57 per cent, respectively, of the tissue sulfur in these plants was accounted for by the 100- and 300-lb. applications of gypsum.

The relation between the amount of percolating water and the downward movement of sulfate was determined with radioactive S^{35} by workers in Oregon. A given amount of sulfate sulfur was added to columns of soil. Different amounts of water were then added to these columns and the distribution pattern of the sulfur was measured. The results are shown in Figure 8-5. The greater the amount of added water, the greater the net downward movement of the sulfate. The soil used in this study contained 29 per cent clay and is classed as a silt or silt loam.

Another factor influencing the loss of sulfates is the nature of the cation population of the soil solution. This is discussed more fully in the following section which deals with sulfate retention, but because the relationship between sulfate retention and sulfate leaching is generally an inverse one it is mentioned here briefly. Leaching losses of sulfate are greatest when monovalent ions such as potassium and sodium predominate; next in order are the divalent ions such as calcium and magnesium; and leaching losses are least when soils are acid and appreciable amounts of exchangeable aluminum and iron are present.

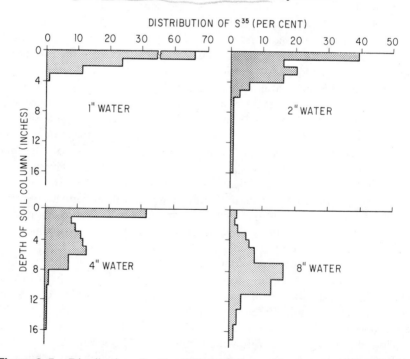

Figure 8-5. Distribution of sulfur (S^{35}) throughout columns of a Willamette soil as a function of the amount of added water. [Chao et al., *SSSA Proc.*, **26**:27 (1962).]

Al + Fe — take up sulphates

ADSORBED SULFATES.   Several years ago in Alabama Ensminger observed that when the subsurface horizons of soils in that state were leached with a solution containing acetate and phosphate ions, large amounts of sulfate sulfur appeared in the extracts. Leaching these same samples with water or a dilute solution of hydrochloric acid failed to extract any sulfate. This work led to the realization that some soils have the capacity to retain sulfates in an adsorbed form, particularly those in which the content of clay and hydrous oxides of iron and aluminum are appreciable. This finding by Ensminger helped to explain the anomalous responses to sulfate fertilization that had been observed. In many field experiments cotton or tobacco plants would exhibit sulfur-deficiency symptoms in the early stages of growth. However, the plants that survived the seedling stage frequently grew out of the deficiency. Further, deep-rooted crops, such as alfalfa and sericea lespedeza, never exhibited the symptoms. It was suggested that the sulfate adsorbed in the subsoil was available to plants and would be utilized by plant roots entering that zone.

Subsequent studies of sulfate adsorption have shown that this phenomenon is influenced by several soil properties. In North Carolina, for example, in a Cecil soil (a type that contains a relatively large proportion of 1:1 clay and hydrous oxides of iron and aluminum) sulfate adsorption increased with the concentration of the sulfate ion in the soil solution and decreased with increasing pH, as illustrated in Figure 8-6.

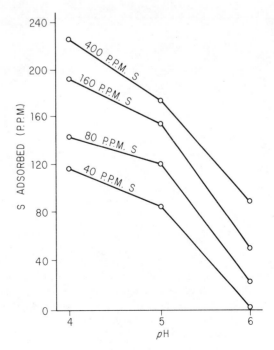

Figure 8–6. Effect of pH on the adsorption of sulfate by a Cecil soil at various solution concentrations of sulfate. [Kamprath et al., *SSSA Proc.*, **20**:463 (1956).]

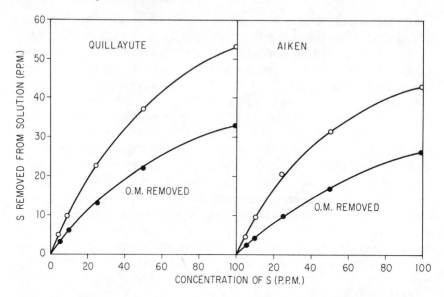

Figure 8–7. The effect of the removal of organic matter on the adsorption of sulfate by Quillayute and Aiken soils. Data were obtained from a soil-solution ratio of 1:10. [Chao et al., *Soil Sci.,* **94:**276 (1962). Reprinted with permission of The Williams & Wilkins Co., Baltimore.]

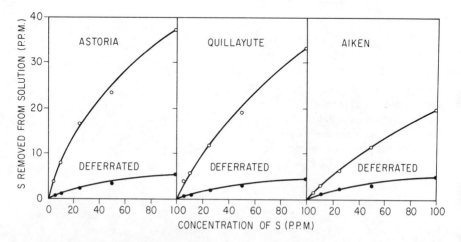

Figure 8–8. The effect of the removal of free iron oxides on the adsorption of sulfate by three retentive soils. Data were obtained by using a soil-solution ratio of 1:10. [Chao et al., *Soil Sci.,* **94:**276 (1962). Reprinted with permission of The Williams & Wilkins Co., Baltimore.]

Workers in Oregon also showed that removal of iron, aluminum, and organic matter significantly reduced the amounts of sulfate retained by soils. The effect of removing organic matter on sulfate retention in two Oregon soils is shown in Figure 8-7. In both samples removal of organic matter decreased the amount of sulfate retained. This is in keeping with the work referred to earlier, in which it was pointed out that appreciable quantities of organically bound sulfate were present in soils.

The effect of removal of iron oxides on sulfate retention is shown in Figure 8-8. Removal of the aluminum oxides from these soils gave similar results. The effect of adsorbed and soil-solution cations on sulfur retention, mentioned in the preceding section, is illustrated by the curves in Figure 8-9. The great effect of pH on adsorption is also illustrated by these data. At any given pH level, however, the amount of adsorption

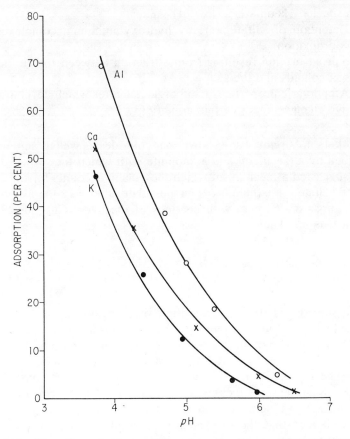

Figure 8-9. The effect of pH on sulfate adsorption from a 0.005N K$_2$SO$_4$ solution by an Aiken soil saturated with the indicated cations. [Chao et al., *SSSA Proc.*, **27**:35 (1963).]

was least when the cation adsorbed on the clay was potassium. It was greater when calcium was the adsorbed ion and greatest when aluminum occupied the exchange sites.

Other work in Oregon showed that the nature of the clay also influenced sulfate adsorption. When the clays were saturated with H^+, the magnitude of sulfate retention was kaolinite > illite > bentonite. When saturated with aluminum, adsorption was about the same for kaolinite and illite but much lower for bentonite.

The results of the studies carried out in Oregon led these workers to suggest several possible mechanisms to explain the observed retention of sulfate in soils.

1. Anion exchange caused by positive charges developed on hydrous iron and aluminum oxides or on the crystal edges of clays, especially kaolinite, at low pH values.
2. Retention of sulfate ions by hydroxy-aluminum complexes by coordination.
3. Salt adsorption resulting from attraction between the surface of soil colloids and the salt.
4. Amphoteric properties of soil organic matter which develop positive changes under certain conditions.

Workers in Virginia have shown that chlorides as well as sulfates were adsorbed by a Cecil soil and a kaolinite with which they worked. They have suggested a mechanism which they claim accounts for the adsorption of sulfate. In explaining their postulate, they assume a homoionic aluminum-saturated clay with coatings of the hydrated oxides R (iron and aluminum). Then

$$yK + Al_x[clay] + yH_2O \longrightarrow Al_x(OH)_y{}^{Ky}[clay] + yH$$
$$SO_4{}^{2-} + R_x(OH)_y[clay] \longrightarrow R_x[(OH)_{y-z}(SO_4)_z]clay + z\ OH^-$$

It is assumed that the K^+ adsorption sites develop from the exchange and/or hydrolysis of aluminum on the clay surface. As a result of this hydrolysis, some H^+ ions go into solution. At the same time $SO_4{}^{2-}$ replaces OH^- ions from R(OH) coatings on clay and substitutes for them. The replaced OH^- ions in turn react with the H^+ ions. Whether the pH of the system increases or decreases depends on the relative rates of the two reactions: hydrolysis and OH^- exchange.

It is claimed that the mechanism proposed explains the several observed phenomena as follows: sulfate adsorption is increased as the pH is lowered because the replaced OH^- ions are more effectively neutralized. Increased cation affinity causes the replacement of more aluminum and results in still further hydrolysis. It has been shown that sulfate adsorption is a function of time as well as of the other factors discussed. In explaining the time effect, it is suggested that the continuation of the aluminum hydrolysis, which produces H^+ ions, neutralizes OH^- ions and thus carries the reaction to completion.

Elemental Sulfur and Sulfides. Elemental sulfur is not found in well-drained upland soils. Under water-logged conditions, in which bacterial reduction is taking place, sulfides are formed and in some instances elemental sulfur is deposited.

In its pure form elemental sulfur is a yellow, inert, water-insoluble crystalline solid. Commercially, it is stored in the open, where it remains unaltered by moisture and temperature. When sulfur is finely ground and mixed with soil, however, it is oxidized to sulfate by soil microorganisms. Because of this property, sulfur has for years been used in the reclamation of alkali soils (which also contain free calcium carbonate).

The oxidation of sulfur in the soil, plus the fact that it is in the elemental form (meaning, of course, that it provides the highest amount of plant nutrient sulfur for the least bulk), has suggested its use as a source of this element in fertilizers. As mentioned in Chapters 5, 6, and 7, many of the higher analysis fertilizers currently being manufactured contain no sulfur. When used for several years on soils low in sulfur, crop deficiencies of this element have appeared. In some areas the deficiency is severe. Work is being carried out to develop fertilizers containing sulfur in the elemental form, and the agronomic suitability of these products is being tested. Sulfur has been successfully introduced into such materials as urea, anhydrous ammonia, CSP, ammonium phosphate, and solid and fluid N-P-K materials. Its usefulness as a plant nutrient depends, of course, on the rate at which it is oxidized to sulfate.

FACTORS AFFECTING SULFUR OXIDATION IN SOILS. Several factors influencing the oxidation of elemental sulfur in soils include the microfloral population of the soil, temperature, moisture, pH, and fineness of the applied sulfur.

Soil Microflora. It has been known for years that elemental sulfur is oxidized in soil by several bacterial species of the genus *Thiobacillus*. The most common form is *T. thiooxidans*, but others are *T. thioparus*, *T. copraliticus*, and *T. ferrooxidans*. These organisms are obligate autotrophic aerobes. In a manner similar to that of the nitrifiers, they obtain their energy by the oxidation of an inorganic material, in this case sulfur, and their carbon from carbon dioxide. A host of organisms takes part in sulfur oxidation, some of which carry the process only a step or two, but the over-all reaction results in the production of sulfuric acid. This reaction requires molecular oxygen,

$$S + \tfrac{3}{2}O_2 + H_2O \longrightarrow H_2SO_4$$

TABLE 8-2. Amounts of Sulfur Oxidized, in Four Different Soils after Incubation at Room Temperature for One-, Two-, and Four-Week Periods*

Soil number	Soil type	pH	Field moisture capacity (%)	Sulfur oxidized (%)		
				1 week	2 weeks	4 weeks
B-51	Plainfield s	4.8	5	1	1	3
B-48	Ella ls	6.8	11	0	4	15
B-62	Kellner ls	6.4	5	3	9	23
B-64	Kellner ls	6.6	5	16	21	57

* Elemental sulfur passing a 230-mesh sieve was mixed with the soil at the rate of 500 pp2m. The soils were maintained at field moisture capacity. Kittams, Ph.D. Thesis, University of Wisconsin, 1963.

and sulfuric acid is produced. Sulfur is one of the most effective agents for increasing soil acidity.

It appears that soil organisms other than the autotrophs are able to oxidize elemental sulfur. The conversion to sulfates apparently can be brought about by facultative autotrophs as well as by heterotrophic organisms.

When finely divided elemental sulfur is mixed with dissimilar soils and incubated under similar conditions of moisture and temperature, it is found that the soils vary considerably in their ability to oxidize the added sulfur. This is illustrated by the data in Table 8-2. The added

TABLE 8-3. Effect of Inoculation with a Soil Known to Be Plentiful in Sulfur-Oxidizing Organisms on the Amount of Sulfur Oxidized in Different Soils after a Two-Week Incubation Period*

Soil type	Sulfur oxidized (%)	
	Not inoculated	Inoculated
Plainfield s	4	11
Ella ls	5	58
Kewaunee l	16	62
Auburndale sil	55	64
Dane sil	40	62
Dane sil	43	68
Dodgeville sil	32	51
Miami sil	18	65
Ontonogan sil	10	58
Ontonogan sic	10	39

* Element sulfur passing 230-mesh was added at the rate of 500 pp2m. The soils were maintained at field moisture capacity. Kittams, Ph.D. Thesis, University of Wisconsin, 1963.

sulfur oxidized at the end of four weeks ranged from 3 to 57 per cent. The textural class and initial pH of three of the soils were similar, but even here the sulfur oxidized ranged from 15 to 57 per cent. It is not unlikely that these differences were in part, at least, the result of variations in the microfloral population.

Some work has been done to show that when the soil or the added sulfur is inoculated with *Thiobacillus* organisms the oxidation of the sulfur is frequently greatly increased.

The effect of inoculating several soils with another of high sulfur oxidizing power is illustrated by the data in Table 8-3. In all cases oxidation at the end of two weeks was greater in the inoculated soil. In some

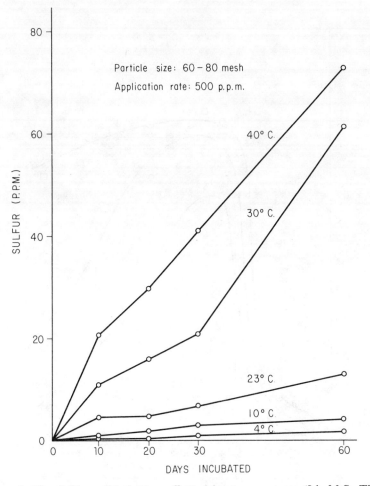

Figure 8–10. Sulfur oxidation as affected by temperature. (Li, M.S. Thesis, University of Minnesota, 1964.)

cases the increases were large indeed. This practice, however, has not been introduced on a commercial farming scale, for as a general rule most soils studied oxidize sulfur at a suitable rate when environmental conditions were satisfactory.

Temperature. As in most biological reactions, an increase in temperature increases the rate at which sulfur is oxidized in the soil. This is illustrated in Figure 8-10. The data indicate an increasing rate of oxidation up to 40°C. Earlier work has suggested that maximum oxidation occurs between 27 and 35° and that at temperatures of 55 to 60°C. the organism was killed. From a practical standpoint it appears that if soil temperatures are above 25°C. appreciable oxidation of added sulfur will take place.

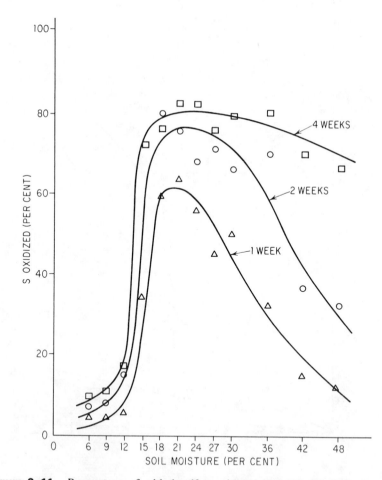

Figure 8–11. Percentage of added sulfur oxidized in a Miami silt loam soil in vials incubated at various moisture levels after one-, two-, and four-week periods. (Kittams, Ph.D. Thesis, University of Wisconsin, 1963.)

Soil Moisture. The effect of soil moisture on sulfur oxidation is shown in Figure 8-11. The most rapid oxidation took place at a moisture level near that corresponding to field capacity. When soils are either excessively wet or excessively dry oxidation of added sulfur will be seriously impeded.

Because the degree of soil aeration is inversely related to soil moisture, at high moisture content oxygen will be a limiting factor. Because the sulfur-oxidizing organisms are aerobic the decrease in the rate of oxidation at high moisture levels is undoubtedly related to the decreased supply of oxygen.

Soil pH. *Thiobacillus thiooxidans* is capable of surviving at extremely low *p*H values. Other sulfur-oxidizing organisms appear to have different *p*H requirements, but in general oxidation of added sulfur proceeds more rapidly on more acid soils.

FINENESS OF APPLIED SULFUR. The particle-size distribution of applied sulfur has a tremendous impact on the rate at which it is converted to sulfates, a consequence, of course, of the surface exposed to the attacking microorganisms. The finer the particle size of a given mass of sulfur, the larger the specific surface and the faster the conversion to sulfate, as illustrated by the data in Table 8-4. Oxidation is slow in the coarsely ground sulfur, 40 mesh and coarser. It is most rapid with the most finely ground material. When elemental sulfur is to be applied to the soil as a source of plant-nutrient sulfur, a general rule of thumb to be followed is that 100 per cent of the material must pass a 16-mesh screen

TABLE 8–4. **Effect of Granule Size on Rate of Oxidation of Elemental Sulfur Added to a Miami Silt Loam Soil at a Rate of 1000 ppm. and Incubated at Room Temperature for Increasing Periods of Time**[*]

Total number	Particle size of sulfur (meshes/in.)	Incubation periods	
		Applied sulfur oxidized (%)	
		2 weeks	4 weeks
1	Check — No sulfur	—	—
2	5–10	0.85	2.27
3	10–20	2.08	5.16
4	20–40	5.27	13.67
5	40–80	15.44	36.07
6	80–120	35.99	68.39
7	120–170	60.52	81.34
8	Less than 230 meshes/in.	79.75	82.44

[*] Unpublished data, O. J. Attoe, University of Wisconsin, 1964.

and that 50 per cent should in turn pass through a 100-mesh screen. Because the grinding of sulfur is an expensive operation, this general rule will serve to minimize grinding costs and at the same time ensure that a significant portion of the added sulfur will be converted to plant-available sulfate.

Sulfides and Polysulfides. Sulfides and polysulfides do not exist in well-drained upland soils. Sulfides of heavy metals and other ions are found in soils under waterlogged conditions. In fact, the deep color of the shore of the Black Sea is caused by the accumulation of iron sulfide.

When sulfates are added to waterlogged soils from which oxygen is excluded, the sulfate is reduced to hydrogen sulfide. If hydrogen sulfide is not precipitated by iron and other similar metals, it escapes to the air aboveground. The effect of waterlogging on the production of hydrogen sulfide in a paddy soil was studied in the laboratory. Its production as affected by time of submergence and added organic matter is shown in Table 8-5. It is obvious that it increased with both time and added organic matter.

In some paddy soils sulfide production is believed to cause Akiochi. The soils on which this disease of rice is observed are old and degraded and low in iron content. On soils containing adequate iron, it does not occur.

In some tidal marsh lands large quantities of reduced sulfur compounds accumulate and the soil pH is raised. When the areas are drained, the sulfur compounds are oxidized to sulfates with a considerable lowering of soil pH. A classic example is found in Kerala State on the southwestern coast of India. During the monsoon season these soils are under water and the pH is about 7.0. When the monsoon season

TABLE 8-5. The Effect of Time and Added Organic Matter on the Production of Hydrogen Sulfide in a Waterlogged Soil*

Time after submergence (days)	Concentration of hydrogen sulfide in soil for three treatments		
	Control (ppm.)	Green manure (ppm.)	Straw (ppm.)
0	nil	nil	nil
7	nil	12.5	10.5
14	3.5	23.8	18.5
21	5.0	32.9	28.8
28	6.8	52.8	40.5
38	8.2	57.6	54.8
51	10.5	60.3	62.5
63	13.0	62.5	65.6
78	15.0	65.0	66.0

* Mandal, *Soil Sci.,* **91**:121 (1961). Reprinted with the permission of The Williams & Wilkins Co., Baltimore.

passes, the soil is no longer inundated. The sulfur compounds are oxidized and the pH drops to around 3.5. This cycle is repeated annually.

Polysulfides are used as fertilizers or soil conditioners. Ammonium polysulfide is a source of fertilizer sulfur as well as a soil conditioner. Calcium polysulfide is applied to a limited extent in the southwestern United States as a soil conditioner. In each case the polysulfide is changed to colloidal sulfur and a sulfide when it is added to the soil or irrigation water, and in this state it is converted to sulfuric acid by soil bacteria. Because of the colloidal nature of the sulfur deposited from polysulfides, its conversion to sulfates takes place quite rapidly.

Some Practical Aspects. Plant roots absorb sulfur almost entirely as the sulfate ion. The concentration of this ion in the soil is important to growing plants. It will be governed largely by the factors affecting its retention in and removal from soils, items which have been discussed in the preceding sections.

Leaching losses of sulfur are greatest on coarse-textured soils under high rainfall. Under such conditions sulfur-containing fertilizers may have to be applied more frequently than on fine-textured soils and under lighter rainfall. In some areas a fertilizer containing both sulfate sulfur and elemental sulfur may be required. This practice is followed in some of the more humid regions of Australia and New Zealand. Elemental sulfur is mixed with ordinary superphosphate and applied to the land. Plant response to sulfur is thus extended over a longer period of time.

Sulfates are retained by certain clays and the hydrous oxides of iron and aluminum usually found in the B or C horizons of some humid region soils. Deep-rooted crops, such as alfalfa, are able to utilize this adsorbed sulfur. Even in these soils, however, the surface horizons may be deficient in sulfur, and a supply of this element should be applied to the young seedling until its roots have penetrated the sulfur-rich zones.

Grasses are better able to utilize sulfate than the legumes. In grass-legume meadows the grasses can absorb the available sulfate at a faster rate than the legumes. Unless an adequate soil level of this element is maintained, the legumes will be forced out of the mixture, for sulfur is required for nitrogen fixation by the *Rhizobia*. This condition has been demonstrated in New Zealand by Walker and his co-workers.

There may be immobilization of added sulfur in some soils because of its conversion to organic forms. This will be particularly true on soils that have large amounts of carbon and nitrogen but only limited quantities of sulfur. On other soils containing larger amounts of sulfur in relation to carbon and nitrogen, there will be the release of sulfate sulfur.

Elemental sulfur and polysulfides must be converted to the sulfate form before they can be absorbed by plants. The rate of conversion of elemental sulfur will be influenced by temperature, moisture, and the particle-size distribution of the applied material. Sulfur applied to dry soils or during cold weather will not be rapidly oxidized. In addition, coarse sulfur (60-mesh and coarser) will be much less rapidly converted to the sulfate form than the more finely divided material (80–100 mesh and finer). For both quick availability and residual action, the applied material should contain both fine and coarse sulfur. If immediate availability to plants is desired, a small amount of sulfate sulfur should also be added.

With the exception of ammonium, aluminum, and iron sulfates, sulfate salts have no effect on soil pH. These three salts, elemental sulfur, and polysulfides are acid forming and their continued use will lower the soil pH. This reduction can be corrected by an adequate liming program. One pound of elemental sulfur will produce an amount of acid which will require 3 lb. of calcium carbonate. Elemental sulfur, aluminum sulfate, and iron sulfate are used to lower soil pH and are considered additionally in Chapter 11 which deals with liming.

Prolonged use of sulfur-free fertilizers will eventually induce a sulfur deficiency of crops. This deficiency will appear sooner on coarse-textured, humid-region soils and in areas in which the amounts of sulfur in the rainfall and atmosphere are low. In addition, the production of high-yielding crops, most of which are removed from the soil, will further aggravate a deficiency of sulfur.

Sulfur-Containing Fertilizers. There are numerous sulfur-containing fertilizer materials, some of which have already been mentioned in Chapters 5, 6, and 7, dealing with nitrogen, phosphorus, and potassium. Some of the more common fertilizer compounds which contain significant quantities of sulfur are shown in Table 8-6 with their content of other plant nutrients.

TABLE 8–6. The Plant Nutrient Content of Some Sulfur-Containing Fertilizer Materials[*]

Material	Plant nutrient content (%)				
	N	P_2O_6	K_2O	S	Other
Aluminum sulfate	0	0	0	14.4	11.4 (Al)
"Ammo-Phos"	11	48	0	4.5	
Ammonia-sulfur solution	74	0	0	10	
Ammonium bisulfite	14.1	0	0	32.3	
Ammonium bisulfite solution	8.5	0	0	17	
Ammonium nitrate-sulfate	30	0	0	5	
Ammonium phosphate-sulfate "Ammo-Phos B"	16.5	20	0	15	
Ammonium polysulfide solution	20	0	0	40	
Ammonium sulfate	21	0	0	24.2	
Ammonium sulfate-nitrate	26	0	0	12.1	
Ammonium thiosulfate (soln.)	12	0	0	26	

TABLE 8-6. **(Continued)**

Material	N	P_2O_6	K_2O	S	Other
				Plant nutrient content (%)	
Basic slag (Thomas)	0	15.6	0	3	
Copper sulfate	0	0	0	11.4	21 (Co)
Copper sulfate	0	0	0	12.8	25.5 (Cu)
Ferrous ammonium sulfate	6	0	0	16	16 (Fe)
Ferrous sulfate	0	0	0	18.8	32.8 (Fe)
Ferrous sulfate (copperas)	0	0	0	11.5	20 (Fe)
Gypsum (hydrated	0	0	0	18.6	32.6 (CaO)
Kainit	0	0	19	12.9	9.7 (Mg)
Langbeinite	0	0	21.8	22.8	
Lime sulfur (dry)	0	0	0	57	43 (Ca)
Lime sulfur (solution)	0	0	0	23–24	9 (Ca)
Magnesium sulfate (Epsom salt)	0	0	0	13	9.8 (Mg)
Manganese sulfate (see "Techmangam")	0	0	0	21.2	36.4 (Mn)
Potassium sulfate	0	0	50	17.6	
Pyrites	0	0	0	53.5	46.5 (Fe)
Sodium bisulfate (niter cake)	0	0	0	26.5	
Sodium sulfate	0	0	0	22.6	
"Sul-Ammo" (paper by-product)	10	0	0	23	
Potassium magnesium sulfate	0	0	26	18.3	
Sulfuric acid (100%)	0	0	0	32.7	
Sulfuric acid (66° Be = 93%)	0	0	0	30.4	
Sulfur	0	0	0	100	
Sulfur dioxide	0	0	0	50	
Superphosphate, normal	0	20	0	13.9	
"Techmangam" (65–80% $MnSO_4$)	2	0	0	15.5	
Urea-gypsum	17.3	0	0	14.8	
Urea-sulfur	40	0	0	10	
Zinc sulfate	0	0	0	17.8	36.4 (Zn)

* Bixby et al., *Adding Plant Nutrient Sulphur to Fertilizers*. Washington, D.C.: The Sulphur Institute, 1964.

Most of these materials are old and established sources of fertilizer sulfur. Their behavior in the soil is determined by the nature of the sulfur, and the reactions they undergo are described in the preceding section. Studies comparing the effectiveness of sources of sulfur are by no means as numerous as are those that compare the agronomic effectiveness of nitrogen, phosphorus, and potassium fertilizers. The available data suggest that when the sulfate itself is considered, one source of sulfate sulfur is generally equal to any other (provided the accompanying cation is not zinc, copper, or manganese, which salts must be applied sparingly) and that the factor determining the selection should be the cost per unit of sulfur applied to the land.

When elemental sulfur is compared with sulfate sulfur, the results will depend on the method of application and the particle-size distribution of the elemental sulfur. When the sulfur is finely ground and mixed with the soil, it is usually just as effective as the sulfate form.

If elemental sulfur is placed on the surface of the soil and compared with a soluble sulfate similarly placed, the sulfate may give initially better responses. Because of its solubility it can move into the root zone with percolating waters. The elemental sulfur must first be oxidized to sulfate, and this is not a rapid process when it is surface applied.

A product consisting of 90 per cent elemental S and 10 per cent bentonite has recently been marketed in the USA. It is nondusty and has good storage and handling properties, and it is of a particle size $(-8, +12M)$ which makes it ideal for blending with solid N, P, and K materials. The bentonite is added to molten sulfur; the molten mass cools and solidifies, and is then crushed and screened. The bentonite imbibes moisture when the product is added to the soil, disintegrating the particles so that oxidation to the sulfate form is facilitated. This material shows promise of becoming an important source of plant nutrient sulfur for high analysis bulk-blend materials.

Ammonia-sulfur and the polysulfides have not been evaluated in many agronomic tests. The limited data available suggest that they are suitable sources of plant-nutrient sulfur. The fact that they are marketed successfully in the sulfur-deficient areas of the Pacific Northwest U.S.A. suggests that they have gained farmer acceptance and that they must therefore be giving satisfactory results.

Ammonium thiosulfate is a comparatively new addition to the list of nitrogen and sulfur fertilizers. It is a clear liquid, with no appreciable vapor pressure, containing 12 per cent N and 26 per cent S. It can be applied directly to the soil or it can be added to irrigation water, both sprinkler and ditch systems. It has the additional advantage of being compatible with all low-pressure or no-pressure nitrogen solutions and with most clear N-P-K solutions. It can also be added to suspension or slurry fertilizers. It is probably the most widely used source of plant nutrient sulfur by the liquid fertilizer industry.

Liquid SO_2 has been used successfully as a source of sulfur for wheat in Washington State. Its acceptance on a commercial farming scale will depend on its cost per pound of sulfur put down in the field. Because it requires pressure equipment and is not compatible with anhydrous ammonia, some difficulties will be encountered in gaining farmer acceptance.

SO_2, produced from elemental sulfur in specially designed field burners, is added to irrigation waters in some areas in California. The SO_2 seems to increase the infiltration rate of the water in some soils, especially when it is added to irrigation waters of a low electrolyte content.

Urea-sulfur has been successfully used in some areas and in others it has not. Its failure to give suitable responses to the sulfur component may be caused by improper placement or distribution within the soil mass.

Ammonium bisulfite solutions are marketed in the Pacific Northwest. They are satisfactory liquid fertilizers but have the drawback of a relatively low content of nitrogen and sulfur. The applied cost of these elements will determine their acceptance.

The addition of finely ground elemental sulfur to water containing 2 to 3 per cent attapulgite clay results in a suspension containing 40 to 60 per cent S. These suspensions can be applied directly to the soil or they can be combined with suspension fertilizers to supply plant nutrient sulfur. They are easy to handle and have the added advantage of being nondusty, a great drawback to the handling of run-of-pile agricultural grade sulfur.

Some granular phosphate-elemental sulfur assemblages have been prepared on a pilot-plant scale by the United States Department of Agriculture Fertilizer Laboratory and by the TVA. They have excellent handling qualities and contain about 17.5 per cent phosphorus (40% P_2O_5) and 10 to 14 per cent sulfur. Limited agronomic tests in the United States indicate that these materials are suitable sources of both elements. They are produced by applying molten sulfur to granular CSP.

THE MICRONUTRIENTS

The need for supplemental additions of fertilizer materials containing the microelements is comparatively recent. As a consequence, the wealth of information concerning the behavior of these elements in soils that exists for the plant nutrients discussed in the preceding chapters of this book has not been accumulated. The reactions in the soil of several of these microelements are considered in the following sections.

Boron. Boron occurs in most soils in extremely small quantities, ranging generally from about 20 to 200 ppm. There are three regions in the United States in which the soils seem to be particularly low in this element: the Atlantic and Gulf coasts from Maine to Texas, some of the north central states, and California and the Pacific Northwest.

Boron does not normally occur in toxic quantities on most arable soils unless it has been added in excessive amounts in commercial fertilizers. There are arid regions, however, in which soils with a high salt content contain this element in toxic quantities, but these areas are few in number and of small agricultural value. Native boron in most humid-region soils is in the form of tourmaline, which is a material quite insoluble and resistant to weathering. It is a borosilicate that contains varying amounts of iron, aluminum, magnesium, manganese, calcium, lithium, and sodium. Release of boron from this mineral is quite slow, and the increasing frequency with which deficiencies of boron occur suggests that native-soil tourmaline cannot supply plant requirements under prolonged heavy cropping.

Most of the available soil boron is held by the organic fraction, and it is retained rather tightly. As the organic matter decomposes, the boron is released, part is taken up by plants, and part is lost by leaching. Some boron is held by clay and its loss from humid-region, fine-textured soils is generally less than from coarse-textured soils.

Some boron is adsorbed by the inorganic fraction of soils. The principal sites for such adsorption are thought to be (a) Fe- and Al-hydroxy compounds present as coatings on or associated with clay minerals; (b) Fe or Al oxides; (c) clay minerals, especially of the micaceous type; and (d) magnesium or hydroxy clusters or coatings that occur on the weathering surfaces of ferromagnesian minerals. It is believed that boron in the soil solution is present essentially as H_3BO_3 in which form it is absorbed by plant roots.

BORON MOVEMENT IN SOILS.   There are several factors associated with the movement of boron in soils. These are soil texture, pH, and moisture.

Soil Texture.   It has generally been found that coarse-textured, well-drained, sandy soils are low in boron and crops with a high requirement, such as alfalfa, respond to applications of 30 to 50 lb./A. of *borax.* Sandy soils with fine-textured subsoils generally do not respond as much to the addition of boron as do those soils with coarse-textured subsoils. Leaching studies have shown that boron added to soils remains soluble and may move out of the upper horizons. The removal of added boron depends on the quantity of water added and the texture of the soil. The finer-textured soils retain the added boron for longer periods than the coarse-textured soils. The effect of soil texture on boron losses from two widely different textural classes of soil is illustrated in Figure 8-12. A water-soluble source of boron had been applied to these two soils for a period of six years. This element moved out of the surface horizons of the sandy soil, although it remained in the surface horizon of the clay soil. Boron accumulated in the 12- to 36-in. depths of the sandy soils. This particular soil type has a concentration of clay in the profile at this depth. Apparently boron is retained in some manner by the clay fraction.

The fact that clays retain boron more effectively than sands does not necessarily imply that plants will absorb this element from clays in greater quantities than from sands when equal concentrations of water-soluble boron are present. In fact, plants will take up much larger quantities of boron from sandy soils than they will from fine-textured soils at

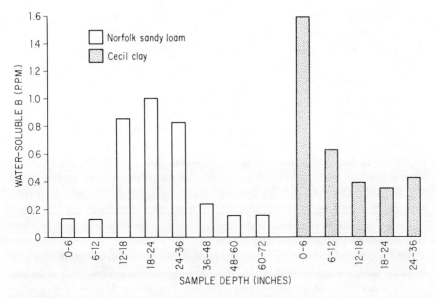

Figure 8–12. Water-soluble boron at various depths after approximately 30 pounds of borax per acre had been applied annually to alfalfa for six years. [Wilson et al., *Agron. J.*, **43**:363 (1951).]

equal concentrations of water-soluble soil boron. This is quite clearly illustrated by the data in Figure 8-13.

Soil pH and Boron Availability. An important factor that influences the availability of boron to plants is the pH or lime level of the soil. The relationship between pH, available calcium, and the boron status of a soil is not well understood. It has been shown repeatedly, however, that

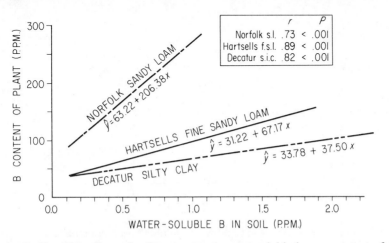

Figure 8–13. The effect of soil texture and water-soluble boron content of the soil on plant uptake of boron. [Wear et al., *SSSA Proc.*, **26**:344 (1962).]

the symptoms of boron deficiency are associated with high pH values, and that boron uptake by plants is reduced by increasing the soil pH. The effect of pH on reducing the plant uptake of boron from soils with similar levels of water-soluble boron is shown in Figure 8-14. A similar reduction in plant uptake was observed on fine-textured soils, though the effect of pH was not so pronounced.

Although some workers have suggested that this reduction is the result of a decreased availability of this element in the soil, others have shown that overliming did not decrease the boron uptake by corn and tobacco. It has been accordingly proposed that the ill effects of over-liming are the result of an unfavorable calcium:boron ratio in the plant. On the overlimed soils high quantities of calcium are absorbed in relation to the boron. It is significant to note that leaching studies have shown that there is less removal of boron in percolating waters from limed soils than from unlimed soils.

The practical implication of these data is, of course, that rates of applied water-soluble boron fertilizer should be less on coarse-textured sandy soils than on fine-textured soils for the same degree of expected plant uptake (and, perhaps, crop response) of boron.

Soil Moisture and Boron Availability. Boron deficiency of many crops is accelerated under extremely dry soil conditions. This behavior, although observed under field and greenhouse conditions, is not completely understood, but it may well be related to the rate of decomposition of organic matter. It may also be related to the rate of root proliferation in the soil which will generally be reduced under extremely dry

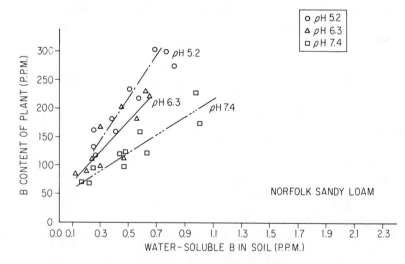

Figure 8–14. The effect of pH and water-soluble boron content of a coarse-textured soil on plant uptake of boron. [Wear et al., *SSSA Proc.*, **26**:344 (1962).]

conditions. As a general rule, boron deficiency is less prevalent on soils that have a high content of organic matter.

In summary, boron deficiency results from excessive leaching (particularly on sandy soils), overliming, and excessively dry weather. The symptoms can be prevented by preplant soil applications of 10 to 50 lb./A. of commercial borax or its equivalent, the amount depending on the soils and the crop to be grown. Care must be exercised to prevent excessive applications of this element, for many crop plants are as adversely affected by too much boron as they are by a deficiency.

Boron Fertilizers. Boron is one of the most widely applied of the microelements. It is required in generally small quantities by alfalfa and certain of the root and cruciferous crops, among which are cabbage, cauliflower, rutabagas, and turnips. Plant species differ in their requirements and tolerances, however, and fertilizers will vary widely in boron content. Alfalfa fertilizer, for example, may contain 4 to 8 lb. of boron per ton, whereas sweet potato fertilizer will contain not more than 1 to 1.5 lb. per ton.

A list of commonly used boron fertilizers is shown in Table 8-7.

TABLE 8–7. Commonly Used B Fertilizers*

Source		Formula	% B (approx.)
Borax		$Na_2B_4O_7 \cdot 10H_2O$	11
Sodium pentaborate		$Na_2B_{10}O_{16} \cdot 10H_2O$	18
Sodium tetraborate:	Fertilizer borate-46	$Na_2B_4O_7 \cdot 5H_2O$	14
	Fertilizer borate-65	$Na_2B_4O_7$	20
Solubor		$Na_2B_4O_7 \cdot 5H_2O +$ $Na_2B_{10}O_{16} \cdot 10H_2O$	20
Boric acid		H_3BO_3	17
Colemanite		$Ca_2B_6O_{11} \cdot 5H_2O$	10
Boron frits		–	2–6

* Mortvedt, et al., Eds., *Micronutrients in Agriculture*, p. 349. Madison, Wisc.: Soil Sci. Soc. Amer., 1972.

Borax ($Na_2B_4O_7 \cdot 10H_2O$) contains 10.6 per cent boron and is a white compound soluble in water. It is generally the most popular of the boron-containing fertilizers. Because it is leached from soils and because of an initially large uptake by the plant, fusion with glass has been employed to reduce its solubility.

BOROSILICATE GLASSES. Several years ago it was found that the salts of boron, and other trace elements as well, could be fused with glass, shattered and applied to soils, into which the salt is slowly released as the glass dissolves. These materials are referred to as *frits* and they have been successfully employed to extend the plant availability of highly soluble and reactive microelement salts. Borosilicate glass is one example.

The boron content of these frits varies, but is generally on the order of 3 to 6 per cent. As would be anticipated in a low water-soluble material of this type, its availability is influenced by the particle-size distribution of the finished product. The more coarsely divided materials will be less effective per unit mass of material (or per pound of contained boron, assuming the same percentage of boron in coarse and fine materials) than the more finely divided materials. This, together with their availability, compared with that of borax, is shown in Figure 8-15. The lower boron content of the alfalfa grown on soils treated with frits is apparent from these curves. The impact of particle size is also obvious. The use of borosilicate glasses offers certain advantages in terms of extended availability not offered by the more readily soluble borax. These advantages are most obvious on sandy soils and under conditions of high rainfall. It is likely that under such soil conditions the use of borosilicate glasses will increase.

BORON IN LIQUID FERTILIZERS. Boric acid (H_3BO_3) and a soluble commercial borate known as *Solubor* can be dissolved in water and sprayed directly onto the leaves of plants. The boron is absorbed by the leaves and utilized by the plant. These compounds are applied in biological control sprays and they can also be added to liquid and slurry fertilizers for application to the soil.

Cobalt. Cobalt, it will be recalled from Chapter 3, has not definitely been shown to be needed by higher plants though several reports of responses to this element by nonleguminous plants have been made. Cobalt, however, is required by *Rhizobia* for the fixation of elemental nitrogen.

CONTENT IN SOILS. The total cobalt content of soils is variable but

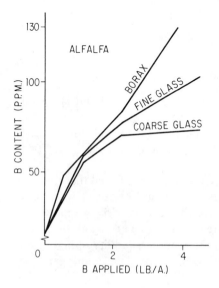

Figure 8–15. The effect of borax and two particle sizes of a boron glass on the uptake of boron by alfalfa. [Holden et al., *J. Agr. Food Chem.*, **10**:188 (1962). Copyright 1962 by the American Chemical Society and reproduced by permission of the copyright owner.]

generally low. Some figures for India list the range of this element as 4 to 78 ppm. total cobalt in the soils studied. Undoubtedly there are soils with lower and higher amounts than these figures. Levels of available cobalt are even lower, being of the order of a few hundredths of a part per million to perhaps 2 or 3 ppm.

FACTORS AFFECTING COBALT AVAILABILITY. Several factors influence the availability of soil cobalt. Work in Russia has shown that increasing the humus content of soil from 3.4 to 16.9 per cent reduced the cobalt content of vetch and oats. Workers in India, however, were unable to show a relationship between soil organic matter and available cobalt.

Polish investigators determined that the nature of the clay had a pronounced influence on the adsorption of cobalt from solutions. The order of adsorption was muscovite > hematite > bentonite = kaolin. Workers at Cornell University obtained similar results which showed that the expanding-lattice clays have a greater capacity for the adsorption of cobalt than has the nonexpanding kaolinite.

Increases in soil pH decrease the availability of cobalt. Yugoslavian workers recently reported that the cobalt content of several grasses decreased with increasing soil pH values. Additional work in Scotland has shown essentially the same relationship, which is illustrated by the data in Table 8-8.

TABLE 8–8. The Effect of Liming on the Cobalt Content of Several Plants Grown on a Granitic Soil*

Soil treatment	Cobalt (ppm.)	Soil pH
A. Mixed Pasture		
Unlimed	0.28	5.4
115 cwt. $CaCO_3$/A.	0.19	6.1
216 cwt. $CaCO_3$/A.	0.15	6.4
B. Red Clover		
Unlimed	0.22	5.4
115 cwt. $CaCo_3$/A.	0.18	6.1
216 cwt. $CaCO_3$/A.	0.12	6.4
C. Rye Grass		
Unlimed	0.35	5.4
115 cwt. $CaCO_3$/A.	0.20	6.1
216 cwt. $CaCO_3$/A.	0.12	6.4

* After Mitchell, as quoted in Underwood, *Trace Elements in Human and Animal Nutrition*, 2nd ed., 1962. Reprinted with permission of Academic Press, Inc., New York.

Cobalt is presently not added to fertilizers. As indicated in Chapter 3, it is administered directly to the animals suffering from a deficiency of this element. It is required in such small amounts by plants that its application as a fertilizer is difficult because of the small quantities of the carrier that are needed. With advances in the technology of fertilizer production, however, it may be possible to introduce this element into fertilizers of the future.

Cobalt is deficient on some soils in the southern United States and in Australia, New Zealand, and Scotland. The deficiency is most pronounced on coarse sandy soils and under conditions of high rainfall.

Copper. Copper deficiencies have been reported in many countries of the world. In the United States they have occurred in the states and on the crops indicated in Figure 8-16. Most of these deficiencies appear on organic soils, but copper deficiency has been found on mineral soils in some countries.

COPPER CONTENT OF SOILS. The total amount of native copper in soils depends on the amount of copper in the parent material. It is usually higher in the soil because of weathering and the concentration of the element in the upper horizons of the soil profile by growing plants. The average copper content of the lithosphere is about 100 ppm., whereas that in soils is reported to range between 2 to 100 ppm.

Copper is found in soils principally as the Cu^{2+} ion adsorbed by clay minerals and as that tied up with organic matter. Smaller amounts of neutral insoluble salts, water-soluble compounds, and copper minerals may also be present.

FACTORS INFLUENCING COPPER AVAILABILITY. The availability of copper to plants is conditioned by several factors: amount of soil organic matter, pH, and the presence of other metallic ions such as iron, manganese, or aluminum.

Organic Matter. As a rule, copper retention in soil increases with an increase in the organic matter content. It is greatest in peats and mucks. When copper is applied to such soils, it is largely retained in the zone of placement. Reports from different research workers suggest that copper-humus complexes vary in their stability. In some cases the copper is retained so tightly that it is not plant available; in others, plants are able to absorb copper from these organic complexes.

Workers in Ireland found that 60 per cent of the copper exchange capacity of the peat studied was accounted for by phenolic hydroxy groups and that 30 per cent was the result of carboxylic groups. Further, as the organic nitrogen or sulfur content of the peat increased, the number of stable copper-peat complexes increased. It was suggested that thiol groups were probably responsible for the formation of some of the stable complexes.

It has been found in other studies with clay-organic matter mixtures

COPPER DEFICIENCIES BY STATES

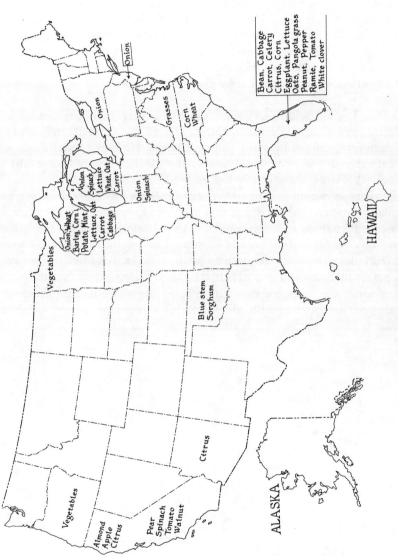

Figure 8-16. Areas of copper deficiency and plants affected in the United States. (McVickar, Bridger, and Nelson, Eds., *Fertilizer Technology and Usage*, 1963. Reprinted with permission of the Soil Science Society of America, Madison, Wisconsin.)

that organic matter will preferentially adsorb copper until its CEC is saturated. After this, the clay fraction will adsorb copper but the clay-copper association is more subject to hydrolysis than the organic-copper complex, meaning, of course, that the copper on the clay is more plant available than the copper held by the organic matter.

Much is to be learned about the nature of the reactions of copper with organic matter. The evidence presently available, however, suggests that those soils high in organic matter are more subject to copper deficiencies than those with smaller amounts. It is fairly obvious, too, that soils of similar organic matter contents will differ in their ability to complex the copper ion.

*Soil p*H. Soil acidity has been found to influence copper availability. Several years ago Peech studied the effect of soil pH on the exchangeable copper in sandy soils in Florida. He found that the amount of exchangeable copper decreased as the pH increased. The results of some of his work are given in Figure 8-17.

Workers at Kentucky found no relationship between soil acidity and available copper. They did find that increasing aluminum concentrations in solution cultures at levels greater than 0.1 ppm. decreased the copper uptake by wheat plants. This is illustrated in Figure 8-18. Even though the Kentucky workers were unable to show a relationship between copper availability and soil pH, it should be remembered that the activity of ionic Cu^{2+} decreases as the pH of the solution increases. In a sandy soil, such as that with which Peech was working, copper solubility

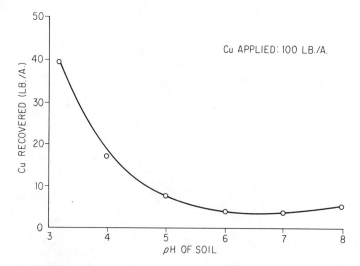

Figure 8–17. The effect of soil acidity on the availability of copper. [Peech, *Soil Sci.*, **51**:473 (1941). Reprinted with permission of The Williams & Wilkins Co., Baltimore.]

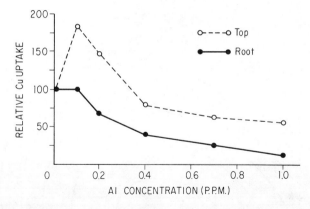

Figure 8–18. Relative uptake of copper by wheat plants in forty-eight hours from a solution containing 0.02 ppm. copper and different concentrations of aluminum. [Blevins et al., *SSSA Proc.*, **23:**296 (1959).]

could have been decreased at the higher pH values. It is entirely probable that both pH and aluminum activity affect the uptake of copper by plants.

Presence of Metallic Ions. The preceding discussion suggests that copper uptake by plants is related to the concentration of aluminum in the rooting medium. The ratio of copper to other metallic ions in the rooting medium and the effect of these ratios on plant growth were investigated by workers in North Carolina. They related the growth of lettuce to the ratio of copper to iron in solution cultures. Different ratios of copper to iron produced different yield values, although some of the yield values were the same for different ratios. Yield curves, constructed by connecting similar yield values, resulted in the series of concentric contour lines illustrated in Figure 8-19. These data suggest that the maximum growth of lettuce is related not to the absolute amounts of copper or iron but rather to the ratio of copper to iron in the rooting medium. In this experiment the maximum yield occurred at a ratio of about 2.5 : 3.0. Additional work by these researchers showed that a generally similar relationship existed among the three elements included in this study: copper, iron, molybdenum.

The studies indicate that the absolute level of a micronutrient in the rooting medium may not be the most important factor in its relation to plant growth. More important may be the amounts of the elements in relation to one another. Thus, the use of absolute quantities of elements in plant and soil diagnostic work may be misleading; these values, especially for the microelements, should be considered in relation to one another. It is further evident that a great deal of research is still needed. If maximum plant growth is indeed the product of some unique ratio

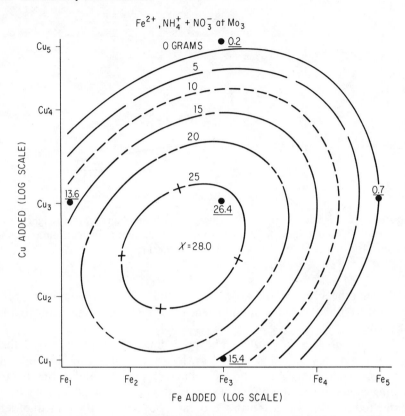

Figure 8–19. Contours of yield of lettuce tops as affected by additions of copper and iron to nutrient solutions containing Fe^{2+}, NH_4^+, NO_3^-, and molybdenum. The observed points are underlined. The point at the center of the contours is the predicted maximum yield. [Moore et al., *SSSA Proc.*, **21**:65 (1957).]

among the various nutrient elements, it is of obvious interest to determine what these ratios are.

COPPER TOXICITY. Like most of the microelements, copper in large amounts is toxic to plants. Excessive amounts depress the activity of iron and may cause iron-deficiency symptoms to appear in plants. Toxicities resulting from these excesses in the soil have been reported in France and in the United States, but it is generally not a problem of serious proportions.

COPPER FERTILIZERS. Copper can be applied by spraying a solution of soluble or slightly soluble salts on the plant leaves or by applying the fertilizer materials to the soil. Copper sulfate is commonly used for both purposes. This salt ($CuSO_4 \cdot 5H_2O$) contains 25.5 per cent copper and 12.8 per cent sulfur. It is quite soluble in water and is compatible with numerous fertilizer materials.

Copper ammonium phosphate can also be used either for soil application or as a foliar spray. It is only slightly soluble in water but can be suspended in water and sprayed on the plants. It contains 30 per cent copper and, like the other metal ammonium phosphates (recall the discussion of magnesium ammonium phosphate in Chapter 5), is a slowly available material.

Copper salts are also contained in frits and in this form are suitable for soil application. Another relatively recent development is that of the chelates of copper and other micronutrients. The nature and development of these compounds are covered in detail in a subsequent section of this chapter. It is sufficient at the moment to indicate that chelates are metal-organic complexes which, though soluble themselves, do not ionize to any degree. They retain the copper (and similar metals) in a soluble form, permitting their absorption by plants, yet preventing their conversion to insoluble forms in the soil.

Copper can be applied as organic compounds in the form of CuEDTA, copper ligninsulfonates, and copper polyflavonoids. These compounds can be applied either to the soil at rates of about 1 to 5 pounds per acre or as a foliar spray at rates considerably less than this.

Iron. Iron is one of the most common of the metallic elements in the earth's crust. Its total content in soils, however, is variable, for it ranges from a low of 200 ppm. to more than 10 per cent. It occurs in soils as oxides, hydroxides, and phosphates, as well as in the lattice structure of primary silicate and clay minerals. Under different soil conditions small amounts of iron are released during the weathering of the primary and secondary minerals and part is absorbed by plants. Total iron content is of no value in diagnosing iron deficiencies; in fact, no suitable test has been developed for this purpose. Iron deficiencies are quite pronounced on some calcareous soils and a high level of soil phosphorus has also been related to iron chlorosis.

IRON-DEFICIENT AREAS. Iron deficiencies have been reported on numerous crops in several states in the United States, most of which have been related to high soil pH, though not always. Crops responding to iron fertilization include rice, small grains, sorghum, soybeans, corn, several grasses, beans, pecans, numerous fruit species, ornamentals, blueberries, and many other vegetable and horticultural crops.

CAUSES OF IRON CHLOROSIS. Iron deficiency, or chlorosis, is believed to be caused by imbalance of metallic ions, such as copper and manganese, excessive amounts of phosphorus in soils, a combination of high pH, high lime, high soil moisture, cool temperatures, and high levels of HCO_3^- in the rooting medium.

Ion Imbalance. The effect of imbalance of ions such as copper, iron, and manganese was covered in the section dealing with copper in soils. The same situation exists for iron. Iron deficiencies observed on many

Florida soils probably result from an accumulation of copper in these soils after long years of application in sprays and fertilizers. Pineapples in Hawaii have exhibited iron chlorosis when grown on soils high in manganese, and other plants growing on soils developed from serpentine have exhibited iron deficiency because of excess nickel. Work carried out by scientists of the United States Department of Agriculture has shown that iron deficiencies on soybeans grown on two soils occurred because of a low ratio in the plants of $Fe:(Cu + Mn)$.

Effects of Bicarbonate, Phosphorus and Calcium. It was observed several years ago that iron chlorosis in orchards was induced by the bicarbonate ion of the irrigation water. This observation led to the suggestion that carbon dioxide fixation by plant roots was responsible for iron chlorosis by inactivating iron within the plant. Iron chlorosis is quite common in many crop plants and ornamentals growing on alkaline and calcareous soils.

Further work has shown that a high HCO_3^- level in the soil increases the solubility of phosphorus and results in a large uptake of this element which interferes with iron metabolism in the plant. United States Department of Agriculture scientists investigated the effect of HCO_3^-, Ca^{2+}, and phosphorus concentrations in the rooting medium on the development of iron chlorosis by two strains of soybeans. One strain developed iron-deficiency symptoms before the other when both were growing on media with the same iron concentration. Using a split rooting-medium technique, these workers established that the effects of the bicarbonate ion on iron chlorosis were only indirectly related to iron uptake. A relationship, however, was found among iron, phosphorus, and calcium concentrations in the rooting medium and the consequent effect on the concentrations of these elements in the plant. High concentrations of phosphorus cause a deposition of iron on the surface of or just inside the root. When iron is supplied in the chelated form, this deposition does not occur, and the element is translocated from the roots to the leaves of the plants.

The differential ability of the two strains of soybeans to utilize iron effectively was suggested as being the result of differences in the quantity or quality of natural iron chelators in the plants themselves. It has also been shown that plant roots will oxidize iron in the rhizosphere. The more effective oxidizers were found to be grasses, timothy, and trefoil, whereas alfalfa was an extremely poor oxidizer. The poor oxidizers had the greatest quantities of iron in the tops of plants; the good oxidizers tended to be iron excluders.

IRON IN WATERLOGGED SOILS. When oxygen is excluded from the soil, as in flooding, ferric iron compounds will be reduced to the ferrous form. Other reducible metals such as manganese will also be reduced. In well-drained soils iron is normally in the ferric state and ferric compounds are largely insoluble. The ferrous compounds are much more soluble, but when added to well-drained soils they are rapidly oxidized to the ferric state.

The behavior of iron under waterlogged conditions was studied recently and the effect of time of submergence and the presence of organic matter on the solubility of iron was measured. In one set of treatments the soil was submerged without added organic matter and in the other set rice straw was mixed with the soil. Soluble and exchangeable iron were determined on these samples at the times indicated in Table 8-9. Two points are made clear by these data. The presence of organic matter greatly hastened the reduction of iron, and the longer the period of submergence, the greater the amount of soluble plus exchangeable iron. The soluble iron reached a peak after about thirty-eight days of submergence with the straw-treated soil, after which it declined. How-

TABLE 8–9. The Effect of Time and Added Organic Matter on the Reduction of Ferric Iron in a Submerged Soil*

Time of submergence (days)	Fe^{2+} in solution (ppm.)	Exchangeable Fe^{2+}
Control		
0	nil	nil
7	nil	nil
14	nil	trace
21	nil	10
28	nil	20
38	nil	31
51	trace	40
63	2.0	46
78	4.0	50
Straw		
0	nil	nil
7	nil	26.0
14	30.0	108.0
21	90.4	162.0
28	132.0	200.0
38	192.0	446.0
51	184.0	450.0
63	128.0	482.0
78	104.0	523.0

* Mandal, *Soil Sci.*, **91**:121. (1961). Reprinted with permission of The Williams & Wilkins Co., Baltimore.

ever, there was a continuous increase in the level of exchangeable iron.

It will be recalled that in the discussion on sulfur in submerged soil it was stated that the hydrogen sulfide released from the reduction of sulfur compounds was precipitated as iron sulfide. If the soil were low in iron, the hydrogen sulfide was not precipitated and was thought to cause Akiochi. This disease of rice occurs on soils that have been in paddy rice for years. The data shown in Table 8-9 suggest what happens to the iron. After years of reduction it is converted to the ferrous form, in which state it is probably slowly leached from the soil.

In general, deficiencies will most likely occur in soils high in pH and/or carbonates. Excessive phosphate fertilization will also induce an iron chlorosis, especially in certain ornamentals and shrubs. Generally, on poorly buffered soils, prolonged applications of copper, manganese, or perhaps zinc salts in biological sprays and dusts or in fertilizers will induce an iron deficiency because the ratio Fe:(Cu, Mn, Zn) may become too narrow in the plant. Plant species differ in their ability to absorb and translocate iron, a characteristic thought to be caused by the differences in internal chelating mechanisms. Iron in waterlogged soils is reduced to the ferrous state, and this is hastened by the presence of organic materials in the soil.

IRON-CONTAINING FERTILIZER MATERIALS. Crop deficiencies of iron can be corrected by applying certain iron compounds to the soil or directly to the foliage in aqueous sprays. In general, soil applications of ionizable ferrous salts, such as ferrous sulfate, have proved inefficient because of their rather rapid oxidation to ferric iron. When these salts are applied as foliar sprays, however, their effectiveness is greatly increased. Injections of dry iron salts directly into the trunks and limbs of several fruit-tree species have been very effective in controlling iron chlorosis.

Iron sulfate ($FeSO_4 \cdot 7H_2O$) is the inorganic salt most commonly used in sprays for controlling iron chlorosis. Solutions containing 4 to 6 per cent iron sulfate are applied at rates of 30 to 50 gal./A., depending on the crop.

Fritted iron can also be employed on acid soils, though frits in general are not suitable for use on alkaline or calcareous soils. Although not presently produced commercially, iron ammonium phosphate can be effective as a spray or soil application.

With the exception of ferrous sulfate, perhaps the most widely used iron compounds are the chelates. These compounds generally contain 6 to 12 per cent iron. Several chelates, all of which are water-soluble, can be applied to the soil or foliage. Other organic sources of iron include iron polyflavonoids and iron lignosulfonates. A list of the compounds supplying plant nutrient iron are given in Table 8-10. Generally iron compounds are more effective when applied as a foliar spray and in this

TABLE 8–10. Some Sources of Fertilizer Fe*

Source	Formula	% Fe (approx.)
Ferrous sulfate	$FeSO_4 \cdot 7H_2O$	19
Ferric sulfate	$Fe_2(SO_4)_3 \cdot 4H_2O$	23
Ferrous oxide	FeO	77
Ferric oxide	Fe_2O_3	69
Ferrous ammonium phosphate	$Fe(NH_4)PO_4 \cdot H_2O$	29
Ferrous ammonium sulfate	$(NH_4)_2SO_4 \cdot FeSO_4 \cdot 6H_2O$	14
Iron frits	Varies	Varies
Iron ammonium polyphosphate	$Fe(NH_4)HP_2O_7$	22
Iron chelates	NaFeEDTA	5–14
	NaFeHEDTA	5– 9
	NaFeEDDHA	6
	NaFeDTPA	10
Iron polyflavonoids	–	9–10
Iron ligninsulfonates	–	5– 8
Iron methoxyphenylpropane	FeMPP	5

* Mortvedt, et al., Eds., *Micronutrients in Agriculture*, p. 357. Madison, Wisc.: Soil Sci. Soc. Amer., 1972.

way they may also be applied at lower rates than are required for soil application.

Altering the soil pH in a band in the rooting zone may also be effective in correcting iron deficiencies. Several sulfur products such as elemental sulfur, ammonium thiosulfate, sulfuric acid, and ammonium polysulfide will lower soil pH and simultaneously act as reducing agents to convert ferric iron to the ferrous form which is absorbed by plant roots.

Workers at TVA have shown that fluid polyphosphate fertilizers may also be effective carriers of iron sulfates for crops growing on Fe-deficient soils.

Manganese. The manganese contained in soils originated from the decomposition of ferromagnesian rocks. The quantities present vary from a trace to several thousand pounds per acre. The amount of this element required by many plant species is small, and when appreciable quantities appear in easily available form a toxicity may develop. The availability of manganese to plants is influenced by several factors, and the range between deficiency and toxicity is relatively narrow.

Manganese deficiencies have been reported in a number of states in the USA. This condition is largely the result of a high soil pH but it can to a lesser extent be induced by an imbalance with other elements such as Ca, Mg, and ferrous iron. Crops that have been found to be manganese deficient include small grains, beans, corn, sorghum, cotton, numerous vegetable and fruit crops, and several ornamentals.

FORMS OF SOIL MANGANESE. Manganese in the soil is generally considered to exist in three valence states: (1) divalent manganese, Mn^{2+}, which is present as an adsorbed cation or in the soil solution; (2) trivalent manganese, which is supposed to exist as a highly reactive oxide, Mn_2O_3; and (3) tetravalent manganese, Mn^{4+}, which exists as the very inert oxide, MnO_2. It is believed by some workers that these three forms exist in equilibrium with one another.

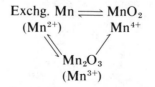

$$\text{Exchg. Mn} \rightleftharpoons MnO_2$$
$$(Mn^{2+}) \qquad\quad Mn^{4+}$$
$$Mn_2O_3$$
$$(Mn^{3+})$$

According to this concept, the exchangeable divalent manganese is in equilibrium with the tri- and tetravalent forms, which are favored by a high pH and oxidizing conditions. The highly stable MnO_2 is the form most likely to occur in soils at pH values greater than 8.0. The trivalent form is presumably favored by pH values near neutrality, whereas the divalent form is found in acid soils. The Mn_2O_3 by breaking down is thought to give rise to both MnO_2 and MnO.

It was shown in Germany that several strains of bacteria, isolated from the rhizosphere of oat plants (a crop that is quite susceptible to manganese deficiency), were capable of fixing manganese in a form that could not be extracted by usual methods. It was suggested that an increased bacterial activity in moor and heath soils with a low content of available manganese would result in a deficiency to crops, particularly if such soils were limed.

Other work in Scotland suggested that biological oxidation of manganese to higher oxides is not the cause of its reduced availability in soils. Rather the formation of complexes of manganese with soil organic matter may account for the observed relationship between manganese availability and soil pH.

Work in the United States has shown that increasing soil pH increased both the chemical and microbial oxidation of manganese. It is apparent from the foregoing that the forms and behavior of this element in soils are not completely understood. Nonetheless, certain fairly well-defined factors seem to be associated with its availability.

SOIL CONDITIONS AND MANGANESE AVAILABILITY. On soils high in organic matter and near neutrality in reaction, crops, particularly legumes and small grains, exhibit varying degrees of manganese deficiency. The presence of considerable organic matter frequently results in the appearance of deficiency symptoms at lower pH values than on soils

TABLE 8–11. The Effect of Added Peat on the Acid-Extractable Manganese Content of a Limed and Unlimed Norfolk Sandy Loam after Three Weeks Incubation[*]

	Acid-extracted manganese (ppm.)	
Treatment	No lime	Limed
54 ppm. Mn	48.0	64.8
54 ppm. Mn + 2% peat	50.2	50.6
54 ppm. Mn + 4% peat	50.0	39.0
54 ppm. Mn + 10% peat	53.9	32.4

[*] Sanchez et al., *SSSA Proc.*, **23**:302 (1959).

with a lower humus content. This has led to the suggestion that certain types of organic matter will form insoluble complexes with divalent manganese, thus rendering it unavailable to plants.

The effect of organic matter, lime, and degree of soil moisture, all factors that have been associated with manganese deficiency, were studied in relation to the ammonium acetate extractable manganese in two North Carolina soils. When peat was added to an unlimed Norfolk soil (low in organic matter), the content of exchangeable manganese was increased. When this soil was limed, however, additions of peat reduced the content of extractable manganese. These relationships are illustrated by the data in Table 8-11.

Different amounts of manganese were added to samples of a Norfolk and a Bladen soil. The Bladen soil has a high content of organic matter. Both soils were incubated for three weeks in moist conditions. Prior to analysis, half of the samples were air-dried and half left in the moist condition. The results are shown in Figure 8-20. Liming decreased the amount of added manganese that could be extracted from both soils at the end of three weeks. Keeping the soils in the moist condition prior to analysis resulted in smaller amounts being extracted.

The effect of soil moisture on manganese availability is somewhat confusing. It is frequently observed in the field that manganese deficiency is most severe on high organic matter soils during the cool spring months when the soils are waterlogged. As the soils dry out and the season warms up, the symptoms disappear. This could be related to changes in soil brought about by increased microbiological activity. It has been found that the pH of these soils is highest in the winter, frequently in excess of 7.0. However, during the summer the pH value of the same soils may be less than 6.5, probably as a result of greater

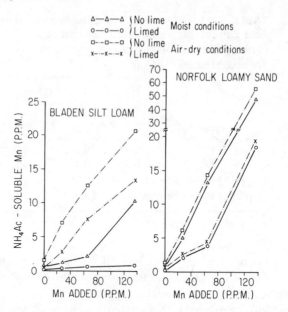

Figure 8–20. The effect of lime and moisture level prior to analysis on the fixation of added manganese in a Bladen and Norfolk soil after three weeks of incubation. [Sanchez et al., *SSSA Proc.*, **23**:302 (1959).]

biological activity. There is a good correlation between the amount of ammonium acetate extractable Mn^{2+} and the pH of these soils, as shown in Figure 8-21.

The oxidation potential for the conversion of Mn^{2+} to MnO_2 is a linear function of the pH between values of 3.2 to 8.0. Whether the pH or the oxidation status of the soil is more closely related to manganese availability has not been definitely established. As already pointed out, certain bacteria are capable of reducing the availability of manganese, but the interrelation between these factors has not been clearly established.

The reduction of MnO_2 to more soluble forms, brought about by sterilization of the soil and by the addition of certain reducing compounds, has led to the suggestion that although pH changes may be well correlated with manganese deficiency the lowering of the pH is not of itself the cause of increased manganese availability.

CORRECTION OF MANGANESE DEFICIENCY SYMPTOMS. Deficiency symptoms induced by high soil pH may be corrected in three ways. To prevent the appearance of the symptoms on soils suspected of lime-induced deficiencies, the pH should be kept below 6.2 to 6.4. This can be accomplished by the use of several materials, such as elemental sulfur, ammonium polysulfide, and ammonium thiosulfate. When these materi-

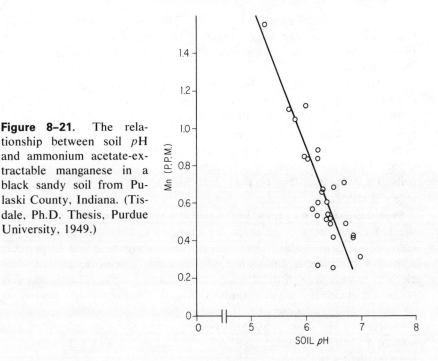

Figure 8–21. The relationship between soil pH and ammonium acetate-extractable manganese in a black sandy soil from Pulaski County, Indiana. (Tisdale, Ph.D. Thesis, Purdue University, 1949.)

als are applied in a band near to but not in contact with the seed, they lower the soil pH and, being oxidized themselves, act as reducing agents converting the oxidized manganese into the plant-available manganous form. Another way in which the deficiency can be prevented is by the addition of soluble manganese salts, such as manganese sulfate at the rate of 30 to 60 lb./A. These salts should be mixed with the starter fertilizer, which, for best results, should be banded. Fritted manganese compounds have also been successfully used.

If a growing crop exhibits manganese-deficiency symptoms, it is generally too late to apply either of these measures. It is possible to correct the deficiency, however, by spraying the plants over an area of one acre with 5 to 10 lb. of manganese sulfate dissolved in 100 gal. of water. Generally within three or four days the plants will begin to lose the symptoms and a normal or nearly normal yield may be expected, provided that the treatment is begun promptly.

A practical solution on a soil that has been limed too high for soybeans is to sow a less sensitive crop such as sweet clover or alfalfa. Corn may also be planted and side-dressed with high quantities of a material such as ammonium sulfate. This fertilizer is quite acid forming, and after a few years the pH should drop.

There have been a few examples in the United States in which manganese deficiency has been the result not of overliming, but of the complete absence of that element in the soil. This condition obtains on sandy soils with very low exchange capacities in areas of high rainfall. Certain parts of Florida, in which the soils are little more than quartz sand, are cases in point.

MANGANESE TOXICITY. From the preceding discussion regarding the reactions of manganese in the soil, it follows that maximum availability of this element occurs on soils at pH values of less than 6.5. On some soils an extremely acid condition may result in a toxicity to crops because of its presence in excessive amounts. Toxicities have been reported on burley tobacco, soybeans, and cotton growing on very acid soils. This condition can be prevented by liming the soil to a more suitable pH value.

MANGANESE FERTILIZER MATERIALS. Manganese sulfate is one of the older and more popular of the manganese fertilizer materials. Its commercial form contains about 26 per cent manganese and 15 per cent sulfur. A manganese ammonium phosphate has been developed. This element is available in the chelated and fritted forms. A comparatively recent product has been manganous oxide, in two forms—one containing 48 per cent manganese and the other 65 per cent. Both are water-insoluble but are slightly soluble in dilute acids. Some slags may contain sizable quantities of manganese. A list of commercially available manganese fertilizers is shown in Table 8-12.

Molybdenum. Molybdenum, which is present in the earth's crust and in soils in extremely small quantities, is required by plants in very small amounts. It has been estimated that the lithosphere contains an average of only 2.3 ppm.; the average for soils is about 2 ppm. Molybdenum is present largely in the crystal lattice of primary and secondary minerals or as an exchangeable anion in soils.

Deficiencies of Mo in the United States occur largely on the acid sandy soils of the Atlantic and Gulf Coasts though responses to this ele-

TABLE 8-12. Commonly Used Mn Fertilizers*

Source	Formula	% Mn (approx.)
Manganese sulfate	$MnSO_4 \cdot 3H_2O$	26–28
Manganous oxide	MnO	41–68
Manganese methoxyphenylpropane	MnMPP	10–12
Manganese chelate	MnEDTA	12
Manganese carbonate	$MnCO_3$	31
Manganese chloride	$MnCl_2$	17
Manganese oxide	MnO_2	63
Manganese frits	—	10–25

* Mortvedt, et al., Eds., *Micronutrients in Agriculture*, p. 363. Madison, Wisc.: Soil Sci. Soc. Amer., 1972.

ment have also been reported in California and the Pacific Northwest, Nebraska, and the States bordering the Great Lakes. Large soil areas in New Zealand and Australia are also deficient in molybdenum.

Crops responding to Mo fertilization include legumes, cruciferous crops, grasses and several vegetable crops.

REACTIONS OF MOLYBDENUM IN SOILS. The reactions of molybdenum in soils are not well understood. It may be present as (1) a part of the crystal lattice of primary and secondary minerals, in which form it is unavailable to plants, (2) as adsorbed MoO_4^{2-}, which is held by clays and which is available to plants, (3) as a part of the soil organic matter, and (4) as water-soluble molybdenum compounds.

Molybdenum in soils is largely unavailable. Some work with an ammonium oxalate extracting solution showed that in soils with a total molybdenum content of 0.24 to 4.45 ppm. only 0.05 to 0.24 ppm. could be extracted. It has been known for some time that the availability of molybdenum increases with the soil pH. This, of course, is the reverse of what is true for most of the other microelements. If MoO_4^{2-} is adsorbed by clays, in a fashion similar to that in which SO_4^{2-} is retained, an increase in its availability with increasing pH values can be explained, in part at least, by the following equation:

$$\text{clay}\left.\begin{array}{c}MoO_4\\MoO_4\end{array}\right] + 2OH^- \rightleftharpoons MoO_4^{2-} + \left[\begin{array}{c}HO\\HO\\MoO_4\end{array}\right. \text{clay}$$

This equation is oversimplified, but it serves to illustrate the oft-observed behavior of molybdenum in the field.

Other workers have suggested that molybdenum occurs in soils as (1) soluble salts, (2) molybdenum oxide (MoO_3), and (3) reduced oxides such as Mo_2O_5 and MoO_2. They have suggested that Mo_2O_5 and MoO_2 are slowly converted to MoO_3 which is in turn very slowly converted to soluble molybdate salts. The conversion of molybdenum oxide to soluble molybdate salts is favored by an alkaline reaction, which could also explain the increased availability of this element in soils with high pH values.

In addition to soil pH, the presence in the soil of the oxides of iron, aluminum, and titanium also increases the adsorption of molybdenum. Work in California has shown that the sorption of MoO_4^{2-} by hydrated ferric oxide is accompanied by the stoichiometric release of two OH^- ions and a molecule of water. This reaction is effected in an acid medium and the compound $Fe_2(MoO_4)_3$ is formed.

Heavy applications of phosphatic fertilizers will increase the molybdenum uptake by plants. Heavy applications of sulfates, on the other hand, have a depressing effect on plant uptake. This could result from ion competition at the root surface, for the MoO_4^{2-} and SO_4^{2-} ions are of similar size and charge. On soils with borderline molybdenum deficiencies, the application of excessive amounts of sulfate-containing fertilizers may induce a molybdenum deficiency in plants. Under such conditions the inclusion of molybdenum in the fertilizer at locally recommended rates may be advisable.

MOLYBDENUM TOXICITIES. Excessive amounts of molybdenum are toxic, especially to grazing animals. Some cases of molybdenum toxicity on cattle or sheep have been reported in the western part of the United States and in Australia, where some of the soils are locally quite high in their content of this element. Molybdenosis, as this disease of cattle is called, is actually caused by an imbalance of Mo and Cu in the diet of the ruminant. This can happen when the Mo content of the forage is greater than 5 ppm. Mo toxicity results in stunted growth and bone deformation in the animal. It can be corrected by oral feeding of copper sulfate, injections of copper glycinate suspensions, or the application of copper sulfate to the soil.

MOLYBDENUM FERTILIZERS. Several compounds available to supply fertilizer molybdenum include ammonium and sodium molybdate and molybdenum trioxide. These materials are normally mixed with the N-P-K fertilizer and are applied at rates equivalent to 2 oz. to 2 lb./A. They can also be applied as foliar sprays.

It has been demonstrated in Australia, New Zealand, and elsewhere that the application of molybdenum to clovers will in some cases produce yield increases equivalent to those obtained from the use of several tons of limestone. In inaccessible areas in which transportation and application of large tonnages are difficult and expensive molybdenum has been of great value.

Several years ago, some Australian workers found that soaking the

TABLE 8–13. Effect of Seed and Foliar Application of Mo on the Yield of Canning Peas[*]

		Yield, kg./ha.	
Treatment[**]		Vines	Peas
Check		1,070	1,520
Seed application:	Slurry	2,150	2,770
	Dust	1,500	1,950
Foliar spray		2,370	2,500
LSD .05		427	410

[*] Hagstrom and Berger, *Soil Sci.,* **100**:52 (1963). Reprinted with permission of The Williams & Wilkins Co., Baltimore.
[**] Molybdenum was applied as $Na_2MoO_4 \cdot 2H_2O$ at a rate of 56 g./ha.

seed of subclover in a solution of sodium molybdate before seeding was just as effective as applying the molybdenum in the fertilizer. Seed treatment is now probably the most common way of correcting Mo deficiencies in the USA and elsewhere. The seed must be soaked in a solution containing the molybdenum salt, for dusting the seed has been proved ineffective. Mo deficiencies can also be corrected by applying foliar sprays containing this element. A comparison of the effectiveness of seed treatment vs. foliar sprays is shown in Table 8-13.

Zinc. The zinc content of the lithosphere has been estimated to be about 80 ppm. Its total content in soils ranges from 10 to 300 ppm., but its presence in the soil is no more a criterion of its availability to plants than is the presence of many of the other plant nutrients.

Zinc deficiencies are fairly widespread throughout the United States and have been reported on several crops, including corn, pecans, sorghum, citrus and deciduous fruit, soybeans, numerous horticultural crops, and ornamentals.

BEHAVIOR OF ZINC IN SOILS. The plant availability of zinc is conditioned by several soil factors: pH, phosphorus level, organic matter content, and adsorption by clays.

Soil pH. Zinc is generally more available to plants in acid than in alkaline soils. Zinc deficiencies do not occur on all alkaline soils by any means, but plant uptake of this element has shown that its availability is a function of soil acidity. This is illustrated by the plant uptake data in Figure 8-22. These data were obtained from a study in which various sources of nitrogen were applied to three crops of milo and an intercrop of clover, all of which were grown in potted soil in the greenhouse. Zinc uptake by the plants was measured. The application of the different sources of nitrogen resulted in different soil pH values, as would be predicted from the discussion of this topic in Chapter 5. Ammonium sulfate had the greatest acidifying effect and sodium nitrate the least. The greatest uptake of zinc, both native and applied, took place at the lowest pH values. As a general rule, most pH-induced zinc deficiencies occur within the range of 6.0 to 8.0.

It should be emphasized again, however, that there have been several cases in which no relation was found between soil pH and zinc uptake by plants or plant response to applied zinc. This was illustrated by work from California in which the zinc status of 53 soils was investigated. It was found that sweet corn, growing on 86 per cent of the soils studied, responded to zinc applications if the soil contained 0.55 ppm. zinc or less, extractable with a dithizone extracting solution. Sweet corn growing on 76 per cent of the soils containing more than 0.55 ppm. extractable zinc did not respond to applications of zinc fertilizer. The pH of these soils ranged from 4.0 to 8.3, but there was no apparent connection between this property and plant response.

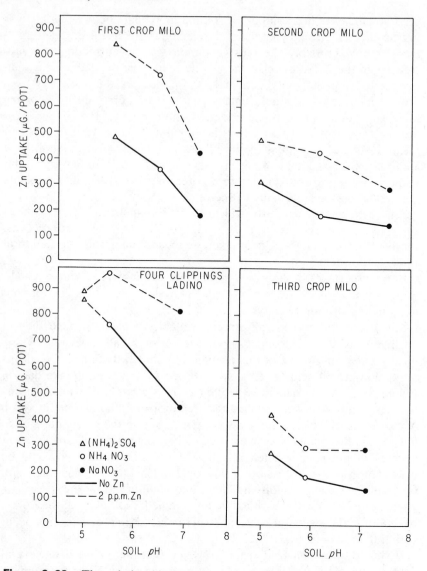

Figure 8–22. The relationship between soil pH and the plant uptake of zinc. [Viets et al., *SSSA Proc.*, **21**:197 (1957).]

Soil Phosphorus Level. Zinc deficiency on high phosphate soils has been observed frequently. In Florida zinc deficiency of citrus was associated with high soil phosphorus. In Tennessee zinc deficiency is common on corn grown on certain high phosphate soils in the central part of the state. In other studies the zinc content of beans was reduced 20 to 30 per cent and that of corn 30 to 50 per cent by the addition per acre of 800 lb. of P_2O_5 (344 lb. of P). Some solution culture studies in which corn was the test crop showed that over the pH range 6.5 to 8.5, the plant uptake of zinc was lower at the higher level of phosphorus in the rooting medium.

A Zn-P interaction has been observed in several plant species. High levels of one of these elements may reduce plant uptake of the other. If the soil is marginally deficient in either element, application of one may induce a deficiency of the other. This situation can usually be corrected by applying both elements to the soil as illustrated by the data from Michigan shown in Figure 8-23.

In pH-zinc relations, however, high soil phosphate is not always associated with zinc deficiencies. Workers in Washington State failed to influence zinc uptake by beans with applications of as much as 400 lb. of P_2O_5 (172 lb. of P) per acre. Workers in Utah were unable to increase zinc deficiency in a mildly zinc-deficient peach orchard even with an application of 1 ton of P_2O_5 (860 lb. of P) per acre.

The nature of phosphate-induced zinc deficiency is not fully understood. It is generally considered that simple precipitation of zinc phosphate is inadequate to explain this phenomenon, even though the solubility of the zinc decreases in high concentrations of the phosphate ion.

Soil Organic Matter. Zinc deficiency has been observed on soils high

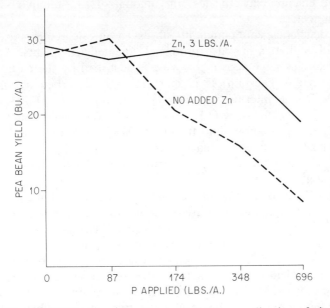

Figure 8–23. The yield of pea-beans as affected by application of zinc sulfate and superphosphate to a Kawkawlin loam. [Judy et al., *Mich. State Univ. Agr. Expt. Sta. Res. Rep. No. 33.* (1965).]

in organic matter, especially that resulting from treatment with animal manures. In other cases sterilization of zinc deficient soils has corrected the deficiency. The same effect was observed whether the soil was sterilized with steam, ether, or formalin. These studies were carried out in the 1930's and generally have not been followed up. If these effects are real, it would suggest that living organisms themselves may be involved in zinc immobilization.

Though the effects of organic matter on zinc availability cannot be separated from the effects of phosphorus and other constituents of the organic matter, the opinion currently held is that organic matter per se probably is not significant in zinc immobilization.

Adsorption by Clay Minerals. Numerous studies have shown that zinc is adsorbed by various clay minerals and by the carbonates of calcium and magnesium. Results of studies carried out in Illinois suggest that zinc retention by soils has the following relation to other cations:

$$H > Zn > Ca > Mg > K$$

Zinc adsorbed by hydrogen-saturated soil systems could be replaced with ammonium acetate extraction. When zinc was added to a calcium system, only a part of that added could be removed by ammonium acetate extraction. Furthermore, the longer the time of contact between the zinc and calcium-clay, the less the amount of zinc removed with ammonium acetate.

Zinc deficiencies are commonly observed on calcareous soils, and the liming of acid soils has also produced zinc-deficient plants. Results of several studies have shown that zinc is adsorbed by the carbonates of calcium and magnesium. It is most strongly adsorbed by magnesite ($MgCO_3$), to an intermediate degree by dolomite [$CaMg(CO_3)_2$], and least of all by calcite ($CaCO_3$). In magnesite and dolomite it appears that zinc is actually adsorbed into the crystal surfaces at the sites in the lattice normally occupied by magnesium atoms. The strong adsorption of zinc on soil minerals may be responsible in part for the low solubility of this element.

The results of numerous studies show that there is a tendency for zinc to form many compounds of low solubility in soils. Precipitation as carbonates, hydroxides, and phosphates can reduce the available zinc in soils to a low level, but present evidence does not indicate that these reactions alone can result in zinc deficiencies in plants. It does suggest, however, that adsorption reactions can reduce the available zinc to deficiency levels, and that these reactions occur on many types of surface, including clay minerals, organic matter, and insoluble carbonates.

Movement of Zinc in Soil. It is obvious from the foregoing that zinc is relatively immobile in most soils. Its movement has been studied by workers in California. Zinc sulfate and zinc oxide were applied to the surface of soil placed in leaching columns. Water was passed through these columns and the distribution of zinc throughout the length of the columns was determined. The results are shown in Figure 8-24. It is apparent that most of the zinc was retained in the surface inch of soil and that only little if any downward movement occurred. These researchers indicated that even though the zinc was retained in a water-insoluble form much of it was removable with a dithizone extractant. Zinc so removed has been well correlated with plant uptake of this element, at least by workers in California.

ZINC FERTILIZER MATERIALS. Zinc sulfate, containing about 36 per cent zinc, has for years been a popular fertilizer material. It can be applied to the soil, usually at a rate of 40 to 80 lb./A., and can also be sprayed over vegetable, field, and fruit crops.

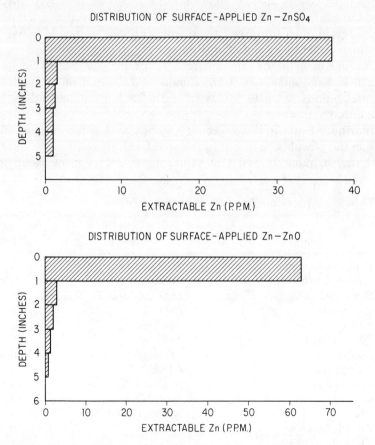

Figure 8–24. Movement of added zinc through a column of soil leached with demineralized water. [Brown et al., *SSSA Proc.*, **26:**167 (1963).]

Zinc can be enveloped in frits and is available in the form of various chelates. A recent development is a zinc-iron-ammonium sulfate compound made by ammoniating spent sulfuric acid containing iron and zinc as impurities. It contains 4 per cent zinc. Zinc ammonium phosphate is another compound. As in other metal ammonium phosphates, it is only slightly soluble in water and will extend the availability of the zinc over a long period of time when applied to the soil. It contains 33.5 per cent zinc. Zinc oxide is an important source also.

Organic zinc compounds have also been used successfully to correct deficiencies of this element. These include several zinc chelates of the EDTA type as well as zinc polyflavonoid and lignosulfonate types. A list of some currently available zinc fertilizers is shown in Table 8-14.

Soil application of the various zinc compounds is the most common way of overcoming deficiencies of this element. Foliar application of zinc compounds in aqueous solution have also been employed successfully, though this method should generally be considered as a temporary or emergency measure. Seed treatment with zinc compounds has also been investigated with only limited success.

Zinc, like many of the other micronutrients, is toxic to the plants when applied in large quantities. Care should be exercised in its application and precautions taken to follow rates that are recommended by local agricultural authorities.

Chlorine. Little is known about chlorine as a plant nutrient. Work done in the greenhouse in California on a California soil and a soil from Western Australia showed that subterranean clover growing on these

TABLE 8-14. Some Sources of Fertilizer Zn*

Source	Formula	% Zn (approx.)
Zinc sulfate monohydrate	$ZnSO_4 \cdot H_2O$	35
Zinc sulfate heptahydrate	$ZnSO_4 \cdot 7H_2O$	23
Basic zinc sulfate	$ZnSO_4 \cdot 4Zn(OH)_2$	55
Zinc oxide	ZnO	78
Zinc carbonate	$ZnCO_3$	52
Zinc sulfide	ZnS	67
Zn frits	(silicates)	Varies
Zinc phosphate	$Zn_3(PO_4)_2$	51
Zinc chelates	$Na_2ZnEDTA$	14
	$NaZnNTA$	13
	$NaZnHEDTA$	9
Zn polyflavonoid	—	10
Zn ligninsulfonate	—	5

* Mortvedt, et al., Eds., *Micronutrients in Agriculture*, p. 371. Madison, Wis.: Soil Sci. Soc. Amer., 1972.

soils developed severe chlorine deficiency. The soils contained 3 to 5 meq. Cl/100 g. A chlorine deficiency of potatoes growing in the field has also been reported by workers in Maine.

Numerous studies have shown that chlorides readily leach from soils. They are generally not considered to be retained to any extent by sorption processes, although workers in Virginia have indicated that limited adsorption of Cl^- by soil clays is possible. This was pointed out earlier in this chapter in the section dealing with sulfate retention in soils. Chlorides will be removed more rapidly by waters percolating through coarse-textured soils than through fine-textured soils, largely because fine-textured soils retain a larger amount of water and there is less leaching with a given amount of rainfall. Workers in North Carolina several years ago found that chlorides tended to accumulate in depression areas in sandy soils. Leaching in such areas was not so pronounced as in adjoining well-drained soils, and chlorides tended to accumulate.

Selenium. This element was not included with the essential microelements, but it is an important factor in crop production in localized areas throughout the world. In some areas soils are so high in this element that animals grazing thereon are poisoned. In other cases the soils are not abnormally high in selenium but are covered with vegetation that tends to accumulate selenium in its tissue. The first recorded case of selenosis in livestock was in Nebraska in 1856. Also known as "alkali disease" and "blind staggers," it has since been found in areas of the Great Plains of the United States and in Canada, Ireland, Israel, Russia, and Australia. Plants classified as selenium accumulators are members of the genera *Sium* and *Astragalus*.

The treatment of seleniferous soils to correct this problem has met with little success. Treatment with gypsum has proved unsuccessful on naturally seleniferous soils because they are often already high in native gypsum. However, when too much selenium may have been added to soils, this condition can be corrected by additions of barium chloride or calcium sulfate. The barium chloride is thought to form an insoluble barium selenate, which is unavailable to plants. The sulfate of the gypsum reduces the uptake of selenium simply because of ion competition.

More recently attention has been given selenium because of its deficiency in certain areas. It was established in 1944 that selenium is an essential element for higher animals. It has not yet been found to be essential for plants. A deficiency of this element in higher animals manifests itself in the form of a disease known as "muscular dystrophy" or "white muscle disease." Cases of white muscle disease in sheep and cattle have been reported in the United States, New Zealand, and Australia. Because of the danger of selenium toxicity, no additions of this element have been made to such soils via the fertilizer, at least in the USA.

Animals so affected are given the element in appropriate quantities either orally or by subcutaneous injections of a selenate salt.

Chelates. In preceding discussion of the microelements, the terms *chelates* or *chelating agents* were frequently employed. Very simply, metal chelates are defined as cyclic structures of a metal atom and an organic component in which the two components are held together with varying degrees of strength, varying from a rather loose bonding force to the strong metal-organic bond typical of metal porphyrins.

Metal chelates are soluble in water. Those commonly used for agriculture dissociate to only a very slight degree. Hence, although they decrease the *activity* of metallic ions in aqueous solution, the *solubility* of these metals is greatly increased in combination with the chelating agent. The term *sequestering agent* is sometimes used in lieu of chelating agent.

Numerous substances have the ability to chelate or sequester metallic ions. Several compounds are commercially important in agriculture, four of which are ethylenediaminetetraacetic acid (EDTA), diethylenetriaminepentaacetic acid (DTPA), cyclohexanediaminetetraacetic acid (CDTA), and ethylenediaminedi(o-hydroxyphenylacetic acid) (EDDHA).

The metallic ions commonly sequestered commercially are iron, copper, zinc, and manganese. Just how these chelated metals are absorbed and utilized by the plant is not known. It was once thought that the intact molecule was absorbed by the root hairs, for some of the early work showed a 1:1 ratio of metal:chelate in the root cells. Later work indicated that the metal and the chelate were not absorbed uniformly and that at low pH values appreciable amounts of the chelates remained in the solution external to the root even though the metal itself was absorbed. It must be admitted that at this writing the mechanism of metal:chelate absorption and utilization by plants is not completely understood.

The chelates without the metals iron, copper, manganese, and zinc exist as acids or sodium salts. When applied to soils, they have the ability to sequester the heavy metals from their insoluble forms in the soil. In fact, it has been possible to correct iron deficiencies simply by applying sodium-chelate directly to the soil around the root zone.

These chelated compounds are quite effective, for only small amounts are required to correct deficiency of the various metals. Chelates can be applied either to the soil or to plant leaves in the form of an aqueous spray.

Summary

1. In most humid region soils a large proportion of the sulfur occurs in organic combination. With soils containing appreciable amounts of hydrous oxides, some adsorbed sulfate sulfur will be found. In arid

regions, of course, soluble sulfate salts accumulate in the soil profile.
2. The mobilization and immobilization of soil sulfur depends on the supply of organic carbon, nitrogen, and phosphorus as well as the activity of soil microorganisms.
3. Sulfate sulfur, which is the form absorbed by plants, is lost to a great extent in leaching waters. With soils high in iron and aluminum oxides and in 1:1 clays, however, appreciable quantities are retained by adsorption.
4. Sulfides, polysulfides, and elemental sulfur are converted to sulfates by soil microorganisms. The speed with which this conversion takes place depends on the temperature, moisture, organisms present, and the soil *p*H. Of great importance, particularly with elemental sulfur, is the fineness of the material.
5. The behavior in the soil of several microelements was covered. The availability of iron and manganese is dependent on soil *p*H, degree of waterlogging of the soil, temperature, and sometimes the amount of organic matter present. Boron is leached readily from sand but is retained fairly well in fine-textured soils. Copper deficiencies occur largely on peat and muck soils, but its availability is also affected by soil *p*H. The availability of zinc is decreased by high *p*H values, the presence of free calcium carbonate in the soil, and by high soil phosphate levels. The availability of molybdenum, unlike that of most of the other microelements, increases as soil *p*H increases.
6. Chelates, which are compounds that form soluble complexes with certain metallic elements, are used as carriers of iron, copper, manganese, and zinc. They maintain these elements in a plant-available form in soils in which the elements would otherwise be rendered unavailable. Other micronutrient fertilizers were discussed, among them frits, which are trace element compounds suspended in a glass matrix. The glass is then finely ground and applied to the soil.

Questions

1. What are the forms of sulfur found in soils?
2. What effect has drying the soil on the availability of organic sulfur?
3. What is the importance of the $C:N:P:S$ ratio to the availability of soil sulfur?
4. What are the soil conditions under which losses of sulfur by leaching would be expected?
5. What are the soil conditions under which little loss of sulfur by leaching would be expected?
6. Describe the soil and climatic conditions under which sulfur deficiencies in the field are most likely to occur.
7. Discuss sulfate adsorption by soils with emphasis on the factors affecting this phenomenon.

8. What are the factors affecting the oxidation of elemental sulfur in soils?
9. What are the soil conditions contributing to the loss of boron by leaching?
10. What are the common forms of fertilizer sulfur that may be added to soils?
11. What is the effect on soil acidity of adding elemental sulfur to soil?
12. How is the availability of copper affected by the level of soil organic matter and soil pH?
13. How is a cobalt deficiency in animals commonly remedied?
14. What are some common copper fertilizers?
15. What are the reasons that have been suggested to explain iron chlorosis of plants?
16. What are the ways commonly employed to overcome iron chlorosis?
17. In what forms is manganese believed to exist in soils?
18. Under what soil conditions has a manganese toxicity been observed?
19. What are some common manganese fertilizer materials?
20. In what way is the behavior of molybdenum in soils different from the behavior of the other microelements?
21. What soil factors influence the availability of zinc in soils?
22. In what ways is selenium important in soils?
23. What are chelates? Which of the microelements are frequently applied as chelates?
24. What are frits? How is their availability influenced by their particle size?

Selected References

Adams, F., and J. I. Wear. "Manganese toxicity and soil acidity in relation to crinkle leaf of cotton." *SSSA Proc.*, **21:**305 (1957).

Anderson, A. J., "Molybdenum deficiency on a South Australian ironstone soil." *J. Australian Inst. Agr. Sci.*, **8:**73 (1942).

Anderson, A. J., "The significance of sulphur deficiency in Australian soils." *J. Australian Inst. Agr. Sci.*, **18:**135 (1952).

Anderson, A. J., "Molybdenum as a fertilizer." *Advan. Agron.*, **8:**164 (1956).

Anon., "Control techniques for sulfur oxide air pollutants." *Natl. Air Poll. Cont. Admin. Pub. No. AP-52*, January 1969.

Atkinson, W. T., M. H. Walker, and R. G. Weir, "The phosphorus and sulphur needs of pastures in New South Wales." *9th Intern. Grassland Confer., Brazil*, January 1965.

Banerjee, D. K., R. H. Bray, and S. W. Melsted, "Some aspects of the chemistry of cobalt in soils." *Soil Sci.*, **75:**421 (1953).

Bardsley, C. E., Jr., and H. V. Jordan, "Sulfur availability in seven Southeastern soils as measured by growth and composition of white clover." *Agron. J.*, **49:**310 (1957).

Barrow, N. J., "A comparison of the mineralization of nitrogen and of sulfur from decomposing organic materials." *Australian J. Agr. Res.*, **11**:960 (1960).

Barrow, N. J., "Studies on the mineralization of sulfur from soil organic matter." *Australian J. Agr. Res.*, **12**:306 (1961).

Barrow, N. J., "Studies on the adsorption of sulfate by soils." *Soil Sci.*, **104**:342 (1967).

Beaton, J. D., S. L. Tisdale, and J. Platou, "Crop responses to sulphur in North America." *Tech. Bull. No. 18*. Washington, D.C.: The Sulphur Institute, December 1971.

Beaton, J. D., "Fertilizers and their use in the decade ahead." *Proc. 20th Ann. Meet., Agr. Res. Inst.*, October 13, 1971.

Biggar, J. W., and M. Fireman, "Boron adsorption and release by soils." *SSSA Proc.*, **24**:115 (1960).

Bingham, F. T., and A. L. Page, "Specific character of boron adsorption by an amorphous soil." *SSSA Proc.*, **35**:892 (1971).

Bixby, D. W., and J. D. Beaton, "Sulphur-containing fertilizers, properties and applications." *Tech. Bull. No. 17*. Washington, D.C.: The Sulphur Institute, December 1970.

Bledsoe, R. W., and R. E. Blaser, "The influence of sulfur on the yield and composition of clovers fertilized with different sources of phosphorus." *J. Am. Soc. Agron.*, **39**:146 (1947).

Blevins, R. L., and H. F. Massey, "Evaluation of two methods of measuring available soil copper and the effects of soil *p*H and extractable aluminum on copper uptake by plants." *SSSA Proc.*, **23**:296 (1959).

Bloomfield, C., Effect of some phosphate fertilizers on the oxidation of elemental sulfur in soil." *Soil Sci.*, **103**:219 (1967).

Boawn, L. C., and F. G. Viets, Jr., "Zinc deficiency of alfalfa in Washington." *Agron. J.*, **44**:276 (1952).

Boawn, L. C., F. G. Viets, Jr., and C. L. Crawford, "Effects of nitrogen carrier, nitrogen rate, zinc rate, and soil *p*H on zinc uptake by sorghum, potatoes, and sugar beets." *Soil Sci.*, **90**:329 (1960).

Bornemisza, E., and R. Llanos, "Sulfate movement, adsorption, and desorption in three Costa Rican soils." *SSSA Proc.*, **31**:356 (1967).

Bridger, G. L., M. L. Salutsky, and R. W. Starostka, "Micronutrient sources, metal ammonium phosphates as fertilizers." *J. Agr. Food Chem.*, **10**:181 (1962).

Brown, A. L., B. A. Krantz, and P. E. Martin, "Plant uptake and fate of soil-applied zinc." *SSSA Proc.*, **26**:167 (1963).

Brown, J. C., "An evaluation of bicarbonate induced iron chlorosis." *Soil Sci.*, **89**:26 (1960).

Brown, J. C., R. S. Holmes, and L. O. Tiffin, "Hypotheses concerning iron chlorosis." *SSSA Proc.*, **23**:231 (1959).

Brown, J. C., "Agricultural use of synthetic metal chelates." *SSSA Proc.*, **33**:59 (1969).

Brown, J. C., and J. E. Ambler, "Further characterization of iron uptake in two genotypes of corn." *SSSA Proc.*, **34**:249 (1970).

Brown, J. C., L. O. Tiffin, A. W. Specht, and J. W. Resnicky, "Iron absorption by roots as affected by plant species and concentration of chelating agent." *Agron. J.*, **53**:81 (1961).

Burleson, C. A., A. D. Dacus, and C. J. Gerard, "The effect of phosphorus fertilization on the zinc nutrition of several irrigated crops." *SSSA Proc.*, **25**:365 (1961).

Burns, G. R., "Oxidation of sulphur in soils." *Tech. Bull. No. 13.* Washington, D.C.: The Sulphur Institute, June 1968.

Chang, M. L., and G. W. Thomas, "A suggested mechanism for sulfate adsorption by soils." *SSSA Proc.*, **27**:281 (1963).

Chao, T. T., M. E. Harward, and S. C. Fang, "Movement of S^{35} tagged sulfate through soil columns." *SSSA Proc.*, **26**:27 (1962).

Chao, T. T., M. E. Harward, and S. C. Fang, "Adsorption and desorption phenomena of sulfate ions in soils." *SSSA Proc.*, **26**:234 (1962).

Chao, T. T., M. E. Harward, and S. C. Fang, "Soil constituents and properties in the adsorption of sulfate ions." *Soil Sci.*, **94**:276 (1962).

Chao, T. T., M. E. Harward, and S. C. Fang, "Cationic effects on sulfate absorption by soils." *SSSA Proc.*, **27**:35 (1963).

Davies, E. B., "Factors affecting molybdenum deficiency in soils." *Soil Sci.*, **81**:209 (1956).

Donald, C. M., and K. Spencer, "The control of molybdenum deficiency in subterranean clover by presoaking the seed in sodium molybdate solution." *Australian J. Agr. Res.*, **2**:295 (1951).

During, C., "Recent research work: Sulphur." *New Zealand J. Agr.*, **93**:549 (1956).

Ensminger, L. E., "Some factors affecting the adsorption of sulfate by Alabama soils." *SSSA Proc.*, **18**:259 (1964).

Ensminger, L. E., "Sulfur in relation to soil fertility." *Alabama Agr. Exp. Sta. Bull. 312* (1958).

Fleming, G. A., "Selenium in Irish soils and plants." *Soil Sci.*, **94**:28 (1962).

Follett, R. F., and S. A. Barber, "Properties of the available and the soluble molybdenum fractions in a Raub silt loam." *SSSA Proc.*, **31**:191 (1967).

Forsee, W. T., Jr., "Conditions affecting the availability of residual and applied manganese in the organic soils of the Florida Everglades." *SSSA Proc.*, **18**:475 (1954).

Fox, R. L., A. D. Flowerday, F. W. Hosterman, H. F. Rhoades, and R. A. Olson, "Sulfur fertilizers for alfalfa production in Nebraska." *Nebraska Res. Bull. 214* (1964).

Freney, J. R., "The oxidation of cysteine to sulfate in soil." *Australian J. Biol. Sci.*, **13**:387 (1960).

Freney, J. R., "Some observations on the nature of organic sulfur compounds in soils." *Australian J. Agr. Res.*, **12**:424 (1961).

Freney, J. R., and K. Spencer, "Soil sulfate changes in the presence and absence of growing plants." *Australian J. Agr. Res.*, **11**:339 (1960).

Fried, M., "The absorption of sulfur dioxide by plants as shown by the use of radioactive sulfur." *SSSA Proc.*, **13**:135 (1949).

Gausmann, H. W., G. O. Estes, and A. Burns, "Chloride deficiency on potatoes under field conditions." *Maine Farm Res. 21–22* (October 1959).

Greenwood, M., "Sulphur deficiency in groundnuts in Northern Nigeria." *Trans. 5th Intern. Congr. Soil Sci.*, **3**:245 (1955).

Guinn, G., and H. E. Joham, "Effect of two chelating agents on absorption and translocation of Fe, Cu, Mn, and Zn by the cotton plant." *Soil Sci.*, **94**:220 (1962).

Hader, R. J., M. E. Harward, D. D. Mason, and D. P. Moore, "An investigation of some of the relationships between Cu, Fe, and Mo in the growth and nutrition of lettuce: I. Experimental design and statistical methods for characterizing the response surface." *SSSA Proc.*, **21:**59 (1957).

Haertl, E. J., "Chelation in nutrition, metal chelates in plant nutrition." *J. Agr. Food Chem.*, **11:**108 (1963).

Hammes, J. K., and K. C. Berger, "Manganese deficiency in oats and correlation of plant manganese with various soil tests." *Soil Sci.*, **90:**239 (1960).

Hemstock, G. A., and P. F. Low, "Mechanisms responsible for the retention of manganese in the colloidal fraction of soil." *Soil Sci.*, **76:**331 (1953).

Hiatt, A. J., and J. L. Ragland, "Manganese toxicity of burley tobacco." *Agron. J.*, **55:**47 (1963).

Hilder, E. J., "Some aspects of sulfur as a nutrient for pastures in New England soils." *Australian J. Agr. Res.*, **5:**39 (1954).

Hodge, J. E., E. C. Nelson, and B. F. Moy, "Chelates in agriculture, metal chelation by glucose ammonia derivatives." *J. Agr. Food Chem.*, **11:**126 (1960).

Hodgson, J. F., R. M. Leach, Jr., and W. H. Allaway, "Micronutrients and animal nutrition, micronutrients in soils and plants in relation to animal nutrition." *J. Agr. Food Chem.*, **10:**171 (1962).

Holden, E. R., and A. J. Engel, "Boron supplements. Response of alfalfa to applications of a soluble borate and a slightly soluble borosilicate glass." *J. Agr. Food Chem.*, **5:**275 (1957).

Holden, E. R., and A. J. Engel, "Boron fertilization, borosilicate glass as a continuing source of boron for alfalfa." *J. Agr. Food Chem.*, **6:**303 (1958).

Holden, E. R., N. R. Page, and J. I. Wear, "Micronutrient glasses, properties and use of micronutrient glasses in crop production." *J. Agr. Food Chem.*, **10:**188 (1962).

Hortenstine, C. C., D. A. Ashley, and J. I. Wear, "An evaluation of slowly soluble boron materials." *SSSA Proc.*, **22:**249 (1958).

Jackson, W. A., N. A. Heinly, and J. H. Caro, "Trace elements in fertilizers, solubility status of zinc carriers intermixed with N-P-K fertilizers." *J. Agr. Food Chem.*, **10:**361 (1962).

Jensen, J., "Some investigations of plant uptake of sulfur." *Soil Sci.*, **95:**63 (1963).

Johnson, C. M., G. A. Pearson, and P. R. Stout, "Molybdenum nutrition of crop plants. II. Plant and soil factors concerned with molybdenum deficiencies in crop plants." *Plant Soil*, **4:**178 (1952).

Jones, L. H. P., and G. W. Leeper, "The availability of various manganese oxides to plants." *Plant Soil*, **3:**141 (1951).

Jones, M. B., W. E. Martin, and W. A. Williams, "Behavior of sulfate sulfur and elemental sulfur in three California soils in Lysimeters." *SSSA Proc.*, **32:**535 (1968).

Jones, M. B., and J. E. Ruckman, "Effect of particle size on long-term availability of sulfur on annual-type grasslands." *Agron. J.*, **61:**936 (1969).

Jordan, H. V. "Sulfur as a plant nutrient in the southern United States." *USDA Tech. Bull. 1297* (1964).

Jordan, H. V., and C. E. Bardsley, "Response of crops to sulfur in Southeastern soils." *SSSA Proc.*, **22:**254 (1958).

Jordan, H. V., and L. E. Ensminger, "The role of sulfur in soil fertility." *Advan. Agron.*, **10**:408 (1958).

Jurinak, J. J., and D. W. Thorne, "Zinc solubility under alkaline conditions in a zinc-bentonite system." *SSSA Proc.*, **19**:446 (1955).

Kamprath, E. J., "Possible benefits from sulfur in the atmosphere." *Combustion*, **44**:16 (October 1972).

Kamprath, E. J., W. L. Nelson, and J. W. Fitts, "The effect of p H, sulfate, and phosphate concentrations on the adsorption of sulfate by soils." *SSSA Proc.*, **20**:463 (1956).

Kamprath, E. J., W. L. Nelson, and J. W. Fitts, "Sulfur removed from soils by field crops." *Agron. J.*, **49**:289 (1957).

Kittams, H. A., "The use of sulfur for increasing the availability of phosphorus in rock phosphate." Ph.D. Thesis, University of Wisconsin, 1963.

Lavy, T. L., and S. A. Barber, "A relationship between the yield responses of soybeans to molybdenum applications and the molybdenum content of seed produced." *Agron. J.*, **55**: 154 (1963).

Leeper, G. W., "Forms and reactions of manganese in soil." *Soil Sci.*, **63**:79 (1947).

Leeper, G. W., and R. J. Swaby, "The oxidation of manganous compounds by microorganisms in the soil." *Soil Sci.*, **49**:163 (1940).

Li, Paulina Y. W., "The oxidation of elemental sulfur in soil." MS Thesis, University of Minnesota, 1964.

Lingle, J. C., and D. M. Holmberg, "The response of sweet corn to foliar and soil zinc applications on a zinc-deficient soil." *Proc. Am. Soc. Hort. Sci.*, **70**:308 (1957).

Liu, M., and G. W. Thomas, "Nature of sulfate retention by acid soils." *Nature (London)*, **192**:384 (1961).

Lobb, W. R., "Sulphur investigations in North Otago." *New Zealand J. Agr.*, **89**:434 (1954).

Löhnis, Marie P., "Manganese toxicity in field and market garden crops." *Plant Soil*, **3**:193 (1951).

Lowe, L. E., and W. A. DeLong, "Aspects of the sulphur status of three Quebec soils." *Can. J. Soil Sci.*, **41**:141 (1961).

Ludwick, A. E., "Manganese availability in manganese-sulfur granules." MS Thesis, University of Wisconsin, 1964.

Lundblad, K., O. Svanberg, and P. Ekman, "The availability and fixation of copper in Swedish soils." *Plant Soil*, **1**:277 (1949).

McClung, A. C., L. M. DeFreitas, and W. L. Lott, "Analyses of several Brazilian soils in relation to plant responses to sulphur." *SSSA Proc.*, **23**:221 (1959).

McKell, C. M., and W. A. Williams, "A lysimeter study of sulfur fertilization of an annual-range soil." *J. Range Management*, **13**:113 (1960).

McLachlan, K. D., "The occurrence of sulfur deficiency on a soil of adequate phosphorus status." *Australian J. Agr. Res.*, **3**:125 (1952).

McLachlan, K. D., "Phosphorus, sulfur, and molybdenum deficiencies in soils from Eastern Australia in relation to the nutrient supply and some characteristics of soil and climate." *Australian J. Agr. Res.*, **6**:673 (1955).

Mandal, L. N., "Transformations of iron and manganese in water-logged rice soils." *Soil Sci.*, **91**:121 (1961).

Martin, W. E., "Sulfur deficiency widespread." *Calif. Agr.*, **11:**10 (1958).

Massey, H. F., "Relation between dithizone-extractable zinc in the soil and zinc uptake by corn plants." *Soil Sci.*, **83:**123 (1957).

Metson, A. J., "Sulphur in forage crops." *Tech. Bull. No. 20.* Washington, D.C.: The Sulphur Institute, January 1973.

Miller, G. W., J. C. Brown, and R. S. Holmes, "Chlorosis in soybeans as related to iron, phosphorus, bicarbonate, and cytochrome oxidase activity." *Plant Physiol.*, **35:**610 (1960).

Mokragnatz, M., and Z. Filipovic, "Further evidence of the influence of soil pH on cobalt content of grasses." *Soil Sci.*, **92:**127 (1961).

Moore, D. P., M. E. Harward, D. D. Mason, R. J. Hader, W. L. Lott, and W. A. Jackson, "An investigation of some of the relationships between copper, iron and molybdenum in the growth and nutrition of lettuce: II. Response surfaces of growth and accumulations of Cu and Fe." *SSSA Proc.*, **21:**65 (1957).

Mortenson, J. C., "Complexing of metals by soil organic matter." *SSSA Proc.*, **27:**179 (1963).

Mortvedt, J. J., P. M. Giordano, and W. L. Lindsay, Eds., *Micronutrients in Agriculture.* Madison, Wisconsin: SSSA, 1972.

Mulder, E. G., "Importance of molybdenum in nitrogen metabolism of microorganisms and higher plants." *Plant Soil,* **1:**94 (1949).

Mulder, E. G., and F. C. Gerretsen, "Soil manganese in relation to plant growth." *Advan. Agron.*, **4:**221 (1952).

Muth, O. H., and J. E. Oldfield, Eds., *Symposium: Sulfur in Nutrition.* Westport, Connecticut: The Avi Publishing Company, 1970.

Nearpass, D. C., M. Fried, and V. J. Kilmer, "Greenhouse measurement of available sulfur using radioactive sulfur." *SSSA Proc.*, **25:**287 (1961).

Newton, J. D., C. F. Bentley, J. A. Toogood, and J. A. Robertson, "Grey-wooded soils and their management." *Univ. Alberta Bull. No. 21,* 5th Ed. Rev. (March 1959).

Nicolson, A. J., "Soil sulfur balance studies in the presence and absence of growing plants." *Soil Sci.*, **109:**345 (1970).

Oliver, S., and S. A. Barber, "Mechanisms for the movement of Mn, Fe, B, Cu, Zn, Al and Sr from one soil to the surface of soybean roots (*Glycine max*)." *SSSA Proc.*, **30:**468 (1966).

Olson, R. A., et al., Eds., *Fertilizer Technology & Use,* 2nd ed. Madison, Wisconsin: Soil Science Society of America, 1971.

Ozanne, P. G., "Chlorine deficiency in soils." *Nature (London)*, **182:**1172 (1958).

Page, E. R., "Studies in soil and plant manganese: II. The relationship of soil pH to manganese availability." *Plant Soil,* **16:** 247 (1962).

Parr, J. F., and P. M. Giordano, "Agronomic effectiveness of anhydrous ammonia-sulfur solutions: 2." *Soil Sci.,* **106:** 448 (1968).

Reddy, K. G., and B. V. Mehta, "Cobalt investigations on Gujarat (India) soils." *Soil Sci.*, **92:**274 (1961).

Reisenauer, H. M., "Relative efficiency of seed- and soil-applied molybdenum fertilizer." *Agron. J.*, **55:**459 (1963).

Reisenauer, H. M., A. A. Tabikh, and P. R. Stout, "Molybdenum reactions with soils and the hydrous oxides of Fe, Al, and Ti." *SSSA Proc.*, **26:**23 (1962).

Roberts, S., and F. E. Koehler, "Extractable and plant-available sulfur in representative soils of Washington." *Soil Sci.,* **106:** 53 (1968).

Sanchez, C., and E. J. Kamprath, "The effect of liming and organic matter content on the availability of native and applied manganese." *SSSA Proc.,* **23:**302 (1959).

Seatz, L. F., A. J. Sterges, and J. C. Kramer, "Crop response to zinc fertilization as influenced by lime and phosphorus applications." *Agron. J.,* **51:**457 (1959).

Soil Sci., Vol. 101, No. 4 (April 1966).

Sorteberg, A., "Copper relationships with oats on adding cultivated to uncultivated peat." *Soil Sci.,* **94:**80 (1962).

Stephens, C. G., and C. M. Donald, "Australian soils and their responses to fertilizers." *Advan. Agron.,* **10:**168 (1958).

Stewart, B. A., L. K. Porter, and F. G. Viets, Jr., "Effect of sulfur content of straws on rates of decomposition and plant growth." *SSSA Proc.,* **30:**355 (1966).

Stout, P. R., W. R. Meagher, G. A. Pearson, and C. M. Johnson, "Molybdenum nutrition of crop plants. I. The influence of phosphate and sulfate ion on the absorption of molybdenum from soils and solution cultures." *Plant Soil,* **3:**51 (1951).

Thorne, W., "Zinc deficiency and its control." *Advan. Agron.,* **9:**31 (1957).

Tisdale, S. L., and B. R. Bertramson, "Elemental sulfur and its relationship to manganese availability." *SSSA Proc.,* **14:**11 (1949).

Underwood, E. J., *Trace Elements in Human and Animal Nutrition,* 2nd ed. New York: Academic, 1962.

Venema, K. C. W., "Some notes regarding the function of the sulfate anion in the metabolism of oil producing plants, especially oil palms. Part I." *Potash Trop. Agr.,* **5** (July 1962).

Viets, F. G., Jr. "Micronutrient availability, chemistry and availability of micronutrients in soils." *J. Agr. Food Chem.,* **10:**174 (1962).

Walker, T. W., "Sulfur responses on pastures in Australia and New Zealand." *Soils Fertilizers,* **18:**185 (1955).

Walker, T. W., "The use of sulphur as a fertilizer." *Intern. Conf. Sulfur on Agriculture,* Palermo, 1964.

Walker, T. W., and A. F. R. Adams, "Competition for sulfur in a grass-clover association." *Plant Soil,* **9:**353 (1958).

Walker, T. W., A. F. R. Adams, and H. D. Orchiston, "The effects and interactions of molybdenum, lime, and phosphate treatments on the yield and composition of white clover grown on acid, molybdenum responsive soils." *Plant Soil,* **6:**20 (1955).

Wallace, A., "Chelation in nutrition, review of chelation in plant nutrition." *J. Agr. Food Chem.,* **11:**103 (1963).

Wallace, A., *Regulation of the Micronutrient Status of Plants by Chelating Agents and Other Factors.* Los Angeles, Calif. 1971.

Walsh, L. M., and J. D. Beaton, *Soil Testing and Plant Analysis* (rev.). Madison, Wisc.: Soil Sci. Soc. of Amer., 1973.

Wear, J. I., "The effect of soil pH and calcium on uptake of zinc by plants." *Soil Sci.,* **81:**311 (1956).

Wear, J. I., and R. M. Patterson, "Effect of soil pH and texture on the availability of water-soluble boron in the soil." *SSSA Proc.*, **26:**344 (1962).

Westfall, D. G., W. B. Anderson, and R. J. Hodges, "Iron and zinc response of chlorotic rice grown on calcareous soils." *Agron. J.*, **63:**702 (1971).

White, J. G., "Mineralization of nitrogen and sulfur in sulfur-deficient soils." *New Zealand J. Agr. Res.*, **2:**225 (1959).

Wilkinson, H. F., J. F. Loneragan, and J. P. Quirk, "The movement of zinc to plant roots." *SSSA Proc.*, **32:**831 (1968).

Williams, C. H., and A. Steinbergs, "Sulphur and phosphorus in some Eastern Australian soils." *Australian J. Agr. Res.*, **9:**483 (1958).

Williams, C. H., E. G. Williams, and N. M. Scott, "Carbon, nitrogen, sulfur, and phosphorus in some Scottish soils." *J. Soil Sci.*, **11:**334 (1960).

Wilson, C. M., R. L. Lovvorn, and W. W. Woodhouse, "Movement and accumulation of water soluble boron within the soil profile." *Agron. J.*, **43:**363 (1951).

9. MANUFACTURE OF NITROGEN, PHOSPHORUS, AND POTASSIUM FERTILIZER

The properties and use of fertilizer materials containing nitrogen, phosphorus, and potassium were discussed in Chapters 5, 6, and 7. It is not the purpose of this book to cover in detail the manufacturing processes of these materials, for this is a topic more properly covered in courses on chemical engineering. Some knowledge, however, of the basic reactions in fertilizer production is of value in better understanding the essential parts assumed by these materials in the over-all picture of soil fertility and its place in the agribusiness enterprise.

In the following sections are discussed the reactions and processes encountered in the manufacture of the basic nitrogen-, phosphorus-, and potassium-containing materials. The production of only a few compounds is necessary for the synthesis of the many discussed in Chapters 5, 6, and 7. The information contained in the present chapter therefore deals with the reactions involved in the manufacture of these fundamental materials.

NITROGEN FERTILIZERS

Before 1940 most of the fertilizer nitrogen consumed in the United States was in the form of natural organic materials; ammonium sulfate, a by-product of the coking industry; and nitrate of soda imported from Chile. The use of natural organics as sources of fertilizer nitrogen in this country has dwindled to almost nothing. The importation of Chilean nitrate has declined in terms of absolute tonnages consumed and greatly in terms of the percentage of total fertilizer nitrogen. This fact is attributed to the great increases in the manufacture of synthetic nitrogen goods. The production of ammonium sulfate is geared largely to the production of steel, a coke-consuming process. More recently large tonnages of by-product ammonium sulfate have become available from the production of caprolactam, an intermediate in the manufacture of nylon. The production of Chilean nitrate and by-product coke-oven ammonium sulfate is touched on briefly and is followed by a discussion of the reactions involved in the production of the more important synthetic nitrogen fertilizers.

Chilean Nitrate of Soda. Tremendous deposits of salts high in sodium nitrate occur in Chile. They lie on the eastern slope of the coastal range in a desertic valley between these mountains and the Andes to the east. The most important are found in an area 450 miles long and 10 to 50 miles wide, the elevation of which is between 4,000 and 7,500 ft. As would be expected from the accumulation of a salt as soluble as sodium nitrate, the region is barren—devoid of vegetation, water, or fuel. To mine the ore all facilities must be carried in.

The Chilean deposits were discovered by Thaddeus Haenke in 1809. The Spaniards are believed to have begun mining operations in 1813, but the first recorded shipment of this material to the United States did not arrive until 1830.

The origin of these deposits is not known. Several theories have been advanced to explain their occurrence, but each has been criticized for one reason or another. Some of these have been (1) fixation of atmospheric nitrogen by nonsymbiotic soil bacteria; (2) fixation by atmospheric discharge of electricity; (3) decomposition of seaweed and other marine flora after a geological uplift had exposed the land; and (4) the accumulation of nitrates from leaching of surface waters into the valley from the surrounding highlands which are populated by large herds of vicunas and llamas. None of these theories seems to be consistent in all respects with the known facts, and the origin of these most unusual deposits remains a mystery.

Basically, the sodium nitrate is manufactured by blasting the ore which is found only a few feet from the surface of the ground. The raw ore is crushed, moved to the refinery, and extracted with a solution of boiling sodium nitrate. This is a stepwise extraction and the refined sodium nitrate ($96+\%$ $NaNO_3$) is recovered, dried, and shipped. It contains small quantities of several trace elements which have enhanced its value when used on many of the coarse-textured sandy soils of the southeastern United States, an area in which it finds its greatest consumption.

Coke-Oven Ammonium Sulfate. Certain types of coal, when heated in the absence of air, release volatile materials, such as ammonia gas. The nitrogen content of coal varies, but the type generally used for coking purposes contains 1 to 2 per cent nitrogen. The recovery of this nitrogen in 1913 amounted to an equivalent of 13 lb. of ammonium sulfate per ton of coal coked, but since that time it has declined because of several factors. The coke produced is used largely by the steel industry, but the by-products are themselves of considerable commercial value.

The basic principle in the production of coke-oven ammonia is the destructive distillation of coal. The coking coal is placed in ovens and heated to 1800°F. in the absence of air. The volatile materials released, which include tar and ammonia, are separated by various processes. The ammonia gas is purified and collected. It can be sold as such, but much of it is reacted with sulfuric acid to produce ammonium sulfate. Some of the ammonia may be dissolved in water and sold as aqua ammonia.

Basic Synthetic Processes. The fixation of atmospheric nitrogen is required for the production of all synthetic nitrogen fertilizer materials. The supply of this element is for all intents and purposes inexhaustible. It constitutes 80 per cent of the atmosphere.

Until the end of the nineteenth century this limitless store of nitrogen was virtually unavailable to man in a combined form. Some, of course, was fixed by symbiotic and nonsymbiotic bacteria and by discharge of electricity in the atmosphere, but large-scale commercial fixation has been undertaken only in recent years.

There are three basic reactions by which the fixation of elemental nitrogen is brought about: (1) direct oxidation of nitrogen, (2) the cyanamide process in which nitrogen is reacted with calcium carbide, and (3) the Claude-Haber process in which nitrogen and hydrogen form ammonia gas.

DIRECT OXIDATION OF NITROGEN. The oxidation of nitrogen has been taking place since the earth's atmosphere, as we know it now, evolved. With every flash of lightning some nitrogen is fixed. It was Cavendish in 1766 who first achieved this combination in a laboratory by passing electric sparks through a mixture of the two gases. The reactions are

$$N_2 + O_2 \longrightarrow 2NO$$
$$2NO + O_2 \longrightarrow 2NO_2$$

When the nitrogen dioxide is dissolved in water, nitric acid is formed:

$$3NO_2 + H_2O \longrightarrow 2HNO_3 + NO$$

The fixation of nitrogen by this method is appropriately termed the *arc process.*

Several attempts have been made in this country to manufacture nitrates by this process, but none has been sufficiently successful to permit continued operations for more than a few years, for it has a large electrical requirement and it cannot compete with the less expensive ammonia fixation. A modification of the arc process was developed by Birkeland and Eyde in Sweden, where, because of cheap hydroelectric power, it has been successful. By this method nitrogen and oxygen are passed through an arc which is expanded in an electromagnet to increase the contact. The gas mixture leaving the furnace contains about 1.3 to 1.7 per cent nitric oxide.

THE CYANAMIDE PROCESS. The cyanamide process was developed in Germany in 1898 by Frank and Caro. Essentially, it requires the reaction at high temperatures of nitrogen gas with calcium carbide to form calcium cyanamide. Calcium oxide is first prepared by heating calcium carbonate:

$$CaCO_3 \xrightarrow{\text{heat}} CaO + CO_2$$

The calcium oxide formed is next reacted with coke at 2200°C. to produce fused calcium carbide:

$$CaO + 3C \xrightarrow{\text{heat}} CaC_2 + CO$$

The calcium carbide is then reacted at 1100°C. with highly purified nitrogen gas to form calcium cyanamide:

$$CaC_2 + N_2 \xrightarrow{\text{heat}} CaCN_2 + C$$

After the residual carbides have been removed and the calcium oxide has been hydrated by treatment with water, the product is granulated or treated with oil to reduce dustiness.

The cyanamide process is limited in the United States because of the need for high power and large amounts of coke. A plant in operation at Niagara Falls, Ontario, manufactures a quantity of this material, some of which is shipped into the United States, but there is no commercial production of fixed nitrogen by this process within the limits of this country. However, it is still manufactured in Germany in appreciable quantities.

SYNTHETIC AMMONIA PRODUCTION. Ammonia production capacity is increasing rapidly throughout the world. A number of obsolete plants have been phased out and have been replaced with high-capacity, lower-cost factories. Manufacture involves three simple materials: natural gas, steam, and air. A basic problem is an adequate supply of natural gas (methane $- CH_4$).

The direct combination of nitrogen and hydrogen to form ammonia was first developed in Germany in 1910 by the efforts of Haber, Nernst, and others. A modification introduced by Claude, a Frenchman, improved the method, and his name is now associated with the *Claude-Haber ammonia synthesis*. The production of synthetic ammonia in the United States has increased tremendously since World War II and con-

stitutes by far the principal original source of all chemical nitrogen fertilizers.

The production of synthetic ammonia is based on the reaction of nitrogen and hydrogen in the presence of a catalyst (originally osmium and
iron) at temperatures that range from 400 to 500°C. The pressure
required varies from 200 to 1,000 atm., depending on the modification
employed. The reaction is expressed by the following equation:

$$3H_2 + N_2 \longrightarrow 2NH_3$$

The nitrogen for the synthesis is obtained in several ways, by one of
which air is passed into the reactor with the hydrogen. This method,
however, is wasteful of hydrogen, and if the cost of hydrogen is high, it
cannot be used. If the water-gas method is employed for the production
of hydrogen, producer gas resulting from the reheating of the coke beds
is used as the source of nitrogen. Tail gases from ammonia oxidation,
containing about 95 per cent nitrogen, are also used, as is the nitrogen
recovered by the fractional distillation of liquid air.

Several sources of hydrogen can be utilized in ammonia synthesis.
The well-known water-gas method, in which incandescent carbon reacts
with superheated steam, is represented by the following equation:

$$C + H_2O \longrightarrow CO + H_2$$

The method most important in the United States treats methane with
steam in the presence of a nickel catalyst. The reaction, shown by the
following equations, takes place at a pressure of about 40 lb./in.^{2} at a
temperature of 315–425°C.:

$$CH_4 + H_2O \longrightarrow CO + 3H_2$$
$$CO + H_2O \longrightarrow CO_2 + H_2$$

By subsequent treatment the carbon oxide content is reduced, and the
gas is rendered suitable for ammonia synthesis.

Naphtha, a partly distilled hydrocarbon, is used in some areas of the
world as a source of hydrogen. It is treated in a way similar to the
methane process.

An important source of hydrogen is the by-product of petroleum-
refining operations, and several refineries maintain adjacent ammonia-
fixation plants to take advantage of this supply. It can be seen that such
an arrangement provides considerable flexibility to the installation of ni-
trogen-fixation plants.

Hydrogen is also a by-product of the electrolytic production of caustic
soda. This is the oldest method of commercial production of hydrogen in

this country, but it now accounts for only a very small fraction of the total supply. In areas in which electric power is cheap, particularly in Italy and Norway, hydrogen is obtained from the electrolytic decomposition of water. Only one such unit, however, is located in North America.

AMMONIA OXIDATION. Much of the ammonia manufactured by the Claude-Haber and coke-oven processes is used directly in fertilizers, either in solution or as anhydrous ammonia. A sizable percentage of the total supply, however, is converted to nitric acid for industrial as well as fertilizer-production purposes. The oxidation of ammonia is accomplished in the presence of a platinum catalyst at 800°C. The process, though developed by Ostwald, is now known as the low-pressure process, because it takes place under pressures of only 50 to 100 lb./in.^{2}. The oxidation of the ammonia proceeds according to the following equation:

$$4NH_3 + 5O_2 \longrightarrow 4NO + 6H_2O + heat$$

The reaction is exothermic and once initiated is self-sustaining. The correct temperature is maintained by controlling the input of the reacting gases. The reactions for the conversion of the nitric oxide to nitric acid are as follows:

$$2NO + O_2 \longrightarrow 2NO_2$$
$$3NO_2 + H_2O \longrightarrow 2HNO_3 + NO$$

These reactions repeat through the absorber.

UREA. Urea, or carbamide, as it is sometimes called, is a nonionic nitrogen material used industrially in the manufacture of plastics, in fertilizers, and as a protein supplement in the feed of ruminant animals. Its preparation is a bit more complicated than that of many fertilizer salts, which essentially require only the neutralization of an acid with ammonia. It is prepared by reacting ammonia and carbon dioxide gas under very high pressure in the presence of a suitable catalyst. The reactions are represented in the following equations:

$$2NH_3 + CO_2 \longrightarrow NH_2COONH_4$$
$$NH_2COONH_4 \longrightarrow NH_2CONH_2 + H_2O$$

Fertilizer-grade urea is a water-soluble, acid-forming material containing about 45 per cent nitrogen. When conditioned, it is noncaking and free-flowing, which are excellent storage and handling qualities. The use of urea as a nitrogen fertilizer is increasing rapidly because of use in bulk blends and nitrogen solutions, and ease of handling for direct application.

OTHER NITROGEN FERTILIZER MATERIALS. Most of the nitrogen fertilizers discussed in Chapter 5 are made by different processes involving the use of the compounds just discussed—ammonia, nitric acid, and urea. Salts such as ammonium phosphate, ammonium sulfate, ammonium nitrate, and their various combinations are manufactured by neutralizing their acids with ammonia. Nitric phosphates are made by acidulating phosphate rock with nitric acid or mixtures of nitric acid with phosphoric or sulfuric acid and ammoniating. Many nitrogen solutions are made by dissolving ammonia gas in water and adding ammonium nitrate and/or urea.

The nitrogen content of the various solid and liquid nitrogen fertilizer materials was discussed in Chapter 5.

PHOSPHATE FERTILIZER MATERIALS

Today nearly all of the industrial and agricultural phosphates originate from phosphate rock. During the nineteenth century and the early part of the twentieth century bones and guano were important sources of fertilizer phosphorus. They no longer are and in the United States have no part whatever in the supply of fertilizer phosphorus.

Guano. Guano is an interesting material, both because of its formation and its importance in the early development of the fertilizer industry. Guano is the excreta and remains of seafowl and contains nitrogen and phosphorus as well as traces of the microelements. It is found in deposits that accumulate, of course, only under arid conditions, particularly along the western coasts of continents in the low latitudes, where a dry, subtropical climate inhibits leaching. The largest deposits of guano are found in vicinities in which marine phosphorites are precipitated. The phosphate supply in the water encourages a luxuriant growth of plankton, which in turn supports a large fish population and leads to a large concentration of seafowl. Guano deposits are greatest along the western coasts of Lower California, South America, and Africa.

The sources of guano are limited. The rich deposits of this material have been exhausted, and it no longer accounts for an appreciable fraction of the fertilizer phosphate tonnage. The phosphorus content is about 9 per cent, most of which is water-soluble, and the nitrogen content may be as high as 13 per cent.

Phosphate Rock. The basic phosphate compound in all commercially important deposits of phosphate rock is apatite. It may be a fluoro-, chloro-, or carbonatoapatite, as indicated in Chapter 6, though most of the apatites are fluorine. The larger deposits are of sedimentary origin, laid down in beds in the ocean and then elevated to land masses. The phosphate is usually in the form of small pellets cemented by $CaCO_3$. It may be loose pebbles or hard rock. There is considerable speculation as to the amount of reserves of phosphate rock and es-

TABLE 9-1.   Estimated Additional World Reserves of Phosphate Rock and Apatite*

	Million tons
Florida Hawthorne	200,000
Western United States	172,000
Southeast United States	50,000
Tennessee	3,000
Australia	200,000
Seafloor	300,000
Colombia	80,000
Saudi Arabia	150,000
Estimated total	1,155,000

In addition, the following areas contain known reserves which cannot be quantified now, but all of which are probably large. All these occur in sedimentary deposits. There are also reserves in igneous phosphates.

Baja California, Mexico	Iran
Zacatecas area, Mexico	India
Alaska	California
Iraq	Utah-Idaho (Mississippian)
Jordan	Spanish Sahara
Turkey	Morocco
Libya	Peru
Syria	People's Republic of
Tunisia	China
Algeria	

---

* Emigh, G. Donald, "World phosphate reserves—are there really enough?" *Engineering and Mining J.,* **173**(4):90–94. Reprinted with permission of *Engineering and Mining Journal,* copyright 1972, McGraw-Hill, Inc., 1221 Avenue of the Americas, New York, New York 10020.

timates are given in Table 9-1. In addition, estimates in Europe are on the order of 8,000 million tons and this and others bring total world estimates by Emigh to about 1,300,000 million tons. Russia also has large deposits of almost pure apatite located on the Kola Penninsula.

Table 9-2 brings in a new basic concept in that as the price per recoverable ton increases, phosphate reserves increase. The amount of phosphate rock in the world has not changed. However, the reserves that may be economically exploited have increased tremendously. In fact, as price per ton increases, industry can afford to mine rock that at a lower price could not be made available in the market place. There is no shortage of phosphate rock in the world. All that is needed for it to be made available is time and money.

In the United States the deposits presently being mined include those

TABLE 9-2.   World Phosphate Reserves as Related to Price per Recoverable Ton* (Millions of short tons)

Area	$8	$12	$20
North America	1,836	5,350	16,340
South America	53	290	930
Europe	829	2,050	4,100
Africa	1,770	8,430	20,500
Asia	335	1,186	4,600
Oceania	120	750	1,300
World	4,943	18,036	47,770

* Turbeville, W. J., Jr., "The phosphate rock situation." Presented at the *Phosphate-Sulphur Symposium*, Tarpon Springs, Florida. January 22–23, 1974.

of Florida, Idaho, Montana, Utah, Wyoming, Tennessee, and North Carolina.

MINING OF PHOSPHATE ROCK.   The working of phosphate rock is accomplished by both strip and shaft-mining techniques. Strip mining is employed in Florida, Tennessee, North Carolina, and in some of the western deposits, and shaft mining in some of the western deposits but not in any of the other fields.

Rock from most of the deposits has to be treated to separate the phosphate-containing fraction from the inert material. In some cases the ore must also be crushed. The phosphate-containing fraction is then separated from the waste material and concentrated by a complex system of washers, screens, classifiers, table agglomeration, and flotation. When dry, the concentrated ores are suitable for the manufacture of processed phosphate fertilizers.

TREATMENT OF PHOSPHATE ROCK.   As pointed out in Chapter 6, the apatite bond in phosphate rock must be broken if the contained phosphate is to be rendered easily available to plants. This can be done either by heat or acid treatment.

Defluorinated Phosphates.   When phosphate rock is heated in a gas- or oil-fired chamber to around 1500 to 1600°C., the fluorine is driven off and the remaining calcium phosphate is of greater plant availability than the phosphorus in the original rock. This defluorination is brought about both by calcination and fusing. Calcination refers to the heating of rock with silica and steam to temperatures below the melting point of the mix (see Chapter 6). Fusion refers to the process in which the mix is heated to a temperature above its melting point so that the furnace charge will run together, or fuse, to form a glassy product.

Regardless of whether fusion or calcining is employed, the reaction believed to take place during the defluorination of rock with heat may be represented as follows:

$$Ca_{10}(PO_4)_6 \cdot F_2 + xSiO_2 + H_2O \longrightarrow$$
$$3Ca_3(PO_4)_2 + CaO \cdot xSiO_2 + 2HF$$

The phosphorus remaining is alpha-tricalcium phosphate, and the fluorine escapes as a gas. Several of the materials discussed in Chapter 6 resulted from one of these two processes, and their manufacture is based on the reaction just described.

Elemental Phosphorus Manufacture. If phosphate rock is heated above 1400°C. in the presence of silica and carbon in a reducing atmosphere, elemental phosphorus is formed. The reaction is complex, but it can be expressed by the following generalized equation:

$$Ca_3(PO_4)_2 + 3SiO_2 + 5C \longrightarrow 3CaSiO_3 + P_2 + 5CO$$

The recovered elemental phosphorus is stored in the liquid state until ready for use. The United States Department of Defense uses elemental phosphorus in incendiary shells, grenades, and other items of warfare, but most of the industrial needs center around a high-purity phosphoric acid made from elemental phosphorus. This element is then burned to form phosphorus pentoxide, which in turn is reacted with water to form phosphoric acid as indicated in the following equations:

$$2P + \frac{5}{2}O_2 \longrightarrow P_2O_5$$
$$P_2O_5 + 3H_2O \longrightarrow 2H_3PO_4$$

By far the greatest tonnage of fertilizer phosphate materials is manufactured by treating phosphate rock with acid rather than with heat. The acids most commonly used are sulfuric, phosphoric, and nitric.

Sulfuric Acid-Treated Rock. Sulfuric acid is basic to the fertilizer industry. It is used in the manufacture of ordinary superphosphate as well as in the production of phosphoric acid. Phosphoric acid is a basic ingredient in the production of triple superphosphate and the ammonium phosphates.

Ordinary Superphosphate. Ordinary superphosphate is manufactured by the simple expedient of mixing gravimetrically equal parts of sulfuric acid and rock phosphate. The reaction, represented by the following simplified equation, gives off a considerable quantity of heat:

$$[Ca_3(PO_4)_2]_3CaF_2 + 7H_2SO_4 \longrightarrow 3Ca(H_2PO_4)_2 + 7CaSO_4 + 2HF$$

Three points are illustrated by this equation. First, the phosphate originally present as apatite is converted to water-soluble monocalcium phosphate. Second, one of the products of the reaction is gypsum, which is intimately mixed in with the monocalcium phosphate. Third, the reaction releases toxic hydrofluoric acid gas, which is usually recovered as a valuable by-product.

In one process, known as a *batch mix,* weighed quantities of rock phosphate and sulfuric acid of a certain concentration are combined. The ingredients are mixed and allowed to react for about a minute, after which the slurry is dumped into a compartment called a den. Here the phosphate may remain for about fifteen minutes after the den is filled. It is then removed, stirred, and stored. However, removal may not take place for twenty-four hours. The acidulated phosphate in the den sets up into a hard block, and removal is accomplished by means of various mechanical excavators which are usually equipped with revolving knives. These knives cut into the block, and the disintegrated superphosphate is stored until needed.

The batch-type process is used to some extent in the United States, but a continuous process of rock acidulation in which the phosphate rock and the acid are added to a mixer is now the major method. Ingenious metering and weighing devices are required for the successful operation of this phase of the process. The mixture is agitated for two or three minutes, and is then discharged onto an endless slat conveyor on which it solidifies. The slat conveyor moves the block of hardened superphosphate toward a revolving cutter, which disintegrates the material. It is then transferred to a storage bin.

Regardless of the mixing process employed, the manufacture of superphosphate is a simple operation. The product is a very satisfactory fertilizer material which contains calcium, sulfur, and 7 to 9 per cent available phosphorus, the lowest content of any of the important sources of fertilizer phosphorus.

Wet-Process Phosphoric Acid. Wet-process phosphoric acid is manufactured by extracting phosphate rock with sulfuric acid. The principal reaction taking place is represented by the following equation:

$$Ca_{10}(PO_4)_6F_2 + 10H_2SO_4 + 20H_2O \longrightarrow$$
$$10CaSO_4 \cdot 2H_2O + 6H_3PO_4 + 2HF$$

This reaction is carried out for about eight hours in a digestion system. The reaction itself is essentially complete within a matter of a few minutes, but the additional time is needed to ensure the formation of gypsum crystals of a size adequate to permit more rapid filtration. The slurry is ultimately filtered and the acid concentrated to the desired strength by heating.

Other Phosphatic Fertilizers. Ordinary superphosphate and phosphoric acid are the two basic phosphate materials manufactured with sulfuric acid. Ordinary superphosphate is used in the production of mixed fertilizers as well as in direct application to build up the phosphorus level of soils low in this element. Ordinary superphosphate, either singly or in combination with the carriers of nitrogen and potassium, is frequently ammoniated. Phosphoric acid is used in the manufacture of triple superphosphate, ammonium phosphates, and liquid fertilizers. (See Chapter 6.)

SULFURIC ACID. Sulfuric acid is literally the workhorse of the fertilizer industry. More than 60 per cent of the total consumption of this industrial acid is accounted for by the fertilizer industry alone, its largest single user.

Sulfuric acid is manufactured either from elemental sulfur or from the sulfur dioxide collected in the roasting of metal sulfides known as pyrites. In either case the sulfur dioxide formed is oxidized to sulfur trioxide and reacted with water to form sulfuric acid.

There are two principal processes in the manufacture of sulfuric acid. One is termed the *contact process*, the other the *lead-chamber process*. Although some lead-chamber plants still produce acid, all of the newer installations are contact plants.

In the lead-chamber process sulfur is burned in an oven to produce sulfur dioxide:

$$S + O_2 \longrightarrow SO_2$$

The reaction is exothermic. The fumes are pulled by a fan-induced draft into a series of large lead tanks, in which the sulfur dioxide is catalytically oxidized to sulfur trioxide by NO_2. This NO_2, known to the industry as niter gas, is recovered and used over and over again; otherwise the process would not be economically possible. The generalized reactions are illustrated by the following equations:

$$2HNO_3 + H_2O + 2SO_2 \longrightarrow 2H_2SO_4 + NO + NO_2$$
$$SO_2 + H_2O + NO_2 \longrightarrow H_2SO_4 + NO$$

As the gases are swept along, the nitric oxide is reconverted to niter gas and recovered in concentrated sulfuric acid in a tank at the end of the system. The sulfuric acid produced is collected in and withdrawn from the various lead tanks or chambers in the system. It is then stored or used immediately, depending on the prevailing situation.

In the contact process a mixture of sulfur dioxide and air is passed through iron tubes containing a finely divided catalyst, usually platinum. In the presence of this catalyst and at 400°C. the sulfur dioxide is rapidly converted to sulfur trioxide.

$$SO_2 + \tfrac{1}{2}O_2 \longrightarrow SO_3 + 22,600 \text{ Calories}$$

The sulfur trioxide is then passed into 98 per cent sulfuric acid, the reason being that sulfur trioxide is not readily soluble in water. The concentration of the acid is maintained at 98 per cent by the constant addition of water as the sulfur trioxide is passed into it.

SOURCE OF SULFUR DIOXIDE. The source of sulfur dioxide for both the contact and lead-chamber processes is either elemental sulfur or SO_2 recovered from the roasting of pyrites.

Dome Sulfur. Elemental sulfur is the principal source of sulfur dioxide in most countries, even in many in which there are substantial deposits of pyrites. Elemental sulfur is produced commercially in several ways. Deposits of the element are found along the Gulf Coast of Mexico, Louisiana, and Texas. These deposits occur in domelike formations, in conjunction with gypsum, several hundred feet below the surface of the earth.

An ingenious engineer named Herman Frash devised a scheme by which this sulfur could be brought to the surface. In essence, his method consists of drilling into the sulfur-gypsum formation and inserting three concentric pipes leading from the deposit to the surface of the earth. Superheated water and compressed air are forced down through two of the pipes into the formation. As the temperature of the superheated water is higher than that of the melting point of sulfur, the sulfur melts. The compressed air forces the molten sulfur to the surface of the ground through the third pipe. It is essentially 99.5 per cent pure as mined. It is either poured into molds, solidified, and stored in this state until ready for use or shipped in molten form. About 90 per cent of all Frash sulfur is handled in molten form.

Sour Gas. Elemental sulfur is also recovered from natural gas containing hydrogen sulfide, or *sour gas*. Before the natural gas can be used as fuel, the hydrogen sulfide must be removed. It is this removal that results in the production of elemental sulfur.

Without going into details, the sulfur is produced by this method essentially as follows. The hydrogen sulfide is stripped from the methane-H_2S gas mixture as it comes from the well. A portion is oxidized to sulfur dioxide, which in turn is allowed to react with the stream of hydrogen sulfide being stripped from the methane. The result is the production of elemental sulfur. The reactions involved are illustrated by the following equations:

$$H_2S + \tfrac{3}{2}O_2 \longrightarrow SO_2 + H_2O$$
$$SO_2 + 2H_2S \longrightarrow \tfrac{1}{2}S_6 + 2H_2O$$

The sulfur produced is handled in the same way as that mined by the Frash process.

Volcanic Sulfur. Local deposits of elemental sulfur throughout the world are associated with volcanic activity. The largest occurs in Sicily, where sulfur is mined by pick-and-shovel methods and marketed. In fact, before the discovery of the sulfur domes on the Gulf Coast of the United States, the Italians and Sicilians monopolized the world market. Today, however, Sicilian sulfur accounts for only a small fraction of the total world production of elemental sulfur.

Pyrites. Pyrites are the sulfides of heavy metals such as iron, lead, copper, and zinc. In recovering the metal from these ores, the sulfur is driven off by roasting or, more properly, by burning. The general reaction can be illustrated by the equation for the roasting of an iron pyrite:

$$2FeS_2 + \frac{7}{2}O_2 \longrightarrow 2SO_2 + Fe_2O_3 + heat$$

Roasting pyrites for their sulfur values is generally not economical. Acid produced by this method can compete in price with acid made from elemental sulfur only when the metal recovered from the pyrite can be profitably marketed. The sulfur dioxide produced from pyrites is converted to sulfuric acid as previously described.

Many of these smelting operations are located at places inaccessible to the market and because of shipping costs, the SO_2 value has not contributed to the supply of sulfuric acid. Legislation aimed at reducing environmental pollution is forcing the smelting industry to recover SO_2, much of which was released into the atmosphere.

Another possible future source of sulfur is the SO_2 which is released by the coal and oil-fired steam generating plants. The SO_2 emissions from the burning of fossil fuels in the United States amounted to the equivalent of about 12.8 million tons of elemental sulfur. Air pollution control legislation has required that most of these emissions be eliminated by the mid to late 1970's. If an economical process for the recovery of this sulfur can be found, this will afford an additional and significant supply.

NITRIC ACID ACIDULATION. For a number of years European manufacturers of phosphate fertilizers have used nitric acid for acidulating phosphate rock. This method, until a few years ago, was not employed in this country. The Tennessee Valley Authority investigated the possibility of developing a process suitable for use in the United States, largely because of a shortage of sulfur that developed near the end of World War II. The processes developed by TVA include the acidulation of phosphate rock with nitric acid, either alone or in combination with sulfuric or phosphoric acid. Four basic processes involve acidulation with (1) nitric and phosphoric acid, (2) nitric and sulfuric acid, (3) nitric acid with potassium sulfate added to the slurry, and (4) nitric acid in which the slurry is ammoniated and carbonated with carbon dioxide. The first two are probably best suited for commercial application, although some European manufacturers have had considerable success with modifications of the fourth process as indicated in Chapter 5.

That nitric phosphates have not been widely accepted in the United States stems in part from two inherent drawbacks in most of the nitric phosphate processes. The first is the additional cost of obtaining a product in which more than 50 or 60 per cent of the contained phosphorus is water-soluble. The second is the limited number of ratios and grades that can be manufactured by a given nitric phosphate process. Not only is the number of grades limited, but it is a difficult and costly process to change from one grade to another. Under present economic conditions an ammonia-phosphoric acid plant combination is capable of producing highly water-soluble phosphates in a large number of grades and ratios to meet specific crop requirements.

The known reserves of elemental sulfur, sour gas sulfur, pyrites and sulfur from fossil fuels are finite, though very large. The supply of elemental nitrogen, although not infinite, is for all intents and purposes so limitless that its shortage should not be considered. In the final analysis the economics of production will govern the part that nitric acid will take in the acidulation of phosphate rock. However, a shortage of natural gas from which hydrogen is obtained may limit future supplies of low-cost NH_3-based fertilizers.

POTASSIUM FERTILIZERS

Like phosphates, potassium fertilizers are obtained from deposits found several hundred to several thousand feet below the earth's surface. Like phosphates, too, the potassium ores must be beneficiated to produce high-grade potassium fertilizers; but, unlike phosphate rock, potassium salts do not require treatment with heat or strong acids to render the contained potassium available to plants, for they are water-soluble.

The word *potash*, which is the trade term commonly applied to potassium-containing fertilizers, was derived from *pot ashes*. During the early days wood and other plant residues were burned in pots to obtain the salts, largely for the manufacture of soap. The ashes containing the salts were extracted with water and the solution was then evaporated. The residue consisted of a mixture of potassium carbonate and other salts. The production of these salts from wood ashes was one of the first important chemical enterprises in the early colonial days in the United States. The first patent was granted in 1790 for the preparation of potassium salts.

Potash Deposits. Potash deposits occur as beds of solid salts at varying depths in the earth's surface and also as brines in dying lakes and seas. The principal potash deposits in the world today are listed in Table 9-3.

TABLE 9-3. Estimated World Reserves of Potassium in Lake Brines and Soluble Salt Deposits*

Country	Million metric tons of potassium
United States	225–360
Canada	18,000
East Germany	3,600–5,400
West Germany	1,800–3,600
USSR	24,000
France	180–225
Spain	83
Poland	8
Sicily	23
Israel-Jordan	544
Great Britain	23–45
Africa	141–150

* Kilmer, V. J., S. E. Younts, and N. C. Brady, Eds., *The Role of Potassium in Agriculture*, p. 6. Reprinted with permission of Am. Soc. of Agron., Crop Sci. Soc. of Am., and Soil Science Society of America, Madison, Wisconsin, 1968.

The largest world reserves of soluble potassium salts are found in Canada and the USSR. There are extensive deposits also in East and West Germany and Israel. Canadian deposits, the mining of which was begun in 1959, constitute an important world source of potassium.

In the United States the largest known deposits occur in the Permian salt basin, which includes southeastern New Mexico, northwestern Texas, and some of the western part of Oklahoma. The center of pro-

TABLE 9-4. Some Commercially Important Potash Minerals*

	Composition	Approximate content (%)	
		K	K_2O
Sylvite	KCl	52.4	63.0
Langbeinite	$K_2SO_4 \cdot 2MgSO_4$	18.8	22.6
Sylvinite	$KCl \cdot NaCl$ mixture	—	—
Carnallite	$KCl \cdot MgCl_2 \cdot 6H_2O$	14.1	17.0
Kainite	$MgSO_4 \cdot KCl \cdot 3H_2O$	15.7	18.9
Niter	KNO_3	38.6	46.5
Polyhalite	$K_2SO_4 \cdot MgSO_4 \cdot 2CaSO_4 \cdot 2H_2O$	12.9	15.5

* Kilmer, V. J., S. E. Younts, and N. C. Brady, Eds., *The Role of Potassium in Agriculture*. Reprinted with permission of Am. Soc. of Agron., Crop Sci. Soc. of Am., and Soil Science Society of America, Madison, Wisconsin, 1968.

duction in this area is around Carlsbad, New Mexico. Mining of extensive potassium salt deposits in central Utah, which are located at depths
of more than 2,000 feet, was begun in 1964. Deposits may also be
present in the Williston basin of North Dakota.

Important sources of lake brines in the United States are Searles
Lake, California, the Great Salt Lake, and the Salduro Marsh in Utah.
Salduro Marsh is a remnant of Lake Bonneville in Utah.

The minerals found in these deposits, together with their approximate
potassium contents, are shown in Table 9-4.

Mining Potassium Salts. The method used to mine potash ores
depends on the nature of the deposit. Shaft mining of the solid ore is
employed in some areas, solution mining in others, and for the surface
lake brines the method varies. A brief summary of these mining operations follows.

SOLID ORE MINING. The problem of mining potassium-containing
ores is complicated by the great depths at which the salts occur. In general, it is difficult to mine profitably at depths greater than 4,000 ft. However, the depth at which the German and Canadian deposits are mined is
considerably greater than that in the Carlsbad area in the United States.

In the room-and-pillar mining pattern rectangular rooms are mined out
and pillars of ore are left for support. The first stage takes about 50 to 60
per cent of the ore. The second stage, *pillar robbing*, may result in
removal of about 90 per cent of the ore.

The first mining in the Carlsbad area was begun in 1931, and coal-
mining methods common at that time were used in the initial effort.
Later the operation was entirely mechanized, and all underground equipment was electrically powered. Continuous mining machines, which
remove the ore continuously from the potassium vein, are largely used
now. The loosened ore is loaded onto shuttle cars or continuous belts
and hauled to the foot of an elevator shaft, where it is crushed to a maximum size of about 6 in. before being carried to the surface for further
processing.

The development of continuous mining machines ranks as one of the
major advances in potash mining. This equipment actually cuts the ore
directly from the mine face and does away with undercutting, drilling,
and blasting operations.

Solution Mining. Considerable attention has been given to the possibility of drilling wells down to the ore bed. The principle of this method
is based on pumping hot solution down to the bed, dissolving the potassium salts, and returning the potassium-laden brine to the surface for
refining. So far this technique is being used by two companies.

BRINES. At Searles Lake wells are driven to a few feet short of the
bottom of the deposit and the brine is pumped a distance of several miles
to the processing plant. Wells may have a life of several years before the
composition of the brine becomes unsatisfactory.

At Great Salt Lake in Utah there is a 14,000-acre complex of evaporation ponds and mineral processing plants. The sun furnishes 90 per cent of the energy in evaporation and time required to convert to brines to harvestible salts is two years. Potassium sulfate, sodium sulfate, magnesium chloride, and common salt are some of the products.

At Salduro Marsh brine is pumped from a network of more than 50 miles of canals, 3 ft. wide and 14 ft. deep.

CEMENT KILN POTASSIUM. In the production of cement some of the potassium-bearing ingredients are given off in flue dusts and gases. These dusts and fumes are collected in a precipitator, and the product is sold as a fertilizer for its potassium and calcium content. Such materials contain about 30 per cent calcium and 5 per cent potassium. Limited quantities of cement kiln dusts are collected in Maryland and California.

Refining of Potassium Ores. As mentioned in Chapter 7, potassium chloride is, in the matter of tonnage, the most important of the potash fertilizers. In the following sections some of the details of the production of this salt are covered and a brief description of the processing of the other potash fertilizers is given.

POTASSIUM CHLORIDE. Recovery of potassium chloride from sylvinite ore is made by the mineral flotation process or by solution of KCl, followed by recrystallization. Flotation is by far the most widely used. Recovery from brines at Searles Lake is by fractional crystallization.

Flotation. Sylvinite ore is a mixture of interlocked crystals of potassium chloride and sodium chloride plus small quantities of clay and other impurities. Flotation is essentially a mechanical separation. A simplified description of the process follows:

1. The ore is ground to separate the crystals and disperse the clay slime.
2. The ground ore is suspended and agitated in a saturated NaCl-KCl brine.
3. The slurry is deslimed to remove the clay. This is important, for clay will increase the requirement for the flotation reagents.
4. The deslimed slurry is conditioned with aliphatic amine acetate salts to film the potassium chloride particles selectively. The sodium chloride particles are not filmed.
5. The conditioned slurry then passes to rougher flotation cells into which air is drawn by agitation. The air bubbles attach to the filmed potassium chloride particles and float them to the surface. The froth is mechanically skimmed off by paddles and passes into cleaner flotation cells for further purification.
6. The froth and brine are centrifuged and the potassium chloride is dried in rotary driers and screened to desired particle size.

At Salduro Marsh the brine is concentrated by solar evaporation in ponds covering 11,000 acres. The potassium chloride is separated from the sodium chloride by a flotation process.

Heavy Media Separation. Because of its coarse crystalline structure, Saskatchewan ore does not need to be crushed as fine as other ores. After screening, the coarse fraction is blended with a mixture of brine and magnetite. Potassium chloride is separated by pumping through hydroclones which act as centrifugal separators. The KCl is then debrined, dried, crushed, and screened to size.

Crystallization. The difference in the temperature-solubility relationships of potassium and sodium chlorides is the usual basis of this method of recovery. The solubility of potassium chloride increases rapidly with a rise in temperature, whereas sodium chloride solubility varies only slightly over a wide temperature range.

Cool brine saturated with both salts is heated and passed over the ore countercurrent to the flow of the ore. Potassium chloride plus small amounts of sodium chloride are dissolved. The slurry is clarified and pumped through vacuum crystallizers to crystallize the potassium chloride, which is filtered out and dried.

At Searles Lake, where product recovery is accomplished by evaporation and fractional crystallization, the brine is concentrated in huge evaporators under vacuum. The twenty-three separate crystallization steps are required to separate potassium chloride and ten other basic chemical products. Among the by-products are sodium chloride, lithium carbonate, soda ash, salt cake, bromine, borax, boric acid, and pyrobor.

POTASSIUM SULFATE. Potassium sulfate is a white salt which contains 41.5 to 44.2 per cent potassium (50.0 to 53.2% K_2O). It is produced commercially by a number of processes.

Langbeinite Process. Production from langbeinite ($K_2SO_4 \cdot 2MgSO_4$) is according to the following equation:

$$K_2SO_4 \cdot 2MgSO_4 + 4KCl \longrightarrow 3K_2SO_4 + 2MgCl_2$$

Trona Process. Burkite ($Na_2CO_3 \cdot 2Na_2SO_4$) is reacted with potassium chloride to form glaserite ($Na_2SO_4 \cdot 3K_2SO_4$) and reacted with potassium chloride brine to give potassium sulfate.

Hargreaves Process. Potassium sulfate and hydrochloric acid are made directly from sulfur and potassium chloride. Sulfur dioxide from a sulfur burner is mixed with water vapor and air and passed over heated beds of potassium chloride to give potassium sulfate and its by-product hydrochloric acid.

Mannheim Process. Although the reactions pass through two stages, the over-all equation is the following:

$$2KCl + H_2SO_4 \longrightarrow K_2SO_4 + 2HCl$$

POTASSIUM MAGNESIUM SULFATE. As sold in the United States, this material contains about 18 per cent potassium (22% K_2O), 10.8 per cent magnesium (18.0% MgO), and a maximum of 2.5 per cent chlorine.

Langbeinite is mixed with sodium and potassium chlorides. The slower rate of dissolution of langbeinite is used as the basis for purification in which the ore is crushed and a countercurrent washing process removes the chloride salts.

POTASSIUM NITRATE. There are several means of producing potassium nitrate. One recently patented method reacts anhydrous liquid nitrogen pentoxide with potassium chloride. Liquid chlorine is a byproduct. A double decomposition reaction has been used for many years but the cost is comparatively high:

$$NaNO_3 + KCl \longrightarrow KNO_3 + NaCl$$

Methods have been proposed in which nitric acid and potassium chloride are the basis of new commercial production:

$$6KCl + 12HNO_3 \longrightarrow 6KNO_3 + 3Cl_2 + 6NO_2 + 6H_2O$$

The chlorine is liquefied and recovered. The nitrogen dioxide is converted to nitric acid and reused. Two grades of potassium nitrate, industrial and agricultural, are crystallized and filtered out.

Potassium Polyphosphates. Commercial processes have been developed to make materials with a wide range in water solubility and concentration. Some of the products are:

	Liquids	Solids
	0-30-11	9-48-16
Orthophosphates	0-20-20 (clear)	5-46-30
	0-25-20 (clear)	0-47-31
Polyphosphates		0-50-40

General reactions are:

Muriate of potash + sulphuric acid $\longrightarrow$ potassium bisulphate + HCl

Sulphuric acid + rock phosphate + potassium bisulphate $\longrightarrow$
$$KH_2PO_4 + H_3PO_4 + CaSO_4$$

The H_3PO_4 reacts with KH_2PO_4 to produce polyphosphates. Solubility and concentration are varied in part by adding K or removing P.

These materials have a low salt index, high analysis, varying solubility, and no chlorine.

OTHER SOURCES. Waste in the manufacture of tobacco products, consisting largely of the stems and ribs of tobacco leaves which are ground and sold for use in the fertilizer industry, contains 4 to 8 per cent potassium and 2 to 4 per cent nitrogen. This product also serves as a good conditioner for mixed goods.

Kelp, a giant seaweed, occurs in large beds extending from lower California to the Alaskan peninsula. The plants grow rapidly and the beds are accessible. It appears that kelp could furnish a considerable amount of potassium annually. The ash, which contains as much as 25 per cent potassium, is also an important source of iodine. During World War I limited use was made of California kelp as a source of potassium.

A vast, untapped source of potassium is sea water, a cubic mile of which contains the equivalent of about 1.6 million tons of potassium. Considerable research has been conducted on the extraction of potassium from this source and, although little progress has been made, Norway has reported some success.

Summary

1. Most of the fertilizer nitrogen consumed is produced from ammonia which is synthesized from hydrogen gas and atmospheric nitrogen. Smaller amounts of Chilean nitrate of soda are still used, as is ammonium sulfate, which is a by-product of the coking of coal and the production of caprolactam from the manufacture of nylon. In Germany some nitrogen fertilizers are produced by the cyanamide process, and in the Scandinavian countries the direct oxidation of nitrogen is practicable because of the inexpensive electric power available in that area.

2. Most of the other synthetic nitrogen fertilizers are manufactured from ammonia. Ammonia as such, or in aqueous solution, is used as a fertilizer. It can be oxidized to nitric acid from which various nitrate fertilizers are then manufactured and can be reacted with carbon dioxide to produce urea, which in turn is reacted with formaldehyde to synthesize slowly available urea-formaldehyde materials. The basic manufacturing processes in the production of the different nitrogen fertilizers were discussed.

3. For all intents and purposes rock phosphate is the original source of phosphatic fertilizers. The phosphorus compound in the rock is apatite, which is a highly insoluble material.

4. Phosphate rock is used to a limited extent as a fertilizer material. It must be very finely ground to be effective and it is of value only with certain cropping systems which include legumes capable of utilizing it or on acid soils. Most of the phosphate rock is processed either by

heat or acid treatment. In both cases the apatite structure is destroyed and a product results in which the phosphorus is more readily available.

5. Heat treatment of phosphate rock is usually effected in gas- or oil-fired chambers. In some cases electric energy is used, particularly if elemental phosphorus is being produced. Rock phosphate can also be treated with sulfuric, phosphoric, and nitric acids. The various processes for treating phosphate rock and their reactions were discussed.

6. Sulfuric acid is the workhorse of the fertilizer industry. It is manufactured by burning elemental sulfur and converting the sulfur dioxide to sulfur trioxide. It can also be made from the sulfur dioxide that results from the burning of pyrites. The reactions and processes in the production of sulfur and sulfuric acid were covered.

7. The sources of potassium fertilizers are deposits of salts containing this element and the brines of dying lakes or seas. The salt deposits are situated in layers or beds several hundred to several thousand feet below the earth's surface and are mined by shaft-mining processes. Mining of these deposits by hot solution pumped into the deposits has also been perfected. Mining of brine is relatively simple. The brine is pumped to a centralized location for refining and the potassium salts are recovered. The techniques of the recovery of potassium salts from deposits and brines were presented.

8. The principal fertilizer-grade potassium salts are potassium chloride, potassium sulfate, potassium-magnesium sulfate, and potassium nitrate. Potassium nitrate, however, is not found to any extent in salt deposits and is made by reacting potassium chloride with nitric acid.

Questions

1. What are the three basic processes for the fixation of elemental nitrogen used in fertilizer manufacture?
2. Illustrate by appropriate equations the reactions of each of these processes.
3. How is the oxidation of ammonia to nitric acid brought about in fertilizer manufacture?
4. How is defluorinated phosphate rock manufactured?
5. Which acid is commonly termed *the workhorse of the fertilizer industry?*
6. What are the three principal sources of the sulfuric acid used in fertilizer manufacture?
7. What is *wet process* phosphoric acid?
8. How is elemental phosphorus manufactured?
9. How does calcination of phosphate rock differ from fusion?
10. How is nitrogen obtained from coal?
11. How is urea produced commercially?

12. Where are the principal deposits of potassium salts in North America?
13. What two processes are commonly employed in mining these salts?
14. In what other form besides solid salt deposits do commercial potash sources occur?

Selected References

"Control techniques for sulfur oxide air pollutants." *NAPCA, Pub. No. AP-52* (1969).

"Fertilizers," *Encyclopedia of Chemical Technology*, 2nd ed., Vol. 9, pp. 25–150. New York: Wiley-Interscience, 1966.

Jacob, K. D., Ed., "Fertilizer technology and resources in the United States." *Agronomy: A Series of Monographs*, Vol. II. New York: Academic, 1953.

Kilmer, V. J., S. E. Younts, and L. B. Nelson, Eds., *The Role of Potassium in Agriculture*. Madison, Wisconsin: Am. Soc. of Agron., Crop Sci. Soc. of Am. and Soil Sci. Soc. of Am., 1968.

Olson, R. A., T. J. Army, J. J. Hanway, and V. J. Kilmer, Eds., *Fertilizer Technology and Use*, second ed. Madison, Wisconsin: Soil Sci. Soc. Am., 1971.

"Phosphorus, properties of the element and some of its compounds." *Chem. Eng. Rept. No. 8*, Tennessee Valley Authority, 1950.

Sauchelli, V., Ed., *Chemistry and Technology of Fertilizers, ACS Mono. Ser. 148*. New York: Reinhold, 1960.

Sauchelli, V. *Manual of Fertilizer Manufacture*, 3rd Ed. Caldwell, New Jersey: Industry Publ., 1963.

Sauchelli, V., Ed., *Fertilizer Nitrogen. Its Chemistry and Technology. ACS Mono. Ser. 161*. New York: Reinhold, 1964.

Striplin, M. M., David McKnight, and T. P. Hignett, "Compound fertilizers from rock phosphate, nitric and sulfuric acid, and ammonia." *Ind. Eng. Chem.*, **44**:236 (1952).

Tuller, W. N., Ed., *The Sulphur Data Book*. New York: McGraw-Hill, 1954.

Turbeville, W. J., Jr., "The phosphate rock situation." *Phosphate-Sulphur Symposium*, Tarpon Springs, Florida, 1974.

10. THE MANUFACTURE AND PROPERTIES OF MIXED FERTILIZERS

For more than one hundred years mixtures of fertilizer materials have been offered for sale in the United States, where the mixed fertilizer industry is more extensive than in any other country.

A wide range of crop needs and soil conditions exists, and numerous fertilizer mixtures have been developed over the years. Often, however, more grades have been formulated than were actually needed. However, in 1971–72 about 200 grades made up 80 per cent of the United States tonnage.

Although the tonnage of nitrogen, phosphorus, and potassium supplied in mixtures has increased, the percentage of the total nitrogen and potassium applied in mixtures has decreased, nitrogen quite rapidly because of the increased emphasis on the direct application of nitrogen materials and the realization by the farmer that savings can be effected.

The total amounts of nutrients both in mixtures and in materials will continue to grow. Materials will increase somewhat faster in view of research which shows the importance of applying corrective applications of a given nutrient to soils low in that nutrient.

The manufacture of mixed fertilizers constitutes a major agricultural industry. Before considering the properties of these fertilizers, a few of the more important terms in common use should be explained.

Percentage of Total Applied in Mixtures

	Nitrogen	Phosphorus	Potassium
1949–1950	49	69	93
1954–1955	41	80	88
1959–1960	37	79	87
1963–1964	32	81	84
1972–1973	27	83	64

TERMINOLOGY

A *fertilizer* is any substance that is added to the soil to supply those elements required in the nutrition of plants.

A *fertilizer material or carrier* is any substance that contains one or more of the essential elements.

A *mixed fertilizer* is a mechanical or chemical combination of two or more fertilizer materials and which contains two or more essential elements.

A *complete fertilizer* contains the three major plant-nutrient elements — nitrogen, phosphorus, and potassium.

The *fertilizer grade* refers to the minimum guarantee of the plant-nutrient content in terms of total nitrogen, available phosphorus pentoxide, and soluble potassium oxide (6-24-24, for example).

The *fertilizer ratio* refers to the relative percentages of nitrogen, phosphorus pentoxide, and potassium oxide (a 6-24-24 grade has a 1-4-4 ratio).

The *fertilizer formula* is an expression of the quantity and analysis of the materials in a mixed fertilizer.

A *filler* is *make-weight* material added to a mixed fertilizer or fertilizer material to make up the difference between the weight of the added ingredients required to supply the plant nutrients in a ton of a given analysis and 2,000 lb.

An *acid-forming fertilizer* is one capable of increasing the acidity of the soil, which is derived principally from the nitrification of ammonium salts by soil bacteria.

A *basic fertilizer* is capable of decreasing the acidity of the soil.

A *nonacid-forming* or neutral fertilizer is one that is guaranteed to leave neither an acidic nor a basic residue in the soil.

Dry bulk blending is the process of mechanically mixing solid fertilizer materials.

Clear liquid fertilizer is one in which the NPK and other materials are completely dissolved.

Suspension liquid fertilizer is one in which some of the fertilizer materials are suspended as fine particles.

Fluid fertilizer is clear or suspension liquid fertilizer.

Compound fertilizer is a term often used in Europe and has about the same meaning as *mixed* in the United States.

The Association of American Plant Food Control Officials is an organization of officers and their deputies charged by law with regulating the sale of fertilizers and of research workers employed by state, dominion, or federal agencies engaged in the investigation of fertilizers. Its object is to promote uniform and effective legislation, definitions, and rulings, and to enforce the laws relating to the control of sale and distribution of fertilizers and fertilizer materials.

ACID, NEUTRAL, AND BASIC FERTILIZERS

The acidifying effect of mixed fertilizers is largely the result of contained ammonium nitrogen, as explained in Chapter 5. Limestone may be added to mixed fertilizer in amounts sufficient to neutralize the expected acidity, and even greater amounts may be added to leave a basic residue. Formerly, some states stipulated that fertilizers must be nonacid forming and required that the potential acidity or basicity of the fertilizer be stated in the guarantee. However, this is a thing of the past.

Neutralizing Agents. Dolomitic limestone is commonly used as a neutralizer, for it is also a source of magnesium. In addition, the dolomite is less likely than calcite to cause reversion of monocalcium phosphate. Caution must be used in adding lime to mixed fertilizer, for large excesses may result in reversion of the soluble form of phosphate to the insoluble or less available forms as shown by the following equation:

$$Ca(H_2PO_4)_2 + CaCO_3 \longrightarrow 2CaHPO_4 + H_2CO_3$$

With dolomite, the water-soluble phosphorus decreases more rapidly in ammoniated than in nonammoniated mixtures because of the formation of magnesium ammonium phosphate.

Excess lime may cause loss of ammonia gas by reaction with the NH_4^+ ion as

$$NH_4^+ + OH^- \longrightarrow NH_4OH \longrightarrow NH_3 \uparrow + H_2O$$

With an inadequate liming program in humid areas the continued use of acid-forming fertilizers will increase soil acidity. Lime added in the fertilizer helps to maintain the content of soil bases. However, with higher-analysis mixed fertilizers, there is less room for lime; and with the gradual trend toward higher-analysis fertilizers, the proportion of acid-forming fertilizer is increasing.

Acid-forming mixed fertilizers may be manufactured for use on alkaline soils. Ammonium sulfate is one of the principal sources of nitrogen for this purpose, and elemental sulfur may also be included as an additional source of acidity.

HIGH-ANALYSIS FERTILIZERS

The cost of a ton of mixed fertilizer is influenced largely by two factors:

1. The cost of the plant nutrients in the materials used to make up the mixed fertilizer.
2. The fixed costs, which include manufacturing, bagging, transporting, and distributing.

For example, for equivalent amounts of nutrients only half as many bags would be required for a 10-20-20 as for a 5-10-10. Obviously, the fixed costs per pound of nutrients decrease with an increase in the nutrient content. Because of these lower fixed costs and in spite of somewhat greater costs of the materials in higher-analysis fertilizer, the cost of nutrients in the higher-analysis mixed fertilizer tends to be reduced.

Beyond a certain point, however, the extra cost of the higher-analysis materials required may offset any saving. For example, for normal superphosphate versus concentrated superphosphate the manufacturing costs of concentrated superphosphate at the plant are generally greater than those of ordinary superphosphate. Hence the savings that may be effected by the use of concentrated superphosphate must result from lower unit costs of manufacturing, packing, and distributing the higher-analysis mixtures.

Of course the greater the transportation distance, the greater the advantage of higher-analysis fertilizers, for transportation costs have risen tremendously in the last ten years.

More Concentrated Materials. The content of plant nutrients in materials from which mixed fertilizers are being manufactured has increased rapidly. Natural organics, which contain 6 to 8 per cent nitrogen and were popular around the turn of the century have been almost entirely replaced by higher-analysis materials, including urea (45 per cent nitrogen). Anhydrous ammonia (82 per cent nitrogen) is also being added to mixed fertilizers, and concentrated superphosphate and ammonium phosphates are replacing the low-analysis superphosphates. Potassium sources have shifted from kainite and manure salts to potassium chloride (50 to 52 per cent potassium). As a consequence of the more concentrated materials, the nutrient content of mixed fertilizers is rising steadily (Table 10-1).

Effect of Concentration on Cost of Nutrients. An example of the effect of concentration on the cost to the farmer is shown in Figure 10-1.

TABLE 10–1. Average Analysis of Mixed Fertilizers in the United States (USDA)

Year	Total N	Available $P(P_2O_5)$	Soluble $K(K_2O)$	Total
		Nutrient content (%)		
1949–1950	4.02	4.78 (10.93)	6.88 (8.29)	15.68 (23.24)
1954–1955	5.24	5.19 (11.86)	8.96 (10.80)	19.39 (27.90)
1959–1960	6.50	5.70 (12.99)	10.00 (12.06)	22.20 (31.55)
1963–1964	7.68	6.57 (14.94)	10.50 (12.65)	24.75 (35.27)
1972–1973	10.32	8.3 (19.02)	10.6 (12.71)	29.2 (42.05)

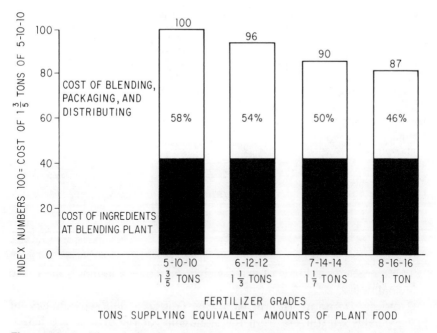

Figure 10–1. The relative farm cost of plant nutrients in fertilizers of different concentrations. Maine. (From Maurice H. Lockwood, "High-analysis Mixed Fertilizers," in K. D. Jacob, Ed., *Fertilizer Technology and Resources in the United States,* Vol. III., *Agronomy: A Series of Monographs.* Copyright 1953 by Academic Press, Inc.)

By going from a 5-10-10, still the leading grade in 4 states in the United States in 1971, to 8-16-16 a considerable saving is effected. The cost of the manufacturing, packaging, and distributing was 58 per cent of the total cost of 5-10-10 and only 46 per cent for the 8-16-16.

In the foregoing comparison 60 per cent more bags are required for a ton of 5-10-10 than for a ton of 8-16-16. Hence the increased cost of bagging, handling, and transportation becomes a sizable factor.

Make-Weight Materials.   In mixed fertilizer there is often a difference between the weight of the materials required to furnish the nutrients in a ton of a given analysis and 2,000 lb. This difference is made up with materials, known as *make-weight* or *filler,* to supply nutrients other than nitrogen, phosphorus, or potassium.

The example in Table 10-2 shows a 10-10-10 in which the weight of the materials containing the primary nutrients would be 1,666 lb. An additional 334 lb. of lime would be added per ton to make the grade. If no extra material were used when diammonium phosphate replaced the superphosphate, the analysis of the mixture would be 15-15-15. It is thus possible to avoid using those materials in mixed fertilizers that bear the make-weight stigma.

TABLE 10–2. Formulation of a Granulated Fertilizer*

	Analysis	10-4.4-8.3(10-10-10) (lb.)	15-6.6-12.5(15-15-15) (lb.)
Nitrogen solution 440	44% N	320	254
Sulfate of ammonia	21% N	320	667
Phosphoric acid	23.6% P	160	280
Superphosphate	8.7% P	588	
Diammonium			
phosphate	18% N		333
	20.3% P		
Muriate of potash	50.2% K	338	506
Limestone		334	
Totals		2000	2000

* R. C. Smith and B. Makower, "Advances in manufacture of mixed fertilizer," Chapter 11. Reprinted from *Fertilizer Technology and Usage*, published and copyrighted by Soil Sci. Soc. Am., 1963, by permission of copyright owners.

There are several kinds of make-weight materials and limestone is the principal one.

Growers Demand Higher-Analysis Goods. The advantages of higher-analysis fertilizers from the standpoint of the grower are listed:

1. Lower cost per unit of plant food.
2. Lower transportation cost.
3. Less storage space required.
4. Less labor in handling.
5. Increased speed of application in the field because of fewer stops.

Some growers may consider it worthwhile to buy high-analysis fertilizer even if the cost per unit of plant food is somewhat higher, because labor, storage, transportation, and/or time required for application may be critical factors.

Agronomic Factors Involved. Fertilizers must be applied according to their plant-nutrient content and not by the number of pounds per acre. For example, if a farmer has been applying 5-10-10 at the rate of 500 lb./A., a 10-20-20 would be applied at the rate of 250 lb. if he expected to have the same amount of nitrogen, phosphorus, and potassium per acre. This may seem to be a simple point, but many of the unfortunate experiences that have resulted from the use of higher-analysis goods have been caused by failure to reduce the rate of application accordingly. This, in turn, is usually due to the effect of improper placement of the larger amount of fertilizer with respect to the seed or the plant.

The other side of the picture, however, is that many crops are inadequately fertilized. The application of higher-analysis fertilizers at the same rate once used for low-analysis goods is, of course, an advantage in that greater quantities of plant nutrients are applied.

One problem, mentioned earlier, is that higher-analysis fertilizers generally contain less of the secondary plant-nutrient elements, calcium, magnesium, and sulfur. Although there is usually little or no room for the addition of dolomitic limestone in such fertilizers, the problem presented is not serious, for the calcium and magnesium requirements can and should be met by broadcast applications of limestone. Although the cost of these broadcast applications is small, it should be considered in evaluating higher-analysis fertilizers.

The sulfur supply is important in many areas of the United States, and the continued use of fertilizers that lack this element will result in a reduction in crop yield. Sulfur deficiencies have been observed on numerous crops, as indicated earlier in Chapter 3, and fertilizer manufacturers are now adding small amounts in mixed, granular, and blended goods. The grower may also add sulfur in the form of gypsum or elemental sulfur or he may choose a nitrogen-sulfur material. Sulfur tends to acidify the soil, but when applied in amounts needed to supply sulfur as a plant nutrient the acidifying effect is minor.

High-analysis fertilizers are likely to contain fewer micronutrients than the lower-analysis goods. However, numerous examples of micronutrient deficiencies appear even when low-analysis materials are used. The most effective insurance against the incidence of these deficiencies is the addition of micronutrients in the form of their respective carriers when the deficiencies are known to exist.

Agricultural experiment stations and extension services support a program of demonstrating and promoting new practices which include the use of high-analysis fertilizers, but the greater part of the burden of developing consumer acceptance of higher-analysis fertilizers must fall on the fertilizer industry. The local dealer also can help considerably by his direct contact with the farmer.

MANUFACTURE OF FERTILIZERS

Many changes in the materials and in the composition of mixed fertilizers have taken place during the last fifty years. Materials with widely different properties are mixed together and physical and chemical changes occur. As few as two or as many as ten different materials may be combined in the manufacture of a complete fertilizer, and, with the trend toward ammonium phosphates and higher-analysis goods, improvement will continue as a direct result of efforts to market a more efficient product.

The present-day fertilizer manufacturer is confronted with a multiplicity of problems of formulation, processing, control of chemical reactions, and the physical condition of the mixed goods.

Three general types of process are used in the manufacture of mixed fertilizers:

1. Ammoniation of phosphorus materials and the subsequent addition of other materials and granulation.
2. Bulk blending of solid ingredients.
3. Fluid mixing.

Chemical Reactions. Chemical reactions in mixed fertilizers are influenced by moisture content, temperature, and particle size, all of which are factors that the manufacturer must control. To reduce these reactions after bagging nongranulated fertilizers, he usually prepares the mixture well in advance and allows it to cure in the storage pile at the factory. Because heat and moisture tend to speed up chemical reactions, the use of high temperature and high moisture during mixing is generally desirable. However, before the mixture is placed in the curing bin, both temperature and moisture are reduced.

DOUBLE DECOMPOSITION. Double decomposition sets in between two compounds, without a common ion, in the presence of moisture. Common examples are the following:

$$CaH_4(PO_4)_2 + (NH_4)_2SO_4 \longrightarrow CaSO_4 + 2NH_4H_2PO_4$$
$$NH_4NO_3 + KCl \longrightarrow NH_4Cl + KNO_3$$
$$(NH_4)_2SO_4 + 2KCl \longrightarrow 2NH_4Cl + K_2SO_4$$

Certain of these reactions may be relatively slow, and the first may require two months or more. It is evident that new compounds, with properties entirely different from those of the original materials, may be formed.

NEUTRALIZATION. The most common reactions in this group, which are those that take place during ammoniation of superphosphate, are described in a subsequent section. In addition, the reaction of lime with free acid is a common occurrence.

$$2H_3PO_4 + CaCO_3 \longrightarrow CaH_4(PO_4)_2 + H_2CO_3$$

HYDRATION. Hydration is an important reaction in that anhydrous forms of certain salts tie up free water and bring about chemical drying in the curing bin. It takes place most rapidly at temperatures of 50°C. or lower.

$$CaSO_4 + 2H_2O \longrightarrow CaSO_4 \cdot 2H_2O$$
$$CaHPO_4 + 2H_2O \longrightarrow CaHPO_4 \cdot 2H_2O$$
$$MgNH_4PO_4 + 6H_2O \longrightarrow MgNH_4PO_4 \cdot 6H_2O$$

Under some conditions certain compounds may
r mixtures.

$$NH_2)_2 + H_2O \longrightarrow 2NH_3 + CO_2$$

of urea is accelerated by higher temperatures, par-
:e of such compounds as monocalcium, dicalcium,
hosphate. This reaction is also favored by high
e.

$$)_2HPO_4 \longrightarrow NH_4H_2PO_4 + NH_3$$

ins calcium sulfate, the liberated ammonia will not
sulfate and dicalcium phosphate will be formed:

$$NH_4H_2PO_4 + CaSO_4 + NH_3 \longrightarrow CaHPO_4 + (NH_4)_2SO_4$$

Ammoniation. Ammoniation is of prime importance to the mixed
fertilizer industry for several reasons:

1. Anhydrous ammonia and ammonia solutions are economical
   forms of nitrogen.
2. Ammonia combines with normal or triple superphosphate and
   phosphoric, nitric, or sulfuric acids.
3. Ammoniation eliminates the need for the large quantities of acids
   that would be necessary if ammonia were converted to certain
   salts, such as ammonium nitrate or sulfate.
4. Ammoniation improves the physical condition of mixed fertilizers.
5. Free acid in superphosphate is neutralized.
6. Higher-analysis goods may be prepared.

The chemistry of the reactions that take place during the ammoniation
of ordinary and concentrated superphosphates is somewhat complex.
For the purposes of this discussion, these reactions are summarized by
the following four equations:

$$Ca(H_2PO_4)_2 \cdot H_2O + NH_3 \longrightarrow NH_4H_2PO_4 + CaHPO_4 + H_2O \quad (1)$$
$$2NH_3 + 2CaHPO_4 + CaSO_4 \longrightarrow Ca_3(PO_4)_2 + (NH_4)_2SO_4 \quad (2)$$
$$NH_4H_2PO_4 + NH_3 \longrightarrow (NH_4)_2HPO_4 \quad (3)$$
$$3CaHPO_4 + 2NH_3 \longrightarrow Ca_3(PO_4)_2 + (NH_4)_2HPO_4 \quad (4)$$

Mixed Fertilizers. Reaction 1 marks the initial stages of the ammoniation of both ordinary and concentrated superphosphates. If the

ammoniation is continued, there is, with ordinary superphosphate, a conversion of the phosphorus to the unavailable tricalcium phosphate and an attendant formation of ammonium sulfate, as indicated by equation 2. Equations 3 and 4 represent the reactions that occur when the ammoniation of concentrated superphosphate is continued.

The ammoniation of phosphoric acid in the formation of mono- and diammonium phosphates was covered in Chapter 6. Calcium metaphosphate and fused tricalcium phosphate do not combine with ammonia. As already mentioned, however, when calcium metaphosphate is hydrolyzed with sulfuric acid, it can be partly ammoniated.

EFFECT OF DEGREE OF AMMONIATION ON PHOSPHATE SOLUBILITY. The ammoniation of ordinary superphosphate to much greater than 3 per cent by weight of ammonia causes a marked reduction in the content of water-soluble phosphorus in the fertilizer, as shown by the data in Table 10-3 and Figure 6-8. If precautions are taken to introduce the correct amount of ammonia, reaction 2 will be minimized.

It will be recalled from Chapter 6 that monocalcium phosphate and ammonium phosphates are largely water-soluble, dicalcium phosphate is citrate-soluble, and tricalcium phosphate is soluble in neither water nor citrate. This last salt, however, is considerably more available to plants than rock phosphate.

Although there is a continuous linear decrease in the water-soluble phosphorus content of ordinary superphosphate with increasing ammoniation, this is not true of concentrated superphosphate. Regardless of the degree of ammoniation, the water-soluble phosphorus does not decrease below about 60 per cent (see Figure 6-8).

The formation of di- and tricalcium phosphates during ammoniation is known as *reversion,* a term frequently used in the trade to refer to a decrease in phosphate availability.

CHLORINE-FREE FERTILIZER. A European fertilizer manufacturer has

TABLE 10–3. Composition of Ammoniated Superphosphate*

Compound	Percentage of composition with percentage of ammonia added as indicated						
	0.0	1.0	2.0	3.0	4.0	5.0	6.0
$CaH_4(PO_4)_2$	25.0	14.3	3.5	0.0	0.0	0.0	0.0
$CaHPO_4$	4.5	9.5	12.0	11.2	5.0	0.0	0.0
$Ca_3(PO_4)_2$	0.0	0.0	0.0	8.2	17.5	25.7	30.6
$NH_4H_2PO_4$	0.0	8.9	14.3	15.5	12.5	9.1	6.0
$(NH_4)_2SO_4$	0.0	0.0	0.0	2.7	8.5	14.3	20.0
$CaSO_4 \cdot 2H_2O$	62.0	62.0	62.0	58.3	51.0	42.0	32.0
Inerts	3.0	3.0	3.0	3.0	3.0	3.0	3.0
Totals	94.5	97.7	94.8	98.9	97.5	94.1	91.6

* F. W. Parker, "The availability of phosphoric acid in precipitated phosphates." *Com. Fertilizer,* 42:28 (1931).

developed a process for making a chlorine-free NPK fertilizer based on ion exchange. Phosphate rock is digested by nitric acid. The solution of calcium, phosphorus and nitrogen is then pumped to the potassium-loaded ion exchanger where the calcium is absorbed. The resulting KC1 solution is pumped to the ion exchanger, the potassium and calcium exchange places, and the chlorine is removed in the form of calcium chloride. Many high-analysis grades are made.

Mixing Procedure. To ammoniate mixed fertilizers a batch of superphosphate, muriate of potash, and conditioner is placed in the mixing unit and sprayed with the ammoniating solution. The reactions of ammonia with superphosphates and acids are exothermic. If this heat is not dissipated fairly rapidly, additional reversion is favored. The ammoniated material must also be handled to prevent overheating in the stored pile. To reduce the heat content of the finished product, it is frequently passed through a rotary cooler in which it is tumbled in a countercurrent stream of air.

A continuous ammoniation process is now in use in many plants. In this method the ammoniating solution is introduced at a constant rate with the other materials. Thus there is no loss of time in recharging the machine.

Materials Used in Manufacture. The number of materials used in the manufacture of fertilizers has decreased as a consequence of the need for regularly available, more uniform ingredients in more complex processes.

Factors influencing the eligibility of a material have been listed by Smith and Makower (1963): (1) nutrient content, (2) cost per pound of nutrient, (3) chemical form, such as nitrate or ammoniacal, (4) absence of toxic compounds, (5) availability, (6) content of other nutrients, (7) moisture content, (8) hygroscopicity, (9) reactions with other ingredients, (10) heat of reaction, (11) particle size, and (12) effect on physical properties.

Suppliers must continually be alert to the characteristics of materials that may be needed by the manufacturer, or they may be left with a product that no one will buy.

A case in point is the standardization of particle sizes for potash materials. Earlier the Fertilizer Industry Roundtable reached agreement on three grades: standard, coarse, and granular, which was a compromise between the needs of the manufacturer and the capability of material suppliers. Unfortunately, the sieve specifications for the three grades vary among manufacturers.

Mixing Processes

NONGRANULATED FERTILIZERS. All of the early fertilizer plants in the United States produced nongranulated fertilizers. Few are still operating.

Ammoniation is practiced because of the economy of the nitrogen source and the physical condition of the product. Standard (fine) solid materials are combined with ordinary and concentrated superphosphate, which are sources of phosphorus, and sulfate of ammonia and nitrogen solutions, which are sources of nitrogen. Muriate of potash is the common source of potassium.

There are two kinds of mixing machines in general—the stationary and the rotary drum. The stationary mixer has a vertical cylinder fitted with a baffle system and the mixing is done by gravity.

The rotary drum, which is a horizontal cylinder equipped with a series of baffles, permits mixing and ammoniation to be done simultaneously. The materials are introduced into the drum, the ports are closed, and the solution is sprayed into the system as the materials are mixed. With high ammoniation a rotary cooler is necessary before the fertilizer is taken to bin storage. Two to four weeks of storage are generally adequate to permit completion of the reactions.

A conditioner added during mixing to serve as a parting agent for the particles tends to reduce crystal knitting during changes in moisture and temperature under pressure in storage.

GRANULATION. Nearly all plants employ this process. It is essentially an extension of nongranulated fertilizer production in which particles of reasonably uniform size and stability are manufactured. The mixture is introduced into a rotary drum until granulation occurs, then dried and sized. The large granules are crushed and the fines are put through the process once again. The final product contains about one per cent moisture. When fertilizers are dry, the prime cause of caking is no longer present.

An example of materials required in the formulation of a granulated fertilizer is given in Table 10-2. Continuous methods of ammoniation mixing are most common.

The two general processes are semisolid and slurry. In the semisolid process ordinary or concentrated superphosphate is used. Ammoniation of the solid particles is slower and there is more danger of local over-ammoniation.

The advantages of the slurry system are these:

1. Its reaction is rapid and easy to control.
2. There is no need for curing the final product.
3. It uses the more economical phosphate rock.
4. It uses the less expensive nitrogen raw materials, nitric acid and ammonia.

These materials tend to offset the additional equipment that would be needed to handle the slurry and the higher cost of removing the water. Nitric phosphates may also be produced with the slurry process.

Nearly all mixed fertilizers in England are granulated. In the United States, except in certain parts of the South, where ordinary superphosphate is still most economical, the trend is toward high-analysis granulated fertilizer.

Granulated products offer many advantages to the manufacturer and the farmer. When properly stored, granulated fertilizers maintain good physical and handling properties for some time. Caking is reduced to a minimum, and the manufacturer will have fewer operations to perform between mixing and bagging. Segregation is almost eliminated, and the grower is assured of a material that can be uniformly applied in the field. In addition, there is less dust to contribute to the physical discomfort of the individuals who work with the fertilizer. However, with the exception of the water-soluble phosphates discussed in Chapter 6, little agronomic advantage from granulation has been noted.

STORAGE. In off seasons the freshly mixed fertilizer is transferred to storage bins in which it is allowed to cool and cure. The fertilizer may be left in the bins for several months. Nongranulated fertilizer is then put through a hammermill, when necessary, and screened before bagging. During the season of peak demand in the late winter and early spring, however, the fertilizers may be packaged directly from the mixing machine.

BAGGING. Packages vary in kind and size. Plastic bags keep the fertilizer in a virtually moistureproof condition. This is quite important, particularly with some fertilizers, and under conditions of high humidity. In addition, the appearance of the product is quite attractive. Paper bags with polyethelyene liners are used for much of the remaining bagged-goods market. A considerable proportion of the fertilizer was once marketed in 200-lb. burlap bags, but there has been a shift to 100, 80, and even 50-lb. paper or plastic bags to make handling easier.

The tightness of the pack depends on the fertilizer. For a material that is likely to cake, a loose pack is preferred, for each time the bag is moved the contents are loosened. On the other hand, a tight pack is preferred for granulated mixtures or for materials that do not cake.

Although there is considerable variation, cost of bags and bagging runs from six to eight dollars a ton. The percentage of mixed fertilizers marketed in bags is gradually decreasing.

Physical Condition. A major problem faced by the fertilizer manufacturer is the physical condition of the product. In the manufacture of nongranulated fertilizer this problem generally becomes more serious as the nutrient content of mixtures is increased. Nothing discourages a grower more quickly than to have lumpy or caked fertilizer. Such materials are spoken of as "tombstones," which may also well describe the situation of the fertilizer manufacturer who continues to put them out. The grower demands a product that will flow easily through the distributing equipment and can be uniformly applied in the field. The principal factors that determine the drillability of fertilizer are its moisture content and its state of subdivision.

HYGROSCOPICITY. The absorption of water from the air is called hygroscopicity. Materials such as calcium nitrate, ammonium nitrate, sodium nitrate, and urea absorb this moisture at fairly low humidity and temperature. On the other hand, superphosphate and potassium sulfate take up water only at a very high relative humidity. Mixtures of certain salts such as ammonium nitrate and urea are more hygroscopic than either material alone, and this complicates the problem. As a consequence, the fertilizer materials which make up a given formula must be selected with considerable care. Many plants limit the use of ammonium nitrate or urea to reduce moisture absorption.

CAKING. The moisture in fertilizers dissolves some of the more soluble compounds to form a saturated salt solution. As the moisture content of the system decreases or the temperature drops, the dissolved salts crystallize and the crystals knit together. Crystal knitting may also occur when a new compound is formed by chemical reaction or when crystals flow together under pressure. Although finely divided particles may cake slightly as the result of cohesion or adhesion, the main source of caking appears to be crystal knitting.

Under ideal curing conditions most of this knitting will take place before bagging. However, a state of equilibrium is seldom attained under conditions of manufacture of nongranulated fertilizer, and again temperature and moisture changes may take place in storage. During the rush season, when it is necessary to bag directly from the mixing machine, there is a good chance of caking if the material is allowed to stand in the bag for any length of time.

CONDITIONERS. Conditioning materials are added to nongranular and granular fertilizers to improve their physical condition or to decrease caking. Their actual purpose is to reduce crystal knitting during changes in moisture and temperature when the fertilizer is under pressure.

In nongranulated fertilizers the conditioners serve as separators between the particles. Ground cocoa shells or ground corn cobs may be used in amounts of about 100 lb./ton, or 5 per cent. A basic material such as lime will react with any free acids to form hydrated salts.

In granulated fertilizers conditioners serve as a coating agent for the granules. Most efficient materials are those with low bulk densities; for example, diatomaceous earth and hydrated silica. Granulation with the accompanying reduction in surface area, drying, and cooling have markedly decreased the need for conditioners and only a few manufacturers now use them.

COATED FERTILIZERS. An interesting development is the covering or encapsulating of granules with water-resistant or impermeable coatings. The purpose is to give a *metered* supply of nutrients, reduce injury from seed-fertilizer contact, reduce leaching losses, and improve physical condition. In fertilizer encapsulated with polyethylene film availability is regulated by the number of pinholes (Ahmed et al., 1963). It is postulated that moisture entered the capsule through the pinholes primarily as a vapor and dissolved the salt. The saturated solution then flowed out by gravity.

With uncoated fertilizer the plant can take advantage of an early and rather high content of nutrients, but in later growth stages the supply is sharply reduced. This is in contrast to the even supply provided by coated fertilizers. Research results show reduced leaching losses of potassium and nitrates and less seedling injury from coated fertilizer. Actual agronomic value is still to be determined. Coated fertilizers have helped to maintain vigor in ornamentals and heavy applications can be safely made.

Various coating substances such as plastics, resins, waxes, paraffin, elemental sulfur, and asphaltic compounds are under investigation. Temperature and thickness of the coating have a marked influence on release, and raising the temperature from 10 to 20°C. almost doubles the rate. However, pH has little effect and the release is not dependent on microbial action.

Sulfur as a coating substance was discussed in Chapter 5. The coating of urea and muriate of potash with sulfur has the advantage of being economical as well as of supplying a necessary element. The sulfur is sprayed on as a melt. Sulfur-coated urea is the most useful coated fertilizer at present.

SEGREGATION. Nongranulated mixed fertilizers are composed of particles that vary considerably in size, shape, and density. The tendency is to segregate during such operations as bagging, transporting, or pouring into the distributor. This segregation results in fertilizers that are not of uniform chemical composition. In addition to the agronomic implications, it is more difficult to obtain representative samples for chemical control work.

BULK BLENDING

Bulk blending is the mechanical mixing of dry, solid fertilizer materials. In the early years mixed fertilizer production was primarily a mechanical operation of dry mixing. Later the trend turned toward ammoniation, nitrogen solutions, and granulation.

Bulk blending has increased rapidly since the late 1950's. And in the 1960's, for example, the number of manufacturing plants in Missouri increased from twenty in 1956 to more than 400 in 1974. In Illinois there were six plants in 1951 and about 700 in 1974. The total number in the Midwest in 1974 was estimated to be 4500. Hence it would appear that the fertilizer mixing plant has completed a cycle.

Numerous factors have created interest in bulk blending. Net savings to the farmer have been estimated at eight to twelve dollars a ton as compared to N-P-K granulated fertilizers. There are several reasons for this saving. The plants are small and the investments relatively low. Additional savings are made when costs of transporting and spreading the material are considered. Many of the costs between the plant and the farm are fixed costs per ton of material. Hence a higher-analysis mixture, characteristic of bulk blends, leads to a lowered cost per ton of nutrients spread on the field. TVA workers made a study of the effect of content of plant nutrients on the cost per unit for several ratios (Figure

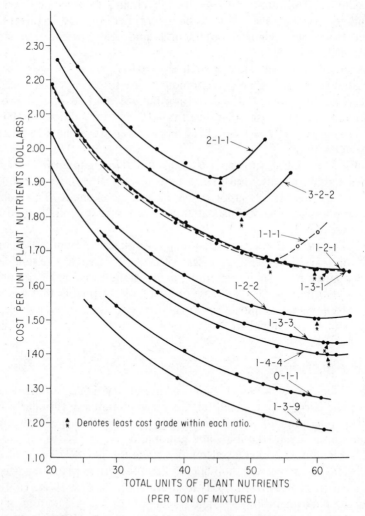

Figure 10-2. Effect of concentration of plant nutrients on cost per unit — 20 lb. [Douglas et al., *Commercial Fertilizer*, **101**(5):23 (1960).]

10-2). For example, for a 1-1-1 ratio a total content of 54 units of $N + P_2O_5 + K_2O$, or an 18-18-18, was most economical. This would be 18-7.9-15 in terms of N-P-K.

Although cost comparisons will depend on the assumptions made, and bases may change rapidly, the *economics* of bulk blending are attractive to the grower.

Materials. In the first attempt at bulk blending muriate of potash was dumped on top of rock phosphate in the spreader, and ammonium sulfate was subsequently blended with the mixture. Gradually the choice of materials moved to ammonium nitrate, urea, muriate of potash, and concentrated superphosphate. The development of supplies of diammonium phosphate gave bulk blending a new impetus and now the material is widely used.

Smith and Makower (1963) report that a bulk fertilizer 15-6.6-12.5 (15-15-15) might contain the following:

Sulfate of ammonia	21% N	786 (lb.)
Urea	45% N	51
Diammonium phosphate	18% N, 20.5% P(46.5% P_2O_5)	658
Muriate of potash	50.2% K (60.5% K_2O)	505
Total		2,000

The difference between the dry mixes of the earlier years and the bulk blends of today is the result in large part of the higher-analysis raw materials now available.

Segregation. One of the problems of bulk blends has been that of segregation, or the separating out of certain components of the blend during transportation. Uniformity of particle size, shape, and density can minimize segregation; particle size, however, is by far the most important. A real effort to obtain closely sized raw materials is being made and many materials now fall in the minus-6-plus-14-mesh range (through a 6-mesh but retained on a 14-mesh).

Introduction of pesticides or micronutrients into the blend merely aggravates the problems of segregation and uniform distribution, again because of particle size. Careful granulation may be the answer. Addition of oil. H_2O or liquid mixed fertilizer helps the micronutrient carriers to adhere to the particles.

Bulk Application of Blended and Manufactured Goods. Blending generally implies bulk handling and spreading of fertilizers. Although by tradition manufactured goods have been handled in bags, more and more of these materials are now being handled in bulk. The growth of shipments of both solid and liquid fertilizers in bulk for retail markets in the United States is as follows (USDA):

	Solids (1,000 tons)		Liquids (1,000 tons)	
Year	Mixtures	Materials	Mixtures	Materials
1953–1954	469	1360	28	560
1960–1961	1357	2490	531	2244
1971–1972	9800	7700	3040	7800

This trend is a reflection of interest in saving six to eight dollars a ton for bags and bagging, as well as labor.

One problem under study has been the matter of uniformity of distribution of fertilizer particles on the field. Critical factors have been size, shape, and density of particles and design and adjustment of spreaders. Much improvement has been made in spreading equipment to allow a more even distribution of fertilizer and its application at the desired rate per acre.

Research in Illinois and Missouri has shown that uneven spreading did not affect crop yields to a measurable degree. Theoretically, however, when crop yields are high, Jensen and Pesek postulate that uneven distribution of plant nutrients will have an adverse effect on yields. This is explained by the diminishing response from higher rates (Figure 10-3). The loss in yield caused by underfertilization (A) is larger than that

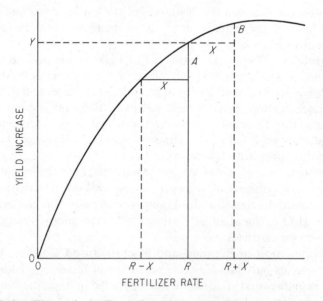

Figure 10-3. Theoretical effect of uneven spreading on yield as shown by decreases or increases in fertilizer rates (R). [Jensen and Pesek, *SSSA Proc.*, **26**:170 (1962).]

gained by overfertilization (B). This holds only for the higher rates of fertilizer near the top of the yield response curve.

To meet the demand for small applications at planting, some blenders sell a part of their product in bags. Others mix standard grades on a registered basis.

Compartmented bulk spreaders designed to blend and spread two or three dry fertilizers are now on the market, and a liquid nitrogen distribution system may be used in conjunction with a two-compartment blender-spreader. However, adoption has been very slow.

FLUID MIXED FERTILIZERS

N-P-K and N-P grades of fluid mixed fertilizers are used in starter solutions and foliar sprays, as water-soluble fertilizers for use in irrigation water, and for direct application to the soil. This discussion deals largely with the first and last of these methods.

Possibly the first liquid fertilizer plant in the United States was built in California in 1923, and the first plant east of the Rocky Mountains in 1954. Fluid mixtures, while only about 14 per cent of the total tonnage of mixtures used by the American farmer in 1972, are growing fast. Ease of handling and less labor contribute to this growth. Too, herbicides and micronutrients can be readily added. Fluid mixtures may be either clear liquids or suspensions. The industry is concentrated in California, Texas, and parts of the Midwest. A typical plant may distribute 1,500 to 3,000 tons in a twenty-five-mile radius.

Fluid fertilizers are essentially of three different types: hot mix, cold mix, and suspensions. A brief description follows:

Manufacture. The hot-mix method has been widely used and is based on the neutralization of different types of phosphoric acids to make such base solutions as 10-15-0 (10-34-0), 11-16-0 (11-37-0) and 12-17-0 (12-40-0). Nitrogen solution is usually used as an additional source of nitrogen and finely divided potassium chloride is the usual source of potassium.

Mixing equipment is relatively simple and a batch system is often employed. The mix tank has a capacity of 4.5 to 18 metric tons (5 to 20 short tons). Considerable heat is generated and the liquid is recirculated through a cooler to prevent excessive boiling in the mix tank.

In contrast to hot-mix plants cold-mix plants are simple and storage tanks, meters and a simple mix tank make up the equipment list. The base solution, 10-34-0 or 11-37-0, urea–ammonium nitrate solution, and potash are mixed together. Heat is not generally evolved. Clear liquids or suspensions can be made.

Materials. The major problem in fluid mixed fertilizers is the cost of the phosphorus. Superphosphoric acid from the wet process, 29–31 per cent P (68–72% P_2O_5), of which about 50 percent is in the polyphosphate form, or furnace process acid, 33–35 per cent P (76–80% P_2O_5), of which 60 to 85 per cent is in the polyphosphate form, may be used. Merchant grade orthophosphoric acid, 23 per cent P (52–54% P_2O_5), is lower in price. However, because of shipping distance the delivered cost of fluid fertilizer made from the superphosphoric acid is often no more than that from the wet process. Generally, merchant grade wet process orthophosphoric acid is becoming more popular because impurities are being removed and it is now a relatively clear solution.

One of the difficulties in the production of clear solutions has been the incorporation of potassium. Potassium chloride is the most economical source, but it has low solubility. Potassium hydroxide or potassium carbonate can be used, for they react with phosphoric acid to form potassium phosphate. TVA reports that a 6-7.9-15 (6-18-18) can be made with potassium hydroxide but only a 3-4-7.5 (3-9-9) with potassium chloride. Unfortunately only the chemical grade potassium compounds, which are currently quite expensive, give high solubility.

Pesticides (herbicides and insecticides) are being added successfully to fluid fertilizers for broadcast application. Such additions save trips over the field. Compatability charts may be available in order to avoid undesirable chemical reactions or precipitation. In case of doubt, farmers are often advised to mix up a sample in a glass jar and observe it.

Micronutrients in various forms may be added readily, particularly with suspensions. Addition of chelating agents favors solubility. A polyphosphate such as 11-37-0 gives higher micronutrient solubility by sequestration. Sulfur may be added in several forms including ammonium thiosulfate and ammonium bisulfite.

Precipitation or salting out may occur in clear solutions at low temperatures. Careful attention must be given to the selection of materials for clear solutions to avoid salting out during storage or application. The general aim is to keep the salting-out temperature below 32°F. Suspensions do not salt out but under cool temperatures viscosity may increase making application difficult.

Suspensions. As mentioned earlier, one of the problems in fluid fertilizers has been low analysis and low potash content. Suspensions help to solve this problem in that they are so concentrated that small fertilizer crystals are suspended in fluid mixtures. Attapulgite clay (up to 2%) is used as a suspending agent. The principle is to keep the crystals from growing to sizes that will clog applicators. Some type of mechanism such as a propellor or air agitation is used to help prevent settling out.

In a 1971 survey, one-third of the fluid fertilizers were suspensions and 58 per cent of the fluid plants were producing suspensions.

Suspensions have a number of advantages, including higher analysis, additives possible, lack of salt-out problems, more economical, and more flexibility in grades produced. For these reasons, suspensions should capture an increasingly greater proportion of the fluid fertilizer market.

Grades. A wide variety of grades are manufactured. A TVA-National Fertilizer Solutions Association survey in 1971 showed the following leading grades (Achorn and Hargett, 1972).

Clear liquids	Suspensions
7-21-7	4-12-24
8-25-3	3-10-30
4-10-10	5-15-30
8-8-8	14-14-14

Interest is growing in N-P materials, particularly in the West, on soils very high in K where potassium is not needed. Where potassium is needed in sizable amounts, broadcast application of potash materials may be used in addition to the fluid fertilizer.

Distribution. Equipment to apply clear and suspension fluid fertilizers is constantly being improved. Much of the clear fluid fertilizer goes on at planting in bands beside the row. The fluid is pumped through a hose from a tank on a truck into the planter fertilizer hopper. The dealer often loans the tank to the farmer.

Suspensions are usually applied broadcast by the dealer. The tank is mounted directly on a truck or on a tricycle-type spreader. Nurse trucks are used to keep the supply of suspensions to the latter. This equipment can cover a strip at least 60 ft. wide and as much as 60 acres can be fertilized in one hour.

Use. Assuming the same type of placement and water solubility of phosphorus, responses to fluid and solid fertilizers are similar. When the water solubility of phosphorus is important, liquids are superior to solids that contain a high amount of water-insoluble phosphorus. In some instances, when zinc in the soil is deficient, the polyphosphates may have a sequestering effect and make the zinc more available to the plant.

Cost of nutrients in fluid fertilizer is somewhat higher than that in bulk blends but convenience is an important factor. Lower labor costs and the fact that liquids can be moved by gravity, air pressure, or pumps are distinct advantages over dry materials.

AGRONOMIC SERVICE

An advisory service including plant and soil analysis is now being provided by the fertilizer industry in many areas of North America.

In many situations the plants are designed to serve a limited area. The managers of these plants, from whom the growers can obtain the exact ratio and amount of fertilizer called for by their soil-test recommendation sheets, can meet their customers on a more personal basis.

The philosophy governing the use of fertilizer, which is to build and maintain soil fertility, leads to more broadcast applications of the actual nutrients needed. The grower calls his dealer, and the fertilizer is custom-applied with no physical effort whatever on the part of the customer.

FERTILIZER-PESTICIDE MIXTURES

Such mixtures combine two operations into one and save labor, time and energy. Hence, one of the important developments in fertilizers in recent years involves their use as pesticide carriers (herbicides, insecticides, and fungicides). This has brought on technical chemistry problems, application questions, and legal requirements. The incorporation of insecticides in fertilizers is not new, for in 1904 a patent was issued in France for such mixtures. Numerous patents have been issued since that time in the United States as well as in foreign countries. Before World War II a sizable tonnage of lead arsenate-fertilizer mixtures was formulated in the Middle Atlantic States.

Manufacture.   A wide variety of formulations of pesticides has been developed. These include sprayable concentrates, granular materials, dusts, solutions, and minor formulations.

Dry mixes may be made up as a simple blend of granular materials, spraying the liquid pesticide on the dry fertilizer, spraying on to a carrier such as vermiculite and mixing with fertilizer, or premixing prior to granulation or blending. Instructions furnished by the pesticide manufacturer must be carefully followed.

Liquid fertilizers are made up in various ways and a given pesticide may react differently depending on the fertilizer materials used. Reactions may affect physical properties as well as performance of the pesticide. For example, paraquat may be adsorbed on colloidal clay in fluid fertilizers and rendered unavailable. Only in a few instances are mixing guides for pesticide-fertilizer combinations provided by manufacturers. Mixing small quantities in a glass jar and observing the results is a good alternative to precipitation in the tank and clogged nozzles.

Insecticides.   The use of chorinated hydrocarbons was quite popular until questions were raised as to insect resistance, persistence in soil, and legal curtailment. Carbamates and organophosphates have become of more interest. Insecticide stability when mixed with fertilizer and stored for any length of time has been a problem. However, use of an insecticide such as aldrin with liquid fertilizer, mixed just before use, has been quite satisfactory. Uniform mixing and stability in solid fertilizers has been more of a problem. Airtight containers help in the latter and use of insecticides in turf fertilizers is acceptable.

Use of insecticide–fertilizer mixtures.   In some areas aldrin and other organic insecticides are recommended to control corn rootworms.

TABLE 10–4. Effect of Insecticide Applied with Starter Fertilizer on Rootworm Control*

	Rate of insecticide (lb./A.)	Lodging at harvest (45° angle) %	Yield of corn (bu./A.)
Fertilizer alone	—	30.0	71.3
Fertilizer plus aldrin	0.25	9.9	93.0
	0.50	11.8	104.5
	1.00	8.6	107.0
Fertilizer plus heptachlor	0.25	10.6	104.0
	0.50	3.3	110.5

* Apple, *Econ. Entomol.*, **50**:28 (1957).

The beneficial effect of aldrin applied with a starter fertilizer for corn-rootworm control increased corn yields and reduced lodging (Table 10-4). One half pound of heptachlor with the starter increased the yield 39 bu./A.

PROBLEMS. Distribution of a given amount of insecticide in the field is a complicating factor, for fertilizers vary greatly in analysis and in the amount required for various crops.

A multiplicity of problems is connected with insect control. The type of insect, the intensity of infestation, and the stage in the life cycle at which the treatment would be most effective must all be recognized in selecting the insecticide, the amount to apply, and the time and method of application.

Residual effects must be considered, for insecticides vary in lasting qualities. For example, toxaphene is easily decomposed in the soil, whereas DDT is fairly stable. An additional problem is that of imparting undesirable flavors to subsequent crops. Benzene hexachloride, for example, has been found to give a distinctly musty taste to peanuts, Irish potatoes, and other food crops. Because of this quality such mixtures must bear labels that specify how, at what rate, and on what crops they may be applied.

Large acreages of crops still receive fertilizer at planting. Often this is in sidebands and the fertilizer-insecticide mixture is quite effective in control of rootworms on corn for example. For other insects, this placement may not be most satisfactory. Systemic insecticides show promise and when these are absorbed by the plant so that insects feeding on the plant are killed, sideband application is satisfactory.

Herbicides. The use of fertilizer-herbicide mixtures is growing rapidly and is commonly called "Weed and Feed." Much work has been done concerning the compatability of herbicides and fluid fertilizers. More herbicides are physically compatible in suspension fertilizers than in clear liquids. Some separation of suspension-herbicide combinations takes place two hours after the last agitation. However, the effectiveness of the herbicide or the fertilizer was not affected by time.

USE OF HERBICIDE–FERTILIZER MIXTURES. Fluid or dry fertilizer-herbicide mixtures may be applied preplant. Some herbicides require soil incorporation but this is compatible with fertilizer usage. Herbicide-nitrogen solution mixtures may be applied preplant, immediately after planting, or after plant emergence as directed sprays to avoid some chemicals burning the plants.

This practice should continue to increase because it is agronomically sound saves time, labor, and energy for the farmer, and it is an area of service and profit for the fertilizer applicator.

However, there is considerable need for additional research on pesticide-fertilizer mixtures. New pesticide chemicals, new fertilizers, and regulations all render the field a challenging one.

FERTILIZER CONTROL AND REGULATION

Laws and regulations governing the sale of fertilizers are necessary because of the opportunity to defraud. This has to do with both the quantity of nutrients and quality of carriers present in the fertilizer. Regulatory laws, which apply to mixed fertilizers and to fertilizer materials, protect the farmer and the reliable fertilizer manufacturer by keeping goods of questionable value off the market.

The fertilizer laws in the various states are similar in many respects, and this, of course, is helpful to fertilizer companies operating across state lines. When adjoining states have different requirements for labeling and guarantees, manufacturing problems are increased.

The Association of American Plant Food Control Officials has been directing considerable attention to a model state fertilizer bill. Although this model is under continual revision, essentially all of the states have adopted it.

Guarantee. The total nitrogen (N), available phosphorus pentoxide (P_2O_5), and the soluble potash (K_2O) must be guaranteed in terms of the percentage of each of these constituents present. For example, in a fertilizer grade such as 6-24-24 the nutrients are guaranteed as 6 per cent total nitrogen, 24 per cent available P_2O_5, and 24 percent soluble K_2O. The elements are listed in that order for mixed fertilizers in all states. In some states it is necessary to guarantee the percentage of water-insoluble nitrogen present, particularly in tobacco fertilizers. Fertilizer material must bear the guarantee of the primary nutrient it is carrying: for example, ammonium nitrate, 33 per cent nitrogen.

The oxide expressions for phosphorus and potassium are actually inaccurate and confusing, for they are based on early practices with which chemists determined the elements by ignition and weighed the oxides. These methods have long since been discarded. Proponents of the change to the elemental expressions for phosphorus and potassium feel that the understanding of fertilizer formulation will be enhanced by the change. In 1963 several scientific societies, including the American Society of Agronomy, the Soil Science Society of America, and the Crop Science Society of America, initiated the practice in their publications of giving the elemental analysis of fertilizers, with the oxide in parentheses. Many states now use the elemental as well as the oxide expressions in extension and research publications. Some fertilizer companies print both the oxide and elemental guarantees for phosphorus and potassium on their bags.

Norway, New Zealand, and South Africa use the elemental basis. Just how soon a change will be made in the United States is not known. However, a real effort is being made by some groups to bring about the change.

There is great variability among states in the guarantees required for nutrients other than nitrogen, phosphorus, and potassium. Because of this lack of uniformity and the growing market for fertilizers containing micronutrients, the committee on Fertilizer Guarantees and Tolerances of the Association of American Plant Food Control Officials, in cooperation with representatives of many other organizations, worked out the following proposed regulation for adoption under state fertilizer laws:

Other plant nutrients, when mentioned in any form or manner shall be registered and shall be guaranteed. Guarantees shall be made on the elemental basis. Sources of the elements guaranteed and proof of availability shall be provided the ____ upon request. The minimum percentages which will be accepted for registration are as follows:

Element	%
Calcium (Ca)	1.00
Magnesium (Mg)	0.50
Sulfur (S)	1.00
Boron (B)	0.02
Chlorine (Cl)	0.10
Cobalt (Co)	0.0005
Copper (Cu)	0.05
Iron (Fe)	0.10
Manganese (Mn)	0.05
Molybdenum (Mo)	0.0005
Sodium (Na)	0.10
Zinc (Zn)	0.05

Guarantees or claims for the above listed plant nutrients are the only ones which will be accepted. Proposed labels and directions for use of the fertilizer shall be furnished with the application for registration upon request. Any of the above listed elements which are guaranteed shall appear in the order listed immediately following guarantees for the primary nutrients of nitrogen, phosphorus and potassium.

A warning or caution statement is required on the label for any product which contains 0.03% or more of boron in a water soluble form. This statement shall carry the word "WARNING" or "CAUTION" conspicuously displayed, shall state the crop(s) for which the fertilizer is to be used, and state that the use of the fertilizer on any other than those recommended may result in serious injury to the crop(s).

Products containing 0.001% or more of molybdenum also require a warning statement on the label. This shall include the word *Warning* or *Caution* and the statement that the application of fertilizers containing molybdenum may result in forage crops containing levels of molybdenum which are toxic to ruminant animals.

The majority of the states have now adopted this regulation. A maximum guarantee of 3 per cent is required for the chlorine content of tobacco fertilizers in some states.

Information Required on Bag or Tag. Although the information required may vary somewhat from one state to another, the following statements are usually made on the bag or tag:

1. Net weight of fertilizer.
2. Brand and grade.
3. Guaranteed analysis.
4. Name and address of manufacturer.

If distribution is made in bulk, a written statement must accompany the delivery to the purchaser.

It is extremely important that farmers become well acquainted with the meaning of these data, particularly the N-P-K guarantee. It must be realized that not just fertilizer is being bought but that the plant-nutrient elements contained in the fertilizer are the items of purchase.

Inspection and Enforcement of Fertilizer Laws. Inspection and analysis are usually carried out by the state department of agriculture or the state agricultural experiment station. The expense of this service is borne by the tonnage tax on the fertilizer. Each manufacturer is required to register each grade annually with the regulatory organization in his state. From time to time during the year the fertilizers are sampled by inspectors who are trained for that particular job, and the samples are sent to the control laboratory for analysis. The findings are then checked with the registration on file, and if they fall outside the allowable tolerance, as specified in the law, a penalty is assessed.

The official methods of sampling and analysis are prescribed by the A.O.A.C. This uniformity is essential to the fertilizer companies operating in several states.

In the manufacture of granulated and nongranulated fertilizers inspection of the product in the bin or in bags is reasonably simple. However, the advent of bulk blending presented a different situation. Some blenders mix standard grades in advance and register them. Others mix to the specific order of the final purchaser and load the mixtures into truck or spreader for immediate transportation to the field. The physical task of getting an inspector on hand to sample a percentage of the loads is difficult, but it is being accomplished.

It is required by law that a bulletin bearing the guaranteed and found analysis of all fertilizer brands inspected be published each year. This sort of publicity is quite effective in minimizing fraud. If a company persists in the manufacture of fertilizers that regularly fall below the guarantee, this fact soon becomes known. Generally, the amounts of nutrients found in most fertilizers are somewhat above the guarantee. Most companies overformulate to decrease the chance of a deficiency in the analysis.

Fertilizers purchased by mail order from outside a state are generally not subject to state inspection. Extravagant claims are sometimes made in local newspapers by these outside companies, but the control agencies have no authority to curb this activity.

UNIFORMITY OF RECOMMENDED GRADES AND RATIOS

Over the years many grades and ratios have been developed. It is often difficult to determine how a given grade originated, but as a rule it was developed for a specific purpose. To manufacturers a large number of grades means a considerable expense in terms of money, time, storage bins, and labor if these grades are to be stocked. To the farmer a long grade list means a confusing array from which to make a choice. To the agronomist these long lists usually represent a needless duplication of grades or unsatisfactory and pointless analyses.

As one example, the variation between certain grades is too small to be detected when it is considered that the fertilizers may be applied at the rate of 200 to 500 lb./A. However, once they have become established in a state and the industry has built up a demand, both farmers and industry dislike dropping them. In some instances industry markets a grade a little different in analysis from existing grades in order to have something new on which the customer cannot readily make price comparisons or to have a distinctive grade.

To complicate matters further, adjoining states may have different recommended grades and ratios, which makes a change difficult if shipments are being made into several states. In view of this situation, various organizations and states over the country have been striving for more uniformity. For example, since 1938 industry and university representatives in the Midwest have been making an effort to establish more uniformity among the states in ratios and minimum grades for each ratio. This represents an outstanding example of cooperative endeavor between state agronomists and members of the industry to eliminate unsuitable grades and to encourage the manufacture of more useful fertilizers.

Certain groups of states in the East are working toward the development of standard recommendations across state lines, and regional soil test work groups have made real strides in ironing out some of the differences.

Some states have ratio and/or grade lists. There is an effort to agree on ratios that will meet the requirements of both crops and soils. Minimum grades are then established, and multiples of each ratio are permitted. This protects the smaller manufacturer who can use only conventional materials and manufacturing methods. It is also advantageous to the larger manufacturers with more modern plants in which new techniques can be adopted and more concentrated fertilizers can be produced.

In bulk blending the materials are mixed together to meet the results shown by soil tests with no regard to grade or ratio. This is developing into a trend away from specific grades for broadcast applications.

University recommendations now are largely in terms of pounds of N, P_2O_5, and K_2O per acre and sometimes give examples of grades to meet these needs. When it is considered that extra nitrogen, phosphorus, and potassium may be applied separately as materials in accordance with recommendations or known needs, a minimum number of grades is really needed.

It is recognized, of course, that special fertilizers may be needed for some crops. This may require the addition of micronutrients to some of the foregoing ratios under certain conditions.

Summary

1. Although the tonnage of nitrogen, phosphorus, and potassium supplied in mixtures has increased, the percentage of the total nitrogen and potassium applied as mixtures has decreased quite rapidly.
2. The greater use of nitrogen, both in mixed fertilizer and as a material, is increasing soil acidity and emphasizes the need for more attention to lime.
3. The fixed costs, which include manufacturing, bagging (in some instances), transporting, and distribution, and the cost of the plant nutrients influence the cost of a ton of mixed fertilizer. Although the cost of the nutrients may be somewhat greater, the lower fixed costs

make higher-analysis fertilizers more economical. The grower appreciates the reduced labor of handling as well as the lower cost.

4. Higher-analysis fertilizers contain less of the micro- and secondary nutrients. Careful attention must be directed toward identifying deficiencies and toward adding supplements as needed.

5. Rapid changes are taking place in the mixed fertilizer industry because of the new materials and processes.

6. A multiplicity of reactions takes place in fertilizers, including double decomposition, neutralization, hydration, decomposition, and ammoniation. The rate at which these reactions proceed is influenced by moisture, temperature, and particle size, all factors over which manufacturers must exercise close control.

7. Ammoniation of phosphorus sources is of greatest importance to the improvement of the product. Reduction in water solubility of phosphorus does occur, however, and ammoniation can be carried out only to a certain point with ordinary superphosphate.

8. Early plants produced nongranulated fertilizer, but now almost all plants granulate. Granulation of fertilizers reduces caking to a minimum, eliminates segregation, and results in a product that can be uniformly applied on the field.

9. Physical condition of the fertilizer, which is of major importance, is influenced by moisture content and state of subdivision. The moisture dissolves certain salts, new compounds form, and crystal knitting occurs. Conditioners such as ground cocoa shells are used to separate the particles in nongranulated fertilizers. Coatings are used in some granulated fertilizers.

10. Bulk blending is the mechanical mixing of solid, granular fertilizer materials in a stationary mixer. This practice is becoming more popular because of the lower cost per unit of plant nutrient spread on the field. The rapid growth in the production of diammonium phosphate has been an important factor. This material is mixed with muriate of potash, and nitrogen sources such as urea and ammonium sulfate.

11. Bulk application of blended and manufactured goods saves six to eight dollars a ton bagging cost as well as considerable labor. Unity of particle size is of major importance in preventing segregation and in producing an even spreading pattern. Improved design of spreader trucks is helping to provide greater uniformity of spreading.

12. Fluid mixed fertilizers make up about 14 per cent of the tonnage of mixed fertilizers in United States. Neutralization of phosphoric acid with ammonia is the basic process and potassium is added as KCl. Cost of the phosphorus has been a major factor. Speed of adoption is influenced by economy and convenience. Suspensions in which

the salts are suspended in their own saturated solutions are means of creating higher-analysis fluid fertilizers.

13. Fertilizer-pesticide mixtures combine two operations into one and save time, labor, and energy. The multiplicity of the chemical reactions of organic pesticides with the fertilizer materials constitutes a challenge to the manufacturer. Care must be used to avoid undesirable physical reactions such as precipitates or the rendering of the pesticide ineffective. New chemicals and regulations make the practice a continually changing one.

14. Fertilizer-insecticide mixtures are being used to a considerable extent in liquid starter fertilizer and in turf fertilizer. Fertilizer-herbicide mixtures or "Weed and Feed" are growing rapidly. The herbicide may be applied in fluid fertilizer preplant, and in nitrogen solutions just after planting or as a directed spray.

15. The control and regulation of fertilizer production is essential to protect both the grower and the reliable manufacturer. Fertilizers are sampled and analyzed systematically by the control agency, and the results are reported to the manufacturer and publicized annually.

16. Guarantees are in terms of nitrogen, phosphorus pentoxide, and potassium oxide for the primary nutrients. Attention is being given to the guarantees of phosphorus and potassium on an elemental basis. The micro- and secondary nutrients are guaranteed as elements on a minimum basis.

17. A large number of fertilizer grades and ratios are being sold annually in the United States. This large number of grades is based on crop requirements, methods of manufacture, and competition. However, just a few N-P-K ratios would meet the requirements of most agronomic and horticultural crops when it is considered that supplements of N-P-K materials can be made directly to the soil.

18. University recommendations are now largely in terms of pounds of N, P_2O_5, and K_2O per acre and sometimes examples of grades to meet needs are given.

Questions

1. Evaluate the practice of including lime in fertilizers.
2. Assume that a 10-20-20 fertilizer sells for 120 dollars a ton. What would the difference be in the cost of a ton of the 10-20-20 and an equivalent amount of nutrients supplied in 5-10-10 at 72 dollars a ton? What costs does this difference represent? What other benefits would the farmer derive from the use of 10-20-20? Why has the percentage of filler in fertilizers been decreasing?
3. What are some of the materials being used to formulate higher-

analysis fertilizers? What are the basic costs that determine whether a higher-analysis fertilizer will be more or less economical than a lower-analysis fertilizer?

4. What deficiencies may occur in crops grown with a higher-analysis fertilizer? Why?

5. Obtain prices on a 16-8-8 and a 12-12-12. How much is the nitrogen costing per pound in the 16-8-8? How does this compare with the price of nitrogen in ammonium nitrate, anhydrous ammonia, nitrogen solutions, urea, and sodium nitrate?

6. Why does the water solubility of phosphorus tend to decrease with continued ammoniation of normal superphosphate but not with concentrated superphosphate? Give equations.

7. What are the three general sizes of particles in fertilizer materials? Why is there a need for some degree of standardization?

8. Why is there a need for drying in the granulation process? What are the advantages of granulated over nongranulated fertilizer?

9. What functions do conditioners serve? On what principle are coated fertilizers based?

10. What reactions may occur during curing? What effect will such reactions have on physical condition? Why may caking be more of a problem in higher-analysis fertilizers than in lower-analysis goods?

11. Explain the difference in unit cost of plant nutrients in bulk blends and in granulated fertilizer.

12. What are the problems in uniform spreading of bulk blends? Will this affect crop yields?

13. List the advantages and disadvantages of fluid mixed fertilizers. What are fertilizer suspensions?

14. (a) The maximum amount of chlorine that can safely be applied to tobacco is 40 lb./A. At planting you apply 1,000 lb. of a fertilizer containing 3 per cent chlorine. Suppose then you want to make an additional N-K side-dressing of 15 lb. of nitrogen and 35 lb. of potassium oxide per acre. How much muriate and sulfate would you mix per acre to keep within the allowable chlorine limit and at the same time keep the cost of potassium at a minimum? The price of muriate is 70 dollars a ton and the price of sulfate of potash 110 dollars a ton. (b) What source of nitrogen would you use to side-dress corn if you could buy anhydrous ammonia at 140 dollars a ton, ammonium nitrate at 100 dollars a ton, and urea at 110 dollars a ton?

15. Why are fertilizer-herbicide mixtures gaining in popularity? What precautions must be taken in mixing?

16. What nutrients must be guaranteed in your state? What nutrients are permitted to be guaranteed?

17. How many fertilizer ratios are sold in your state? In your opinion how many are needed?

Selected References

Achorn, Frank P., and Thomas R. Cox. "Production, marketing and use of solid, solution and suspension fertilizers," in R. A. Olson, T. J. Army, J. J. Hanway, and V. J. Kilmer, Eds., *Fertilizer Technology and Use,* 2nd edition, p. 381. Madison, Wisconsin: Soil Sci. Soc. Am., 1971.

Achorn, Frank P., and Norman L. Hargett, "NFSA and TVA survey of suspensions in 1971." Tennessee Valley Authority, Muscle Shoals, Ala. (1972).

Ahmed, I. U., O. J. Attoe, L. E. Engelbert, and R. B. Corey, "Factors affecting the rate of release of fertilizer from capsules." *Agron. J.,* **55**:495 (1963).

Apple, J. W., "Reduced dosage of insecticides for corn-rootworm control." *J. Econ. Entomol.,* **50**:28 (1957).

Douglas, J. R., John I. Bucy, and Robert M. Finley, "Bulk blending with linear programming." *Com. Fertilizer,* **101** (5), 23 (1960).

Gribbins, M. F., "Conversion of ammonia to fertilizer materials," in K. D. Jacob, Ed., Fertilizer Technology and Resources in the United States. *Agronomy,* Vol. III, pp. 63–84, New York: Academic, 1953.

Hignett, T. P., "General considerations on operating techniques, equipment, and practices in manufacture of granular mixed fertilizers," in V. Sauchelli, Ed., *Chemistry and Technology of Fertilizers,* ACS Monograph No. 148, pp. 269–298, New York: Reinhold, 1960.

Hill, W. L., "Problem of evaluating nutrient power of phosphorus in mixed fertilizers produced by different manufacturing techniques." *USDA, B.P.I.S.A.E., Res. Rept. 251* (1952).

Jacob, K. D., Ed., "Fertilizer Technology and Resources in the United States." *Agronomy,* Vol. III. New York: Academic, 1953.

Jensen, D. and J. Pesek, "Inefficiency of fertilizer use resulting from nonuniform spatial distribution: I." *SSSA Proc.,* **26**:170 (1962).

Kapusta, E. C., "Potash use in liquid fertilizers." *Solutions,* **5** (5) 6 (1963).

Knudsen, K. C., "Superfos pure plant food." *Fertilizer Industry Roundtable* (1973).

Lawton, K., "Coated fertilizer in the future?" *Better Crops Plant Food,* **45** (2), 18 (1961).

Lockwood, M. H., "High analysis mixed fertilizer," in K. D. Jacobs, Ed., Fertilizer Technology and Resources in the United States. *Agronomy,* Vol. III, pp. 393, New York, Academic, 1953.

Lunt, O. R., J. J. Oertli, and A. M. Kofranek, "Coated fertilizers 'meter out' plant nutrients." *Crops Soils,* **14** (6), 14 (1962).

Murphy, L. S., L. J. Meyer, O. G. Russ and C. W. Swallow, "Evaluating fluid fertilizer-herbicide combinations." *Fertilizer Solutions,* **17** (4):26 (1973).

Nelson, W. L., and G. L. Terman, "Nature, behavior, and use of multinutrient (mixed) fertilizers." *Fertilizer Technology and Usage,* Madison, Wisconsin: Soil Sci. Soc. Am., 1963, pp. 379–427.

Parker, F. W., "The availability of phosphoric acid in precipitated phosphates." *Com. Fertilizer,* **42**:28 (1931).

Petty, H. B., O. C. Burnside, and John P. Bryant, "Fertilizer combinations with herbicides or insecticides", in R. A. Olson, T. J. Army, J. J. Hanway and V. J. Kilmer, Eds., *Fertilizer Technology and Use,* 2nd edition, pp. 495–515, Madison, Wisconsin: Soil Sci. Soc. Am., 1971.

"Fertilizers." *Encyclopedia of Chemical Technology,* 2nd ed., Vol. 9, pp. 25–150, New York: Wiley-Interscience, 1966.

Slack, A. V., "Liquid fertilizers," in V. Sauchelli, Ed., *Chemistry and Technology of Fertilizers,* ACS Monograph No. 148, pp. 513–537. New York: Reinhold, 1960.

Smith, R. C., "Plant practices in the manufacture of nongranulated mixed fertilizers," in V. Sauchelli, Ed., *Chemistry and Technology of Fertilizers.* ACS Monograph No. 148, pp. 403–433, New York: Reinhold, 1960.

Smith, R. C., and B. Makower, "Advances in manufacture of mixed fertilizer." *Fertilizer Technology and Usage,* pp. 341–378, Madison, Wisconsin: Soil Sci. Soc. Am., 1963.

Tennessee Valley Authority, *TVA Bulk Blending Conference Proc.* (1973).

11. LIMING

Liming, as the term applies to agriculture, is the addition to the soil of any calcium or calcium- and magnesium-containing compound that is capable of reducing acidity. Lime correctly refers only to calcium oxide (CaO), but the term almost universally includes such materials as calcium hydroxide, calcium carbonate, calcium-magnesium carbonate, and calcium silicate slags. This chapter is devoted to a discussion of the theory of soil acidity and the practice of liming and to some of the materials employed for this purpose.

WHAT IS ACIDITY?

An acid is a substance that tends to give up protons (hydrogen ions) to some other substance. Conversely, a base is any substance that tends to accept protons. This concept of acids and bases was developed by Brönsted and Lowry and is generally accepted as describing the behavior of materials that yield or gain protons in all liquid media, including water. The earlier theory of Arrhenius described the behavior of compounds in water, and an acid was accordingly defined as any substance yielding hydrogen ions (protons) when dissolved in water. A base was defined as any substance yielding hydroxyl ions when dissolved in water. The second definition is a special case covered, of course, by the broader concept of an acid and base defined by the Brönsted-Lowry theory. Soil acids are aqueous systems, and therefore the terms hydrogen and hydroxyl ions rather than proton, proton donor, and proton acceptor are employed in this book.

Active and Potential Acidity. The characteristics of acid solutions are based on the activity of the hydrogen ion (H^+). An acid when mixed with water dissociates or ionizes into hydrogen ions and the accompanying anions, as represented by the hypothetical acid HA:

$$HA \xrightleftharpoons{H_2O} H^+ + A^-$$

The H^+ ions to the right indicate *active* acidity, and the more the reaction tends toward that direction, the greater the activity of the H^+ and the stronger the acid is said to be. The HA on the left side of the equation is the *potential* acidity.

Strong and Weak Acids. Acids are arbitrarily classified according to the extent to which they dissociate in water. If the dissociation is great, the acid is said to be strong; well-known examples are nitric, sulfuric, and hydrochloric. Acids that dissociate to only a slight extent, examples of which are acetic, carbonic, and boric, are termed weak acids.

Expressions of Acid Concentration. The total acidity of a solution is the sum of the concentrations of the active and potential acidities. As an example, suppose that a solution of an acid is $0.099M$ with respect to its active acidity and $0.001M$ with respect to its potential acidity. The total acid concentration is $0.100M$ and would certainly be classed as strong. The foregoing is illustrative of most strong acids in dilute solution. The activity of the H^+ is so nearly equal to the concentration of the total acidity that an expression of the latter is for all practical purposes an automatic expression of the former. In other words, there is little need for a separate designation of the active and total acidity of dilute solutions of strong acids.

With weak acids, however, a knowledge of the total concentration is of no value in predicting the activity of the H^+ ions unless the ionization constant is known. Many weak acids are dissociated to less than 1 per cent. Assume that the total concentration of a weak acid HA is $0.1M$ and that it is 1 per cent dissociated. This means that the activity of the H^+ in such a solution is $0.1 \times 0.01 = 0.001M$. Obviously a measure of the total acidity gives no indication whatever of the active acidity.

The pH Concept. There are means of determining the H^+ ion activity of solutions, and this activity can be expressed as is the case with the strong acids. But with extremely weak acids this mode of expression becomes inconvenient and the H^+ ion activity is generally stated in terms of pH values. This connotation was developed a number of years ago by Sorenson, a Swedish chemist, and has now been almost universally adopted for describing the H^+ ion activity of very dilute acid solutions. The pH is defined as the logarithm of the reciprocal of the H^+ ion activity, or symbolically

$$pH = \log \frac{1}{A_{H^+}}$$

where A_{H^+} is the hydrogen ion activity in moles per liter. A solution with an H^+ ion activity of $0.001M$ will have a pH of 3.0; one with an H^+ ion activity of $0.0001M$, 4.00, and so on. The pH connotation is an integral part of the terminology employed in soil fertility and should be thoroughly understood.

Neutralization. Quite important in any consideration of acids and bases is neutralization, the reaction of an acid with a base to form a salt and water. This reaction is represented by the following equation in which HA and BOH are the hypothetical acid and base:

$$HA + BOH \longrightarrow BA + HOH$$

If a given quantity of acid is titrated with a base and the pH of the solution is determined at intervals during the titration, a curve is obtained by plotting pH values against the amounts of base added. Titration curves so obtained for strong and weak acids differ markedly, as can be seen in Figure 11-1. Neutralization of soil acidity is brought about by liming.

Buffer Mixtures. Buffers or buffer systems are compounds which can maintain the pH of a solution within a narrow range when small amounts of acid or base are added. The term buffering, then, defines the resistance to a change in pH. An example of a commonly known buffer system is acetic acid and sodium acetate. The salt is essentially dissociated and

$$CH_3COOH \rightleftharpoons H^+ + CH_3COO^-$$
$$CH_3COONa \rightleftharpoons Na^+ + CH_3COO^-$$

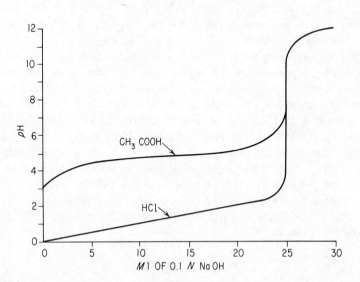

Figure 11-1. The titration of $0.10N$ CH_3COOH and $0.10N$ HCl with $0.10N$ NaOH.

gives rise to a large concentration of acetate ions, which, by mass action, suppress the dissociation of the acetic acid because it tends to form the undissociated acetic acid. If an acid such as hydrochloric is added to this system in small quantities, the excess of acetate ions forms molecular acetic acid, with the net result that there is little change in the active acidity of the solution. Conversely, if sodium hydroxide is added, the hydroxyl ions are neutralized by the hydrogen ions forming water. Because of the large supply of undissociated acetic acid, the equilibrium shifts to the right, replacing the hydrogen ions removed by the preceding neutralization. The net effect is, again, only a slight change in the active acidity of the solution, hence only a small change in the pH. The resistance offered by this system to changes in pH, even though acids or bases may be added, is termed buffering. This brief review of weak acids and buffering has been made only because soils behave in many respects like buffered weak acids. In the soil, however, humus and aluminosilicate clays, with their ability to retain aluminum and other cations, act as the buffer medium.

SOIL ACIDITY

Soil acidity has been considered seriously by soil scientists for more than fifty years, and numerous concepts have been advanced to explain its observed patterns and behavior. In fact, these explanations have almost completed a cycle, a point made very clearly by Professor Hans Jenny of California in an excellent review paper, "Reflections on the Soil Acidity Merry-Go-Round." This article is included in the list of references at the end of the chapter and is recommended reading for those interested in the development of ideas relating to this important soil property.

Nature of Soil Acidity. According to generally held views, acidity in soils has several sources: humus or organic matter, aluminosilicate clays, hydrous oxides of iron and aluminum, exchangeable aluminum, soluble salts, and carbon dioxide.

Humus. As mentioned in Chapter 5, soil organic matter or humus contains reactive carboxyl, phenolic, and amino groups which are capable of bonding H^+ ions. Such H^+-saturated groups will behave as weak acids and the covalently bound H^+ will dissociate, depending on the dissociation constant of the acid formed. The heterogeneity of soil organic matter is such that it varies from place to place, and its contribution to soil acidity will vary accordingly. However, it is a significant factor, particularly in peat and muck soils and in mineral soils containing large amounts of organic matter.

Aluminosilicate Clay Minerals. These minerals are the two- and three-layer clays typified by kaolinite and montmorillonite. The charge on these clays originates from the isomorphous substitution in the crystal lattice of a cation of lower valence for a cation of higher valence. Charges on clays may also originate from the dissociation of hydrogen ions from hydroxyl groups or from bound water of constitution, both of which are structural components of the crystal lattice.

It has been shown that the total charge on soil colloidal materials can be separated into two categories, one of which is called the permanent charge and is responsible for the electrostatic bonding of Al^{3+} and other ions. This charge presumably results from isomorphous substitution. The other is termed the pH-dependent charge and results in the covalent bonding of hydrogen and other ions. This type is illustrated by the charge on soil organic matter which is due to the carboxylic and phenolic groups already mentioned. It also originates in aluminosilicate clays from structural OH^- groups at the corners and edges of the clay lattice which may dissociate H^+ ions within the slightly acid to alkaline pH range. Amorphous aluminum and iron hydroxy compounds may also coat the aluminosilicate clays, and at higher pH values they may hydrolyze and unblock exchange sites on the minerals.

During the early investigations of soil acidity aluminum was believed to be responsible for the replacement of the adsorbed basic cations, such as sodium, calcium, and magnesium, and the consequent increase in soil acidity. This idea fell into disrepute and attention was turned to H^+ ions and their adsorption by clays. This newer focus was given added impetus when the potentiometric determination of soil acidity and Sorenson's pH concept were developed.

Today, however, most soil scientists recognize that the presence of exchangeable Al^{3+}, along with the loss of basic cations such as calcium, magnesium, and potassium, is responsible for the development of acid soils. As Al^{+3} ions which are displaced from clay minerals by cations hydrolyze in the soil solution, the hydrolysis products are readsorbed by the clay minerals causing further hydrolysis. H^+ ions resulting from the hydrolysis of Al and Fe compounds react with and dissolve or decompose soil minerals. Further, the addition of salts, such as those contained in fertilizers, to sesquioxide-coated, interlayered minerals increases the hydrolysis of nonexchangeable Fe and Al resulting in an increase in the H-ion concentration of the soil solution, and hence a lower pH.

If a base is added to this soil, the H^+ ions will be neutralized first. When more of the base is added, the aluminum hydrolyzes with the production of H^+ ions in amounts equivalent to the aluminum present. At low pH values most of the aluminum is the hexahydrated Al^{3+} ion. At pH values above 5.0, hydrated hydroxyaluminum ions exist in exchangeable form. These reactions are illustrated by the following equations:

$$Al^{3+} + H_2O \longrightarrow Al(OH)^{2+} + H^+$$
$$Al(OH)^{2+} + H_2O \longrightarrow Al(OH)_2{}^+ + H^+$$
$$Al(OH)_2{}^+ + H_2O \longrightarrow Al(OH)_3 + H^+$$

The hydroxy aluminum ions described tend to polymerize to produce much more complex systems than those represented by these simple equations.

When acid aluminosilicate minerals are leached with a neutral unbuffered salt such as potassium chloride, the leachate contains both H^+ and Al^{3+} ions. It will also contain other cations such as calcium, magnesium, iron, and potassium. Coleman and his co-workers in North Carolina leached soils with neutral 1*N* KCl and then with a solution of barium chloride and triethanol amine (TEA) buffered at *p*H 8.2. Their results are shown in Table 11-1. The data in the column headed Al are the amounts of this ion displaced by the potassium chloride solution. For those soils with *p*H values near 6 the amounts of aluminum extracted are small. Coleman et al. considered that the sum of the Al^{3+} plus the basic cations extracted by this neutral salt leaching represented the effective cation exchange capacity of the clay, values which are shown in the column labeled ΣM^+.

The column labeled H shows the negative charges that developed when the soils first extracted with potassium chloride were next extracted with $BaCl_2$-TEA. They considered this to be the *p*H-dependent charge that reflects the covalently bound hydrogen released from the

TABLE 11-1. Cations Replaced from B Horizons of Several Southeastern Piedmont Soils on Leaching with 1*N* Potassium Chloride*

	Exchangeable cations (meq. per 100 g.)						Per cent saturation	
Soil	*p*H	Al	ΣCa, Mg, K	ΣM+	H	CEC	I	II
Granville	4.8	12.1	1.3	13.4	5.7	19.1	9.7	6.8
Mayodan	4.9	8.9	1.8	10.7	4.3	16.0	15.4	11.2
White Store	4.6	17.9	4.5	22.4	7.3	29.7	20.0	15.1
Iredell	6.3	0.4	15.8	16.2	7.5	23.7	97.5	68.3
Mecklenburg	5.8	0.8	9.0	9.8	8.3	18.1	92.0	49.7
Davidson	5.9	0.2	4.5	4.7	7.6	12.3	95.8	36.6
Cecil	5.6	0.8	1.7	2.5	5.0	7.5	68.0	22.6
Georgeville	5.3	2.0	1.5	3.5	6.7	10.2	42.9	14.7
Appling	5.2	1.6	1.1	2.7	4.2	6.9	40.7	16.0
H-Peat	4.1	5.0[a]	0	–	89.0	94.0	–	–
Ca-H-Peat	5.4	1.2[a]	46.3	–	47.7	94.0	–	–

* Coleman, et al., *Advan. Agron.*, **10**:475 (1958). Reprinted with permission of Academic Press, Inc., New York.

functional groups on the organic matter and that not bound by permanent charge on the aluminosilicate clays. The magnitude of this pH-dependent charge is quite high, especially so for the two peat samples and several of the mineral soils.

The column labeled CEC contains the sum of the metal cations and the exchangeable H^+. These workers point out that the figures in the CEC column would be those obtained by a CEC determination, using $BaCl_2$-TEA. Ammonium acetate solutions are commonly employed in CEC determinations, but they will give a value somewhere between those for the effective CEC and the CEC developed at pH 8.2. The exchange capacities at pH 9 or 10 would have been correspondingly larger because more pH-dependent charge develops in the more alkaline reaction.

The two columns labeled I and II show the percentage of base saturation, calculated on the basis of the effective CEC (I) and the CEC (II). For the Cecil soil this figure is 68 per cent on the basis of effective CEC but only 22.6 per cent on the basis of CEC. It was generally believed that a pH of about 6 corresponded to the saturation of the effective CEC with the basic cations. A pH value below 5.5 suggests the presence of appreciable amounts of aluminum, whereas pH values below 5 are indicative of a high degree of aluminum saturation.

Subsequent work has shown that the CEC of soils is a continuous function of pH, with the CEC values increasing as the pH increases. Coleman and his associates later modified their views about "permanent charge" and "pH-dependent charge," stating that they had been under the misapprehension that there was a permanent charge on the mineral soil colloid that was analogous with lattice charge and that was invariant below pH 6.0. While the term "effective CEC" is a useful one, its original association with the "permanent charge" did not take into account the interactions between layer silicates and sesquioxide coatings, which also contribute exchangeable Al and hence to the "effective CEC" when present in soil clays.

Even though the original interpretation of the data presented in Table 11-1 may have been in error, they do nonetheless illustrate quite well the dependence of the CEC and the percentage base saturation upon the nature of the soil clay, its pH, and the pH of the extracting solution and for that reason is included in this discussion.

HYDROUS OXIDES. Hydrous oxides, which are principally hydrated oxides of aluminum and iron, may occur in amorphous or crystalline colloidal form as coatings on other mineral particles or as interlayers between crystal lattice structures. When the soil pH is lowered, these oxides may be brought into solution and by stepwise hydrolysis will release H^+ ions.

SOLUBLE SALTS. The presence of salts—acid, neutral, or basic—in the soil solution is accounted for by mineral weathering, organic-matter decomposition, or their addition as fertilizer compounds. The cations of these salts will displace adsorbed aluminum and cause an increase in the acidity of the soil solution, which is easily measured in soil pH determinations. Divalent cations usually have a greater effect on lowering soil pH than monovalent metal cations.

The addition of fertilizer in a band will result in a high soluble salt concentration in and immediately surrounding this band. This in turn will bring Al and Fe into solution, resulting, through hydrolysis, in a lowering of the soil pH. With high rates of band-applied fertilizer, this effect on soil pH in the band can be appreciable.

CARBON DIOXIDE. In soils near neutrality or those containing appreciable quantities of carbonates or bicarbonates the pH is influenced to a great extent by the partial pressure of the carbon dioxide in the soil atmosphere. The pH of a soil containing free calcium carbonate and in equilibrium with carbon dioxide at the pressure of normal aboveground air is 8.5. If the carbon dioxide pressure of the soil atmosphere in such a system increases to 0.02 atmo., the pH will drop to about 7.5.

THE SOIL AS A BUFFER

The foregoing discussion suggests that soil acidity is influenced by numerous factors, not the least of which is the presence of aluminum ions on the exchange complex. It was also pointed out that the soil behaves like a buffered weak acid and that it will resist sharp changes in pH accordingly. This buffering mechanism is explained in the following somewhat oversimplified discussion.

For the purpose of this discussion, it can be assumed that the clay micelle is a large acid radical which behaves as a weak acid when saturated predominantly with Al^{+3} ions. The adsorbed Al ions will maintain an equilibrium with Al ions in the soil solution, the hydrolysis of which gives rise to H^+ ions in the solution as indicated in the following equation:

$$Al^{+3} + H_2O \longrightarrow AlOH^{2-} + H^+$$

If these H^+ ions are then neutralized by the addition of small amounts of a base and the Al ions in solution are precipitated as $Al(OH)_3$, the equilibrium of the system, hence the pH, will tend to be maintained by the movement of adsorbed Al ions to the soil solution. These Al ions in turn hydrolyze, producing H^+ ions, and the pH tends to remain as it was before the addition of the base.

As more and more base is added to the system, however, the above-given reaction continues with more and more of the adsorbed Al being neutralized and replaced on the soil colloid with the cation of the added base. As a result, there is a gradual increase in the pH of the system rather than the abrupt change in pH that is characteristic of the neutralization of unbuffered systems or of strong acids such as HCl.

The reverse of the reaction described above also obtains. As acid is continually added to a neutral soil, OH^- ions in the soil solution are neutralized. Gradually, as these OH^- ions are consumed by the added H^+ ions, the $Al(OH)_3$ dissolves, enters the soil solution, and gradually replaces the basic cations held on the soil clay. As this change takes place, there is a continual but slow decrease in soil pH as the Al^{+3} replaces the adsorbed basic cations.

The total amount of clay and organic matter in a soil and the nature of the inorganic clay will determine the extent to which soils are buffered. Soils containing large amounts of mineral clay and organic matter are said to be highly buffered and require larger amounts of added lime to increase the pH by any given number of units than do soils with a lower buffer capacity. Sandy soils with small amounts of clay and organic matter are poorly buffered and require only small amounts of lime to effect a given change in pH. As a general rule, soils containing large amounts of the $1:1$ type clays (Ultisols and Oxisols) are generally less strongly buffered than soils in which the predominant clay minerals are of the $2:1$ type (Alfisols and Mollisols).

DETERMINATION OF ACTIVE AND POTENTIAL ACIDITY IN SOILS

Active Acidity. Active soil acidity can be determined in several ways. Indicator dyes are frequently employed, particularly in rapid soil test methods. Unless they are handled by a skilled operator, however, considerable error can result. Currently, the most accurate and probably the most widely used method requires a pH meter, or glass electrode potentiometer, as it is sometimes called.

Soil pH is measured by placing a suspension of soil and distilled water in contact with the glass electrode of a pH meter and reading the result on the dial. In soil-water suspensions a phenomenon known as the suspension effect is sometimes observed. If the soil-water suspension is stirred and a reading is taken, the needle will drift to a higher pH value as the suspension settles. As soil colloids behave as weak acids, in which the unionized phase is made up of solids containing surface-adsorbed acidity, the presence of a solid phase may be expected to give a lower pH value when in intimate contact with the electrode. This is exactly what is shown, for the pH of the suspension is always lower than the pH of the supernatant liquid.

Not all workers agree that the suspension effect is produced by the mechanism just suggested. Some feel that it is caused by a junction potential resulting from the salt bridge used with the electrodes attached to the pH meter. The successful measurement of pH by a glass electrode potentiometer requires that the transference of K^+ and Cl^- ions from the salt bridge take place at the same rate. If the salt bridge were in contact with the clay, there would be a more rapid diffusion of K^+ than Cl^-, for the K^+ would be attracted to the negatively charged clay. This would give rise to a potential difference above that resulting from the H^+-ion activity. The thicker the suspension around the electrode and the salt bridge, the greater the junction potential. To get around this difficulty the glass electrode may be placed so that it will come in contact with the soil suspension and the salt bridge will remain in the supernatant liquid.

The presence of soluble salts influences soil pH, and to obtain a true estimate of the acidity of the soil these salts must be removed before a pH determination. This is usually done by leaching with distilled water and then measuring the pH of the salt-free soil, a time-consuming operation when routine determinations on many samples are required.

Another approach used in correcting the acidifying effects of salts is to add a salt solution to the soil instead of water. At first this may seem to be an absurd approach—eliminating the effect of salt by adding a salt solution! The trouble with salt, however, is that it can never be determined just how much is present in the soil and whether the pH measured reflects the true acidity of the soil. Therefore a 0.01 M KCl or $CaCl_2$ solution is added to the soil and the pH is measured. Such pH values, of course, are lower than those made with water, but the salt content of the soil solution is negligible contrasted with that of the added salt solution. Therefore the effect of the former will be of negligible effect in comparison to the effect of the latter on the pH of suspension. This means that differences in soil pH caused by differences in the salt concentration of the soil solution will have no effect on the pH measured in the added soil-salt solution suspension, which is then a more precise estimate of the acidity status of the soil than that measured in a soil-water suspension. Of course, the soil could be leached with distilled water, but it is simpler to measure the pH of the soil in the added salt solution.

Workers in Great Britain, who used this method, have calculated a lime potential for soils expressed by the formula

$$pH - \frac{1}{2}p(Ca + Mg)$$

It has been shown that this value for a particular soil is fairly constant over a wide range of soil-salt levels.

Potential Acidity. Soil pH is an excellent single indicator of general soil conditions. However, as a measure of active acidity in a medium that behaves as a weak acid, it gives no indication of the amount of lime to be applied. Potential acidity must then be considered. Some method of relating a change in soil pH to the addition of a known amount of acid or base is necessary and this is termed a lime-requirement determination.

The lime requirement of a soil is related not only to the soil pH but also to its buffer or cation exchange capacity. As indicated previously, some soils are more highly buffered than others, and a lime requirement determined for one soil will in all probability not be the same as that determined for another. The buffer or exchange capacity is related to the amount of clay and organic matter present: the larger the amount, the greater the buffer capacity. Hence soils classed as clays, peats, and mucks have higher buffer capacities and, if acid, will have a high lime requirement. Coarse-textured soils with little or no organic matter will have a low buffer capacity and, even if acid, will have a low lime requirement.

The implications of the statements made in the preceding paragraph are of tremendous practical importance. The indiscriminate use of lime on coarse-textured soils could lead to excessively alkaline conditions and to serious consequences, such as deficiencies of iron, manganese, and other microelements. Conversely, the application to an acid clay soil of the amount of lime detrimental to crops growing on sandy soils may well be insufficient to raise the pH the desired amount. Adequate liming recommendations are based on a knowledge of the buffer capacity of a soil. Since the pH value alone is no criterion of the amount of lime that should be added to a soil, the indiscriminate use of pH test kits by persons unfamiliar with the rudiments of soil chemistry is undesirable.

Determining the Lime Requirement of Soils. Many state experiment stations and soil-testing laboratories have determined the lime requirement of the major soil series and types in the areas they serve. Once this has been done, a knowledge of the pH and the soil type will make possible an immediate liming recommendation. If the lime requirement has not been determined and a recommendation must be made, the texture and organic matter content must also be taken into consideration.

The lime requirement of a soil can be determined by several different methods. One, which is not used much because it is time consuming, is discussed here to illustrate the mechanics of the determination. It is a simple, straightforward technique which illustrates quite clearly the meaning of buffer capacity and lime requirement. It can also be used to determine the amount of acid or sulfur needed to lower soil pH.

Such data are obtained by adding to a series of small beakers or flasks a known quantity of soil. To each beaker is then added a given amount of acid or base. The base usually employed in calcium hydroxide and the acid is hydrochloric. Water is added to equalize the volume of liquid in all beakers, and the samples are allowed to equilibrate. pH determinations are made, and the values obtained are plotted against the milliequivalents of acid or base added. A buffer curve is then constructed. From

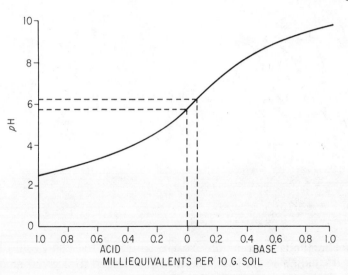

Figure 11–2. The lime requirement curve for a Maumee sandy loam.

these data it is a simple matter to determine the amount of lime to be added to an acre of land. The data obtained from such a determination are illustrated in Figure 11-2.

In an example of how these data may be used the pH of the soil to which no acid or base was added is 5.75. Suppose that we wish to add sufficient lime to raise the pH to 6.25. The milliequivalents of added base or acid corresponding to the untreated soil, of course, are zero. Commencing at the pH value of 6.25 on the Y-axis, draw a line parallel to the X-axis until the line crosses the lime requirement curve. At this point of intersection drop a line to the X-axis parallel to the Y-axis. This perpendicular strikes the X-axis at 0.067 meq. of base. Since the weight of soil added to each beaker was 10.0 g., 0.067 meq. of a base must be added to each 10 g. of the soil in question to change the pH from 5.75 to 6.25. If it were decided to use finely divided calcite, the amount required would be 0.067×0.05, or 0.00335 g. Calculated on an acre basis, the amount of calcite required would be 670 lb., assuming that the weight of an acre furrow slice of this soil is 2 million lb.

An even more time-consuming method involves adding known amounts of $CaCO_3$ to measured quantities of soil and incubating the mixtures for several months to allow the reaction to go to completion. pH values on the samples are then measured and a buffer curve constructed as in the previous case, relating soil pH at the end of the incubation period to amount of $CaCO_3$ added to the sample. An example of the results of one such study conducted in Ohio is shown in Figure 11-3.

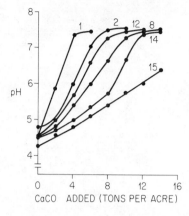

Figure 11–3. Titration curves for representative soils from Ohio after incubation with $CaCO_3$ for 17 months. [Shoemaker et al., *SSSA Proc.*, **25**:274 (1961).]

Other methods in common use are based on the change in pH of a buffered solution when a sample of soil is added to a given amount of solution. When a sample of acid soil is added to a measured quantity of the solution, the pH is depressed in proportion to the original soil pH and its buffer capacity. By calibrating pH changes of the buffered solution which accompany the addition of known amounts of acid, the amount of lime required to bring the soil to some prescribed pH can be calculated.

Several buffer methods have been used in various parts of the country over the years. Schofield was one of the first to employ the buffer solution, using paranitrophenol as the buffer in a displacement and titration procedure. Mehlich in North Carolina developed a method using triethanolamine, while Brown suggested the change in the pH of an ammonium acetate buffer as a suitable technique. Woodruff of Missouri used the change in pH of a Ca-acetate-paranitrophenol buffer to estimate lime requirement. Shoemaker of Ohio and his associates developed a buffer solution technique employing a change in pH of the buffer solution as the criterion of the lime requirement. Theirs was a dilute mixture of triethanolamine, paranitrophenol, potassium chromate, and calcium acetate. Several other workers have developed modifications of these procedures. All of the buffer procedures, especially those in which a pH change is the criterion of the lime requirement, are rapid, easily made, and are suitable for the processing of large numbers of soil samples as would be encountered in the operation of a soil testing service.

Several of the buffer solution methods were recently compared by McLean and his associates at Ohio. Not only were large numbers of soils from Ohio evaluated in these studies, but eight different soils from the North Central Region and thirty-eight soils from twelve states in the United States were included. These latter included Ultisols and Oxisols as well as Alfisols and Mollisols.

It was concluded from this study that the buffer solution technique of Shoemaker, et al., gave the most uniformly acceptable results, being relatively rapid, accurate, and especially well adapted for soils requiring > 4,000 pounds of lime per acre, having pH values < 5.8, containing < 10 per cent organic matter, and having appreciable quantities of exchangeable Al.

SOIL pH FOR CROP PRODUCTION

For many years the pH of soils for best crop production has been considered to be between 6.5 and 7.0. Liming recommendations made by the various agricultural agencies throughout the USA have generally been based on the amounts of lime needed to bring soil pH values within this range.

Work carried out over the past several years, however, has cast doubt on the validity of liming all soils in the USA to the pH 6.5 level for best growth of agricultural crops. Coleman, Thomas, Kamprath, and other workers in North Carolina have proposed that liming of the Oxisols and Ultisols of the warm, humid Southern USA to pH values greater than 6.0 or 6.2 may be not only unnecessary but harmful. It is the opinion of this group, and other scientists as well, that lime sufficient to neutralize the exchangeable Al is all that need be added to the red and yellow soils of the humid, warm areas of the USA. Work by other scientists in tropical countries has generally confirmed this. Liming of most soils that are high in the hydrous oxides of Al and Fe sufficient only to neutralize most of the exchangeable Al will bring the pH to about 5.6 or 5.7 and the exchangeable Al to less than 10 per cent of the effective CEC.

The North Carolina workers have presented data which show that the liming of Ultisols and Oxisols to pH values approaching 7.0 does result in deleterious effects. Kamprath showed that liming red and yellow soils to pH 7.0:

1. reduced water percolation;
2. reduced the growth of legumes and nonlegumes;
3. reduced plant uptake of phosphorus; and
4. reduced the micronutrient uptake by plants.

Workers in the North Central States, notably McLean and his associates at Ohio, believe that for the soils of the Alfisol and Mollisol types liming to pH 6.5 to 6.8 will give the most desirable results. McLean points out that *if* all nutrients are supplied in adequate amounts and *if* no elements are present in toxic concentrations, adequate plant growth can be obtained at pH values considerably lower than 6.5. Because these conditions do not prevail in most soils, liming to pH values of 6.5 to 6.8 is preferred on the Alfisols and Mollisols in the

Midwestern USA. These workers state that in most of the Midwestern USA there appears to be no generally adverse effect from liming to a near-neutral pH value where crops depend for most of their nutrients on the various release mechanisms of the soil. They further state that:

1. At higher base saturation, Ca and Mg are adsorbed at pH-dependent sites favoring hydrolysis reactions which increase the plant availability of these two nutrients. Among other things this will increase the movement of these cations into the lower horizons of the soil profile which may otherwise be devoid of basic cations for utilization by deep-rooted crops.
2. At pH values higher than those required to inactivate Al, Mn, and Fe, the availability of soil supplies of Ca, Mg, P, K, S, B, Cu, and Zn is greater. As most Midwestern soils do not require the addition of many of these nutrients, the maintenance of a soil pH value favoring their availability seems justified.
3. At a higher pH value, biological activity is more intense, more nitrogen is fixed by soil microorganisms, and component elements are released by the more rapid decomposition of plant residues.

The chemical and physical properties of Oxisols and Ultisols are regulated to a large degree by hydroxy-Al and hydroxy-Fe coatings on the clay fraction while Mollisols have clay fractions that are only very infrequently coated with hydroxy-Fe and -Al. It is therefore quite possible that the pH best for crop production in the Southern USA is lower than that for best plant growth in the Midwestern and other areas of the USA. Certainly the arguments of the proponents of each side of the question have merit. It must be kept in mind that these arguments hold only for general field crops such as corn, small grains, soybeans, alfalfa, clovers, and similar crops. Crops such as tobacco, potatoes, and others which may be affected by diseases or micronutrient deficiencies associated with high or low pH values would be excepted and their individual pH requirements should be met.

LIMING MATERIALS

The materials commonly used for the liming of soils are the oxides, hydroxides, carbonates, and silicates of calcium or calcium and magnesium. The presence of these elements alone does not qualify a material as a liming compound. In addition to these cations, the accompanying anion *must* be one that will reduce the activity of the hydrogen and hence aluminum in the soil solution.

The mechanisms controlling the reaction of liming materials with acid soils are complex. The rates of neutralization and the final reaction products are not known with certainty, although such factors as the amount of soil acidity and particle size and reactivity of the added limestone have been studied extensively. It is known, however, that the reaction of all of the liming materials listed begins with the neutralization of H^+ ions in the soil solution by either OH^- or $SiO_3^=$ ions furnished by the liming material. The basic reaction of a liming material when added to the soil can be illustrated with the case for calcium carbonate. In water, $CaCO_3$ behaves as follows:

$$CaCO_3 + H_2O \longrightarrow Ca^{+2} + HCO_3^- + OH^-$$
$$H^+(\text{soil soln}) + OH^- \longrightarrow H_2O$$

The rate of the above-given reaction, and thus of the solution of $CaCO_3$, is directly related to the rate at which the OH^- ions are removed from solution. As long as sufficient H^+ ions are in the soil solution, Ca^{+2} and HCO_3^- ions will continue to go into solution. When the H^+ ion concentration is lowered, however, solution of the Ca^{+2} and HCO_3^- ions is reduced.

In acid soils, the concentration of the H^+ ions in solution is related to the hydrolysis of Al^{+3} or hydroxy-Al or hydroxy-Fe^{+3} ions. Their hydrolysis in turn is influenced by the amount of clay and organic matter in the system. The continued removal of H^+ from the soil solution will ultimately result in the precipitation of the Al and Fe ions and their replacement on the adsorption sites with Ca and/or Mg and other basic cations. When the exchangeable Al and the hydroxy-Al and hydroxy-Fe^{+3} have been precipitated as $Al(OH)_3$ and $Fe(OH)_3$, the soil acidity which remains arises from the other sources mentioned earlier in this Chapter.

As the neutralization of soil solution H^+ by a material is necessary for it to be classed as a liming agent, gypsum ($CaSO_4 \cdot 2H_2O$) and other neutral salts cannot qualify as such. In fact, as has been mentioned previously on several occasions, the addition of neutral salts will actually lower soil pH. Their addition, especially as in band placement, results in replacement of adsorbed Al^{+3} in a localized soil zone, sometimes with a significant lowering of the pH in this region.

Calcium Oxide. This is the only material to which the term *lime* may be correctly applied. Calcium oxide (CaO), known also as unslaked lime, burned lime, or quicklime, is a white powder, quite disagreeable to handle. It is manufactured by roasting calcitic limestone in an oven or furnace. The carbon dioxide is driven off, leaving calcium oxide. The purity of the burned lime depends on the purity of the raw material. This product is shipped in paper bags because of its powdery nature and its caustic properties. When added to the soil, it reacts almost immediately.

When unusually rapid results are required, either this material or calcium hydroxide should be selected. Complete mixing of calcium oxide with the soil may be difficult, however, for immediately after application absorbed water causes the material to form flakes or granules. These granules may harden because of the formation on their surfaces of calcium carbonate, and in this condition they may remain in the soil for long periods of time. Only by very thorough mixing with the soil at application time can this caking be prevented.

On a pound-for-pound basis, calcium oxide is the most effective of all the liming materials commonly employed, for the pure material has a neutralizing value or calcium carbonate equivalent (CCE) of 179 per cent, compared with pure calcium carbonate. The full significance of this last statement will become apparent when the neutralizing value of the various materials is discussed in a subsequent section of this chapter.

Calcium Hydroxide. Calcium hydroxide [$Ca(OH)_2$] is frequently referred to as slaked lime, hydrated lime, or builders' lime. Like calcium oxide, it is white, powdery substance, difficult and unpleasant to handle. Neutralization is rapidly effected, as it is with calcium oxide.

Slaked lime is prepared by hydrating calcium oxide. Much heat is generated, and on completion of the reaction the material is dried and packaged in paper bags. The purity of the commercial product varies, but the chemically pure compound has a neutralizing value of 136, making it pound-for-pound the second most efficient of the commonly used liming materials.

Calcium and Mixed Calcium-Magnesium Carbonates. The carbonates of calcium and magnesium occur widely in nature and in a number of different forms.

Crystalline calcium carbonate ($CaCO_3$) is termed calcite or calcitic limestone. Crystalline calcium-magnesium carbonate [$CaMg(CO_3)_2$] is known as dolomite when the calcium carbonate and magnesium carbonate occur in equimolecular proportions. In other proportions they are said to be dolomitic limestones. Metamorphosis of these high-grade limestones produces marble. Deposits of high-grade limestone are widespread in the United States.

Limestone is most often mined by open-pit methods. First the overburden of soil and undesirable rock is removed, after which holes are drilled in the exposed limestone and filled with an explosive, which is then detonated. The blast breaks out the rock, generally in sizes that can be accommodated by the quarrying and crushing equipment. The loosened material is crushed to sizes of 1 in. or less, and the limestone is ready for grinding and pulverizing. The pulverized material is classified by passing it through one or more screens of some specified size. This is required particularly of agricultural-grade limestone, specifications for which are quite rigid. Having been thus processed, it is generally stored in the open in piles and may be shipped either in bulk or in bags.

The quality of crystalline limestones depends on the degree of impurities they contain, such as clay. The neutralizing values usually range from 65 to 70 per cent to a little more than 100 per cent. The neutralizing value of chemically pure calcium carbonate has been established arbitrarily at 100 per cent, and, theoretically, chemically pure dolomite may have a neutralizing value of nearly 109 per cent. As a general rule, however, the neutralizing value or CCE of most agricultural limestones is between 90 and 98 per cent because of impurities.

Marl. Marls are soft, unconsolidated deposits of calcium carbonate. They are frequently mixed with earth and usually quite moist. Marl occurs in many states in the eastern part of the country and is quite easily mined. The deposits are generally thin, though the layers have been known to range up to 30 ft. in thickness. Marl is recovered by dragline or power shovel after the overburden has been removed. The fresh material is stockpiled and allowed to dry before being applied to the land.

Marls are almost always low in magnesium. Their value as liming materials depends on the amount of clay they contain. Their neutralizing value usually lies between 70 and 90 per cent, and their reaction with the soil is the same as that of calcite.

Slags. Several types of material are classed as slags, three of which are important agriculturally.

Blast-Furnace Slag. Blast-furnace slag is a by-product of the manufacture of pig iron. In the reduction of iron the calcium carbonate in the charge loses its carbon dioxide, and calcium oxide then combines with the molten silica to form a slag that is tapped off and either air-cooled or quenched with water. The cooled product is ground, screened, and shipped in open cars or trucks.

As a liming material, slag behaves essentially as calcium silicate. Metasilicic acid, which is formed when the slag is added to acid soils, is weakly dissociated, and the pH of the soil is raised. The neutralizing value of blast-furnace slags ranges from about 75 to 90 per cent. These slags usually contain appreciable amounts of magnesium. Results of field tests indicate that, when applied on the basis of equivalent amounts of calcium and magnesium, they are just as effective in producing crops as ground limestones.

Basic Slag. A second type of slag is known as basic or Thomas slag and was discussed in Chapter 6. This slag is a by-product of the basic open-hearth method of making steel from pig iron, which, in turn, is produced from high-phosphorous iron ores. The impurities in the iron, including silica and phosphorus, are fluxed with lime and slagged off. The slag is cooled, finely ground, and usually marketed in 80- or 100-lb. bags. In addition to its phosphorous content, basic slag has a neutralizing value of about 60 to 70 per cent. It is generally applied for its phosphorus content rather than for its value as a liming material, but because of its neutralizing value it is a good material to use on acid soils. Its selection in relation to other sources of phosphorous is determined by economic factors.

ELECTRIC-FURNACE SLAG. A third type of slag results from the electric-furnace reduction of phosphate rock in the preparation of elemental phosphorus. The slag is formed when the silica and calcium oxide fuse, and the product is thought to be largely calcium silicate. It is drawn off and quenched with water. The slag is a waste product which is marketed at a low price and usually only within a limited radius of the point of production. It contains 0.9 to 2.3 per cent P_2O_5 and is not ground. The neutralizing value ranges from 65 to 80 percent. Its reaction with the soil is similar to that indicated for blast-furnace slag.

Neutralizing Value or Calcium Carbonate Equivalent (CCE) of Liming Materials. Liming materials differ markedly in their ability to neutralize acids. The value of limestone for this purpose depends on the quantity of acid that a unit weight of the material will neutralize. This property, in turn, is related to the molecular composition of the liming material and its purity; in other words, its freedom from inert contaminants such as clay. Pure calcium carbonate is the standard against which other liming materials are measured, and its neutralizing value is considered to be 100 per cent. Calcium carbonate equivalent (CCE) is defined as the acid-neutralizing capacity of an agricultural liming material expressed as a weight percentage of calcium carbonate.

The molecular constitution is the determining factor in the neutralizing value of chemically pure liming materials. Consider the reactions illustrated by the following equations:

$$CaCO_3 + 2HCl \longrightarrow CaCl_2 + H_2O + CO_2 \qquad (1)$$
$$MgCO_3 + 2HCl \longrightarrow MgCl_2 + H_2O + CO_2 \qquad (2)$$

In each of these equations the molecular proportions are the same; that is, one molecule of each of these carbonates will neutralize two molecules of acid. However, the molecular weight of calcium carbonate is 100, whereas that of magnesium carbonate ($MgCO_3$) is only 84. In other words, 84 g. of magnesium carbonate will neutralize the same amount of acid as 100 g. of calcium carbonate. How much more effective then is 100 g. of magnesium carbonate than the same quantity of calcium carbonate in neutralizing an acid? This is demonstrated quite easily by the following simple proportion:

$$\frac{84}{100} = \frac{100}{x}$$
$$x = 119$$

TABLE 11–2. The Neutralizing Value (CCE) of the Pure Forms of Some Commonly Used Liming Materials

Material	Neutralizing value (%)
CaO	179
Ca(OH)$_2$	136
CaMg(CO$_3$)$_2$	109
CaCO$_3$	100
CaSiO$_3$	86

Therefore magnesium carbonate on a weight basis will neutralize 1.19 times as much acid as the same weight of calcium carbonate; hence its neutralizing value or CCE in relation to CaCO$_3$ = 100 is 1.19/1 × 100, or 119 per cent. The same procedure is used to calculate the neutralizing value of other liming materials. The neutralizing values for several compounds are shown in Table 11-2. Their use makes possible the simplest and most straightforward comparison of one liming material with another in regard to neutralizing properties. These values are one guarantee which may be made for commercial liming agents. There are other methods of expressing the value of limestones, and these are discussed next.

Calcium Magnesium. The composition of liming materials is sometimes expressed in terms of the elemental content of calcium and magnesium. Chemically pure calcite, for example, contains 40 per cent calcium and chemically pure magnesium carbonate contains (24/84)100, or 28.6 per cent magnesium. Obviously, to convert the percentage of calcium to its calcium carbonate equivalent it is necessary to multiply by a factor of 2.5; and to convert the percentage of magnesium to the percentage of magnesium carbonate, it is necessary to multiply by a factor of 84/24, or 3.5. If a limestone carries the guarantee in terms of the element, it is a simple matter, by applying the appropriate factor, to calculate the percentage in terms of the carbonate to which the elemental percentage is equivalent.

Calcium and Magnesium Oxide Content. The quality of a limestone may also be expressed in terms of its calcium or magnesium oxide equivalent. As an example, pure calcite is 100 per cent calcium carbonate and contains 40 per cent calcium. Suppose that we wished to express these quantities, not in terms of either of the constituents, but rather in terms of the oxide (CaO). Calcium oxide has a molecular weight of 56, which means of course that 16 g. of oxygen are combined with 40 g. of calcium. Therefore, if the calcium present in calcium carbonate were expressed as the oxide, it would contain (56/100)100, or 56 per cent calcium oxide equivalent. Thus to convert the percentage of calcium to percentage of calcium oxide, we need only to multiply the calcium by 56/40, or 1.4; and to convert the percentage of calcium carbonate to the percentage of calcium oxide, we have but to multiply the percentage of calcium carbonate by 56/100, or 0.56. Similar figures may be derived for the magnesium-containing limestones.

Total Carbonates. Another expression of the quality of limestones is that of total carbonates. This is a summation of the percentages of the carbonates contained in a given liming material. For example, assume that a limestone contains 78 per cent calcium and 12 per cent magnesium carbonate. The total carbonate content would be 90 per cent.

Conversion Factors for the Various Methods of Expression. The conversion figures for a few transformations, such as the percentage of calcium to that of calcium oxide, have been given. It is on occasion desirable to convert the percentage of magnesium oxide or magnesium carbonate to the calcium carbonate equivalent. This is illustrated in the following example. Assume that a limestone contains this guarantee:

35 per cent CaO

15 per cent MgO

Assume further that we wish to express this analysis in terms of the calcium carbonate equivalent. The conversion may be obtained if it is remembered that 56 g. of calcium oxide are equivalent to 100 g. of calcium carbonate. The percentage of calcium oxide is multiplied by 100/56, and the calcium carbonate equivalent in this sample is 35 per cent × 1.785, or 62.5 per cent.

How is the conversion of magnesium oxide to calcium carbonate to be handled? A glance at their molecular composition shows that 1 mole of each will neutralize the same quantity of acid. A mole, or 1 g. mol. w. of

TABLE 11–3. Limestone Conversion Factors

Per cent		Per cent		Factor
Ca	to	CaO	multiply by	1.40
Ca	to	Ca(OH)$_2$	multiply by	1.85
Ca	to	CaCO$_3$	multiply by	2.50
Mg	to	MgO	multiply by	1.67
Mg	to	Mg(OH)$_2$	multiply by	2.42
Mg	to	MgCO$_3$	multiply by	3.50
Mg	to	Ca	multiply by	1.67
Mg	to	CaCO$_3$	multiply by	4.17
MgO	to	CaCO$_3$	multiply by	2.50
MgCO$_3$	to	CaCO$_3$	multiply by	1.19

the magnesium oxide is 40, whereas that of calcium carbonate is 100. In other words 40 g. of magnesium oxide will neutralize the same amount of acid as 100 g. of calcium carbonate. Therefore all that remains is to multiply the 15 per cent magnesium oxide in the sample by the factor 100/40, or 2.5. The result, 37.5 per cent, is added to the 62.5 per cent, and the total neutralizing value, or calcium carbonate equivalent, for the limestone in question is 100 per cent.

Conversion factors which make possible an expression of the value of a limestone may be determined in any way desired, provided that the content of one of the constituents is given. A few of these factors are listed in Table 11-3. The student should be able to derive them as well as others not shown.

FINENESS OF LIMESTONE

Molecular constitution and freedom from inert impurities are not the only properties that limit the effectiveness of agricultural limestones. The degree of fineness is equally important, because the speed with which the various materials will react is dependent on the surface in contact with the soil. Materials such as calcium oxide and calcium hydroxide are by nature powdery, so that no problem of fineness is involved, but the crystalline limestones are an entirely different matter.

When a given quantity of crushed limestone is thoroughly incorporated with the soil, its reaction depends upon the size of the individual particles. If they are coarse, the reaction will be slight, but if they are fine, the reaction will be extensive. This is strikingly illustrated by the relative slopes of the curves in Figure 11-4. These curves are self-explanatory. It is obvious that the efficiency of the limestone does not increase appreciably when ground to finer than 60–80 mesh. It should be emphasized that these data were obtained with sized samples, that is, only those particles falling between the size limits indicated, which eliminated the fines from the coarser materials.

The pH-time relationship that resulted from the addition to soils of limestones of different particle sizes is shown graphically in Figure 11-5. The curves are of particular interest. The coarse materials, 4–8 and 8–20 mesh, were for all intents and purposes without effect on the soil pH over the eighteen-month period of the experiment. The 30–40 mesh material increased the pH by less than one unit at the end of twelve months. The remaining materials became more effective as the particle size decreased. The response to the 100-mesh material was quite rapid and of the same order as that expected from the application of calcium oxide or calcium hydroxide. Its maximum effectiveness was greater and was reached more quickly than were the remaining materials.

The data shown in Figures 11-4 and 11-5 suggest that the fine agricultural liming materials should give rapid results when added to the soil.

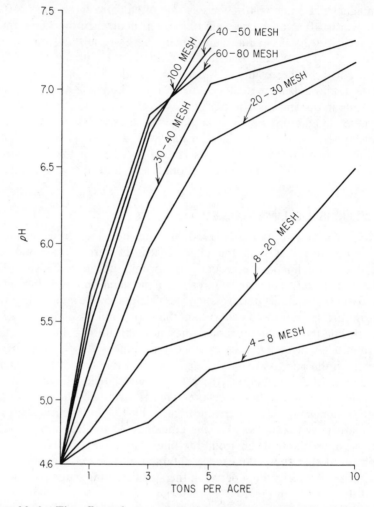

Figure 11–4. The effect of rate and particle size of dolomitic limestone on the *p*H of a Canfield silt loam eighteen months after application. [Meyer et al., *Soil Sci.*, **73**:37 (1952). Reprinted with permission of The Williams & Wilkins Co., Baltimore.]

The cost of limestone increases with the fineness of grinding. What is needed is a material that requires a minimum of grinding, yet contains enough fine material to effect a *p*H change rapidly. As a result, agricultural limestones contain both coarse and fine materials. Many states require that 75 to 100 per cent of the limestone pass an 8- to 10-mesh screen and that 20 to 80 per cent pass anywhere from an 8- to 100-mesh screen. In this way there is fairly good distribution of both the coarse and fine particles.

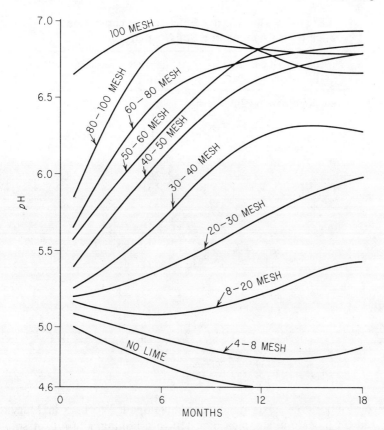

Figure 11–5. The effect of dolomitic limestone of different particle sizes on the *p*H of a Canfield silt loam at various times after application. [Meyer et al., *Soil Sci.*, **73**:37 (1952). Reprinted with permission of The Williams & Wilkins Co., Baltimore.]

RAPID METHODS FOR EVALUATING AGRICULTURAL LIMESTONES

The best criterion for evaluating agricultural limestones is their performance in field soils. Their reaction with soils in small containers in the laboratory is also an effective method of evaluation. Both methods, however, require too much time, for limestone takes two to six months to react with the soil. Its reactivity is determined not only by its purity and particle-size distribution but also by its hardness and magnesium content. The rapid evaluation of large numbers of limestone samples is desirable from the standpoint of the commercial utilization of these materials, for some knowledge of their reactivity is needed for the development of sound promotional and sales programs and for the protection of the purchaser.

TABLE 11–4. The Relative Efficiency of Limestone in Producing pH Changes in a Soil Suspension[*]

Commercial limestone samples	Relative efficiency	Change in pH[†] (%)
Pure calcite passing 200 mesh	100	100
1	100	97
2	101	95
3	92	93
4	97	90
5	84	84
6	78	82
7	79	79
8	78	77
9	74	74
10	73	73
11	71	71
12	64	64
13	59	61
14	41	58
15	36	45
16	19	8

[*] El Gibaly et al., *SSSA Proc.*, **19**:301 (1955).
[†] Total change in pH studied was from 3.90 to 7.00.

Several laboratory tests have been developed. Workers in Maryland reported on limestone samples digested in a boiling solution of sodium EDTA. The efficiency of the limestones was then calculated from the results of similar analyses made on pure calcite. These efficiency values were compared with the effectiveness of these same materials in changing the pH of soil samples. Their results are shown in Table 11-4. It is obvious that with most of them the chemical method gave a good estimate of the effectiveness of the liming materials.

Another method, developed by Shaw and Robinson at the University of Tennessee, is based on the dissolution of limestones in ammonium chloride. Comparison of results by this method with results from incubating limestones with soil showed that the ammonium chloride treatment gave a very good estimate of the agricultural value of the limestones tested.

Using the method just described, Shaw and his colleagues determined the reactivity of different size separates of a calcite and a dolomite. Results of this study are shown graphically in Figure 11-6. Two points are brought out by these data. The coarser the size separate, the less reactive the material. This is true for both dolomite and calcite. The sec-

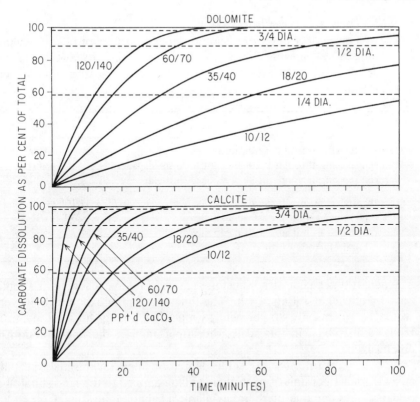

Figure 11–6. Reaction rates of separates of calcite and dolomite of different sizes in boiling 5*N* NH₄Cl solution. [Shaw et al., *Soil Sci.*, **87**:262 (1959). Reprinted with permission of The Williams & Wilkins Co., Baltimore.]

ond point is that dolomite is less reactive than calcite for each of the size separates.

The reactivity of limestones is related in part of the fineness with which they are ground. The greater the surface of the limestone exposed, the faster its reaction with the soil. However, apparent surface, calculated on the basis of the size of solid, smooth particles, may not represent total surface. Limestone particles may be spongy, as found by workers at the United States Department of Agriculture, who used a technique of surface adsorption of an inert gas. They found that the total surface measured by this gas adsorption technique was much greater than simple surface measurements based on particle size determinations. They also found that their method produced results that were closely related to the reactivity of limestone measured by an ammonium chloride technique.

SELECTION OF A LIMESTONE

The selection of a limestone should depend on its neutralizing value, its content of magnesium, its degree of fineness, and its reactivity. An example of evaluation of degree of fineness according to methods used in Ohio follows:

	Efficiency rating
Material passing through a 60-mesh sieve	100
Material passing through a 20- but not a 60-mesh sieve	60
Material passing through an 8- but not a 20-mesh sieve	20
60% through a 60-mesh sieve	60% × 100 = 60
30% passes a 20- but not a 60-mesh sieve	30% × 60 = 18
10% passes an 8- but not a 20-mesh sieve	10% × 20 = 2
Total efficiency rating	80

When 100 per cent of a limestone passes a 60-mesh screen, it should have an efficiency rating of 100. The limestone in the example, however, would be worth only 80 per cent as much, assuming the same neutralizing value. Thus, in this scale, the importance of the finer fraction is emphasized.

As a general rule, for the same degree of fineness, the material that costs the least per unit of neutralizing value applied to the land should be selected. Assume that there are available a calcitic limestone, CCE = 95 per cent, and a dolomitic limestone, CCE = 105 per cent, each with the same fineness or mechanical analysis. Assume also that they both cost 6 dollars a ton applied to the land. Based on the neutralizing value, the first will cost 105/95 × 6, or $6.63 a ton, as compared with the dolomite at 6 dollars a ton. In addition, the dolomite supplies magnesium, which is a nutrient not overly abundant in many humid-region soils.

Whenever possible, some measure of the reactivity of the limestone, based on one of the rapid chemical methods, should be obtained. When not possible, the purchaser should be guided by the fineness of the material, its neutralizing value, its magnesium content, and the cost per ton applied to the land.

USE OF LIME IN AGRICULTURE

The application of lime to soils in many areas of the USA produces striking increases in plant growth. The areas in the country in which the need for lime application is greatest are the humid regions of the East, South, Middle West and Far Western States. Lime, of course, is not needed in those areas where rainfall is low and leaching is minimal, such as parts of the Great Plains States and the arid, irrigated saline-alkali soils of the Southwest, Intermountain, and Far Western States.

When crop responses are obtained from the application of materials carrying the major plant nutrients, nitrogen, phosphorus, and potassium, it is assumed, and usually correctly, that the response was the direct result of overcoming a deficiency of one of these nutrient elements. Responses from the application of lime, however, may not always be attributed to the plant-nutrient value of the calcium or magnesium.

Direct Benefits. Perhaps the greatest single direct benefit of liming acid soils is the reduction in the activity or solubility of Al and Mn. Both of these ions in anything other than very low concentrations are toxic to most plants. The poor growth of crops that is observed on acid soils is due largely to the large concentration of these two ions in the soil solution. When lime is added to acid soils, the activity of the Al and Mn is reduced and they are removed from solution as indicated earlier in this Chapter.

Not only are these ions toxic to plants, the presence of increasing amounts of Al in the soil solution also decreases the uptake of calcium by the plant as is indicated in Figure 11-7.

It has been found that different crops and even different varieties of the same crop will differ widely in their susceptibility to Al toxicity. Foy and his coworkers at the USDA have done quite a bit of work on this subject showing that different varieties of crops such as soybeans, wheat, and barley show wide ranges in their tolerance to high concentrations of Al in the soil solution. An example of this differential effect is shown in Figure 11-8.

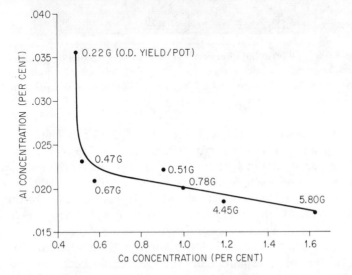

Figure 11-7. Relationship between Al and Ca concentration in cotton tops from nonleached subsoil. [Soileau et al., *SSSA Proc.*, **33**:919 (1969).]

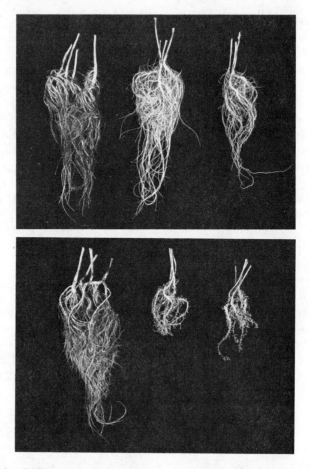

Figure 11–8. Differential effects of Al on root growth of Perry (top) and Chief (bottom) soybean varieties grown in ⅕ Steinberg solution containing 2 ppm. Ca. *Left to right:* 0, 8, 12 ppm. Al added. [Foy, Fleming, and Armiger, *Agron J.,* **61**:505 (1969).]

Indirect Benefits. Although the importance of calcium and magnesium in plant nutrition is not questioned, the scope of benefits derived from the application of lime where needed is much broader than would be expected from a simple response to the addition of a deficient nutrient element. Many of the indirect benefits have been covered in this text in conjunction with other topics. They are summarized here, however, and a few additional advantages are listed.

EFFECT ON PHOSPHOROUS AVAILABILITY. The relationship between the available soil phosphorus and the soil *p*H has been covered. At low *p*H values and on soils high in aluminum and iron, phosphates are rendered less available because of their reaction with these compounds.

The addition of a liming agent to these soils will inactivate the iron and aluminum, thus increasing the level of plant-available phosphorus.

If the soil *p*H is greatly increased by the addition of excessive amounts of lime, phosphate availability will again be decreased because of precipitation as calcium or magnesium phosphates. A liming program should be planned so that the *p*H can be kept between 5.5 and 6.8 to 7.0 if maximum benefit is to be derived from the applied phosphate. As a usual practice, soils are not limed much above 6.5 because of the possibility of decreasing the availability of certain microelements, as discussed in a previous section.

MICROELEMENT AVAILABILITY. The effect of soil *p*H on microelement availability was covered in Chapter 8. With the exception of molybdenum, the availability of the microelements increases with a decrease in *p*H. This can be detrimental because of the toxic nature of many of the elements in anything other than minute concentrations. The solubility of aluminum, iron, and manganese increases with increasing acidity. In addition to toxic effects, their presence may interfere with the absorption of calcium, magnesium, and other basic cations. The addition of adequate lime causes their inactivation, and soil at a *p*H value of 5.6 to 6.0 is usually most satisfactory from the standpoint of minimum toxicity and adequate availability of these elements. The effect of high *p*H on the availability of manganese and iron has been discussed.

Molybdenum deficiency generally decreases with an increase in the *p*H. A deficiency of this element occurs least on those soils that are limed to *p*H values in excess of 7.0, but because of the difficulties mentioned, liming to this value should be discouraged on most crops in humid areas.

NITRIFICATION. Most of the organisms responsible for the conversion of ammonia to nitrates require large amounts of active calcium. As a result, nitrification is enhanced by liming to a *p*H of 5.5 or 6.5. Decomposition of plant residues and breakdown of soil organic matter are also more rapid in this *p*H range than in acid soils.

NITROGEN FIXATION. The process of nitrogen fixation, both symbiotic and nonsymbiotic, is favored by adequate liming. With legumes, the growth of the plant is increased because of the greater amount of nitrogen fixed, and larger quantities of organic matter and nitrogen are thus returned to the soil. With the nonsymbiotic, nitrogen-fixing organisms the greater fixation of atmospheric nitrogen taking place in adequately limed soils makes possible the more rapid conversion to humus of carbonaceous crop residues, such as those of corn and small grains. The increased level of soil nitrogen means a higher content of stable organic matter and a general increase in the fertility status of the land.

SOIL PHYSICAL CONDITION. The structure of fine-textured soils may be improved by liming. This is largely the result of an increase in the organic matter content and to a lesser extent to the flocculation of calcium-saturated colloids. However, the overliming of Oxisols and Ultisols (liming in excess of that needed to neutralize the exchangeable Al) can result in the deterioration of soil structure with the consequent decrease in water percolation through such soils. This was mentioned in an earlier section of this Chapter.

PLACEMENT OF LIME

Particles of limestone cannot move in the soil, and consequently they must be placed where they are needed. In addition, it should be apparent from the indirect effects of lime on plant growth that complete mixing of this material with the plow layer of soil is necessary. This point is emphasized by the results of pot tests. Lime was added to different zones of soil, and the root distribution of the plants grown on these treatments was measured. The results suggest the importance of adding lime to the entire region of soil in which root growth is expected. For deep-rooted crops this would become an almost impossible task with the equipment currently available. Attempts have been made, however, to make deep placement of lime and phosphate. Results of this work have been inconsistent, however. On acid soils a combination of plowing down of part of the lime and disking in a second application is employed. On less acid soils all of the lime may be applied at one time and either plowed down or disked in.

Workers in Virginia found that lime applied to the surface of a silt loam in no-till corn was more effective than lime applied to conventionally tilled corn (each spring cover crop plowed under). In conventional tillage, an occasional lack of available soil moisture probably prevented corn from taking advantage of the increased pH that resulted from liming.

EQUIPMENT

Lime may be spread on the land in any one of several ways. If the grower plans to apply the lime himself, a lime spreader is a useful piece of equipment with which to do the job. Manure spreaders or end-gate seeders may also be adapted to this purpose. In recent years the bulk application of lime by specially constructed trucks has become increasingly popular. Because of the scarcity of farm labor, bulk application by the supplier who hauls lime to the farm is most efficient. The spinner truck spreader, which throws the lime in a semicircle from the rear of the truck, is often used. Uniform spreading is more difficult with this equipment than with the kind that drops the lime from a covered hopper or conveyor.

Regardless of the method employed, care should be taken to ensure uniform application of the lime. An examination of the distribution pattern at the start of the operation is helpful in correcting nonuniform spreading and providing proper lapping. Nonuniform distribution can result in excesses and deficiencies in different parts of the same field and corresponding nonuniform crop growth.

FACTORS DETERMINING THE SELECTION OF A LIMING PROGRAM

It is obvious that a number of factors will determine the selection of a liming program.

1. Lime requirement of the crop to be grown.
2. Texture and organic matter content of the soil as well as the pH.
3. Time and frequency of liming.
4. Nature and cost of the liming material.
5. Depth of plowing.

Liming Requirement of the Crop. Plants differ widely in their response to added lime. The nature of this response, as previously pointed out, is not always known, but it is a matter of common observation that certain plants will grow well in acid soils, whereas others will not. In considering the liming program for a given soil, the type of crop to be grown ranks first in importance. Plants such as blueberries, cranberries, azaleas, and camellias do best on soils that are distinctly acid, whereas plants such as sweet clover, alfalfa, and sugar beets make their best growth on neutral to slightly alkaline soils. On Oxisols and Ultisols crops such as corn and wheat grow well at pH values of 5.5 to 5.8. On Mollisols and similar soils that are not as highly weathered as the Oxisols and Ultisols, best growth may be obtained in the pH range of 6.0 to 6.4.

Texture and Organic Matter Content of the Soil. The importance of this point has already been covered. If a soil is coarse-textured and has a low organic matter content, the quantity of lime to be applied will certainly be less than required to effect the same pH change in a fine-textured soil or one high in organic matter. The overliming of coarse-textured soils is not uncommon, but a knowledge of the basic chemistry of soils can prevent it. In states in which a soil-testing service is available advice should be sought. In the absence of such an organization a soil-test kit in the hands of an experienced person will suffice.

Time and Frequency of Liming Applications. For rotations that include leguminous crops lime should be applied three to six months before the time of seeding. This is particularly important on very acid soils. Liming just a few days before seeding alfalfa under such conditions may produce disappointing results, for the lime may not have adequate time to react with the soil. If clover is to follow fall-seeded wheat, the lime is best applied when the wheat is planted. The caustic forms of lime [CaO and $Ca(OH)_2$] should be spread well before planting to prevent injury to germinating seeds. With bulk spreading there is a greater tendency to add the lime whenever convenient. For example, during the summer and fall after hay harvest or on pastures is a good time to get the truck out on the fields and it also helps to distribute the peak demands on lime dealers.

The frequency of application generally depends on the texture of the soil, nitrogen applied, crop removal, and the amount of lime applied. On sandy soils frequent light applications are preferable, whereas with fine-textured soils larger amounts may be applied less often. The type of limestone added will also determine to a certain extent the frequency of application. The finely divided materials react more quickly, but their effect is maintained over a shorter period than is that of materials containing appreciable amounts of coarse particles.

General statements concerning the frequency of applying lime are perhaps not too safely made, because a number of factors will influence each choice. The most satisfactory means of determing reliming needs is by soil tests. Samples should be taken every three to five years — sandy soils somewhat more frequently.

Liming Material to Be Used. The fineness of the limestone on which recommendations are based varies among states. For example, in Indiana the basis is 25 per cent through a 60-mesh sieve. In North Carolina it is 100 per cent through a 10-mesh sieve and 40 per cent through a 100-mesh sieve. If the individual grower has a much coarser material, he must increase the rate of application accordingly. Similarly, if he plans to use slaked lime and the recommendations call for ground limestone, he must reduce the amount of lime he will apply to his land.

The magnesium content of limestone should also be considered. Many soils are deficient in this element, and the use of dolomitic lime is to be encouraged.

For limestones that are ground to meet local specifications and which contain roughly the same amount of magnesium, an excellent criterion of selection is the cost per unit of neutralizing value applied to the land, the calculation of which has already been covered. As indicated in Chapter 15, the returns per dollar spent on lime are phenomenal.

Depth of Plowing. Recently more and more farmers have been increasing the depth of plowing from the customary six inches to seven to ten inches. Lime recommendations are presently made on the basis of a six-inch furrow slice. When land is plowed to a depth of ten inches, the lime recommendations should be increased by at least 50 per cent.

ACIDULATING THE SOIL

It is occasionally necessary to increase soil acidity. Acidification may be needed when land is inherently high in carbonates, as in the arid western regions of this country. Farmers in humid regions may overlime, or dust from limestone-graveled roads may blow onto field borders, causing a localized and excessively high pH. In other areas moderately acid soils may need further acidification for the growth of such plants as potatoes, azaleas, rhododendrons, or camellias.

The fundamental soil chemistry of the acidification of soils is the same as that of liming soils. The pH is decreased, however, and different materials are employed. The agents used to reduce soil pH are elemental sulfur, sulfuric acid, aluminum sulfate, iron sulfate, and ammonium polysulfide. Ammonium sulfate, ammonium phosphate, and similar compounds, though primarily nitrogen fertilizers, are also quite effective in decreasing the soil pH, as pointed out in the chapter on nitrogen fertilization.

Elemental Sulfur. Pound for pound, elemental sulfur is the most effective of the soil acidulents. Its conversion to sulfuric acid in moist, warm, well-aerated soils by the autotrophic bacteria was discussed in Chapter 8. In calculating the amount to apply to the soil, reference must be made to the buffer curve of that soil. If it is assumed that all of the sulfur is to be converted to sulfuric acid, the calculation of the amount needed is a simple matter, for 1 meq. of sulfuric acid, 0.049 g., will be formed by the oxidation of 1 meq. of sulfur, 0.016 g. Theoretically, then, 1000 lb./A. of limestone could be neutralized by 320 lb. of elemental sulfur if it were completely transformed to sulfuric acid by the *Thiobacillus* organisms.

Ordinary ground sulfur can be applied broadcast and disked in several weeks before planting the crop, for the initial velocity of the reaction, particularly in cold alkaline soils, may be somewhat slow. Under some conditions it may be advisable to acidulate a zone near the plant roots to increase water penetration or increase micronutrient availability. Both of these conditions frequently need to be corrected on the irrigated saline-alkali soils of the Southwestern, intermountain, and Western states of the USA. Elemental sulfur can be applied in bands either as dry ground sulfur or the granular sulfur bentonite mentioned in Chapter 8. Sulfur suspensions can also be employed. When elemental sulfur is applied in a band, much smaller amounts are required than are needed when it is applied broadcast.

Sulfuric Acid. This material can be added directly to the soil, but it is unpleasant to work with and requires the use of special acid-resistant equipment. It can be dribbled on the surface or applied with a knife-blade applicator, similar to the way in which anhydrous ammonia is treated. It can also be applied in ditch irrigation water. Sulfuric acid has the advantage of reacting instantaneously with the soil. In some areas it can be applied by custom suppliers who have the equipment necessary for handling this acid.

Aluminum Sulfate. Aluminum sulfate is a popular material among floriculturists for acidulating the soil in which azaleas, camellias, and similar acid-tolerant ornamentals are grown. When this material is added to water, it hydrolyzes as follows:

$$Al_2(SO_4)_3 + 6H_2O \longrightarrow 2Al(OH)_3 + 3H_2SO_4$$

This solution is quite acid. When the salt is added to the soil, in addition to hydrolysis in the soil solution, the aluminum replaces any exchangeable hydrogen on the soil colloid and drives the pH even lower:

$$Al_2(SO_4)_3 + [clay]\begin{matrix}4H\\Ca\end{matrix} \longrightarrow \begin{matrix}Al\\Al\end{matrix}[clay] + CaSO_4 + 2H_2SO_4$$

Aluminum sulfate, which is largely a specialty material, is not widely used in general agriculture.

Iron Sulfate. Iron sulfate ($FeSO_4$) is also applied to soils for acidification. Its behavior is similar to that of aluminum.

Ammonium Polysulfide. Liquid ammonium polysulfide, described in Chapter 8, is also used to lower soil pH and to increase water penetration in irrigated saline-alkali soils. It can be applied in a band 3 or 4 inches to the side of the seed or it can be metered into the ditch irrigation systems. Band application would be more effective in correcting micronutrient deficiencies than application through irrigation water. The polysulfide decomposes into ammonium sulfide and colloidal sulfur when applied. The sulfur and sulfide are subsequently oxidized to sulfuric acid.

Summary

1. An acid is a substance that tends to give up protons (hydrogen ions) to some other substance. A base is any substance that tends to accept protons.
2. Acids range all the way from strong to weak. Those that are strong are those that dissociate to a great degree, whereas those that are weak dissociate only slightly. The strength of acids and the methods of expressing it were discussed. The pH concept, which is a convenient means of describing the activity of weak acids, was explained. Neutralization of acids was covered and illustrated with appropriate equations and the principle of buffering and buffered systems was explained.
3. In many ways soils are similar to buffered weak acids. This and other current ideas on the fundamental nature of soil acidity were discussed. Soil acidity is affected by the nature and amount of humus

and clay colloids present, the amount of hydrous oxides of iron, especially aluminum, the content of soluble salts in the soil, and the level of carbon dioxide in the soil atmosphere. The importance of aluminum in soil acidity was stressed.

4. Soil *p*H, or active acidity, is registered by a *p*H meter on a sample of soil and water. Potential acidity is determined by adding known amounts of acid and base to a series of soil samples and measuring the *p*H change. The curve plotted from such data is termed a buffer curve and is used in calculating the lime requirement of field soils. Lime requirements can also be measured using various strongly buffered solutions.

5. Several materials, which include dolomitic and calcitic limestones, burnt lime, hydrated lime, marl, and slags, are employed commercially in the liming of soils. Their properties and reactions in soil were discussed.

6. The neutralizing value, or CCE, of limestones is a measure of their effectiveness in neutralizing soil acidity. Several methods of determining this property were discussed.

7. Lime is one of the most important of the production inputs in the farming system. Its effect on phosphate and microelement availability, nitrification, nitrogen fixation, and soil structure influences crop production in many ways.

8. The selection of a liming program to be followed is determined by the lime requirements of the crop, the *p*H, texture, and organic matter content of the soil, the liming material to be used, and the time and frequency of the lime applications.

9. Soils frequently have to be acidified. This can be effected by several materials, among which are elemental sulfur, sulfuric acid, aluminum sulfate, and iron sulfate. The use of these materials and their reactions in the soil were discussed.

Questions

1. The term *agricultural* lime usually refers to what material?
2. Chemically speaking, lime refers to what compound?
3. Distinguish between active acidity and potential acidity. Which of these two forms is measured when a *p*H determination is made?
4. What is meant by the terms *buffer* and *buffer capacity?*
5. What soil properties determine its buffer capacity?
6. How is the buffer capacity of a soil related to the lime requirement of that soil?
7. Define the term *lime requirement.*
8. How is the lime requirement of a soil determined?
9. From the lime requirement curve shown in Figure 11-2, determine the amount of elemental sulfur that would be required to reduce the

pH of this soil from 6.75 to 4.75 in a 10-acre field. Assume complete oxidation of the sulfur and assume also that the weight of an acre furrow slice of soil is 2 million lb.

10. Calcium chloride ($CaCl_2$) contains 53 per cent calcium. Express this as an equivalent percentage of calcium oxide. Can calcium chloride be used as a liming agent? Why?

11. What three types of slag can be used as effective liming materials?

12. What are marls? What determines their value as liming materials?

13. Define *neutralizing value* or *calcium carbonate equivalent* as it refers to liming materials.

14. What is the calcium carbonate equivalent of sodium carbonate?

15. You analyze a limestone and find that it has a neutralizing value of 85 per cent. How many tons of this limestone would be equivalent to 3 tons of chemically pure calcium carbonate?

16. In addition to its purity and neutralizing value, what other property of crystalline limestones is important with respect to their value as agricultural liming materials?

17. What are several indirect benefits of adding lime to soil?

18. A solution has a pH value of 6.5. To what hydrogen ion activity does this correspond?

19. What are several factors that will determine the frequency and rate of liming?

20. You have two fields, A and B, that need liming. The characteristics of the soils in each of these fields are the following:

	Field	
Soil property	A	B
Organic matter (%)	0.8	3.1
Clay (%)	10.0	38.1
Sand (%)	74.0	51.2
Type of clay	1:1	1:2
pH	5.2	5.2

You have lost the liming recommendations sent to you by the soil laboratory but you do recall that 3 tons/A. was recommended for Field B. Because the pH is the same in both fields, you apply 3 tons to Field A as well. Have you acted wisely? Why?

21. For what reasons do soils become acid?

22. What materials may be used to acidulate soils?

23. Solution A has a pH of 3.0. Solution B has a pH of 6.0. The active acidity of solution A is how many times greater than that of solution B?

24. Assume that you have available three dolomitic limestones of equal neutralizing value but of the following mechanical analyses:

	Limestone		
	A	B	C
Coarser than 10 M	20	5	0
Coarser than 50 M	70	30	20
Coarser than 100 M	95	60	50

(a) Which limestone would you not buy? (b) Which one would you select for the quickest results? (c) Under what circumstances could you afford to buy limestone B?
25. With what type of soil would the ammonium acetate method give a fairly good approximation of the effective CEC of soil? On what types of soil would it not give a good estimate? Would the estimate be high or low? Why? What method would give a better estimate of the CEC of such soils?
26. In what parts of the USA is liming a needed and very important practice? Where is it relatively unimportant? In your own particular case, is liming a needed practice? Why?
27. Define the term "effective CEC." How and why does it differ from the CEC as determined by the ammonium acetate method?

Selected References

Abruna-Rodriguez, F., J. Vicente-Chandler, R. W. Pearson, and S. Silva, "Crop response to soil acidity factors in Ultisols and Oxisols: I. Tobacco." *SSSA Proc.* **34:**629 (1970).

Adams, F., and R. W. Pearson, "Neutralizing soil acidity under Bermudagrass sod." *SSSA Proc.*, **33:**737 (1969).

Adams, F., and R. W. Pearson, "Differential response of cotton and peanuts to subsoil acidity." *Agron. J.*, **62:**9 (1970).

Andrew, C. S., and D. O. Norris, "Comparative responses to calcium of five tropical and four temperate pasture legume species." *Australian J. Agr. Res.*, **12:**40 (1961).

Assoc. Amer. Pl. Fd. Cont. Offic., Model Agricultural Liming Materials Bill. (Tentative 1970.)

Barrows, H. L., A. W. Taylor, and E. C. Simpson, "Interaction of limestone particle size and phosphorus on the control of soil acidity." *SSSA Proc.*, **32:**64 (1968).

Bixby, D. W., and J. D. Beaton, "Sulphur-containing fertilizers, properties and applications." *Tech. Bull. No. 17.* Washington, D.C.: The Sulphur Institute, December 1970.

Brown, I. C., "A rapid method of determining exchangeable hydrogen and total exchangeable bases of soils." *Soil Sci.*, **56:**353 (1943).

Buckman, H. O., and N. C. Brady, *The Nature and Properties of Soils*, 7th Ed. New York: Macmillan, 1969.

Burns, G. R., "Oxidation of sulphur in soils." *Tech. Bull. No. 13.* Washington, D.C.: The Sulphur Institute, June 1968.

Coleman, N. T., E. J. Kamprath, and S. B. Weed, "Liming." *Advan. Agron.*, **10**:475 (1958).

Collins, J. B., E. P. Whiteside, and C. E. Cress, "Seasonal variability of pH and lime requirements in several Southern Michigan soils when measured in different ways." *SSSA Proc.*, **34**:56 (1970).

El Gibaly, H., and J. H. Axley, "A chemical method for the rating of agricultural limestones as soil amendments." *SSSA Proc.*, **19**:301 (1955).

Eno, C. F., "The relationship of soil reaction to the activities of soil microorganisms. A review." *Soil Crop Sci. Soc. Florida, Proc.*, **17**:34 (1957).

Evans, C. E., and E. J. Kamprath, "Lime response as related to percent Al saturation, solution Al, and organic matter content." *SSSA Proc.*, **34**:893 (1970).

Fisher, T. R., "Crop yields in relation to soil pH as modified by liming acid soils. *Mo. Agr. Sta. Res. Bull.*, 947 (1969).

Fleming, A. L., and C. D. Foy, "Root structure reflects differential aluminum tolerance in wheat varieties." *Agron. J.*, **60**:172 (1968).

Foy, C. D., A. L. Fleming, and W. H. Armiger, "Aluminum tolerance of soybean varieties in relation to calcium nutrition." *Agron. J.*, **61**:505 (1969).

Foy, C. D., A. L. Fleming, G. R. Burns, and W. H. Armiger, "Characterization of differential aluminum tolerance among varieties of wheat and barley." *SSSA Proc.*, **31**:513 (1967).

Heddelson, M. R., E. O. McLean, and N. Holowaychuck, "Aluminum in soils: IV. The role of aluminum in soil acidity." *SSSA Proc.*, **24**:91 (1960).

Hunter, A. S., H. Kinney, C. W. Whittaker, J. H. Axley, M. Peech, and J. E. Steckel, "Reproducibility of ratings and correlations with chemical and physical characteristics of materials (lime)." *Agron. J.*, **55**:351 (1963).

Hutchinson, F. E., and A. S. Hunter, "Exchangeable aluminum levels in two soils as related to lime treatment and growth of six crop species." *Agron. J.*, **62**:702 (1970).

Jenny, H., "Reflections on the soil acidity merry-go-round." *SSSA Proc.*, **25**:428 (1961).

Kamprath, E. J., "Exchangeable aluminum as a criterion for liming leached mineral soils." *SSSA Proc.*, **34**:252 (1970).

Kamprath, E. J., "Potential detrimental effects from liming highly weathered soils to neutrality." *Soil Crop Sci. Soc. Florida, Proc.*, **31**:200 (1971).

Lee, C. R., "Influence of aluminum on plant growth and tuber yield of potatoes." *Agron. J.*, **63**:363 (1971).

Lee, C. R., "Influence of aluminum on plant growth and mineral nutrition of potatoes." *Agron. J.*, **63**:604 (1971).

Lin, C., and N. T. Coleman, "The measurement of exchangeable Al in soils and clays." *SSSA Proc.*, **24**:444 (1960).

Lindsay, W. L., M. Peech, and J. S. Clark, "Determination of aluminum ion activity in soil extracts." *SSSA Proc.*, **23**:266 (1959).

Long, F. L., and C. D. Foy, "Plant varieties as indicators of aluminum toxicity in the A_2 horizon of a Norfolk soil." *Agron. J.*, **62**:679 (1970).

Love, J. R., R. B. Corey, and C. C. Olsen, "Effect of particle size and rate of application of dolomitic limestone on soil *p*H and growth of alfalfa." *Trans. 7th Intern. Congr. Soil Sci.*, **3**:293 (1960).

Love, K. S., and C. W. Whittaker, "Surface area and reactivity of typical limestones." *J. Agr. Food Chem.*, **2**:1268 (1954).

Mathers, A. C., "Effect of ferrous sulfate and sulfuric acid on grain sorghum yields." *Agron. J.*, **62**:555 (1970).

McLean, E. O., S. W. Dumford, and F. Coronel, "A comparison of several methods of determining lime requirements of soils." *SSSA Proc.*, **30**:26 (1966).

McLean, E. O., and E. J. Kamprath, "Letters to the Editor." *SSSA Proc.*, **34**:363 (1970).

Moschler, W. W., G. D. Jones, and G. W. Thomas, "Lime and soil acidity effects on alfalfa growth in a red-yellow podzolic soil." *SSSA Proc.*, **24**:507 (1960).

Moschler, W. W., D. C. Martens, C. I. Rich, and G. M. Shear, "Comparative lime effects on continuous no-tillage and conventionally-tilled corn." *SSSA Proc.*, **65**:781 (1973).

Motto, H. L., and S. W. Melsted, "Efficiency of various particle size fractions of limestone." *SSSA Proc.* **24**:488 (1960).

Nye, P., D. Craig, N. T. Coleman, and J. L. Ragland, "Ion exchange equilibria involving aluminum." *SSSA Proc.* **25**:14 (1961).

Olson, R. A., et al., Eds., *Fertilizer Technology & Use*, 2nd ed. Madison, Wisconsin: Soil Science Society of America, 1971.

Pearson, R. W., and F. Adams, Eds., *Soil Acidity and Liming*. Madison, Wisconsin: American Society of Agronomy, 1967.

Pionke, H. B., R. B. Corey, and E. E. Schulte, "Contributions of soil factors to lime requirement and lime requirement tests." *SSSA Proc.*, **32**:113 (1968).

Plucknett, D. L., and G. D. Sherman, "Extractable aluminum in some Hawaiian soils." *SSSA Proc.*, **27**:39 (1963).

Pratt, P. F., "Phosphorus and aluminum interactions in the acidification of soils." *SSSA Proc.*, **25**:467 (1961).

Pratt, P. F., and F. L. Bair, "Buffer methods for estimating lime and sulfur applications for *p*H control of soils." *Soil Sci.*, **93**:329 (1962).

Reid, D. A., et al., "Differential aluminum tolerance of winter barley varieties and selections in associated greenhouse and field experiments." *Agron. J.*, **61**:218 (1969).

Rixon, A. J., and G. D. Sherman, "Effects of heavy lime applications to volcanic ash soils in the humid tropics." *Soil Sci.*, **94**:19 (1962).

Rysler, G. J., G. R. Gist, and G. W. Volk, "Equivalent amounts of liming materials." Mimeo., Agron. Dept., Ohio State University.

Schollenberger, C. J., and C. W. Whittaker, "The ammonium chloride-liming reaction." *J. Assoc. Offic. Agr. Chemists*, **36**:1130 (1953).

Schollenberger, C. J., and C. W. Whittaker, "A comparison of methods for evaluating activities of agricultural limestones." *Soil Sci.*, **93**:161 (1962).

Shaw, W. M., "Rate of reaction of limestone with soils." *Tenn. Univ. Agr. Exp. Sta. Bull.*, 319 (1960).

Shaw, W. M., and B. Robinson, "Chemical evaluation of neutralizing efficiency of agricultural limestone." *Soil Sci.*, **87**:262 (1959).

Shoemaker, H. E., E. O. McLean, and P. F. Pratt, "Buffer methods for determining lime requirement of soils with appreciable amounts of extractable aluminum." *SSSA Proc.*, **25**:274 (1961).

Soileau, J. M., O. P. Engelstad, and J. B. Martin, Jr., "Cotton growth in an acid fragipan subsoil: II. Effects of soluble calcium, magnesium, and aluminum on roots and tops." *SSSA Proc.*, **33**:919 (1969).

Thorup, J. T., "pH effect on root growth and water uptake by plants." *Agron. J.*, **61**:225 (1969).

Tisdale, S. L., "The use of sulphur compounds in irrigated arid-land agriculture." *Sulphur Inst. Jour.*, **6**:2 (1970).

Turner, R. C., and W. E. Nichol, "A study of the lime potential: 1. Conditions for the lime potential to be independent of salt concentration in aqueous suspensions of negatively charged clays." *Soil Sci.*, **93**:374 (1962).

Turner, R. C., and W. E. Nichol, "A study of the lime potential: 2. Relation between lime potential and per cent base saturation of negatively charged clays in aqueous salt suspensions." *Soil Sci.*, **94**:58 (1962).

Walsh, L. M., and J. D. Beaton, Eds., *Soil Testing and Plant Analysis* (rev. ed.). Madison, Wisconsin: Soil Science Society of America, 1973.

White, R. P., "Effects of lime upon soil and plant manganese levels in an acid soil." *SSSA Proc.*, **34**:625 (1970).

Whittaker, C. W., J. H. Axley, M. Peech, J. E. Steckel, E. O. McLean, and A. S. Hunter, "Relationship of ratings and calcium carbonate content of limestones to their reactivity in the soil." *Agron. J.*, **55**:355 (1963).

Whittaker, C. W., C. J. Erickson, K. S. Love, and D. M. Carroll, "Liming qualities of three cement kiln flue dusts and a limestone in a greenhouse comparison." *Agron. J.*, **51**:280 (1959).

Wolf, B., "Evaluation of calcined magnesite as a source of magnesium for plants." *Agron. J.*, **55**:261 (1963).

Woodruff, C. M., "Determination of the exchangeable hydrogen and lime requirement of the soil by means of the glass electrode and a buffered solution." *SSSA Proc.*, **12**:141 (1948).

Yuan, T. L., "Some relationships among hydrogen, aluminum and pH in solution and soil systems." *Soil Sci.*, **95**:155 (1963).

12. SOIL FERTILITY EVALUATION

Historically, crop production has been based on the use of plant nutrients already in the soil. Although the addition of plant nutrients has increased since 1950, the majority of crops continues to be grown on the basis of mining the soil for some or most nutrients (Figure 12-1). Soils, of course, vary greatly in how long they can be cropped without yield reduction before a given nutrient must be added. Diagnostic techniques, including identification of deficiency symptoms as well as soil and plant tests, are helpful in determining when additions are needed.

The selection of the proper rate of plant nutrients is influenced by a knowledge of the nutrient requirement of the crop and the nutrient-supplying power of the soil on which the crop is to be grown. When the soil does not furnish adequate quantities of the elements necessary for normal development of plants, it is essential that the required amounts be supplied. This necessitates finding a method that will permit the determination of those deficient elements. Obviously, looking at a given soil will tell little about its nutrient-supplying power. Red, gray, or black soils may all be deficient in nitrogen, phosphorus, potassium, or other nutrients.

Diagnosis of the needs of plants is comparable in many ways to diagnosis of human ills. The medical doctor observes the patient, obtains all the information possible with his questions, and then makes the appropriate tests, all of which are helpful in diagnosing the case. Similarly, the grower or agricultural worker observes the plants, obtains information on past management, and may make tests on the soil or the plant. The success of his diagnosis depends on his understanding of the fundamentals of plant and soil science and on a correct interpretation of the facts at hand.

Diagnostic measurements of the ailing plant or soil are often classed

Figure 12-1. The soil on the right has been "mined" by continuous cropping with little fertilizer addition for many years. Diagnostic techniques such as soil testing (*left*) are helpful in determining how much of a nutrient is needed. (Courtesy of the Potash Institute of North America.)

as trouble shooting. They can and are being used for this purpose, but *a more important application is in preventive measures;* for by the time a plant has shown deficiency symptoms a considerable reduction in the potential yield will already have occurred and the grower will have lost considerable money. By the time potassium deficiency symptoms appear in potatoes, yield reduction may be as much as 50 per cent.

APPROACHES EMPLOYED

The problem of predicting plant-nutrient needs has been under study for many years. In 1813 Sir Humphrey Davy stated that if a soil is unproductive the cause of sterility can be determined by chemical analysis. This has not always been the case, but much work has been done on soil analysis and other techniques, and a gradual improvement in methods of predicting the fertility status of the soil has become evident.

In contrast to chemical soil analysis, which depends on chemical reagents for the determination of available plant nutrients, biological methods make use of plants as extracting agents to achieve the same purpose. Generally speaking, biological soil tests are of two general types—those employing higher plants and those employing lower plants, such as bacteria and fungi. The criterion of treatment will vary with the method and may be expressed in a number of ways, including yield in bushels of grain, amount of a given nutrient extracted, quality, disease resistance, standability, or diameter of a mycelial growth.

In a consideration of the merits of chemical and biological tests it should be understood that to be of value as a basis for making lime and fertilizer recommendations the results must be correlated with crop responses in the field. With no prior knowledge of the relationship of the test results to crop response, the tests themselves are of little practical use.

Several techniques that are commonly employed will give an indication of the fertility status of a soil:

1. Nutrient-deficiency symptoms of plants.
2. Analyses of tissue from plants growing on the soil.
3. Biological tests in which the growth of either higher plants or certain microorganisms is used as a measure of soil fertility.
4. Chemical soil tests.

NUTRIENT-DEFICIENCY SYMPTOMS OF PLANTS

Many of the methods for evaluating soil fertility are based on observations of or measurements on growing plants. These methods have considerable merit because the plants act as integrators of all growth factors and are the products in which the grower is interested.

An abnormal appearance of the growing plant may be caused by a deficiency of one or more nutrient elements. If a plant is lacking in a particular element, more or less characteristic symptoms may appear. This visual method of evaluating soil fertility is unique in that it requires no expensive or elaborate equipment and can be used as a supplement to other diagnostic techniques.

Occurrence of Symptoms. Nutrient-deficiency symptoms may be classified as follows:

1. Complete crop failure at seedling stage.
2. Severe stunting of plants.
3. Specific leaf symptoms appearing at varying times during the season.
4. Internal abnormalities, such as clogged conductive tissues.
5. Delayed or abnormal maturity.
6. Obvious yield differences, with or without leaf symptoms.
7. Poor quality of crops, including unseen chemical composition differences, as in protein, oil, or starch content and in keeping or storage quality.
8. Yield differences detected only by careful experimental work.

In addition, nutrient deficiencies have a marked effect on extent and type of root growth (Figure 12-2). The underground portion of the plant has not received much attention because of the difficulty of making observations. However, when we consider that the roots are the main avenue of entry for nutrients, the importance of this aspect of plant development looms large.

Deficiency of an element does not directly produce symptoms. Rather, the normal plant processes are thrown out of balance, with the result that there is an accumulation of certain intermediate organic compounds and a shortage of others. This leads to the abnormal conditions recognized as symptoms and has a definite relation to shortages of elements. For example, diamine putrescine forms in some potassium-deficient plants and causes characteristic symptoms. Actually, a plant containing adequate potassium will show symptoms when injected with this compound.

Each symptom must be related to some function of the element in question. However, a given element may have several functions to perform, and this makes it difficult to explain the physiological reason for a particular deficiency symptom. For example, when nitrogen deficient, the leaves of most plants tend to become pale green or light yellow. When the quantity of nitrogen is limiting, chlorophyll production is reduced, and the yellow pigments, carotene and xanthophyll, show

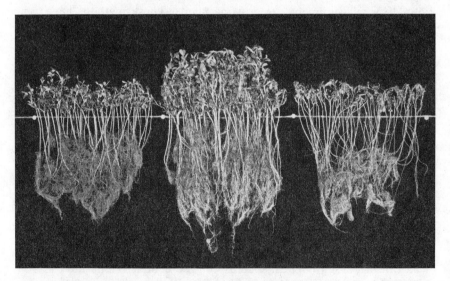

Figure 12-2. Omitting phosphorus (*left*) or potassium (*right*) reduced the growth of alfalfa roots as well as tops in the spring after seeding. There is evidence of alfalfa heaving above the groundline marked by the string. This soil tested low in phosphorus and potassium. (Courtesy of the Potash Institute of North America.)

through. Any one of a number of nutrient deficiencies, however, may produce pale green or yellow leaves, and the difficulty must be further related to a particular leaf pattern or location on the plant.

Deficiencies are actually relative, and a deficiency of one element implies adequate or excessive quantities of another. For example, manganese deficiency may be induced by adding large quantities of iron, provided the manganese supply is close to the critical point. This was discussed in Chapter 8. In addition, a sufficient supply under one condition may become deficient as other elements become more abundant. At a low level of nitrogen supply, the corn plant may not require much phosphorus, but with an adequate level of nitrogen the same phosphorus supply may become critical. In other words, once the first limiting factor is eliminated, the second limiting factor will appear.

Precautions. In the field it is often difficult to distinguish between the deficiency symptoms. It is not infrequent that disease or insect damage will resemble certain minor element deficiencies. An example is the confusion of leaf hopper damage with boron deficiency in alfalfa. It has been observed that boron deficiency is accompanied by a red coloration of the leaves near the growing point when the plant is well supplied with potassium. On the other hand, when the potassium content is low, yellowing of the alfalfa leaves occurs.

A symptom is a secondary effect and may be the result of more than one cause. For example, accumulated sugar in corn may combine with flavones to form anthocyanins (purple, red, and yellow pigments). Sugar accumulation may be based on several factors, such as an insufficient supply of phosphorus, cool nights and warm days, insect damage to the roots, nitrogen deficiency, or transverse creasing of the leaves. A short supply of two or more elements will also complicate the picture.

Nutrient-deficiency symptoms as a means of evaluating soil fertility represent an excellent example of closing the door after the horse has left the barn. These symptoms appear only after the supply of an element is so low that the plant can no longer function properly. In such cases it would have been profitable to have applied fertilizer long before the symptoms appeared.

If the symptom is observed early, it can be corrected during the growing season. This may be true with nitrogen, potassium, and certain micronutrients. Of course, the principal objective is to get the limiting nutrient into the plant as quickly as possible. With some elements and under some conditions this may be accomplished with foliar applications; otherwise, side-dressings must be used. Usually the yield is reduced below the quantity that would have been obtained if adequate nutrients had been available at the beginning. However, if the trouble is properly diagnosed, the deficiency can be fully corrected the following year.

The points just discussed in relation to deficiency symptoms are raised so that the diagnostician may be aware of some of the pitfalls. It should be emphasized that the wise use of deficiency symptoms in conjunction with other methods of diagnosis, such as plant or soil analyses, can do much to promote proper fertilization. For a detailed description the reader is referred to *Hunger Signs in Crops*. David McKay, 1964.

HIDDEN HUNGER

Hidden hunger refers to a situation in which a crop needs more of a given element, yet has shown no deficiency symptoms (Figure 12-3). The content of an element is above the deficiency symptom zone but still considerably below that needed to permit the most profitable crop performance. Because agriculture has changed from a way of life to a business, the grower will make a determined effort to avoid obvious deficiency in his crops. However, he may not add quantities large enough to obtain the most profitable yield. With most nutrients on most crops, significant responses can be obtained even though no recognizable symptoms have appeared.

In the beginning stages of use of a plant nutrient in an area, deficiency symptoms point toward first recognition of trouble. However, as use of the nutrient increases and higher yields are desired, deficiency symptoms are of less value and can be classified as a problem of the marginal farmer.

The question then is how best to eliminate hidden hunger (Figure

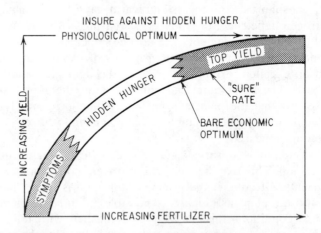

Figure 12–3. Hidden hunger is a term used to describe a plant that shows no obvious symptoms, yet the nutrient content is not sufficient to give the top profitable yield. Fertilization with the "sure" rate rather than the bare economic optimum for an average year helps to obtain the top profitable yield. (Courtesy of the Potash Institute of North America.)

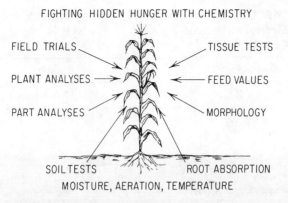

FIGHTING HIDDEN HUNGER WITH CHEMISTRY

FIELD TRIALS — TISSUE TESTS

PLANT ANALYSES → ← FEED VALUES

PART ANALYSES — MORPHOLOGY

SOIL TESTS — ROOT ABSORPTION

MOISTURE, AERATION, TEMPERATURE

Figure 12–4. Detecting hidden hunger in crops is an increasing problem as yield goals rise and higher profits are sought. In this zone, with no symptoms to guide us, we must turn to more diagnostic chemistry to evaluate needs more accurately. Many diagnostic tools are available. (Courtesy of the Potash Institute of North America.)

12-4). Tests on the plant help toward a plan for a plant nutrient program for next year, and tests on the soil help to eliminate the problem for the coming crop. In both approaches careful consideration must be given to past management practices.

Seasonal Effects. Nutrient shortages in the soil may be intensified by abnormal weather conditions. Nutrients may be present in sufficient quantities when conditions are ideal, but in drought, excessive moisture, or unusual temperature the plant may not be able to obtain an adequate supply. For example, under cooler temperatures less nitrogen, phosphorus, and potassium were taken up by tomatoes (Table 12-1).

Likewise, moisture stress influences nutrient uptake. In Iowa, as moisture stress increased, the concentrations of NPK in corn leaves

TABLE 12–1. Effect of Temperature on the Content of N-P-K in Tomato Leaves*

Age of plants (days)	Dry matter (%)					
	12°C			20°C		
	Nitrogen	Phosphorus	Potassium	Nitrogen	Phosphorus	Potassium
36	3.27	0.15	2.12	4.92	0.38	4.23
50	4.11	0.37	3.11	4.78	0.44	4.40
60	4.62	0.35	1.70	6.05	0.47	3.12
110	4.40	0.43	4.95	4.15	0.62	4.20

* From Z. I. Zurbicki, "Dependence of mineral composition of plants on environmental conditions," in Walter Reuther, Ed., *Plant Analysis and Fertilizer Problems*, p. 432. Copyright 1960, by American Institute of Biological Science.

TABLE 12–2. Influence of Applied N, P, and K and Moisture Stress on Per Cent N, P, and K in Corn Leaves*

Nutrients applied			NPK concentration	
N	P	K	No stress days	Maximum stress
—— kg./ha. ——			———— % N ————	
0	78	47	2.0	1.5
179	78	47	2.9	2.2
			———— % P ————	
179	0	47	0.26	0.12
179	78	47	0.32	0.18
			———— % K ————	
179	39	0	1.1	0.7
179	39	93	1.6	1.2

* From Voss. R. D., *Proc. 22nd Ann. Fert. Ag. Chem. Dealers' Conf.* **14**:1, Iowa State Univ., 1970.

decreased (Table 12-2). Application of these nutrients reduced the effects of moisture stress, but concentrations were still below the optimum in stress years.

Home owners buy fire insurance for their homes, hoping never to collect. It is worthwhile to consider the insurance feature and add enough nutrients to meet seasonal variations (Figure 12-5). To eliminate plant nutrients as a limiting factor the nutrient content of the plant must

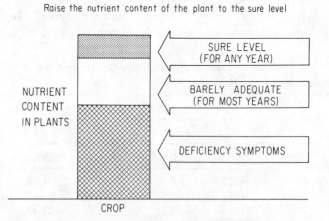

Figure 12–5. Nutrient shortages may be intensified by abnormal weather conditions. Enough nutrients must be added to ensure against seasonal variations in need. (Courtesy of the Potash Institute of North America.)

be raised to the sure or insurance level rather than to the bare economic optimum (adequate in some years). Fertilizing to this level helps to take advantage of a good season and leaves nutrients in the soil for the succeeding crop.

PLANT ANALYSES

Two general types of plant analysis have been used. One is the tissue test which is customarily made on fresh tissue in the field. The other is the total analysis performed in the laboratory with precise analytical techniques.

Plant analyses are based on the premise that the amount of a given element in a plant is an indication of the supply of that particular nutrient and as such is directly related to the quantity in the soil. Since a shortage of an element will limit growth, other elements may accumulate in the cell sap and show high tests, regardless of supply. For example, if corn is low in nitrate, the phosphorus test may show high. This is no indication, however, that if adequate nitrogen were supplied to the corn the supply of phosphorus would be adequate.

Critical levels have been suggested for a number of nutrient elements in a number of plants. Many definitions of critical level have been proposed, but the one that appears to be most meaningful for the efficient grower is the content of an element below which the crop yield or performance is decreased below optimum. For example, in corn around 0.3 per cent phosphorus in the leaf opposite and below the uppermost ear at silking time is considered the critical point. However, it is difficult to choose a specific level because the content of other nutrients in the plant may affect the critical point for a particular element. In corn the critical nitrogen, phosphorus, or potassium level includes a wide range of values, depending on balance of other nutrients and on yield level. With boron the critical point will be higher when the calcium level in the plant is high than when the calcium level is low.

Tissue Tests. Rapid tests for the determination of nutrient elements in the plant sap of fresh tissue have found an important place in the diagnosis of the needs of growing plants. In these tests the sap from ruptured cells is tested for unassimilated nitrogen, phosphorus, and potassium. Tests for other elements such as magnesium and manganese are also used. The results are read as very low, low, medium, or high. The purpose is not to split hairs but to assess general levels.

The plant roots absorb the nutrients from the soil, and these nutrients are transported to other parts of the plant where they are needed. The concentration of the nutrients in the cell sap is usually a good indication of how well the plant is supplied *at the time of testing*. The cell sap of the conducting tissues might be compared with the conveyor belts in a factory. If the factory is to operate at full capacity, all the belts bringing in raw materials must be running on schedule. If one raw material is short, its belt will run empty, the other raw materials will pile up, and production will be drastically reduced. An alert factory superintendent will make sure that there are no shortages. Likewise, an alert farmer or agricultural worker will make certain that no nutrient is limiting crop growth and that the supplies of nutrients are in the proper balance.

GENERAL METHODS. The release in 1926 of Purdue Agricultural Experiment Station Bulletin 298, "Testing Corn Stalks Chemically to Aid in Determining Their Plant Food Needs," by G. N. Hoffer, did much to lay the groundwork for tissue testing. Since that time numerous developments have taken place.

In one test the plant parts may be chopped up and extracted with reagents. The intensity of color developed is compared with standards and used as a measure of the supply of the nutrient in question. This is the system on which the Purdue Soil and Plant Test Kit, the Spurway Kit, and many others are based.

In another more rapid test plant sap is transferred to filter paper by squeezing the plant with pliers. The tests for nitrogen, phosphorus, and potassium, which are then made with various reagents, are simple and easy to perform. Semiquantitative values for the nitrogen, phosphorus, and potassium status of a plant can be obtained in about a minute. Researchers at Pennsylvania State University and the University of Illinois developed the methods. The equipment is shown in Figure 12-6.

Tissue tests are gaining in popularity because of ease of handling and the small amount of equipment needed. Because a number of tests can be made in a few minutes, it is unpardonable to guess at the nutritional status of a plant when a large part of the guesswork can be eliminated. These methods have the advantage over those of the laboratory. The lab tests require more time for an answer, and as a result there is a tendency to let the diagnosis go with a guess rather than send samples to the laboratory.

PLANT PARTS TO BE TESTED. It is essential to test that part of the plant which will give the best indication of the nutritional status. Considerable work is still needed on this point, but certain principles are fairly well established.

As the supply of nitrogen decreases, the upper part of the plant, in which maximum utilization of plant nutrients is in progress, will show a low test for nitrates first. In the case of phosphorus and potassium the reverse is true, and the lower part of the plant will become deficient first. The part of the plant to use for testing, as suggested for work with the Purdue Kit, is given in Table 12-3. Young leaves should not be tested.

The best part to use for testing is generally that showing the greatest range of levels as the nutrient goes from deficient to adequate levels (Figure 12-7).

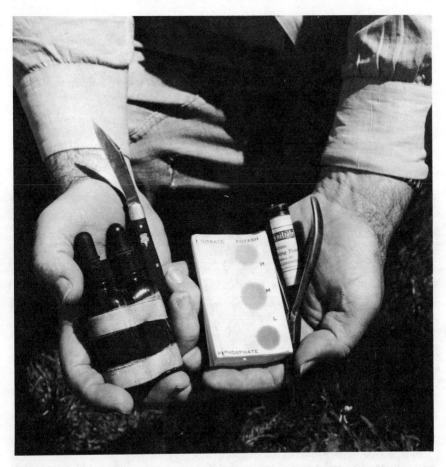

Figure 12–6. Tests on fresh tissue, using the filter-paper technique, are rapid and simple to perform. (Courtesy of the Potash Institute of North America.)

TABLE 12–3. Part of Plant Used for Plant Tissue Tests*

Plant	Nitrogen	Phosphorus	Potassium
Corn	Main stem or leaf midribs	Leaf midribs near ear	Blade tissue or midrib near ear
Soybeans		Petioles in upper third of plant	Petioles
Small grain	Main stem	Leaf tissue near center of plant	Leaf tissue near center of plant
Alfalfa		Upper third of stem and petioles	Upper third of stem and petioles
Potatoes Tomatoes	Main stems or petioles	Leaf petioles in lower third of plant	Petioles

* A. J. Ohlrogge, *Purdue Univ. Agr. Exp. Sta. Res. Bull. No. 635*, revised 1962.

Figure 12–7. Selection of sugarbeet leaves for analysis. A leaf stalk from any one of the recently matured, fully expanded leaves marked *A* may be included in the sample. The small leaves in the center or the old leaves should be avoided. (Courtesy of Dr. Albert Ulrich, and the Potash Institute of North America.)

TIME OF TESTING. The stage of maturity is of considerable importance in tissue testing. The average farm crop grows for a period of 100 to 150 days or longer, and its nutritional status will change during that period. Plants testing high in nutrients when small might test lower later on. However, if deficiencies were expected and the plants were tested early, there would be an opportunity to correct the difficulty.

In general, the most critical stage of growth for tissue testing is at the time of bloom or from bloom to early fruiting stage. During this period the utilization of nutrients is at its maximum, and low levels of nutrients are more likely to be detected. In corn the leaf opposite and just below the uppermost ear at silking is sampled. Although nothing as a rule can be done about the current crop, the tests will indicate any shortage in plant nutrients, and needed adjustments can then be made for future crops.

Forage crops lend themselves well to tissue tests and can be analyzed after considerable growth has been made in the spring. If deficiencies are found, top-dressings of the required nutrients are effective in later growth.

Time of day has an influence on the nitrate level in plants, for nitrates are usually higher in the morning than in the afternoon if the supply is short. They accumulate at night and are utilized during the day as carbohydrates are synthesized. Therefore tests should not be made early in the morning or late in the afternoon.

A few keys points are listed:

1. It is ideal to follow the uptake of nutrients through the season by testing the field five or six times. Nutrient levels should be higher in the early season when the plant is not in stress.
2. The plants' greatest need for nutrients generally comes at the time when they are preparing to make seed (i.e., flowering stage). If the field is to be checked only once a season to determine the adequacy of the fertilization program, this will be the time.
3. Comparison of plants in a field is helpful. Test plants from deficient areas and compare with plants from normal areas.
4. Plants vary. Test ten to fifteen plants and average the results.

USE. Tissue tests and plant analyses are made for the following reasons:

1. To aid in determining the nutrient-supplying power of the soil. They are employed in conjunction with soil tests and management history.
2. To help identify deficiency symptoms and, even more important, to determine nutrient shortages days or weeks before they appear.
3. To aid in determining the effect of fertility treatment on the nutrient supply in the plant. This is helpful in measuring the effect of additional fertilizers even though yield responses are not available. In some cases added plant nutrients may not be assimilated because of improper placement, dry weather, leaching, fixation, or poor aeration.
4. To study the relationship between the nutrient status of the plant and crop performance.
5. To survey large areas.
6. To interest more people in sound soil testing programs.

Interpretation. The plant diagnostician must be well acquainted with the physiology of the plant with which he is working. Some of the more important factors that should be considered before making a decision are these:

1. General performance and vigor of the plant.
2. Level of other nutrients in the plant.
3. Incidence of insects or disease.
4. Soil condition, such as poor aeration.
5. Soil moisture.
6. Climatic conditions.
7. Time of day.

If a plant appears to be discolored or stunted and gives a high test for nitrogen, phosphorus, and potassium, it is not necessarily evidence that these nutrients are present in adequate amounts. It suggests, however, that some other factor is limiting growth to that level. Only after this condition has been corrected can tissue tests be expected to reveal which of the major plant nutrients may be a limiting factor in growth.

Generally, low to medium tests for nitrogen, phosphorus, or potassium in the early part of the growing season mean that a plant will yield considerably less than optimum. At blooming time a test of medium to high is adequate in most crops. Much more work is needed on calibration, however.

Total Analysis. Total analysis is performed on the whole plant or on plant parts. Precise analytical techniques are used for measurement of the various elements after the plant material is dried, ground, and ashed. The spectrograph determines several elements simultaneously and atomic absorption is becoming increasingly important. Electron microprobes and microscopes also are being developed. Results may be reported by computer.

Of course, much smaller differences can be detected by such quantitative methods than by tissue tests. Both assimilated and unassimilated nutrients are included. With total analysis many elements, such as nitrogen, phosphorus, potassium, calcium, magnesium, sulfur, manganese, zinc, boron, copper, iron, molybdenum, cobalt, silicon, and aluminum can be determined. As in tissue tests, the plant part selected is of first importance. Evidence so far indicates that recently matured material is preferable (Figure 12-7).

It has been suggested that in some crops the relationship of potassium content in the lower leaves to potassium content in the upper leaves is an indication of deficiency or sufficiency. If the potassium content of the lower leaves is below that of the upper leaves, the plant is deficient. However, if the potassium content of the lower leaves is equal to or greater than that of the upper leaves the plant is not deficient.

For some purposes plant tissue tests on green material are thought to be more valuable than total analysis. For example, if the nutrient supply had just been exhausted, the difficulty would be more likely to be found by use of the tissue tests on the cell sap. Both tissue tests and total anal-

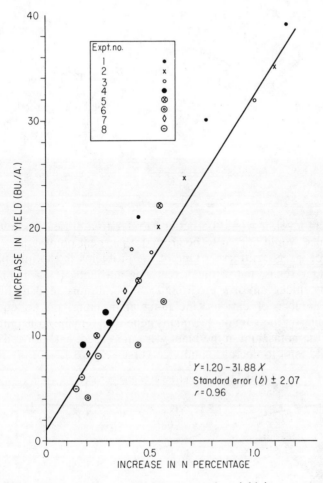

Figure 12–8. With nitrogen applied on corn, the yield increase was directly proportional to the increase in percentage of nitrogen in the corn leaf. [Hanway, *Better Crops with Plant Food,* **46**(3):50. Copyright May–June 1962 by the Potash Institute of North America.]

ysis, however, have been employed to considerable advantage in following the nutrient status of the plants through the growing season.

INCREASE IN YIELD WITH INCREASE IN NUTRIENT CONTENT. Up to a given point, increasing the rates of a nutrient, for example, nitrogen, will increase the elemental content of the plant as well as the yield. An example is shown in Figure 12-8 in which the applied nitrogen increased the percentage of nitrogen in the corn leaf in direct proportion to the increase in yield.

The relationship between corn grain yield and percent potassium in the leaf is shown in Figure 12-9. It is suggested that rather than referring

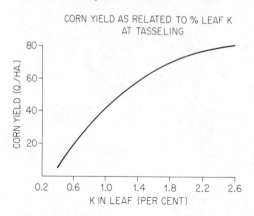

Figure 12–9. Corn grain yields increased with increasing levels of potassium in the leaf sampled at silking. [Loué, *Fertilite,* **20** (Nov.–Dec. 1963).]

to a critical level, it would be better to refer to a critical zone which in this instance would appear to be around 2 per cent potassium.

BALANCE OF NUTRIENTS. One of the problems in the interpretation of plant analyses is that of balance among nutrients. It has been shown that plants under uniform environmental conditions tend to take in a constant number of cations, including ammonium, on an equivalent basis. Similarly, the sum of the anions generally remains constant. If, for example, the potassium in the plant were increased, calcium and magnesium would tend to decrease and vice versa (Figure 12-10).

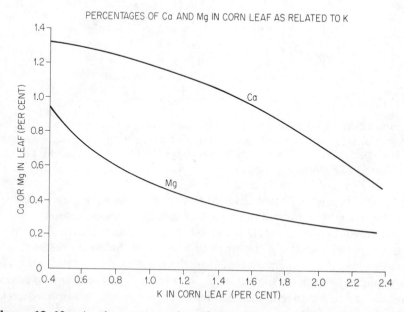

Figure 12–10. As the concentration of one element is increased, in this example potassium, other elements may decrease as shown here with calcium and magnesium [Loué, *Fertilite,* **20** (Nov.–Dec. 1963).]

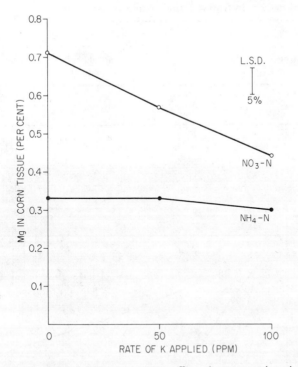

Figure 12–11. The NH_4 ion has a greater effect than potassium in decreasing Mg in corn (Fincastle silt loam). [Claassen and Wilcox, *Better Crops with Plant Food*, **57**(4):10. Copyright 1973-74 by the Potash Institute of North America.]

In recent years considerable attention has been directed to the problem of grass tetany, a disease of cattle. This basically is related to low magnesium in the blood and is influenced by many factors. The effect of the NH_4 ion in depressing magnesium uptake is illustrated in corn 37 days old (Figure 12-11). These examples serve to illustrate the problem of nutrient balance in using the actual quantity of a given element as an indication of adequacy or deficiency.

Another problem occurs when a plant is low in nitrogen, and phosphorus and potassium accumulate to show high values. With nitrogen additions, phosphorus and potassium may drop drastically (Table 12-4). This in part may be a depressing effect and in part dilution.

TIME OF SAMPLING. The percentage of certain plant nutrients may drop rapidly from the early to the late stages of growth. The rapid drop in potassium content in potatoes is shown in Figure 12-12. Hence stage of growth for sampling must be carefully selected and identified.

SURVEYS. Collection of plant samples from many fields, with subsequent analysis by the spectrograph, gives a general indication of the levels of nutrients. To permit interpretation these levels, of course, must

TABLE 12–4. Nitrogen Decreases Phosphorus and Potassium in Sugar Cane—Age Ten Months*

	8–10 internodes		
Nitrogen added (lb./A.)	Nitrogen (ppm.)	Phosphorus (ppm.)	Potassium (ppm.)
0	229	131	1160
300	463	57	340

* From G. O. Burr, in Walter Reuther, Ed. *Plant Analysis and Fertilizer Problems*. p. 336. Copyright 1960, by American Institute of Biological Sciences.

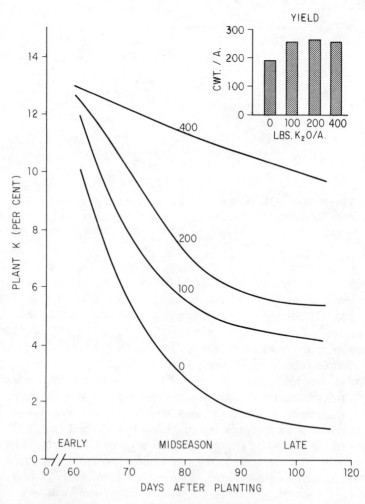

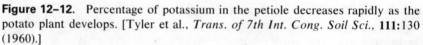

Figure 12–12. Percentage of potassium in the petiole decreases rapidly as the potato plant develops. [Tyler et al., *Trans. of 7th Int. Cong. Soil Sci.*, **111**:130 (1960).]

be compared with critical levels observed in controlled plots. This method has been particularly useful in obtaining preliminary information on elements such as zinc, boron, cobalt, and copper.

ROUTINE USE. Quantitative plant analyses are employed extensively in research to obtain another measure of the effect of treatment. However, crops on a commercial scale, such as sugar cane and pineapples, are analyzed periodically in many areas. A commercial analytical service is often available for tree crops. Such a service is available in Michigan. Several public and private organizations maintain a plant analysis service on agronomic and horticultural crops. California workers have established safe levels and deficiency ranges for nitrogen, phosphorus and potassium in a long list of crops for use by growers (*Western Fertilizer Handbook*).

Crop variety has an effect on plant analyses in some instances. When one looks at an average of many fields the effect may not be apparent but when one is evaluating a given situation the influence may be significant.

Plant analysis is another helpful tool in evaluating the nutrient status of the plant. It must be considered along with soil testing and crop-management practices in diagnosing problems. Its use in crop logging is a case in point.

CROP LOGGING. An excellent example of the use of plant analyses in crop production operations is the crop logging carried out for sugar cane in Hawaii. The crop log, which is a graphic record of the progress of the crop, contains a series of chemical and physical measurements. These measurements indicate the general condition of the plants and suggest changes in management that are necessary to produce maximum yields.

During the growing season plant tissue is sampled every thirty-five days and analyzed for nitrogen, sugar, moisture, and weight of the young sheath tissue. Analyses are made for phosphorus and potassium at critical times, and adjustments in management practices are introduced as needed (Figure 12-13). Knowledge of the percentage of moisture makes it possible to regulate irrigation, particularly during the ripening period. It has been found, for example, that the moisture content should drop gradually to around 73 per cent when the crop is ripe, and irrigation is regulated accordingly. More and more plantations in Hawaii are using crop logging, for it has been found that yields gradually increase under this system of record keeping. One important factor is that, periodically, key personnel are able to take a close look at the fields, and this examination is helpful from the standpoint of observing signs of trouble.

A-VALUE TECHNIQUE. The radiochemical analyses of plants grown on soils which have been treated with fertilizers containing elements such as radioactive phosphorus may be used to calculate the phosphorus supply of the original soil (A = available). This is based on the fact that

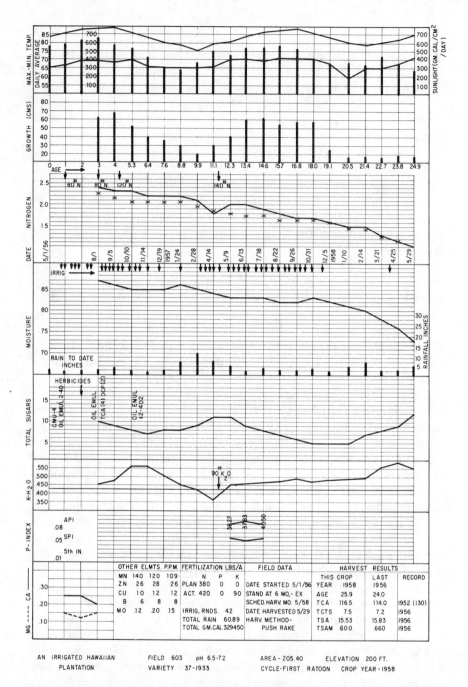

Figure 12–13. A completed crop log for an irrigated Hawaiian plantation. This approach has been valuable in a complete diagnostic approach. (From Clements, in Walter Reuther, *Plant Analysis and Fertilizer Problems,* p. 132. Copyright 1960 by the American Institute of Biological Sciences.)

when plants are given two sources of phosphorus, namely the soil and the fertilizer, they will absorb phosphorus from each in direct proportion to the amounts available.

This relationship has the following mathematical consequence:

$$A = \frac{B(1 - y)}{y}$$

where A is the amount of available soil phosphorus in pounds per acre, B is the amount of added fertilizer phosphorus in pounds per acre, and y is the fraction of the phosphorus in the plant which is derived from the fertilizer. For example, if 50 lb./A. of this element were applied and 20 per cent of the phosphorus in the plant came from the fertilizer, the A value would be 200 lb./A.

An additional development related to the A-value method is an analysis of the total uptake of phosphorus by plants at several rates of applied phosphorus. When these uptake values are plotted against the rates of phosphorus and extrapolated back to the X-axis, a value approximating the A value is determined. This method is an alternate possibility when the equipment necessary for determining radioactive samples is not available. It is also useful for a study of elements such as calcium, sulfur, zinc, manganese, and potassium.

BIOLOGICAL TESTS

Use of the growing plant understandably has much appeal in the study of fertilizer requirements, and much attention has been devoted to this method for measuring the fertility status of soils.

Field Tests. The field-plot method is one of the oldest and best-known of the biological tests. The series of treatments selected depends on the particular question the experimenter wishes to have answered. The treatments are then randomly assigned to an area of land, known as a replication, which is representative of the conditions. Several such replications are used to obtain more reliable results and to account for variations in soil and management.

These experiments are helpful in the formulation of general recommendations. When large numbers of tests are conducted on soils that are well characterized, recommendations based on such studies can be extrapolated to other soils with similar characteristics. Field tests are expensive and time consuming, and one is unable to control climatic conditions and other limiting factors. They are valuable tools, however, and are widely used by experiment stations, although they are not well adapted for use in determining the nutrient status of large numbers of soils. Rather, they are used in conjunction with laboratory and greenhouse studies as a final proving ground and in the calibration of soil and plant tests.

Strip Tests on Farmers' Fields. Strips of fields are being treated with fertilizer by extension, industry, and growers alike to check on recommendations based on soil or plant tests. Although the results of these tests must be interpreted with caution if only one replication is used, careful observations have yielded excellent information. Replication by location is helpful.

Laboratory and Greenhouse Tests. Simpler and more rapid biological techniques which still involve higher plants but utilize small quantities of soil have been developed. These methods have met with wide acceptance on the European continent and have frequently been employed in the United States.

MITSCHERLICH POT CULTURE. In this method oats are grown to maturity in pots containing 6 lb. of soil. A total of ten pots is used. The yields of the N-P and N-K treatments are expressed as a percentage of the yield from the complete N-P-K treatment. For example, if the N-P-K treatment yielded 80 g. and the N-K treatment, 60 g., the yield would be 75 per cent. With these percentage yields, the plant-nutrient reserve in the unfertilized soil can be read in pounds per acre from yield tables prepared by Mitscherlich, and from these same tables predictions of the percentage increases in yield expected from the addition of given amounts of nutrients can be obtained. The growth equations derived from these studies were discussed in Chapter 2.

NEUBAUER SEEDLING METHOD. The Neubauer technique is based on the uptake of nutrients by a large number of plants grown on a small amount of soil. The roots thoroughly penetrate the soil, exhausting the available nutrient supply within a short time. The nutrients removed are usually determined quantitatively by chemical analysis of the entire plant. In some procedures, however, the tops and the roots are harvested and analyzed separately. Tables have been set up to give the minimum values for satisfactory yields of various crops. The Neubauer method has been used for the availability of several nutrients including phosphorus, potassium, calcium, micronutrients, or fertilizer materials.

SUNFLOWER POT CULTURE TECHNIQUE FOR BORON. The sunflower method is based on vigorous extraction of boron from a small amount of soil by massive plant growth. Several modifications have been used. The soil receives a complete nutrient solution, minus boron, and five sunflower seeds are planted. The criterion of boron deficiency is the number of days required for the first of the five plants to exhibit symptoms. The soil is classified as having marked deficiency if the symptoms appear in less than twenty-eight days, moderate deficiency if they appear in twenty-eight to thirty-six days, and little or no deficiency if the elapsed time is greater than thirty-six days. Sand cultures with increasing quantities of boron added are set up as a standard, thus permitting an estimate of the absolute amounts of available boron in the soil.

Short-Term Method. The short-term method helps to bridge the gap between chemical extraction and greenhouse pot methods. Plants deficient in the element under study are grown in sand contained in a cardboard carton with the bottom removed. A dense mat of roots is formed at the bottom in two to three weeks. The roots are then placed in contact with soil, or soil plus fertilizer, contained in a second carton. An uptake time of about one week is allowed, after which the plants are analyzed. Uptake is generally well correlated with uptake in conventional pot tests that require several weeks.

Microbiological Methods. Winogradsky was one of the first to observe that in the absence of mineral elements certain microorganisms exhibited a behavior similar to that of higher plants. It was shown that the growth of *Azotobacter* served to indicate the limiting mineral nutrients in the soil, especially calcium, phosphorus, and potassium, with greater sensitivity than chemical methods. Since that time several techniques which employ different types of microorganisms have been developed and a few are described. In comparison with methods that utilize higher plants, microbiological methods are rapid, simple, and require little space.

SACKETT AND STEWART TECHNIQUE. The Sackett and Stewart technique is based on Winogradsky's work and was used to study the phosphorus and potassium status of Colorado soils. A culture is prepared of each soil, phosphorus is added to one portion, potassium to another, and both elements to a third portion. The cultures are then inoculated with *Azotobacter* and incubated for seventy-two hours. The soil is rated from very deficient to not deficient in the respective elements, depending on the amount of colony growth.

ASPERGILLUS NIGER. To determine phosphorus and potassium small amounts of soil are incubated for a period of four days in flasks containing the appropriate nutrient solutions. The weight of the mycelial pad or the amount of potassium absorbed by these pads is used as a measure of the nutrient deficiency. Mehlich devised a more refined technique in which the mycelial pad is also analyzed for potassium. An example of the criteria used is tabulated.

Weight of four pads	Potassium absorbed by *Aspergillus niger* per 100 g. of soil	Degree of potassium deficiency
< 1.4 g.	< 12.5 mg.	Very deficient
1.4–2.0 g.	12.5–16.6 mg.	Moderate to slight deficiency
> 2.0 g.	> 16.6 mg.	Not deficient

A modification of the *Aspergillus niger* test was made by Mulder to determine copper and magnesium. A rather unusual approach determines the degree of deficiency by using the color of the mycelia and

Figure 12-14. Growth of *Cunninghamella* on soil plaques. Mycelial growth is proportional to amount of available soil phosphorus. (Courtesy of Dr. Adolph Mehlich, North Carolina Agricultural Experiment Station.)

spores as a measure of the amounts of copper or magnesium present. This organism is also employed for the determination of other nutrients such as molybdenum, calcium, and manganese.

MEHLICH'S CUNNINGHAMELLA-PLAQUE METHOD FOR PHOSPHORUS. The soil is mixed with the nutrient solution, a paste is made, spread uniformly in the well of a specially constructed clay dish, inoculated on the surface in the center of the paste, and allowed to incubate for four and a half days. The diameter of the mycelial growth on the dish is used to estimate the amount of phosphorus present (Figure 12-14).

SOIL TESTING

A soil test is a chemical method for estimating the nutrient-supplying power of a soil. Although biological methods for evaluating soil fertility have certain advantages, most of these tests have the disadvantage of being time consuming, hence not well adapted for use with large numbers of samples. A chemical soil test, on the other hand, is much more rapid and has the added advantage over deficiency symptoms and plant analyses in that one may determine the needs of the soil *before* the crop is planted.

A soil test measures a part of the total nutrient supply in the soil. The values are of little use in themselves. To employ such a measurement in predicting nutrient needs of crops the test must be calibrated against nutrient rate experiments in the field and in the greenhouse.

Chemical tests are used widely in the United States as well as in other countries. All states offer soil-testing services by state-operated agencies. In some areas the service is free, whereas in others a nominal charge is made. Large numbers of soil tests have been made; in 1973, 1.0 million samples were reported to have been analyzed in state-operated agencies. There are many independent soil-testing laboratories in the United States and about the same as the above number were analyzed.

Many other countries, including Ireland, Denmark, the Netherlands, Scotland, England, Germany, France, and India, have well-developed soil-testing programs.

Objectives of Soil Tests. Information gained from soil testing is used in many ways.

1. *To maintain fertility status of a given field.* An attempt is made to extract some portion of the nutrients for calibration with the capacity of the plant for taking nutrients from the soil. When one considers the wide range in soil characteristics and the many crops grown, it is not surprising that much effort has been expended over the last century to improve soil-testing methods.
2. *To predict the probability of obtaining a profitable response to lime and fertilizer.* Although response to applied nutrients will not always be obtained on low-testing soils because of other limiting factors, the probability of a response is greater than on high-testing soils.
3. *To provide a basis for recommendations on the amount of lime and fertilizer to apply.* These basic relations are obtained by careful laboratory, greenhouse, and field studies.
4. *To evaluate the fertility status of soils on a county, soil area, or state-wide basis by the use of soil-test summaries.* Such summaries are helpful in developing plans for research and educational work.

Expressed simply, the objective of soil testing is to obtain a value that will help to predict the amount of nutrients needed to supplement the

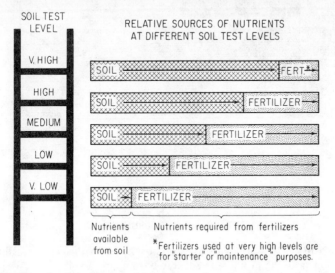

Figure 12–15. As the soil tests higher in a plant nutrient, the amount needed from fertilizers becomes less. The purpose of soil testing is to determine the levels of nutrients. (Courtesy of the Potash Institute of North America.)

supply in the soil. For example, with a high test value the soil will not require so much supplementing as it will at a low test value (Figure 12-15). It must be kept in mind that soil fertility is only one of the many factors affecting plant growth.

Sampling the Soil.   One of the most important aspects of soil testing is the matter of obtaining a soil sample that is representative of the area. Usually a composite sample of only 1 pint of soil (about a pound) is taken from a field. In a 10-acre field there are about 20 million lb. of surface soil, and with this extremely small sample there is a considerable opportunity for error. Hence it is quite important that the sampling in-

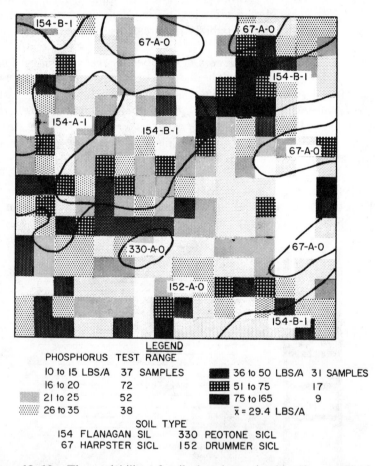

LEGEND
PHOSPHORUS TEST RANGE

10 to 15 LBS/A	37 SAMPLES		36 to 50 LBS/A	31 SAMPLES
16 to 20	72		51 to 75	17
21 to 25	52		75 to 165	9
26 to 35	38		$\bar{x}$ = 29.4 LBS/A	

SOIL TYPE

| 154 FLANAGAN SIL | 330 PEOTONE SICL |
| 67 HARPSTER SICL | 152 DRUMMER SICL |

Figure 12-16.   The variability of soil phosphorus in this 40-acre Illinois soil helps to illustrate the problem in accurately sampling a soil. (Peck, T. R. and S. W. Melsted, "Field sampling for soil testing," in Leo M. Walsh and James D. Beaton, Eds., *Soil Testing and Plant Analysis*, p. 67. Madison, Wisconsin: Soil Sci. Soc. Am., 1971.)

structions outlined by agricultural agencies be followed carefully, for a poor sample is worse than none at all. The soil-testing laboratory analyzes the soil in the sample box. If the sample does not represent the field, it is impossible for a good recommendation to be made. The error in sampling a field is generally greater than the error in laboratory analyses. The variability which may occur in a field is shown in Figure 12-16.

Tools. There are two important requirements of a sampling tool: first that a uniform slice be taken from the surface to the depth of insertion of the tool and, second, that the same volume of soil be obtained from each area. Soil tubes in general meet these two requirements very well (Figure 12-17), and they are becoming more and more popular. An additional advantage is that the volume they take is small enough that fifteen to twenty cores can be placed in a pint container. These tubes have 5 to 15 in. cut away on one side, except at the cutting head which has 1 in. of solid tube of a smaller bore. These tubes work well under most conditions, except in dry or gravelly ground. Other tools used are trowels, augers, spades, or power driven samplers.

Areas to Sample. The size of the area from which one sample may be taken varies greatly but usually ranges from 5 to 20 acres or more. Areas that vary in appearance, slope, drainage, soil types, or past treatment should be sampled separately (Figure 12-18), and small areas that

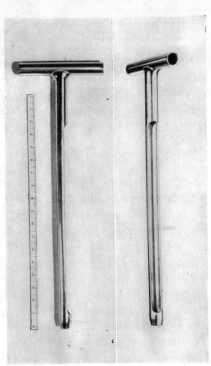

Figure 12-17. The Hoffer soil tube is a convenient sampling tool. (Courtesy of the Elano Corp., Xenia, Ohio.)

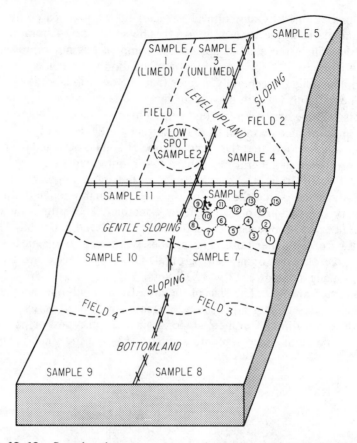

Figure 12–18. Samples that are representative of the field to be fertilized are important. The sampling pattern recommended by the various agricultural agencies should be followed. (Courtesy of the Nebraska Agricultural Extension Service.)

cannot be treated separately by lime and fertilizer applications might well be omitted from the sample. On the other hand, with the trend toward higher and higher production and more uniform crop growth, many growers are directing attention to these small spots in their fields and treating them as needed. Hence separate samples from localized areas of poor crop growth are helpful.

Number of Spots to Sample for the Composites. Each soil sample is a composite consisting of the soil from cores taken at several places in the field. The purpose of this procedure is to minimize the influence of any local nonuniformity in the soil. For example, in fields in which lime or fertilizer applications have been made in the last two or three years the plant nutrients may be incompletely mixed with the soil. In addition, there may be spots in which the fertilizer or lime was spilled or dumped or on which plant refuse was burned. A sample taken entirely from such an area would be completely misleading. Consequently, most recommendations call for taking borings at fifteen to twenty locations over the field for each composite sample (Figure 12-18). If more than a pint is obtained, the soil is mixed well and a subsample of the mixture is taken for the analysis.

As larger amounts of fertilizer are applied in the row, careful attention must be given to sampling between visible rows. The tendency toward heavier broadcast applications will increase the difficulty of obtaining representative samples unless the soil has been plowed and worked at least twice.

Depth of Sampling. For cultivated crops samples are ordinarily taken to a plow depth of 6 to 9 in. In some areas, however, growers are plowing as deep as 12 in., and this fact should be taken into consideration by sampling to this depth. The operations of land preparation tend to mix previous lime and fertilizer applications with the whole plow layer. When lime and fertilizer are broadcast on the surface for established pastures and lawns, a sample from the upper 2 in. is most satisfactory. It is recognized that plants obtain nutrients from below the plow layer, but little information is available for interpreting the analyses on such subsoil samples. Workers in some states have characterized the subsoils of major soil series in respect to levels of phosphorus and potassium and adjust their recommendations accordingly. Obviously the fertility level of the subsoil will not change appreciably over a five-to-ten year period.

An exception is sampling two to six feet in depth in some of the lower rainfall areas in the United States in order to measure nitrate in the profile. This amount is then considered in the recommendations.

Time of Sampling. Samples can be taken any time the soil conditions permit. A key point is that they be taken in time to obtain the needed lime and fertilizer. This is the problem in waiting until spring. On some soils the potassium level is lower during the summer. This has been found in West Virginia and in Illinois. In Illinois it is recommended that, depending on the soil type, 30 to 60 lb. be subtracted from the soil test for samples taken before May 1 and after September 30.

There is a basis for taking the soil sample when the crop is growing. In this way the nutrient content of the soil is determined while and under the conditions plants are drawing nutrients from the soil.

Most recommendations call for testing each field about every three years, with more frequent testing on the lighter soils. In most instances this is often enough to check on the lime level in the soil and to determine whether the fertilization program is adequate for the crop rotation. For instance, if the phosphorus level is decreasing, the rate of application can be increased. If it has risen to a satisfactory level, application may be reduced to maintenance rates.

Copies of the soil test results from previous samplings are usually maintained in the office of the agricultural leader or the fertilizer dealer. When the grower decides what crop he is to grow for a given year, he may check back with these men to get the recommendation.

Analyzing the Soils.   Any chemical soil test should be designed to permit an estimate of the amount of plant nutrients contributed by the cation exchange fraction of the soil, by the fraction retaining the phosphorus, and under some conditions by the decomposition of organic matter. The major plant-nutrient cations available for use by plants are held in exchangeable form. Among those anions that are of major importance in soils the phosphates are retained most strongly, the sulfates less strongly, and the nitrates not at all. The method of retention of many of the micronutrients has not been established, but some of them are held in the exchangeable form.

Several kinds of extracting solutions have been employed in an effort to correlate soil test results with plant growth. However, when it is considered that the soil sample will be in contact with the extracting solution for only a few minutes, whereas the plant will absorb from the soil during the entire growing season, the complexity of the task becomes apparent.

Organic matter contributes to plant nutrition. Certain fractions retain cations in exchangeable form; other fractions are decomposed or mineralized by microorganisms, and nitrogen, phosphorus, and sulfur as well as other nutrients are released.

Soil acidity is an important characteristic and is a good index to a number of conditions. It gives an indication of the base saturation and a lead to possible toxicity or deficiency of certain elements.

CATIONS.   The principle underlying the determination of cations is the replacement of all or a proportionate amount of the cation from the exchange complex. Cations such as H^+, Ba^{2+}, NH_4^+ (if NH_4^+ is not to be determined), or sodium may be used, but ammonium acetate is a common extractant for potassium, calcium, and magnesium. Generally, the soils are dried before extracting for chemical analysis. However, evidence has been obtained that on some soils potassium uptake by plants is better correlated with the exchangeable potassium determined in undried rather than air-dried samples because of the release or fixation that occurs in the drying process. Hence, in 1964 Iowa State University adopted procedures to test undried samples.

In some soils, particularly in South American countries (Andean soils), nonexchangeable potassium has been found to be important. In these soils a sulfuric acid technique is used to measure the exchangeable and a readily available portion of the nonexchangeable potassium.

The percentage base saturation refers to the percentage of exchange capacity made up of exchangeable bases that include ammonia but not hydrogen and aluminum. Rather than the actual amounts, some laboratories are basing recommendations on percentage base saturation. The concept is based on the principle that the availability of a given cation to a plant is influenced by the concentration of other cations present.

It has been found that at a given level of potassium in the soil, a sandy soil will have a greater proportion of the exchangeable potassium in solution than will a heavier soil. Hence, this may be part of the reason why percentage base saturation may be useful in helping to predict needs.

PHOSPHORUS. Extracting solutions, ranging from water, alkalies, and weak acids all the way to relatively strong acids containing ammonium fluoride, have been used for the extraction of phosphorus. In a study conducted by the Soil Test Work Group on a wide range of soils all over the United States the Bray No. 1 method, which employs $0.025N$ $HCl + 0.03N$ NH_4F, gave especially good correlation with A-values in the greenhouse and with crop responses. Olsen's method of employing $0.5N$ $NaHCO_3$ has been satisfactory on alkaline soils. Strong acids cannot be used in alkaline soils because the tricalcium phosphate, which is not available to crops, is dissolved out. Mehlich's $0.05N$ $HCl +$ $0.025N$ H_2SO_4 and the Morgan sodium acetate extractants have been popular in some areas.

Once the nutrients are extracted quantitative equipment such as the flame photometer, photoelectric colorimeters, atomic absorption, etc., are generally used to measure the amounts in the extract.

MICRONUTRIENTS. A number of laboratories make soil tests for micronutrients, mostly on a special basis. Several extractants are used but chelating agents such as DTPA appear most promising for Zn, Cu, Mn and Fe. The pH of the extraction can be controlled and chemical attack on lime and mineral fractions is minimized. Analyses for boron are usually made by the hot-water method. Two major problems stand out with micronutrient soil testing. One is interpretation and the other is laboratory control.

ORGANIC MATTER AND NITROGEN. A knowledge of the organic matter content is helpful in estimating the cation exchange capacity and the nitrogen-supplying power of the soil. Organic matter is usually determined by wet combustion methods in which the soil is subjected to treatment with sulfuric acid and potassium dichromate. Certain of the less resistant fractions, known as "easily oxidizable organic matter," are oxidized.

Available nitrogen is also being determined by chemical oxidation in which the soil is digested with sodium carbonate and potassium permanganate for five minutes to reduce the nitrogen to the ammonium form. In some laboratories nitrification tests are being made. The soil is incubated under optimum moisture and temperature conditions for two weeks, at the end of which time the nitrates are leached out and determined.

Consistent results from nitrogen tests are complicated by the fact that nitrogen availability depends on decomposition of organic matter. Environmental conditions such as moisture and temperature affect decomposition, hence seasonal variations may be great. Past cropping practices are widely used in predicting needs.

SULFUR. Determination of the sulfur needs by soil tests is complicated by the various forms and methods by which sulfur is held in the soil. The organic matter contains sulfur and many factors influence retention of sulfate on the inorganic soil fraction. A number of laboratories now determine sulfur with water or $Ca(H_2PO_4)_2$ as the most common extractants. A turbidimetric measurement of $BaSO_4$ is usually used.

In general, plant tests for assessing sulfur needs of crops have been somewhat more successful than soil tests. Absolute levels of total sulfur vary with species, but for some crops, especially alfalfa, levels have been set which fairly well describe the need for sulfur. For many crops, however, such levels have not been determined but the N:S ratio (% total N: % total S) has been used as an indication of the need for sulfur. N:S ratios of 14 to 16:1 are generally considered to be satisfactory, but ratios in excess of 17:1 suggest the need for sulfur fertilization.

SOIL ACIDITY AND LIME REQUIREMENT. The soil acidity determination is accomplished by the use of the pH meter which is standard equipment in most laboratories. Color indicator dyes are employed in some county laboratories and in field kits but must be used by relatively experienced operators if correct results are to be obtained. These dyes are helpful in pointing out gross needs, and if an acid soil is found a lime requirement test in the laboratory is essential.

The soil pH, along with a consideration of the organic matter and amount and type of clay, is the basis for many lime recommendations. This method, however, is subject to a good bit of estimation and human error. Many laboratories use a method involving a buffer in which hydrogen is replaced from the exchange complex and the depression in pH is read and related to lime requirement.

LABOR-SAVING EQUIPMENT. Soil testing on a service basis involves the handling of large numbers of samples. Special labor-saving equipment has been developed in order that the volume of work may be dealt with rapidly and accurately (Figure 12-19). For example, extraction racks may hold twelve bottles and can be manipulated as a unit; extraction solutions may be added to twelve samples at a time from specially built dispensers, and electrically powered automatic pipettes dispensing 1 to 5 ml. at each stroke may be used. A few seconds saved in each of

Figure 12–19. Large numbers of samples are analyzed in soil-testing laboratories in which rapid and accurate procedures must be used. Labor-saving equipment is essential. (Courtesy of Dr. J. B. Jones, Soil Testing Laboratory, Georgia Agricultural Experiment Station.)

several manipulations is extremely important. In addition, as the operation is made easier and more routine, the laboratory technician is less subject to fatigue, hence less subject to errors. Some laboratories have equipment that translates the instrument readings to punch cards or types out the results on a form.

Detailed analytical procedures are used in research work. More rapid but carefully performed tests are used for the routine soil samples. In many soil-testing laboratories, however, the accuracy of the routine soil tests approaches that of the detailed procedure. The use of well-trained personnel who understand the chemical reactions and the operation of special equipment contributes much toward more reliable results.

PHYSICAL PROPERTIES OF SOILS. Rapid estimates of soil structure have been rather difficult to make. The soil tube described in this chapter is an effective tool for visual examination of the soil profile to a depth of 18 in. Shaving the exposed side of the core has helped to detect compact layers, and a suspension of precipitated chalk in water dropped on the cut surface is also useful. Soil layers in good tilth will absorb both lime and water. Compact layers will absorb the water but not the lime, and the intensity of the lime spot increases with compactness.

Physical condition of the soil becomes increasingly important at higher yield levels. In the future more effort will probably be diverted to the examination and characterization of the physical properties of the root zone as affected by management.

Central Laboratories and County Laboratories. Most states use a central or regional laboratory system, but a few maintain laboratories in many counties in addition to a central control laboratory. There are advantages to both systems.

Analyses are probably more accurately performed in central laboratories because of more skilled technicians, better equipment, and better control. It is also somewhat easier to make uniform recommendations; and if changes are made, these changes can be put into effect quickly. The lime and fertilizer recommendations may be made at the laboratory, or the results may be sent out to regional representatives or county agents for handling. The latter procedure reduces the criticism that individuals unfamiliar with local conditions are making recommendations. Many laboratories are now using the computer to make recommendations. The program is set up to consider soil tests along with many management factors and the recommendations are printed directly on a sheet. The regional representative, be he public or private, may choose to make adjustments based on his knowledge of the managerial capacity of the farmer, the productivity of the soil, and other factors.

Commercial Laboratories. Commercial laboratories are important in all areas of the country. Some which are connected with the fertilizer industry provide tests as a service to customers. Others offer soil testing as a part of a wide range of analytical services. In any event, there is a marked trend toward the use of commercial laboratories.

Most important in commercial laboratories, as in those operated by state agencies, is the correct interpretation of results. Therefore these laboratories must have access to check samples from the state agencies as well as tables showing the interpretation of the results. In some states a system of approved laboratories has been developed.

Growers may be suspicious of recommendations made on tests in a laboratory supervised by a fertilizer company. In some cases this suspicion may be justified, but the rapid trend toward service by the industry, the employment of trained personnel, and the competitive need for a program that will be profitable to the grower should lead to reliable recommendations.

Calibrating Soil Tests. Although the chemical analysis of soils presents some difficulties, perhaps the greatest problem in a testing program is the calibration of the tests. It is essential that the results of soil tests be calibrated against crop responses from applications of the nutrients in question. This information is obtained from field and greenhouse fertility experiments conducted over a wide range of soils. Yield responses from rates of applied nutrients can then be related to the quantity of available nutrients in the soil.

Much additional work in calibration is needed in the United States as well as in other countries. In certain European countries, such as Holland and Denmark, large numbers of experiments have been used in calibration studies. With the increased interest in fertilizers in the developing countries, soil testing is receiving more emphasis. Unfortunately, in these countries as in the early days in the United States, few calibration data are available.

Many of the testing laboratories in the United States classify the fertility level of soils as very low, low, medium, high, or very high, based on the results of chemical tests. Some, however, report results in terms of pounds per acre or ppm. of phosphorus and potassium. In general, the very low to very high classification seems to be somewhat more easily understood by the grower, but crops vary in their requirements and what is low for potatoes may be high for small grain; what is low for a clay loam may be high for a sandy loam. In any case, it is important that the grower know the meaning of the results reported.

Some states such as Alabama have adopted a fertility index. This is the relative sufficiency expressed as percentage of the amount adequate for top yields. The percentage figures can be converted to pounds per acre.

	Fertility index (%)		Fertility index (%)
Very low	0–50	High	110–200
Low	60–70	Very high	210–400
Medium	80–100	Extremely high	410 up

A simplified approach for developing countries simply plots percentage yield and the soil test value. A critical level is established at about 75 per cent.

The probability of a response to fertilization as related to soil-test results has been emphasized. As already stated soil fertility is only one of the factors influencing plant growth, but in general there will be a greater chance of obtaining a response from a given element if the soil test is low in that element. This concept is presented diagrammatically in Figure 12-20. For example, more than 85 per cent of the fields testing very low may give a profitable increase; in the low range 60 to 85 per cent may give increases, whereas in the very high range less than 15 per cent would respond. These values are arbitrary, but they illustrate the idea of expectation of response. The better informed farmers who are

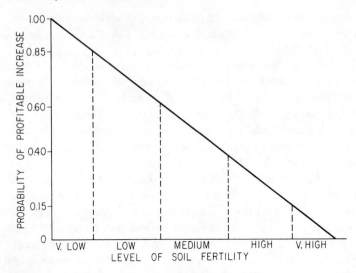

Figure 12–20.   There is a greater probability of obtaining a profitable response from fertilization on soils testing low in an element than from soils testing high in that element. [Fitts, *Better Crops with Plant Food,* **39**(3):17. Copyright March 1955 by the Potash Institute of North America.]

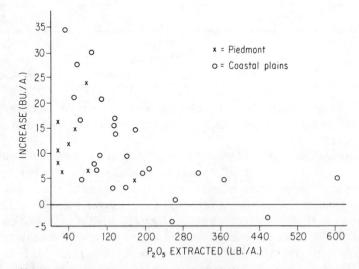

Figure 12–21.   The response of corn to applied phosphorus is related to the soil phosphorus level. [Fitts, *Better Crops with Plant Food,* **39**(3):17. Copyright March 1955 by the Potash Institute of North America.]

using good management practices will be more likely to obtain a response in the high range.

The variable yield response of corn to applied phosphorus obtained in different experiments related to levels of soil phosphorus is well illustrated in Figure 12-21. On soils containing 20 lb./A., the response to applied phosphorus ranged from 5 to 30 bu./A. However, even with this variable response there is more of an opportunity for gain from phosphorus applications at low soil phosphorus levels than at levels of 100 lb. or more.

An example of the calibration of the soil test for zinc with crop response is shown in Figure 12-22. Eighty-six per cent of the soils testing below 0.55 ppm. zinc responded to zinc sulfate, whereas 77 per cent above this level did not respond.

In a North Central regional experiment on corn, responses to potassium were calibrated with soil test values. As the exchangeable potassium increased, the amount of potassium needed to obtain the most economic return was reduced.

The calibration of soil tests is complicated by the fact that many factors other than fertility level influence the response obtained. Temperature, water, soil properties, stand, cultural practices, and pests are more readily controlled in the greenhouse than in the field. However, it is essential to conduct field trials after the preliminary greenhouse tests have been completed.

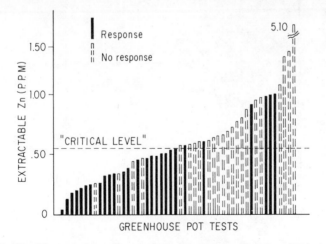

Figure 12-22. The response of sweet corn to zinc as related to dithizone-extractable zinc in fifty-five California soils. Each solid line represents a soil that responded; each broken line shows no response. [Brown et al., *Calif. Agric.* **15**:15 (1961).]

Variety may have a marked effect on response to applied fertilizer as shown by data on paddy rice from the Philippines.

	No potassium ton/ha.	300 kg./ha. K$_2$O ton/ha.
Native variety	1.7	1.9
Improved variety	1.4	4.8

A major problem in soil tests is that they are sometimes calibrated at a yield level too low to be useful to commercial farmers. Many of the calibration data are out of date because of the development of new technology to control more of the limiting factors which results in higher yields. The reason for the emphasis on higher yields in calibration trials is simple. Higher yields use more nutrients, the opportunity for responses is greater, and results are more meaningful for the progressive grower.

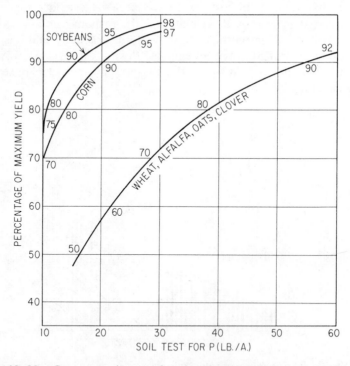

Figure 12-23. Crops vary in need for phosphorus. For example, with a 20-lb. phosphorus test the percentage of the maximum yield is 57, 90, and 94 per cent for wheat, corn and soybeans, respectively. [Bray, *Better Crops with Plant Food*, **45**(3):18. Copyright June–July 1961 by the Potash Institute of North America.]

The Changing Philosophy of Interpretation of Soil Tests. Much progress has been made in soil testing in measuring the available nutrients in soils. However, the big problem is interpreting the results in terms of fertilizer needed. The degree of accuracy depends on several factors including knowledge of the soil, the yield level expected, level of management, and weather. When a soil test is first developed in an area we simply want to know the critical level and if a particular nutrient should be added. As knowledge is gained in a given area the soil test may be divided into three or more categories and the fertilization rate adjusted to the soil fertility level.

The percentage yield concept is based on the idea that the expected yields, as a percentage of the maximum yield, are predicted from the phosphorus and potassium analyses of the soil (Figure 12-23). Sufficient amounts are added to bring yields to 95 per cent or more of the maximum. While this concept fits a number of broad situations, interactions may cause considerable deviation. When this concept was developed by Bray he indicated that it applies only when the same pattern and rates of planting were used and with about the same soil and seasonal conditions. For example, in Indiana on the same soil it was found that in dry years potassium per acre increased corn yields 39 bu., in average years 8 bu., and in wet years 48 bu.* In Kentucky the following information was obtained:

Corn plants/A.	Response to 100 lb. P_2O_5 (bu./A.)	Response to 200 lb. K_2O (bu./A.)
15,700	2	21
24,500	22	39

Hence, considering the weather effects and improved practices of progressive farmers, the importance of individual attention to each situation looms large.

RECOMMENDATIONS FOR DIFFERENT GENERAL YIELD LEVELS. Soil test interpretation involves an economic evaluation of the relation between the soil test value and the fertilizer response. However, the potential response may vary due to several factors among which are the soil, weather, and management ability of the farmer (Figure 12-24).

Some laboratories may vary recommendations with expected yield level (Table 12-5). The amount of N recommended depends on previous crop and yield goal.

Many laboratories make one recommendation assuming best production practices for the region and the area representative may make ad-

* S. A. Barber, Purdue University, personal communication.

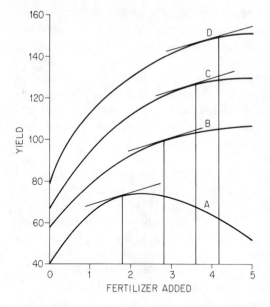

Figure 12–24. The yield response to fertilizer depends on the potential yield level with A being the poorest and D the greatest potential. (Barber, S. A., "The changing philosophy of soil test interpretations," in Leo M. Walsh and James D. Beaton, Eds., *Soil Testing and Plant Analysis,* p. 203. Madison. Wisconsin: Soil Sci. Soc. Am., 1971.)

justments as necessary. However, some base them on average production practice and this tends to discourage progressive farmers from using the laboratory.

As technology and management practices improve or with increased economic incentives, yield potential increases and recommendations increase. *For the commercial grower the goal is to maintain the plant nutrients at a level for sustained top profits per acre, which means that nutrients should not be a limiting factor at any stage, from plant emergence to maturity.* This approaches the "sure" rate (Figure 12-3).

TABLE 12–5. P_2O_5 and N Recommended for Corn (Ohio Agronomy Guide, 1974–75)

Soil test – P_1 lb./P/A.	Yield goals – bu./A.			
	100–124	125–149	150–174	175+
	Annual application of P_2O_5 – lb./A.			
0– 9	70	80	90	100
10–19	60	70	80	90
20–29	50	60	70	80
30–59	40	50	60	70
60–99	30	40	50	60
100+	20	20	25	30
	Annual application of nitrogen			
Continuous corn	140	180	220	260

TYPES OF RECOMMENDATIONS. In general, there are four courses of action if the soil is low in phosphorus or potassium.

1. *Buildup or basic treatment.* Corrective applications are added to bring the soil up to the desired level. A rule of thumb is that about 10 lb. P_2O_5 will raise the P_1 test one lb. and about 3 lb. K_2O will raise the K one lb. However, in Michigan the amount of P_2O_5 was found to vary from 4.5 to 11.5 lb. and K_2O to vary from 2 to 4.5 lb., depending on soil texture. The soil is retested in two to three years to see if additional corrective applications are needed. Maintenance amounts are then added to replace losses by crop removal, erosion, leaching, and fixation.

2. *Annual application.* Phosphorus and potassium are added to each crop in the rotation. This may be in amounts to effect a gradual buildup or merely to maintain the soil test. This approach demands well-calibrated tests and there is considerably more chance of error. When capital is limited, the fertilizer-using area is new, or the land is being rented, the maintenance program is probably the preferred method. Yield will not be so high and profit per acre will be lower, but returns per dollar spent will be higher than in the buildup method.

3. *Rotation.* This method is perhaps the most widely used and a gradual buildup approach may be used. In a rotation involving soybeans, such as corn–soybeans, many farmers just fertilize the corn and supposedly add enough to take care of the soybeans.

However, consider the pounds per acre removed just in the grain.

	N	P_2O_5	K_2O
150 bu. corn grain	135	60	40
50 bu. soybean grain	200*	40	70
Total:	335 N	100 P_2O_5	110 K_2O

* Legumes get much of their N from the air.

It is sometimes difficult to impress on farmers and recommenders the high amounts of nutrients removed in their ever-increasing yields. Removals in crop, other losses and some buildup in the soil must be evaluated in the recommendations.

In rotational fertilization certain points must be considered when deciding how to recommend the fertilizer among the crops in the rotation:

a. Apply the broadcast plowdown before most responsive and profitable crops.
b. Apply row phosphorus to corn, particularly in northern areas.
c. Forage crops remove large amounts of potassium. Annual applications are essential to maintain yields.
d. Soybeans respond more to a high soil fertility level than to direct

application of fertilizer. However, on soils medium or less in fertility apply fertilizer directly for soybeans.

Double-cropping or two crops in one year is practiced in some areas. This may be small grain–soybeans or small grain silage–soybeans. In such instances sufficient phosphorus and potassium is recommended to be applied for both crops before the small grain is planted. Again the removals, losses, and any buildup must be considered in making the recommendations.

4. *Replacement system.* As soils are built up to what is considered to be an adequate level in phosphorus and potassium, amounts may be recommended to replace removals on the basis of expected yield. For example, if 50 bu. of soybeans were expected and at ¾ lb. P_2O_5 and 1.4 lb. K_2O per bu. about 40 lb. P_2O_5 and 70 lb. K_2O would be suggested. With an eight-ton yield of alfalfa and a removal of 15 lb. P_2O_5 and 60 lb. K_2O per ton, 120 lb. P_2O_5 and 480 lb. K_2O would be suggested.

There are several aspects which must be considered in such a system:

a. On soils with a high supplying power only 50 per cent of removal might be recommended.
b. What is an adequate level in the soil?
c. Will the farmer wish to increase the level as yield potential increases?
d. The content of phosphorus, potassium, and other elements in grain may vary some and in forage a good bit.
e. Will adding just what is removed maintain soil fertility level? This will depend on fixation and release in the soil and certain losses, but generally it will not.
f. If 100 lb. of P_2O_5 or K_2O is added can it be expected that the crop is 100 per cent efficient in absorbing it?
g. In some soils the amount added over removal may need to be increased 10 to 25 per cent.

In view of these points, if this system is used, the soil must be monitored periodically to determine if the soil fertility level is decreasing or increasing.

NITROGEN. Recommendations depend on many factors including amount of nitrate in the soil profile in the lower rainfall areas, previous crop, yield goal (Table 12-5) and rate of nitrogen on the previous crop.

Nitrogen credits might be as follows and subtracted from the normal recommendations:

Good legumes	60–80 lb.
Average to poor legumes	20–40 lb.
Soybeans for grain	0–20 lb.
Manure	5 lb. per ton

Nitrogen recommended for corn following a heavy small grain crop would be increased because of the straw while recommendations on small grain following a heavily fertilized corn crop would be reduced. With continuous corn, if the desired yield level had not been reached, more nitrogen would be applied than if it had been reached. Nitrogen use efficiency decreases as yields increase. For example, at 100 bu. one lb. of N/bu. might be adequate but at 150–175 bu. two to three lb. per bu. might be needed.

As potential yield increases, nitrogen needs increase considerably because it is a mobile nutrient. This is in contrast to phosphorus which is an immobile nutrient and need does not increase as much. Potassium is more mobile than phosphorus, so need would increase somewhat more than phosphorus as potential yield increases.

PRESCRIPTION METHOD. The prescription method for making recommendations is based on the idea that plants can secure certain percentages of nitrogen, phosphorus, and potassium contained in the soil, manure, and fertilizer. When the approximate number of pounds per acre required to produce a given yield is known, the amount of supplemental manure and/or fertilizer is calculated. The principle behind this method is to formulate fertilizer recommendations fitted to needs, which are determined by the rotation followed, crop management, soil analysis, and immediate crop to be grown.

Several different methods have been used. An example of one approach is given in Table 12-6. The estimated percentages of nitrogen, phosphorus, and potassium available to a crop such as corn in any one year are listed.

Similar calculations have been made for other crops, including small grain, legumes, potatoes, tomatoes, and sugar beets.

It is recognized that this method has limitations. The problem of developing soil tests which will give an accurate measure of the amounts of nutrients available has been discussed. Tests vary somewhat among the states, and a given prescription must be related to the soil-test method being used. The percentages of elements available from manure

TABLE 12–6. Estimated Percentages of Nitrogen, Phosphorus, and Potassium in Soil, Manure, and Fertilizer Available to a Crop Such as Corn During One Season*

Sources	Percentages obtained during one season		
	Nitrogen	Phosphorus	Potassium
Soil (available)	40	40	40
Manure (total)	30	30	50
Fertilizer (available)	60	30	50

* K. C. Berger, *Farm Chemicals.* **117**:47–50 (1954).

and fertilizer are approximations for they change with crop, soil, and climatic conditions.

In spite of these limitations, such methods serve to point out the nutrient needs of crops and that higher yields will demand greater quantities. This emphasizes the attention that is being given to eliminating as many of the limiting factors in crop production as possible.

Lime. This subject has been covered in detail in Chapter 11. It is mentioned here only to emphasize that limestone is the workhorse in the soil. If a soil is acid, the first step in a buildup fertility program is *adequate* lime application. Applications often do not raise the soil *p*H to the desired level. Testing after three years may reveal the need for reapplication.

Recommendations on Electronic Computers. The computer is fast becoming the farmer's electronic hired hand. Many farmers are already depending on computerized farm management and dairy production records. Now this exceedingly fast, tireless, and accurate brain is being put to work on soil test recommendations and crop production practices.

Many handwritten recommendations consider only the soil test results. On the computer many other factors, such as soil type, yield potential, crop sequence, moisture, and managerial ability of the farmer, may be considered. Many samples can be processed in a minute, yet the farmer gets more information than ever before possible. The recommendations are printed on the soil test report.

Importance of Soil Tests to the Farmer and the Lime and Fertilizer Industries. It should be recognized that soil testing is not an infallible guide to crop production. The problems of representative samples, accurate analyses, correct interpretation, and environmental factors which influence crop responses all enter in. However, the soil test helps to reduce the guesswork in fertilizer practices. The soil test may be used to monitor the soil periodically to determine general soil fertility levels. Rates of lime and fertilizer are then applied to supply adequate quantities to the current crop or crops.

The lime and fertilizer industries can be, and in many cases are being, helpful in state or commercial soil-testing programs. Agronomists, salesmen, and dealers can encourage their customers to have soils tested or can take samples for their customers, and dealers can make a point of stocking the recommended fertilizers. If the grower applies what his crops need, he is much more likely to be able to pay his bills in the fall. On the other hand, if he applies phosphorus and potassium when lime or nitrogen happens to be limiting, his bills may go unpaid. It has been known that a fertilizer dealer will refuse to sell fertilizer to a customer until his soils have been properly limed.

The difficulty in getting growers to take samples is great, and many dealers or private laboratories provide a service for which the grower may pay.

Industry classifies the two most important functions of soil testing as providing a basis for more profitable recommendations and opening the door to customer counseling.

Farmers have been hearing about soil tests and fertilizer for years, and many farmers are convinced of their value, but they want to know how much to apply. This takes the soil test out of the category of an educational tool and into that of a profit-guaranteeing service.

More and more fertilizer dealers are recognizing the importance of soil tests, as well as other diagnostic techniques, in helping to predict the plant nutrients that growers will need on their fields. The fertilizer dealer is selling service along with his product in a personalized approach to sound development of lime and fertilizer use.

The statement is often made that *a soil test means a lime or fertilizer sale*. This is attractive to the dealer and a service to the farmer. As mentioned earlier, it is essential that recommendations be keyed to the changing needs and goals of the commercial grower.

SOIL TEST SUMMARIES

The purpose of soil testing is to give the individual grower dependable information regarding the fertilizer and lime needs of his fields. However, not all growers will have their soils tested, and neither will all growers taking soil samples include all the fields on their farms. Better use of lime and fertilizer may be attained on most of these untested fields by drawing on the information gained from the fields tested. This should lead to the best average recommendation.

To accomplish this purpose, summaries of the results of soil tests have been made in most states by county or on the basis of soil association, type of farming, or crop.

A summary map will serve to alert agricultural workers and industry personnel to the sectional problems of the state. This map can be helpful in the over-all direction of a program and in indicating where more intensive educational efforts can be made.

It should be stressed that although the content of lime, phosphorus, or potassium in the soils of a given county or area may, on the average, be high the soils in individual fields may test all the way from very low to very high. This serves to direct attention to the need for specific soil samples from each field for the most intelligent use of plant nutrients.

The soil phosphorus levels as related to certain crops in North Carolina are listed in Table 12-7. The large percentages of samples in the high and very high ranges for tobacco, cotton, Irish potatoes, and peanuts are related closely to the use of fertilizer. Tobacco, for example, has been receiving the equivalent of 50 to 60 lb. of phosphorus annually, from which not more than 10 lb. are removed. High rates also are applied to cotton and potatoes. Peanuts receive little phosphorus directly but are often grown in rotation with cotton.

TABLE 12-7. Phosphorus Level Related to Crop to Be Grown, Beaufort County, North Carolina*

Crop	Percentage of samples testing				
	VL	L	M	H	VH
Ladino-grass seeding	13	31	38	9	9
Small grains	1	22	26	21	30
Tobacco	1	3	14	19	63
Corn	3	20	30	19	28
Soybeans	3	20	40	19	18
Potatoes	0	7	0	9	84

* C. D. Welch and E. J. Kamprath, "Fertility Status of North Carolina Coastal Plain Soils," North Carolina Department of Agriculture, March 1961.

Another summary by a soil management group is shown in Table 12-8. Such summaries can be readily made when soil testing data and site information are punched on cards.

It is interesting that in the organic soils 47 per cent test more than 300 in potassium. On the other hand, loamy sands and sands tested 31 and 46 per cent, respectively, less than 110. Although there will be considerable variation within each group, these data are helpful to agricultural workers. Other summaries may be prepared by soil areas in the state, such as in Iowa.

A summary by general subsoil fertility groups has been made in some states. An example, for Wisconsin, is shown in Figure 13-24. Subsoil fertility will affect response to fertilizer, and a knowledge of this fertility will contribute to a satisfactory interpretation of soil tests. Few states suggest routine subsoil samples.

TABLE 12-8. Summary of Potassium Tests by Soil Management Group, Michigan*

Soil management group	Pounds potassium per acre					
	<110	110–159	160–209	210–249	250–299	>300
Organic	12.5%	14.2%	14.4%	11.7%	9.8%	47.5%
Clay and silty clay	10.9	24.2	28.1	14.1	14.1	8.7
Clay loam or loam	12.8	29.6	25.9	12.2	7.8	11.7
Sandy loam	23.6	28.7	21.1	11.1	7.2	8.4
Loamy sand	31.3	27.2	18.7	9.5	5.4	7.9
Sand	46.5	24.4	15.7	5.3	3.6	4.6

* E. C. Doll, Michigan State University, personal communication.

Periodic Summaries. Several states make periodic summaries. Some of these show that phosphorus and/or potassium levels are gradually increasing as the result of higher rates of fertilizer in the past decade. For example, the average phosphorus test in Ohio increased from 18 to 35 lb. in the period from 1961 to 1971. The potassium test increased from 165 to 237 lb. in the same period. There is some concern that levels may be too high on a few soils and fertilizer practices may need to be reevaluated. This may be true for a few soils and these should be studied carefully. However the following should be kept in mind:

a. Yields are higher.
b. Yields will continue to increase.
c. As other elements become limiting, they should be added.
d. High levels help plants to endure stress periods.
e. Nutrients are in soil and can be used in case application cannot be made—for example, because of adverse weather, fertilizer shortages, or lack of funds.

A point might be mentioned in regard to accuracy of a summary from routine soil samples. These may be coming from the more knowledgeable growers. Hence, the summary may be showing fertility levels to be higher than they actually are when all farms are considered. Also, few samples come from forage fields and these are generally low in fertility.

Summary

1. The selection of adequate lime and fertilization practices depends on the requirement of the crops, weather, and the soil characteristics and necessitates finding a method that will reveal the deficiencies in the soil.
2. Although diagnostic approaches are used in trouble shooting, they are more important as preventive measures.
3. Deficiency symptoms are helpful guides in new fertilizer-using areas. In high-use areas they are a sign of mediocre farming practices and are often difficult to interpret because of their complication with many other problems.
4. Hidden hunger is insidious, but careful plant and soil tests will help to avoid it.
5. The plant integrates all factors in the environment into itself, and tests can be highly revealing. Quick tissue tests in the field on growing plants are useful, but careful interpretation is essential.
6. As a nutrient is added, the percentage of that nutrient in the plant increases. It is important to identify the point at which there is no further economic yield increase.

7. Plant analyses are of real value in making surveys of incipient micronutrient problems in a given area.

8. Balance among nutrients in the plant may be just as important as actual amounts. The relationships among Ca, Mg, K, and NH_4, Mn and Fe, and Zn and P are examples.

9. There are many biological short-term methods, in which both higher and lower plants are used for determining nutrient needs. Eventually all of these methods must be related to field responses for calibration.

10. The principle of soil testing is to obtain a value that will help to predict the amount of nutrients needed to supplement the supply in the soil. Soil tests are of little use in themselves. They must be calibrated against nutrient rate experiments in the field and in the greenhouse. Too, soil fertility is only one of many factors influencing crop production.

11. The physical properties of soils become more and more important as the top profitable yields are approached, but much more work is needed to identify favorable and unfavorable physical conditions.

12. Soil tests can be classified on the basis of probability of response. For example, in a low test for phosphorus there would be a high probability of response to this element, although in some cases it might not be obtained because of other limiting factors.

13. Most of the calibration experiments need to be rerun, using the new technology in crop production, to get higher yields and to provide greater opportunity for response. These results will have more meaning for the commercial grower.

14. Two or more levels of recommendation or provision for higher yields, made by some laboratories, help to provide benefits from soil testing for the leading growers. The goal is to *maintain soil fertility at a level for top profit yields.*

15. In general there are four approaches in recommendations: (a) *buildup* with heavy broadcast rates plus maintenance; (b) *annual application* with modest additions to each crop in the rotation; (c) *rotational fertilization,* a combination of the first two; (d) *replacement* to replace nutrients in crops removed.

16. Prescription and debit and credit methods, in which the contributions of the soil, crop residues, fertilizer, and manure are related to crop needs, are useful. Although the values obviously are influenced by many factors, the methods serve to point out broad needs and possible adjustments.

17. The lime and fertilizer industries can utilize soil and plant tests in a complete service program to provide better identification of the nutrient needs of farmers' fields. The statement *a soil test means a lime or fertilizer sale* is attractive to the dealer and a service to the grower.

18. Summaries of soil tests help to determine the most needed fertilizer ratios and corrective applications. It is possible to develop more realistic general fertilizer recommendations for use by the growers, the vast majority of whom still do not have their soils tested.
19. Phosphorus and potassium levels are being built up in some soils. Excessive buildup must be studied carefully. However, increased fertility levels encourage higher yields and aid the plant in periods of stress. Nutrient balance must be continually evaluated.

Questions

1. For what reason may a plant develop an unusual red or purple color? What factors encourage this change in color? Distinguish between nitrogen- and potassium-deficiency symptoms in corn.
2. What factors must be taken into consideration in interpreting tissue tests?
3. What is the critical level? How does cation balance influence the interpretation of plant analysis for a given cation? What is the difference between tissue analysis and plant analysis?
4. Why cannot just any part of the plant be used for tissue testing?
5. Why must soil tests be calibrated with crop response? How would you set up a series of experiments to determine the calibration of the phosphorus test for corn soils in your state?
6. Explain why a response to phosphorus would be more generally expected in the northern United States than in the South.
7. Corn plants treated with phosphorus at rates of 20, 40, and 60 lb./A. absorb 20, 25, and 30 lb. P/A., respectively. Using the extrapolation technique to obtain the *A* value, what would be the content of available phosphorus in the soil?
8. Why can the growth of *Cunninghamella* be used as an indication of the amount of phosphorus a soil will supply for higher plants?
9. From the standpoints of the grower and the agencies making the tests, what are the greatest problems in the soil-testing program in your state? What can be done to remedy the situation?
10. Would you apply a given nutrient if there were a 50 per cent chance of obtaining a response? A 25 per cent chance? Why or why not?
11. Of what value to a county agent is a summary by crops of the soil-test results? How can he prepare such a summary?
12. Name some advantages and disadvantages of deficiency symptoms, tissue tests, and soil tests for detecting plant-nutrient needs.
13. What is crop logging?
14. Ten per cent of a grower's field is black low-land soil and the remainder is light-colored upland soil. How should the field be sampled? How soon should it be resampled?
15. What complicates the securing of good correlation of soil tests for nitrogen with response to nitrogen in the field?

16. Explain the percentage maximum yield concept. Under what conditions might it not be accurate?
17. What are four general types of recommendations? List advantages and disadvantages of each in your state.
18. If you were a fertilizer dealer what kind of a crop production service would you have for your customer? (Remember it costs money.)
19. Are soil fertility levels increasing in your state? What are the advantages? Disadvantages?

Selected References

Barber, S. A., "The changing philosophy of soil test interpretations," in L. M. Walsh and J. D. Beaton, Eds., *Soil Testing and Plant Analysis*, p. 201. Madison, Wisconsin: Soil Sci. Soc. Am., 1971.

Barrett, W. B., C. F. Engle, and R. M. Smith, "Factors influencing the levels of exchangeable potassium in Gilpin and Cookport soils." *West Virginia Bull.* **622T** (1973).

Bartholomew, W. V., "Soil nitrogen." International Soil Fertility Evaluation and Improvement Program, *Tech. Bull. No. 6,* North Carolina State University, Raleigh, North Carolina (1972).

Berger, J. C., "Soil tests and fertilizer prescriptions." *Farm Chemicals,* **117:**47 (1954).

Bray, R. H., "You can predict fertilizer need with soil tests." *Better Crops with Plant Food,* **45**(3):18 (1961).

Brown, A. L., and B. A. Krantz, "Zinc deficiency diagnosis through soil analysis." *Calif. Agr.,* **15:**15 (1961).

Cate, Robert B., and Larry A. Nelson, "A rapid method for correlation of soil test analysis with plant response data." International Soil Testing Program, *Tech. Bull. No. 1,* Raleigh, North Carolina, North Carolina State University (1965).

Claassen, Maria, and Gerald E. Wilcox, "Another factor affecting your magnesium level." *Better Crops with Plant Food,* **57** (4):10 (1973–4).

Clements, H. F., "Crop logging of sugar cane in Hawaii," in Harry Reuther, Ed., *Plant Analyses and Fertilizer Problems*, p. 131. Washington, D.C.: Am. Inst. Biol. Sci., 1960.

Cope, J. T., "Fertilizer recommendations and the computer programs key." *Auburn University Agr. Exp. Sta. Cir. 172* (1972).

Diagnostic Techniques for Soils and Crops. Washington D.C.: Potash Institute of North America, formerly American Potash Institute, 1948.

Dumenil, L., "N and P composition of corn leaves and corn yields in relation to critical levels and nutrient balance." *SSSA Proc.,* **25:**295 (1961).

Fitts, J. W., "Using soil tests to predict a probable response from fertilizer application." *Better Crops with Plant Food,* **39**(3):17 (1955).

Fitts, J. W., and J. J. Hanway, "Prescribing soil and crop nutrient needs." in R. A. Olson, T. J. Army, J. J. Hanway and V. J. Kilmer, Eds., *Fertilizer Technology and Use*, p. 57, Madison, Wisconsin: Soil Sci. Soc. Am., 1971.

Follett, R. H., and J. F. Trierweiler, *"Ohio soil test summary 1971–72."* *Ohio Extension Bul. 561* (1973).

Hanway, J. J., "Plant analysis guide for corn needs." *Better Crops with Plant Food*, **46**(3):50 (1962).

Hanway, J. J., "Test undried soil samples." *Better Crops with Plant Food*, **48**(5):1 (1964).

Hauser, G. F., "Guide to the calibration of soil tests for fertilizer recommendations." *FAO Soils Bul. 18* (1973).

Hipp, Billy W., and Grant W. Thomas, "Method for predicting potassium uptake by grain sorghum." *Agron. J.*, **60**:467 (1968).

Hoffer, G. N., "Soil aeration and crop response to fertilizers." *Better Crops with Plant Food*, **31**(12):6 (1947).

Jones, J. B., "Soil testing—changing role and increasing need." *Comm. in Soil Sci. and Plant Analysis*, **4**(4):241 (1973).

Kemmler, G., "Response of high yielding paddy varieties to potassium: Experimental results from various rice growing countries." *International Symposium on Soil Fertility Evaluation*, p. 391. New Delhi: Indian Society of Soil Science, 1971.

Loúe, A., "A contribution to the study of inorganic nutrition in maize with special attention to potassium." *Fertilite* **20** (Nov.–Dec. 1963).

Mehlich, A., "Uniformity of expressing soil test results—a case for calculations in a volume basis." *Comm. in Soil Science and Plant Analysis*, **3**(5):417 (1972).

Mengel, K., "Potassium availability and its effect on crop production." *Japanese Potash Symposium*, p. 141. Tokyo: Yokendo Press, Kali Kenkyu Kai. 1971.

Ohio Agronomy Guide 1974–75. Cooperative Extension Service, Ohio State University, Columbus, Ohio.

Ohlrogge, A. J., "The Purdue soil and plant tissue tests." *Purdue Univ. Agr. Exp. Sta. Res. Bull. 635* (revised 1962).

Parker, F. W., W. L. Nelson, E. Winters, and I. E. Miles, "The broad interpretation and application of soil test information." *Agron. J.*, **43**:105 (1951).

Potash Institute of North America, "Fight hidden hunger with chemistry." *Better Crops with Plant Food*, **48**(3):1 (1964).

Reuther, W., Ed., *Plant analysis and fertilizer problems*. Washington, D.C.: Am. Inst. Biol. Sci., 1960.

Soil Test Work Group, "Soil tests compared with field, greenhouse and laboratory results." *North Carolina Tech. Bull. 121* (1956).

Sprague, H. B., Ed., *Hunger Signs in Crops*. New York: David McKay, 1964.

Sullivan, L. J., "What motivates U.S. farmers in use of fertilizer." *Phosphorus in Agriculture*, **61**:19 (1973).

Syltie, P. W., S. W. Melsted, and W. M. Walker, "Rapid tissue tests as indicators of yield, plant composition, and soil fertility for corn and soybeans." *Comm. in Soil Sci. and Plant Analysis*, **3**(1):37 (1972).

Terman, G. L., D. R. Bouldin, and J. R. Webb, "Evaluation of fertilizers by biological methods." *Advan. Agronomy*, **14**:265 (1962).

Tyler, K. B., F. W. Fullmer, and O. A. Lorenz, "Plant and soil analysis for potatoes in California." *Trans. 7th Intern. Cong. Soil Sci.*, 130 (1960).

Voss, R. D., "P—most limiting nutrient for corn in Iowa." In *Proc. 22nd Ann. Fertilizer Ag. Chem. Dealers Conf.*, **141**:1, Iowa State University, Ames, Iowa (1970).

Walsh, Leo M., and James D. Beaton, Eds., *Soil Testing and Plant Analyses.* Madison, Wisconsin: Soil Sci. Soc. Am., 1973.

Welch, C. D., and E. J. Kamprath, "Fertility status of North Carolina Coastal Plain Soils," North Carolina Department of Agriculture, March 1961.

White, R. P., and E. C. Doll, "Phosphorus and potassium fertilizers affect soil test levels," *Michigan State Research Report 127* (1971).

Wilcox, G. E., and R. Coffman, "Simplified plant evaluation of potassium status." *Better Crops with Plant Food,* **56**(1):8 (1972).

13. FUNDAMENTALS OF FERTILIZER APPLICATION

Crops are fertilized to supply the nutrients that are not present in sufficient quantities in the soil. The purpose of an adequate fertilization program is to supply year in and year out the amounts of fertilizer that will result in sustained maximum net return. In other words, this means the most efficient use of fertilizer and other inputs.

The major factors influencing the selection of the rate and placement of fertilizer are the crop characteristics, soil characteristics, expected yield, and the cost of the fertilizer in relation to the sale price of the crop. The economics of lime and fertilizer use are dealt with in Chapter 15.

CROP CHARACTERISTICS

Nutrient Utilization. The approximate amounts of nitrogen, phosphorus, and potassium in certain crops are shown in Table 13-1. The values do not include the quantities contained in the roots. Although uptake will vary considerably, depending on a number of factors, including yield level, nutrient supply in the soil, fertilization, and rainfall, these data indicate the comparative uptake among crops. Method of harvest of the crop, of course, determines the amounts of nutrients actually removed from the field. In crops from which just the grain is removed much lower quantities of nutrients are lost than if the entire aboveground portion were harvested.

In a legume-hay crop such as alfalfa about six times more potassium than phosphorus is removed. The nitrogen content is about equal to that of potassium. If the legume seed has been properly inoculated and adequate lime, phosphorus, and potassium have been applied, the legumes will fix a large portion of this nitrogen from the air on soils low in organic matter or nitrogen. At planting, however, a small quantity of ni-

TABLE 13-1. **Approximate Utilization of Nutrients by Selected Crops**[*]

Plant	Yield per acre	Nitro-gen (lb.)	P_2O_5 (lb.)	K_2O (lb.)	Magne-sium (lb.)	Sulfur (lb.)
Alfalfa	8 tons	450	80	480	40	40
Orchard grass	6 tons	300	100	375	25	35
Coastal Bermuda	10 tons	500	140	420	45	45
Clover-grass	6 tons	300	90	360	30	30
Corn (grain)	180 bu.	170	70	48	16	14
(stover)	8,000 lb.	70	30	192	34	16
Sorghum (grain)	8,000 lb.	120	60	30	14	22
(stover)	8,000 lb.	130	30	170	30	16
Corn silage	32 tons	240	100	300	50	30
Cotton (1,500 lb. lint, 2,250 lb. seed)		94	38	44	11	7
(Stalks, leaves and burrs)		86	25	82	24	23
Oats (grain)	100 bu.	80	25	20	5	—
(straw)		35	15	125	15	—
Peanuts (nuts)	4,000 lb.	140	22	35	5	10
(vines)	5,000 lb.	100	17	150	20	11
Potatoes, Irish (tubers)	500 cwt.	150	80	264	12	12
(vines)		102	34	90	20	12
Potatoes, sweet (roots)	400 bu.	53	26	126	5	—
(vines)		50	14	84	6	—
Rice (grain)	7,000 lb.	77	46	28	8	5
(straw)	7,000 lb.	35	14	140	6	7
Soybeans (grain)	60 bu.	252	49	87	17	12
(straw)	7,000 lb.	84	16	58	10	13
Tobacco, flue-cured						
(leaves)	3,000 lb.	85	15	155	15	12
(stalks)	3,600 lb.	41	11	102	9	7
Tobacco, burley (leaves)	4,000 lb.	145	14	150	18	24
(stalks)	3,600 lb.	95	16	114	9	21
Tomatoes (fruit)	40 tons	144	67	288	10	28
(vines)	4,400 lb.	88	20	175	26	26
Wheat (grain)	80 bu.	144	44	27	12	5
(straw)	6,000 lb.	42	10	135	12	15
Barley (grain)	100 bu.	110	40	35	8	10
(straw)		40	15	115	9	10
Sugar beets (roots)	30 tons	125	15	250	27	10
(tops)	16 tons	130	25	300	53	35
Sugar cane (stalks)	100 tons	160	90	335	40	54
(tops and trash)		200	66	275	60	32

[*] *Better Crops with Plant Food,* **56**(1). Copyright 1972 by the Potash Institute of North America.

trogen may be applied to the small-seeded legumes to help the plants get started on light-colored soils. It usually takes a little time for the nitrogen-fixing bacteria to become established and start functioning. On the other hand, a properly fertilized nonlegume-hay crop may contain a high amount of nitrogen, compared with phosphorus but about the same as potassium, and the fertilization program must be radically different from that used for legumes.

Corn is a crop that takes up large quantities of nitrogen and potassium in relation to phosphorus. This nitrogen and potassium must come from the soil, manure, or fertilizer. Much potassium is contained in the stover, and if corn is harvested for silage large quantities of this element, in addition to the nitrogen, will be removed.

The peanut is a unique plant in that the nuts, the top portion, and many of the roots are removed during harvest. In a 2-ton crop of nuts and a 2.5-ton crop of hay about 39 lb. of P_2O_5 and 185 lb. of K_2O are removed. Oddly enough, the crop responds only slightly to direct applications of phosphorus and potassium unless the soil is low in these nutrients.

Although figures showing the nutrient removal cannot be used as an accurate guide to the amount of fertilizer to apply, they do show the differences that exist in nutrient needs among crops. In addition, they are

TABLE 13-2. Percentage of Total Nutrient Requirement Taken Up at Different Growth Stages*

	Corn – growth periods (days)				
	0–25	26–50	51–75	76–100	100–115
N	8	35	31	20	6
P_2O_5	4	27	36	25	8
K_2O	9	44	31	14	2

	Soybeans – growth periods (days)				
	0–40	41–80	81–100	101–120	121–140
N	3	46	3	24	24
P_2O_5	2	41	7	25	24
K_2O	3	53	3	21	20

	Sorghum – growth periods (days)				
	0–20	21–40	41–60	61–85	86–95
N	5	33	32	15	15
P_2O_5	3	23	34	26	14
K_2O	7	40	33	15	5

* Basic data on soybeans and sorghum from North Carolina and Kansas, respectively. Corn, soybean, and sorghum data appeared in *Better Crops* **56**(2) (1972), **55**(2) (1971) and **57**(4) (1973). Copyright by Potash Institute of North America.

an indication of the rate at which the reserve or *storehouse* nutrients in the soil are being depleted. Some soils have sufficient reserves to draw on for an extended period, but eventually one or more nutrients become limiting. Soil tests help to establish this point. It is revealing to calculate the amounts of nutrients being removed per rotation and to compare them with the nutrients being added.

The percentage of total uptake during various growth stages for corn, soybeans, and sorghum is shown in Table 13-2. During the first two growth stages all three crops take up a higher percentage of their total potassium than of nitrogen or phosphorus. This illustrates the importance of an adequate supply of potassium early in the life of the plant. In the last two growth periods the percentage uptake of nitrogen and phosphorus was greater than that for potassium. An important key is to have enough nutrients in the right proportions in the soil to supply crop needs during the entire growing season.

Root Characteristics. The growth and appearance of the above-ground portion of plants vary considerably from one species to another and even within species, depending on environment. It might be expected that root systems would also vary greatly in rapidity and extent of development. Since the roots are the principal organs through which plant nutrients are absorbed, understanding of the characteristic rooting habits and relative activity should be helpful in developing fertilization practices. Root systems are usually classified as fibrous or tap. The fact that crops may be annuals, biennials, or perennials is also important.

SPECIES DIFFERENCES. A knowledge of the early rooting habits of various crop plants is helpful in determining the most satisfactory method of placing the fertilizer. If a vigorous tap root is produced early, applications may best be placed directly under the seed. If many lateral roots are formed early, side placement may be best. If the rooting habits during the period of rapid growth are known, it should be possible to determine the most effective placement of fertilizer to be used by the plant during this period of growth. Only limited data are available, however, for rooting habits are difficult to study.

A technique has been employed in which small quantities of radioactive phosphorus (P^{32}) are injected at various places in the soil with respect to the plant. The plant is analyzed for P^{32}, and thus it is possible to study rate of root development and activity within given soil zones. A diagram of the extent of root development of several crops two and three weeks after planting is shown in Figure 13-1. At two weeks the corn root system is more extensive than that of tobacco or cotton. The root development of tobacco and cotton would suggest that at least for early absorption of nutrients the presence of the nutrients under the plants is important. At three weeks corn has the most extensive root system and cotton has the most restricted. These differences tend to persist as long as three months after planting.

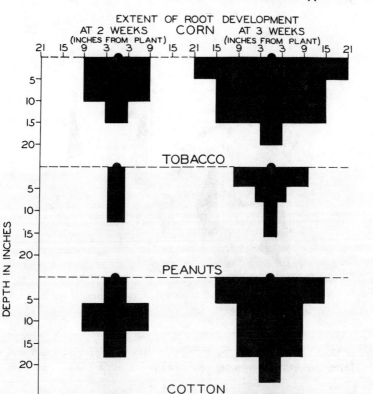

Figure 13-1. The root development at two and three weeks of corn, tobacco, peanuts, and cotton. [*North Carolina Agr. Expt. Sta. Tech. Bull. 101* (1953).]

Carrots show considerable root activity at 33 in. (Figure 13-2) and much more than onions, peppers, and snap beans. The reduction in activity at 13 in. marks the beginning of the mottled compact subsoil in the Blount soil. There was a marked effect of soil, and activity was much less at 33 in. in peat.

Many observations have shown that corn depends heavily on phosphorus near the roots early in the season. From the knee-high stage on, however, corn develops a very extensive root system and has a great capacity for utilizing the nutrients distributed through a large soil zone. The root system of corn, and to a lesser extent that of soybeans, exploits the soil rather thoroughly in contrast to potatoes (Figure 13-3). At later

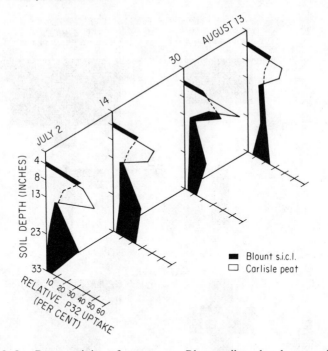

Figure 13-2.   Root activity of carrots on Blount silty clay loam and Carlisle peat. [Hammes et al., *Agron. J.*, **55**:329 (1963).]

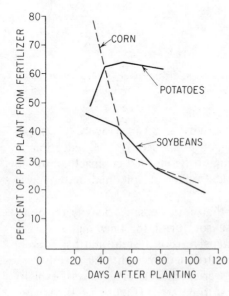

Figure 13-3.   Percentage of phosphorus in corn, soybeans, and potatoes derived from the fertilizer. [Krantz et al., *Soil Sci.*, **68**:171 (1949). Reproduced with permission of The Williams & Wilkins Co., Baltimore.]

stages corn and soybeans utilize only small quantities of fertilizer phosphorus as compared with potatoes. Potatoes have a limited root system, often being confined by the hilled row, and the roots penetrate only a small volume of soil.

In some areas potatoes are grown during a cooler time of year than corn and have a shorter growing season. These factors plus the differences in rooting habits help to explain the greater dependence on the fertilizer and less on the soil phosphorus.

By far the greatest root growth of corn, and for that matter most crops, is in the surface 6 to 9 in. (Figure 13-4). The root weights were

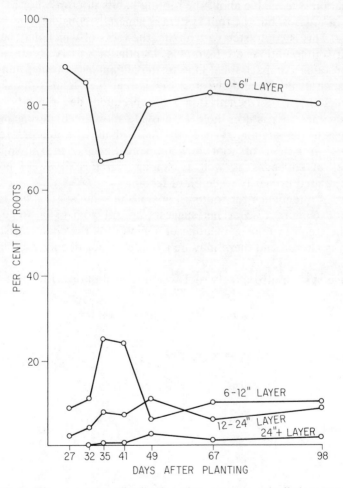

Figure 13-4. The percentage distribution of corn roots vertically in a gray-brown podzolic loam — 150 lb. nitrogen, 105 lb. phosphorus, 200 lb. potassium applied per acre. [Foth et al., *Mich. Quarterly Bul.*, **43**(1):2: (1960).]

obtained by sampling cores of the soil and the total weight was 1,186 lb./A. The soil was an imperfectly drained gray-brown podzolic loam. On a sandy soil in Nebraska corn roots reached a depth of more than 8 ft. and completely extracted readily available soil moisture down to 6 ft. Similarly, corn root weights of 3,000 to 4,000 or more lb./A. have been found. Hence there is as much or even more variation in root growth as in top growth.

Small grain such as wheat may be expected to have a fairly extensive root system that compares rather favorably with corn. The early response of small grains to phosphorus placed near the seed even on soils reasonably well supplied with this element is characteristic. Little information is available about the rooting habits of crops such as Ladino clover or grasses, but the root system is generally considered to be fairly shallow. This is contrasted with alfalfa, the roots of which may penetrate 25 ft. if soil conditions are favorable. Depths of 8 to 10 ft. are common even on quite compact soils. One of the advantages of deep tap-rooted crops such as alfalfa and sweet clover is that they tend to loosen compact subsoils by root penetration and subsequent decomposition. Also, such legumes in pastures help to provide more feed during moisture shortages in the summer than do the more shallow-rooted grasses. Differences in extent of root development have been shown among varieties of soybeans as well as among varieties of sweet potatoes. This is related primarily to depth of feeding.

There is a tendency for root systems of the same species not to interpenetrate (Figure 13-5). This suggests an antagonism or toxic effect. With the trend to thicker planting of some crops the characteristic root pattern is altered and there may be a tendency for deeper rooting if soil conditions permit.

Tillage system also affects root weight distribution with depth (Table

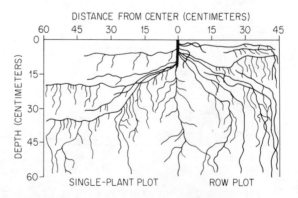

Figure 13–5. Illustration of contrasting rooting patterns of soybeans in single-plant plots and row plots. [Raper and Barber, *Agron. J.*, **62**:581 (1970).]

TABLE 13-3. Effect of Tillage Treatment on Corn Root Weight Distribution with Depth (1969)*

Depth (cm.)	Conv.	Conv. No residues	Chisel	Roto-till	No-till
			(mg./100 cc. of soil)		
0–5	29	49	26	69	137
5–10	38	52	104	136	100
10–15	96	218	110	137	74
15–30	85	111	68	88	73
30–45	73	61	47	52	28
45–60	61	59	35	52	37

* Barber, *Agron. J.*, **63**:724 (1971).

13-3). When the soil was plowed annually (conventional) corn roots developed more extensively than with no-till. Roto-till and chisel were in between. Also, when residues were removed there was greater root growth in the surface 15 cm. Thus it appears that residue decomposition products were inhibiting root growth.

FEEDING POWER. As discussed in Chapter 4, differences exist in cation exchange capacity of roots among various plants. The exchange capacity of the roots of dicotyledonous plants is much higher than that of monocotyledonous plants. Cation exchange values for Ladino clover and alfalfa have been found to be at least twice as great as those for grasses. The magnitude of the exchange capacity affects the absorption of cations, those with a high capacity absorbing relatively more divalent cations such as calcium and less of the monovalent cations such as potassium. In contrast, plants with a low exchange capacity absorb less of the divalent and more of the monovalent cations. This may help to explain the greater potassium absorption by grasses in a legume-grass mixture and the difficulties of maintaining legume stands without an adequate supply of this element. Similarly, plant roots having a high exchange capacity could use calcium phosphates more effectively.

The relative absorption of cations and anions by the root will determine whether H^+ or HCO_3^- will be released by the root. If the NH_4^+ ion is absorbed, H^+ will be released making the soil more acid. If NO_3^- ion is absorbed, HCO_3^- is released by the root and the pH rises. This change in pH of the rhizosphere affects the solubility of such elements as phosphorus, boron, zinc, and manganese. Calcium phosphates may accumulate on the root surface. Soils may vary greatly as to the effect of the pH change on nutrient availability and salt accumulation around the root. Also, phosphorus availability might vary greatly when fertilized with NH_4^+ as compared to NO_3^-.

Plants vary in extent of roots, and the proximity of the absorbing plant roots to the soil surfaces increases nutrient uptake. Even then it has been estimated that roots reach a maximum of 3 per cent of the available nutrients in the soil.

Mychorrizal fungi are associated with the roots of some plants and may help to mobilize some of the nutrients.

SOIL EFFECTS. Soil characteristics have a pronounced influence on the depth of root penetration, and soils with compact B horizons are highly restrictive. Effects of soil series in Illinois on corn root development are shown in tabular form: a number of factors were operating, including bulk density, organic matter content, acidity and plant nutrient content, old root channels, and oxygen supply.

	Flanagan	Muscatine	Clarence	Wartrace	Cisne
Corn root weights (lb./A.)	1846	2008	1758	3136	2647
Depth of penetration (in.)	60	66	38	48	60
Water-holding capacity to rooting depth (in.)	12.8	17.4	6.4	17.1	17.3

The yield of a crop is often directly related to the availability of stored water in the soil. This amount is related to soil characteristics. Illinois reported the following yields of corn (bu./A.):

Available H_2O in profile (in.)	Indiana Lafayette	Illinois Urbana	Iowa Ames
4	79	85	88
8	121	128	129
12	130	136	135

Basically the soil is a rooting medium and a storehouse for nutrients and water. Hence, it is essential that the roots penetrate the soil in order that the plants can not only obtain nutrients but also root deeply enough to obtain water to help the plant better withstand periods of water stress.

In drought-prone areas in the Delta and if the soil is below pH 5.5, annual subsoiling to 18 inches and liming is suggested for soybeans. A crop such as soybeans has only limited ability to penetrate even moderately compacted soil layers, and soybeans have responded to deep tillage in other areas in the South. An additional barrier in the subsoil may be the acidity and the accompanying high degree of Al saturation. Aluminum is toxic to the roots and restricts water uptake.

Soil compaction will become an increasing problem on some soils as other limiting factors are removed, higher yields sought, and heavier equipment used. However, in many instances in the past, attempts to loosen plowpans or heavy subsoils have not been too successful. The operation is most effective when the subsoil is dry so that shattering of the soil occurs. However, in most cases there is a rapid resealing of the subsoil. A practice called vertical mulching has received some attention. Chopped plant residues blown into the slit behind the subsoiler serve to keep the channel open and improve water intake.

PLANT-NUTRIENT EFFECTS.   Heavy fertilization of surface soil and proper management are important in encouraging deeper rooting. Proper nutrition encourages not only greater top growth but also a more vigorous and extensive root system.

Concentration of nutrients in localized zones in infertile soils tends to encourage concentration of roots in that zone (Figure 13-6). The proliferation of roots in the fertilizer band is related to the buildup of high concentrations of nitrogen and phosphorus in the cells that hasten division and elongation. This favors branching and is accompanied by an increase in growth regulator auxins.

Plants absorb nutrients only from those general areas in the soil in which roots are active. It is also well known that plants cannot absorb nutrients from a dry zone. Hence root systems modified by shallow applications of fertilizer may be less effective in time of drought. It ap-

Figure 13-6.   Development of a single soybean root growing through a nitrogen and phosphorus fertilized zone. [Wilkinson and Ohlrogge, *Agron. J.*, **54**:288 (1962).]

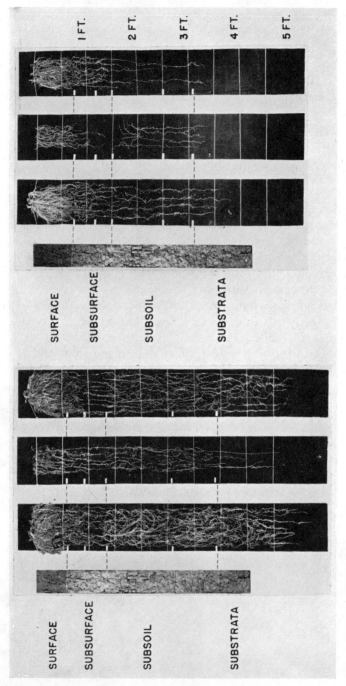

Figure 13–7. Soil treatment affects root growth. A rotation including corn, small grains, and legumes was followed. The corn roots on the left, however, were grown in soil receiving adequate lime, phosphorus, and potassium. Those on the right were grown in soil receiving no fertilizer or lime. (Fehrenbacher et al., *Soil Sci.*, 17:281. Copyright 1954 by The Williams & Wilkins Co., Baltimore.]

pears that the bulk of the less mobile nutrients should be distributed through the plow layer.

Another aspect is that since oxygen is needed when nutrients are absorbed, intense absorption from a fertilizer band may result in a temporary deficiency of oxygen. Of course, oxygen deficiency may be caused by deep placement in wet seasons on poorly drained soils. Also, the acidity may be very high in the fertilizer band because of salt effects and residual acidity from the nitrogen.

Hence it is quite apparent that soil chemistry, fertilizer chemistry, and plant physiology are all important in the fertilizer band. The anions, such as NO_3^-, $H_2PO_4^-$, and Cl^- and cations such as calcium, magnesium, NH_4^+ and potassium are competing with one another for entrance into the plant. New compounds are being formed and some may actually be toxic to the plant roots. Much, however, is yet to be learned about these interactions.

An example of the effect of proper management on growth of corn roots in a Cisne silt loam is shown in Figure 13-7. The Cisne has a clay-pan and is low in native fertility. The use of adequate plant nutrients with proper cropping systems, including legumes, was effective in the development of a much deeper root system.

Root weight of Coastal Bermuda was increased from 6,765 lb. with no nitrogen to 8,098 pounds with 1,600 lb. of nitrogen. Eighty-five to 90 per cent of the roots were in the surface 24 in. on a fine sandy loam.

The effect of fertility treatments on root growth of newly established alfalfa on a Cisne silt loam was shown in Figure 12-2. The influence of plant nutrition on resistance to winter killing is important. By the addition of adequate plant nutrients to the soil, alfalfa is now being grown on many soils in which growth was once thought to be impossible. Similar effects of plant nutrients on extending root growth of wheat have been observed (Figure 13-8).

Winter killing is generally due to the following conditions:

1. Heaving. Plants are lifted from the soil and roots are broken by alternate freezing and thawing.
2. Smothering. Ice-sheet formation causes internal accumulation of toxic products of aerobic and anaerobic respiration.
3. Physiological drought. A frozen soil is like a dry soil.
4. Direct effect of low temperature. As plant tissues freeze and ice forms, the cells rupture and dry out.

The effect of fertilizer treatments on the winter killing of alfalfa is given in Table 13-4. High levels of lime, phosphorus, and potassium increased sugars, soluble proteins, and retention of water, all of which are directly associated with less winter killing.

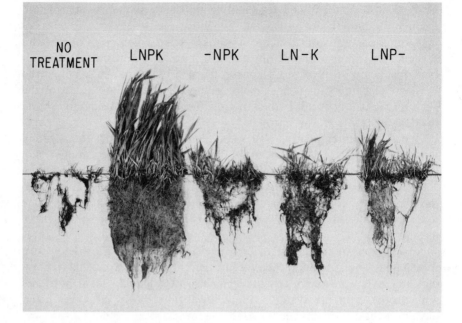

Figure 13–8. Balanced fertility aids winter survival of wheat. Early spring vigor means more stooling and more yield. (Brownstown Agronomy Research Center, Illinois.)

TABLE 13–4. Effect of Plant Nutrients on Winterkilling and Composition of Alfalfa*

Treatments				Starch plus sugar (%) (Nov. 26)	Soluble protein (%) (Nov. 26)	Moisture retained on drying (%) (50°C.)
Lime (tons)	Phosphorus (lb.)	Potassium (lb.)	Winterkill (%)			
0	0	0	90	14.68	10.36	2.70
5	0	0	50	15.53	16.20	3.09
5	132	250	>20	–	–	3.82
10	132	250	>20	19.74	15.37	4.18
5	264	0	50	17.90	16.55	3.85
5	0	500	>20	18.32	14.99	3.46
5	264	50	>20	19.74	17.10	4.45

* L. C. Wang, O. J. Attoe, and E. Truog, *Agron. J.*, **45**:381 (1954).

SOIL CHARACTERISTICS

A most important soil factor in the determination of rate and placement of fertilizers is the amount of soil nutrients that will be needed by the crop during the growing season. Soil and tissue tests, history of management, and observations of plant growth are of great help in evaluating the nutrient-supplying power of the soil. This was covered in Chapter 12. The effects of fixation capacity, certain physical characteristics that influence water movement through the soil, and climatic factors on placement of the fertilizer are discussed in appropriate sections of this chapter.

FERTILIZER PLACEMENT

An important item in the efficient use of fertilizers is that of placement in relation to the plant. Determining the proper zone in the soil in which to apply the fertilizer ranks in importance with choosing the correct amount of plant nutrients. Fertilizer placement is important for at least three reasons:

1. Efficient use of nutrients from plant emergence to maturity. A fast start and continued nutrition are essential for sustained maximum profit. Merely applying fertilizer does not ensure that it will be taken up by the plant. It is usually important to place some of the fertilizer where it will intercept the roots of the young plant and to place the bulk of the nutrients deeper in the soil where they will be more likely to be in a moist zone the greater part of the year.
2. Prevention of salt injury to the seedling. Soluble nitrogen, phosphorus, potassium, or other salts close to the seed may be harmful. An important rule is that there should be some fertilizer-free soil between the seed and the fertilizer band.
3. Convenience to the grower. With a premium on labor saving, speed, and timeliness, placement methods assume additional importance. Commercial growers will want most of the fertilizer plowed down in advance of planting when time is less critical and a minimum amount of fertilizer at planting, properly placed for most efficient plant growth. Under some conditions none will be applied at planting. In the Corn Belt yields are reduced a bushel for each day of delay in planting after the first week of May. The statement has been made that a farmer's time is worth $100 or more an hour at corn planting time.

The importance of adequate quantities of plant nutrients is well recognized. However, full return from their increased use is sometimes limited by improper placement, particularly from the standpoint of injury from fertilizer salts, an important factor in poor stands of crops. When large quantities are applied, maximum returns may not be obtained if most of the nutrients are concentrated near the plant.

It should be stressed that as the soil fertility increases the benefit from applications at planting is generally decreased. The important point is that adequate amounts of nutrients be available to the crop.

Methods of Placement

1. *Broadcast,* in which the fertilizer is applied uniformly over the surface of the land before planting the crop. This method is growing rapidly. The fertilizer is sometimes disked, though preferably plowed down. A drill may be used for shallow placement. With no-till planting in humid areas spreading on the surface may be adequate. Similarly, a special applicator is used to apply anhydrous ammonia or pressure nitrogen solutions below the soil surface.

2. *Sideband,* in which the fertilizer is applied in bands to one or both sides of the seed or plant. Special equipment places the bands of fertilizer 2 to 3 in. to the side and 1 to 2 in. below the seed or transplant (Figure 13-9). It is essential that careful atten-

Figure 13–9. Band application of fertilizer to the side and somewhat below the seed helps to avoid injury to the plant and permits more efficient use of fertilizer. (Courtesy of the Potash Institute of North America.)

tion be directed to the proper adjustment of these machines and that the placement be checked frequently to make certain that the distributor is not out of adjustment. Disastrous results have been reported because of poor adjustment of equipment.

3. *"In-the-row"* applications cover a number of methods. The fertilizer may be run down the same spout with the seed, as in a small grain drill, or it may be placed in the furrow with the seed. As higher rates of higher-analysis fertilizers are used, the emergence of seedlings is delayed and in some cases yields are reduced (Figure 13-10). The fertilizer may also be placed at the bottom of a rather deep furrow. The soil is then bedded back on the row before planting.

4. *Top-dressed or side-dressed,* in which the fertilizer is applied to the crop after emergence. The term *top-dressed* usually refers to broadcast applications on crops such as small grain or forage. The term *side-dressed* refers to fertilizer placed beside the rows of a crop such as corn or cotton.

Results of many experiments might be cited to show the effect of different methods of fertilizer placement on crop stands and yields. The

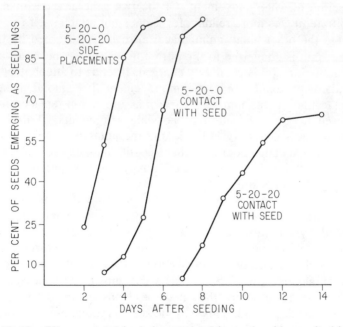

Figure 13–10. Placement 1.5 in. below and 1.5 in. to the side resulted in faster emergence of wheat seedlings under greenhouse conditions in contrast to placement in contact with the seed. [Lawton et al., *Agron. J.,* **52**:326 (1960).]

correct placement of fertilizers on vegetable crops is of particular importance. Large quantities of fertilizers are applied, but any retardation in the rapidity of growth is undesirable. These two factors combine to make sidebands quite beneficial. Results of work with potatoes will be cited. A summary of results from several states in which 1,000 to 2,000 lb./A. of fertilizer were used showed that when the fertilizer was mixed in the row with the seed a yield of 304 bu./A. was obtained. When the fertilizer was placed 2 in. to the side and 2 in. below, a yield of 348 bu. was obtained.

MOVEMENT OF FERTILIZER

Soluble salts are dissolved in the soil solution surrounding the zone of fertilizer application, which thus becomes quite concentrated. The rate and distance of movement of the salts from point of application depends on the nature of the salts, the character of the soil, and climatic conditions.

Phosphorus moves very slowly from the point of placement, for the phosphate ion is almost immobile in the soil. Although water-soluble phosphorus sources move short distances, for all practical purposes the phosphates are not important in salt movement.

The nitrogen salts move up and down in the soil solution, depending on direction of water movement. Of the two general types of nitrogen salts, nitrate moves more readily, for it does not attach itself to soil particles. On the other hand, ammoniacal nitrogen is adsorbed by the soil colloids. As it is converted to the nitrate, it, of course, becomes mobile.

The potassium ion is positively charged. It tends to attach itself to the colloidal complex and is restricted in movement. The anionic make-up of ammoniacal and potassium salts as well as the cation replaced must be considered in any evaluation of possible salt effects. If the accompanying anion is sulfate or chloride and the potassium replaces magnesium on the exchange complex, there is still a soluble salt present. On the other hand, if calcium is replaced, a much less soluble salt is formed if the added anion is sulfate.

Soil water movement is largely in a vertical direction, and for a given set of weather conditions, the extent of movement up and down will be influenced by soil texture. In sandy soils there is usually a greater freedom of soluble salt movement both upward and downward.

As the soils dry out, the concentration of the soil solution is increased, soil water moves upward by capillary movement, and the salts move with it. In some instances they may be deposited on the surface just above the fertilizer band. This may be seen as a white or a light brown deposit coming from dispersed organic matter. Rain immediately after planting followed by a long dry period is conducive to solution of the salts and upward salt movement. With considerable rain the soluble salts tend to move downward again. It would be expected that on soils having a relatively low water-holding capacity the increase in concentration of the soil solution would be greater than on heavier soils having a larger water-holding capacity.

In view of possible movement, the concentration of fertilizer in proximity to the seed is generally hazardous. Excessive concentration of soluble salts in contact with the roots or the germinating seeds causes injurious effects through plasmolysis, restriction of moisture availability, or actual toxicity. The term "fertilizer burn" is often used. The plant will lose water and dry out just as effectively as if it had been placed in an oven.

Certain of the nitrogen-bearing compounds contribute more to damage of germinating seeds than appears to be explained by the osmotic effect. Evidence is that free ammonia is a toxic factor and can move freely through the cell wall, whereas NH_4^+ cannot. Urea, diammonium phosphate, ammonium carbonate, and ammonium hydroxide may cause more damage than materials such as monoammonium phosphate, ammonium sulfate, and ammonium nitrate. This was discussed in Chapter 5. Placement to the side and below the seed is an effective method of avoiding the problem.

SALT INDEX

Fertilizers increase the salt concentration of the soil solution. The salt index of a fertilizer is a measure of this phenomenon and is determined by placing the material under study in the soil and measuring the osmotic pressure of the soil solution. Osmotic pressure is expressed in atmospheres. Salt index is actually the ratio of the increase in osmotic pressure produced by the material in question to that produced by the same weight of sodium nitrate, based on a relative value of 100.

Fertilizer salts differ greatly in their effect on the concentration of the soil solution. Mixed fertilizers of the same grade may also vary widely in salt index, depending on the carriers from which they are formulated. It would be well to emphasize that the higher-analysis fertilizers will generally have a lower salt index per unit of plant nutrients than water-soluble, lower-analysis fertilizers because they are usually made up of higher-analysis materials. For example, to furnish 50 lb. of nitrogen, 250 lb. of ammonium sulfate would be required, whereas with ammonium nitrate 150 lb. would be required. Hence the higher-analysis fertilizers have less of a tendency to produce salt injury than equal amounts of plant nutrients in the lower-analysis fertilizers.

The salt index per unit, 20 lb., of plant nutrient for several materials is shown in Table 13-5. Nitrogen and potassium salts have much higher

TABLE 13-5.   Salt Index per Unit of Plant Nutrients Supplied for Representative Materials*

Material	Analysis†	Salt index per unit of plant nutrients
Nitrogen carriers		
Anhydrous ammonia	82.2	.572
Ammonium nitrate	35.0	2.990
Ammonium sulfate	21.2	3.253
Monammonium phosphate	12.2	2.453
Diammonium phosphate	21.2	1.614
Nitrogen solution 2A	40.6	1.930
Potassium nitrate	13.8	5.336
Sodium nitrate	16.5	6.060
Urea	46.6	1.618
Phosphorus carriers		
Superphosphate	20.0	.390
Superphosphate	48.0	.210
Monoammonium phosphate	51.7	.485
Diammonium phosphate	53.8	.637
Potassium carriers		
Manure salts	20.0	5.636
Potassium chloride	50.0	2.189
Potassium chloride	60.0	1.936
Potassium nitrate	46.6	1.580
Potassium sulfate	54.0	.853
Potassium magnesium sulfate	21.9	1.971

* L. F. Rader, L. M. White, and C. W. Whittaker, *Soil Sci.*, **55**:201. Copyright 1943 by The Williams & Wilkins Company.

† By analysis is meant the percentage of nitrogen in nitrogen carriers, of P_2O_5 in phosphorus carriers, and of K_2O in potassium carriers.

salt indices and are much more detrimental to germination than phosphorus salts when placed close to or in contact with the seed. From this information the relative salt index of a mixed fertilizer can be calculated if the formulation of the fertilizer is known. The comparison between the 5-4.4-8.3 (5-10-10) and 10-8.8-16.6 (10-20-20) is shown in Table 13-6. In fertilizers of these formulations the salt index for equal quantities of plant nutrients is 38.09 and 28.45, respectively. This then is another advantage of higher-analysis fertilizers.

In considering the placement of fertilizers at planting it is important that the differences in salt effects among fertilizers be kept in mind. Obviously, if higher amounts per acre of higher-analysis fertilizers are to be used, the problem becomes increasingly important.

TABLE 13–6. Comparative Salt Index of 5-4.4-8.3 and 10-8.8-16.6

	5-4.4-8.3 (5-10-10)		10-8.8-16.6 (10-20-20)	
	Pounds	Salt index	Pounds	Salt index
Ammoniating solution	148	5.79		
Diammonium phosphate			373	6.37
Urea			260	9.71
Ammonium sulfate	195	6.51		
Treble superphosphate			417	2.10
Superphosphate	1,000	3.90		
Muriate of potash (41% K)	400	21.89		
Muriate of potash (50% K)			667	38.72
Conditioner	100		110	
Filler	157		180	
	2,000	38.09	2,000	56.90

GENERAL CONSIDERATIONS

Band Applications. Early stimulation of the seedlings is usually advantageous, and it is desirable to have N-P-K near the plant roots.

The early growth of the plant top is essentially all leaves. In a crop such as corn leaf growth is completed in about sixty days. Since the photosynthesis is carried on in the leaves, the number of leaves produced in this period will influence the grain produced in the next forty-five days. It is important to have a small amount of nutrients near the very young plants to promote early growth and the formation of large healthy leaves (Figure 13-11). Starter response to N-P-K is often independent of fertility level. Under cool temperatures the early available nutrient supplies may be low because of slow release of nitrogen and phosphorus from the soil organic matter or because of low absorption of phosphorus, potassium, and other nutrients by the plant. Localized applications of fertilizer at planting are commonly referred to as starter or planting fertilizers.

The advantage of early stimulation depends on the crop and seasonal conditions. Some of the factors that might be considered are the following:

RESISTANCE TO PESTS. Under adverse conditions a fast-growing young plant is usually more likely to resist insect and disease attacks.

COMPETITION WITH WEEDS. Weed control is facilitated when the young plants start off rapidly. Herbicides are more likely to be effective and/or cultivation can be started before the weeds become established, and the number of cultivations may be reduced. If one cultivation is eliminated, this saving should be considered in calculating cost of fertil-

Figure 13–11. A readily available supply of nutrients near the young plant helps to insure rapid early growth and the formation of large leaves essential in photosynthesis. (Courtesy of the Potash Institute of North America.)

ization. Vigorous early growth of uncultivated crops such as seedling forage plants is also important in reducing weed competition.

EARLY CROPS. Particularly with vegetables, an early crop is in most cases of paramount importance. A delay of only three or four days may make the difference between a good price on an early market and a break-even proposition. In this same category it should be mentioned that with corn in the areas of shorter growing seasons, early stimulation and rapid growth are important if frost damage in the fall is to be avoided and if high yields are to be realized. On the other hand, in the southern United States, where corn usually matures a month or two before frost, the corn may show an early response to applied phosphorus. This response tends to disappear in some cases, however, before the end of the growing season. If the soil is reasonably well supplied with phosphorus, there may be no effect on the yield. Under such climatic conditions the plants have a sufficiently long growing season to catch up. If the soil is very low in phosphorus, however, yields will be increased.

Broadcast Applications. Broadcast applications usually imply large amounts of nutrients for soil buildup. Phosphorus and potassium as well as some secondary or micronutrients are usually disked or plowed down, although plowing is preferable because the nutrients can be placed deeper. Nitrogen may be plowed down, injected into the soil, or placed on the surface. There are several points in favor of broadcasting the major amount of nutrients.

1. Application of large amounts of fertilizer is accomplished without danger of injuring the plant.
2. Distribution of nutrients throughout the plowlayer encourages deeper rooting. Placement of P-K in a band near the seed tends to encourage concentration of roots in the rope of fertilizer and less complete exploration of the soil for water and for other nutrients takes place. The exception would be with small amounts of nutrients.
3. Labor is saved during planting. The fertilizer marketing season is also spread out.
4. As a consequence of (3), the application of adequate amounts of nutrients is made easier.
5. This method is the only practical means of applying maintenance fertilizer to established stands of forage.

PHOSPHORUS. Since this element tends to move very little, the phosphorus should be placed in the zone of root development. Obviously, surface applications after the crop is planted will not be in the zone of root activity and will be of little value to row crops in the year of application.

An exception to the inefficiency of surface application is with forage-crop fertilization. Top-dressed phosphorus for maintenance purposes is an efficient method of placement. Some of the phosphorus is absorbed by crowns of the plant as well as by very shallow roots. In addition, such applications come in contact with less soil than applications disked in, and there is less opportunity for fixation. The difficulty of placing phosphorus in the soil for a growing sod crop is obvious.

Another exception is with no-till planting where a row crop such as corn is planted directly in sod killed with herbicides or crop residues with no previous tillage. The no-till planter makes a slit in the soil and fertilizer may be applied at planting or broadcast before planting. Many investigators have found broadcast phosphorus to be effective under such conditions. With the surface residues there is more moisture near the surface and roots grow in this area. Under very low fertility levels or drier areas surface-applied phosphorus or potassium with no-till may not be as effective as plowdown. If the soil is low in fertility and the soil permits, the fertility level should be brought up to medium or high before initiating no-till planting. Plowing every four or five years to distribute the phosphorus and potassium accumulated in the surface is desirable. Steep slopes may not allow this, however.

Surface application of phosphorus using chisel tillage appears to be as satisfactory as placement beside the row or at chisel depth for corn.*

Zone fertilization by banding P_2O_5 or K_2O on the surface and plowing down was found to be more effective than banded beside the row or broadcast plowdown on a low phosphorus soil (Table 13-7). Bands of P_2O_5 on surface and plowed down was superior at all rates. Similar results were obtained with potassium. Less fixation and more concentrated zones of nutrients are possible reasons. This would vary with the soil.

The question of localized applications in the row versus broadcast application is one of considerable importance. Fixation as well as proximity to the plant must be considered. Progressive fixation of added phosphorus by certain fractions in the soil tends to diminish its utilization by crops. The mechanism of phosphorus fixation was discussed in Chapter 6.

The placement of water-soluble phosphorus in bands tends to reduce contact with the soil and should result in less fixation than broadcast application. With broadcasting and thorough mixing, the phosphorus comes into intimate contact with a large amount of soil. Results of radioactive phosphorus studies indicate that with the more soluble materials such as monocalcium phosphate there is a greater uptake from band placement. With the less soluble sources such as dicalcium or tricalcium phosphate there is greater uptake when the materials are mixed in the soil. The latter method allows more root-fertilizer contact, and this is apparently important with the less soluble forms. With water-soluble forms granules are more effective, whereas with the less soluble forms finely divided materials are most satisfactory.

Crops during any one season are generally able to recover only a

TABLE 13–7. Effect of Method of Phosphorus Application on Corn Yields (5-yr. ave.)*

	P_2O_5 lb./A.		
	30	60	120
	Bushels per acre		
Banded near row	115	115	115
Broadcast—plowdown	118	121	122
Narrow bands on surface 30 in. apart and plowed down	128	132	133

* S. A. Barber, Purdue University, personal communication.

* L. F. Welch, University of Illinois, personal communication.

Figure 13-12. Forage plants on the left resulted from broadcasting both seed and fertilizer. Plants on the right resulted from drilling seed and banding fertilizer 1 in. below the seed. Both plots were planted September 17 and photographed March 31. [Courtesy of Wagner et al., *Natl. Fertilizer Rev.*, **29**:13 (1954).]

small fraction of the added fertilizer phosphorus, usually less than 25 per cent. This is in marked contrast to the recovery of applied nitrogen and potassium, which may be 50 to 75 per cent. Even with band placement, fertilizer phosphorus is rather inefficiently used. On a Bladen silt loam fairly low in phosphorus, and with a band application of 11 lb. P/A., corn used only 15.2 per cent, soybeans 10.4 per cent, and potatoes 27.2 per cent of the fertilizer phosphorus. Broadcast application would probably have resulted in still lower utilization.

In Illinois it was found that the relative efficiency of broadcast phosphorus as compared to banded for corn on three soils ranged from 0.49 to 1.23. On two of the soils higher yields were obtained by a combination of banded and broadcast. It was emphasized that as the fertility of the soil is increased the advantage for band application would be expected to decrease.

Drilling the phosphorus with small grain seed requires only half as much to produce a given increase in yield as when the material is broadcast. Placing the phosphorus directly under the drill row for forage plants is also superior to broadcast applications or to placement one in. to the side (Figure 13-12). This practice is called band seeding and has been particularly successful in helping to ensure stands. The seeds are dropped behind the disks of the drill and directly over the fertilizer. The beneficial effect of the placement of phosphorus directly under tomato seed instead of 1.5 in. to the side is shown in Figure 13-13. With such seedlings the initial roots grow directly down. Workers in Michigan have shown that phosphorus should be placed directly under the onion sets.

It is generally true that the most efficient use of limited quantities of phosphorus at planting and the highest return from each dollar spent will be obtained by band or localized applications. However, if one considers the profit per acre and has sufficient capital, a different decision may be reached for some crops and under some conditions. On soils low in phosphorus or other nutrients it is usually difficult to produce top yields by row applications. There is some advantage in building up soil fertility, hence productivity, in a long-term fertilizer program. Some research suggests that the soil be built up to a medium to high level in phosphorus by broadcast applications. Maintenance can then be taken care of by periodic broadcast applications and/or by relatively small applications at planting on the most responsive crops in the rotation. In a corn-soybean-wheat-corn rotation the phosphorus might be applied on the corn and wheat. The effects of phosphorus applied to build up soil levels and of row application on a black silt loam are shown in Table 13-8. Row application is quite effective on corn and wheat at the low phosphorus level. With corn and soybeans, the medium and high soil phosphorus levels gave higher yields, although corn gave no response to row applications. In this study only corn and wheat received row application.

Figure 13–13. At five weeks of age 20 lb. P/A. 2 in. directly under the tomato seed (*center*) produced much better growth than 1.5 in. to the side and the same depth (*right*). (Courtesy of G. E. Wilcox, Purdue University.)

When high acre-value crops are to be grown on soils low in phosphorus, it is good insurance to increase the soil level of this element. Flue-cured tobacco, a crop selling for 2,000 dollars or more per acre, is a good example, as are many vegetable crops. An example of establishing basic fertility levels for phosphorus as well as other nutrients is shown in Table 13-9. Annual applications of phosphorus and potassium are then made to correspond roughly with crop removal. Exchange capacity affects desirable levels.

A crop may be grown on a soil high in phosphorus and not respond appreciably to an added quantity. If maintenance applications were recommended, localized placement would probably not be so important, for little response would be obtained. There would tend to be less fixation on a soil high in phosphorus because of the high degree of saturation. This minimizes the importance of localized placement, a point discussed in Chapter 6.

NITROGEN. In contrast to phosphorus, the nitrate salts are mobile and move vertically or horizontally within the soil as the water moves. While ammoniacal nitrogen is held to the clay and organic matter fractions, a mobile salt is produced as soon as it is converted to the nitrate. In fine-textured soils the movement of nitrogen is restricted. A compari-

TABLE 13–8.   The Effect of Soil Test Levels and Row Applications of Phosphorus and Potassium on Yields in a Corn, Soybean, Wheat, and Corn Rotation, Indiana.*

Soil level (lb./A.)	Applied in four years row (lb./A.)	Corn-1 (bu./A.)	Soybeans (bu./A.)	Wheat (bu./A.)	Corn-2 (Tons/A.)
Phosphorus†					
8	0	130	42	37	125
	18	141	46	62	142
	44	151	49	65	150
	88	151	50	68	153
30	0	154	52	63	150
	18	150	50	66	149
	44	152	51	67	151
80	0	151	52	65	148
	18	151	51	68	151
	44	147	51	66	149
Potassium**					
90	0	110	40	63	101
	83	138	47	67	140
130	0	147	49	68	138
	83	147	50	66	147
250	0	151	51	68	150

  * S. A. Barber, Purdue University, personal communication. Results shown are for 1963–1972 inclusive. Experiment started in 1952 with a corn-soybeans-wheat-hay rotation.
  † Row applications applied only for corn-1 and wheat, ¼ on corn-1 and ¾ on wheat. Broadcast applications of 88 and 264 lb. P/A. were applied initially to the 30-lb. and 80-lb. levels, respectively. Additional broadcast applications every four years were 44 lb. and 132 lb./A. until 1959 and 88 lb. and 176 lb./A. since then.
  ** The medium level received 165 lb. K/A. initially and an annual application before corn of 83 lb. in 1955–1957, 126 lb. in 1958, 165 lb. since then. The high level received 500 lb. of potassium initially, 330 lb. every four years until 1961, and 500 lb. since then.

TABLE 13–9.   Basic Fertility Levels for Vegetables*

	P_2O_5 (Bray) (lb./A.)	K (lb./A.)	Mg (lb./A.)	Ca (lb./A.)
Sandy or gravelly loam (8–11 meq./100 g.)	125–300	325–400	200–225	2000–3000
Medium silt loam (12–16 meq./100 g.)	130–150	425–550	275–400	3000–4500
Heavy loam and clay (above 16 meq./100 g.)	165–186	500–650	475–650	5000–7500

  * Lambeth, *Better Crops, Plant Food*, **42:**14. Copyright 1958 by the Potash Institute of North America.

son of soil nitrogen losses from sodium nitrate and ammonium sulfate applied to a Norfolk sandy loam and Cecil clay was made in Alabama. In the Norfolk series 88 per cent of the nitrate and 11.5 per cent of the ammoniacal nitrogen was lost. In the Cecil series comparable losses were 57.8 per cent and 8.1 per cent, respectively.

Under conditions in which large quantities of a wide C:N ratio of organic matter are present in the soil, some of the nitrogen may be temporarily absorbed by the microorganisms and rendered unavailable to the growing plant. Under such conditions fall applications hold fairly well in the soil. The nitrogen may be absorbed by the microorganisms decomposing the organic matter, or the reducing conditions brought about by the actively decomposing organic matter may discourage nitrification of the ammoniacal nitrogen. One characteristic of anhydrous ammonia placed deep in certain soils is that the lack of aeration or decomposing residues delays nitrification and some NH_4-N may stay in the soil a good part of the season. There is increasing evidence of fixation of ammonium by the clay minerals, at least temporarily, as discussed in Chapter 5.

Small amounts of nitrogen are important in starting plants off, but because of its mobility, particularly of the nitrate salts, the fertilizer must be kept well away from the plant roots. On sandy soils extreme care must be taken to avoid damage to the plants when considerable nitrogen is used in the fertilizer.

The addition of ammoniacal nitrogen to the fertilizer at planting has beneficial effects on absorption of phosphorus by the plant (see Chapter 6). Intimate association of nitrogen and phosphorus in the band is essential. Various reasons for the effect of nitrogen on phosphorus uptake are listed:

1. Nitrogen increases root growth and foraging capacity for phosphorus.
2. Nitrogen may affect plant metabolism and the ability of roots to absorb phosphorus.
3. Top growth is increased thus increasing need for phosphorus.
4. Nitrogen compounds may have salt effects on phosphorus solubility. Ammonia sources are generally more effective than nitrates.
5. Residual acidity may increase phosphorus availability.

It is usually undesirable to apply all nitrogen in the row at planting because of possible injury to the crop. A more rational procedure is to apply the bulk either before planting or as a top- or side-dressing after the crop is growing. Water movement carries the dissolved nitrate salts down to the plant roots. If ammoniacal nitrogen, as in ammonium sulfate, is used, it must be nitrified before it moves down in appreciable quantities. An exception is on soils of very low exchange capacity in which some of the ammoniacal nitrogen may move down directly.

POTASSIUM. Potassium salts are much less mobile than the nitrates but more mobile than phosphorus. Concern is sometimes expressed, however, over leaching losses of potassium. Although some losses on sandy soils may occur, losses from most soils are negligible. Studies on several Illinois silt loams over a period of years has shown small annual losses of 2 to 5 lb. K/A.

The lack of movement of potassium is attributed to the fact that as a cation it is adsorbed on the base exchange complex in the soil. Hence amount and type of clay and amount of organic matter will influence movement.

Potassium is loosely held in organic soils. They often flood and the potassium may be moved out of the plow layer or root zone. As much as 60 per cent of the available potassium has been found to be lost from the 0-6 inch layer between fall and spring soil samplings. Potassium will move less in a properly limed soil than in an acid soil, for it can replace calcium more easily than hydrogen and aluminum (Figure 13-14).

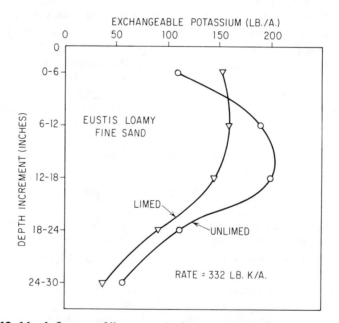

Figure 13–14. Influence of lime on reducing movement of potassium applied at at rate of 332 lb. K/A. in Eustis loamy fine sand. [Adapted from Lutrick. Reprinted from *J. Agr. Food Chem.,* **11**(3):195 (May-June 1963). Copyright 1963 by the American Chemical Society and reprinted by permission of copyright owner.]

Potassium is subject to some degree of fixation on certain soils, particularly those with a high mica content. Such fixation is generally a reversible reaction. Except on soils very low in potassium, fixation is probably not of great importance in permanently immobilizing this element. However, the application of high corrective rates to build up the soil level is usually avoided in high fixing soils.

Because of solubility, potassium salts cannot be placed in contact with the seed in any great quantity. Although the tendency for movement is much less than that of nitrate salts, fertilizers high in potassium should be placed in a band to the side and below the seed or transplant.

Starter responses from potassium similar to those from nitrogen and phosphorus are recognized on many crops and soils. It is generally accepted that potassium should be added to help promote early growth and large leaves except on soils testing high in this element. However, barley responds to potassium fertilizer at planting in North Dakota even on soils testing as high as 600 to 800 lb. exchangeable potassium. This is related in part to the cool temperatures at seeding. The trend to plant crops earlier, under cooler and more stress conditions, improves the probability of responses even on high-testing soils.

The effects of potassium applied to build up soil levels and as a row application on a black silt loam are shown in Table 13-8. Row applications of potassium were effective on corn at the low potassium level. An increase in this level raised the yields of corn and soybeans.

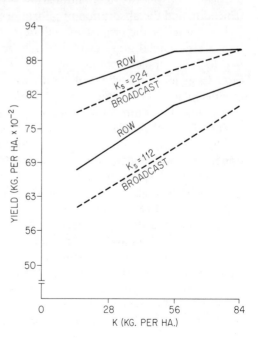

Figure 13–15. The effect of row or broadcast K upon corn yield at two potassium soil test levels, 112 and 224. [Parks and Walker, *SSSA Proc.*, **33**:427 (1969).]

Broadcast potassium is usually less efficient than banded. In Illinois it was found that the relative efficiency ranged from 0.33 to 0.88 on soils low to medium in potassium. As soil test level increases there is generally less difference, however (Figure 13-15). Thus, the importance of placement decreases as higher rates of fertilizer are used over the years.

With no-till and crop residues on the surface in humid areas, broadcast potassium is satisfactorily absorbed by the corn plant because of the roots near the surface.

Workers in Wisconsin showed that alfalfa recovered more of the potassium added to the surface of a silt loam than when applied at various depths. Recovery of the potassium was 41, 29, 19, 16, 10, 15 and 11 per cent from the surface, 7.5, 22.5, 37.5, 52.5, 67.5 and 82.5 cm. soil depths, respectively.

TIME OF APPLICATION

The time at which to apply a fertilizer depends on the soil, climate, nutrients, and crop. With respect to the soil factor, soils differ greatly in the speed with which water will move through them. They also vary greatly in their capacity to fix plant nutrients.

Climate is important in any consideration of fertilizer application. The amount of rainfall between the time of application and the time of utilization by the plant will influence the efficiency of a material. Temperature also affects the availability of certain elements; for example, release of nitrogen, phosphorus, and sulfur from organic matter. It also affects nitrification and the absorption of phosphorus and potassium by the plants.

The nature of the crop itself will determine the need for split applications. A perennial grass crop, if it is receiving fertilizer nitrogen, needs two to four supplements during the year rather than one large application. In general, the effect of a nitrogen application lasts for about two months or less. On the other hand, with a fast-growing plant such as the radish, which matures in forty days, one application should be adequate.

Year-Around Application. This is the process of the farmer and/or fertilizer dealer planning his crop fertilization to take advantage of opportunities that exist throughout the year. The bulk of the fertilizer goes on in the spring in many areas. However, with the trend to plant earlier, the increasing use of fertilizer, and the decline in transportation capacity to deliver the fertilizer to the farmer on time, there is a real need and opportunity to spread any time during the year when soil, weather, and crop conditions permit.

The summer offers opportunity for applications on forages or after small grain is harvested. Considerable interest has been shown by farmers, agronomists, and the fertilizer industry in the application of fertilizers in the fall or winter. It would be economically advantageous to the fertilizer manufacturer and to those engaged in bulk spreading if the period of fertilizer application and use could be extended. The farmer, however, must have a satisfactory answer to his question "How will it benefit me?"

There are a number of factors which point to more fall and winter spreading:

1. The farmer wants to plant earlier and any operations which can be moved to an earlier time are a must. Timeliness is the main key in crop production.
2. Fertilizer usage is increasing and transportation systems are less able to meet requirements.
3. The farmer is more apt to get the fertilizer he needs. In a wet spring he is inclined to plant without all the fertilizer that is required.
4. Custom application by the dealer is increasing so spreading can begin as soon as a field is harvested. The farmer's labor is not involved.
5. Soil compaction is becoming of increasing importance and soils are likely to be drier in the fall than in early spring.

The heavy, relatively level soils which often remain wet in spring and delay broadcasting fertilizers by truck or pull-type spreaders are just the soils in which fall plowdown of fertilizer is agronomically feasible. Phosphorus or potassium applied in the fall will be held safely by the clay and organic matter. Application of nitrogen requires special precautions to be mentioned later.

Sandy soils, which drain more rapidly, are usually less of a problem in the spring because they will support fertilizer spreaders ahead of time to plow and plant. Also it is not advisable to apply nitrogen and/or potassium fertilizers in the fall on sandy or organic soils because of the danger of leaching.

Winter application of phosphorus and potassium is feasible in the warmer areas. In colder areas fertilizers can be applied up to a 4 to 5 per cent slope with a good cover of residues present. With a thin covering of snow the fertilizer pellets melt through and react with the soil.

PHOSPHORUS. In theory phosphorus should not be applied very far in advance of seeding because soluble forms revert to less available forms. The magnitude varies greatly with the fixing capacity of the soil. In actual practice the time of application is adjusted to labor available and other field operations.

Phosphorus can be applied in the fall for a spring planted crop without danger of loss by leaching. On soils of low to moderate fixing capacity broadcasting on the surface and plowing under in the fall is one of the most effective methods. On soils medium to high in phosphorus, the time and method become of less importance (Table 13-8). On such soils proportionately larger applications every two to four years may be recommended. However, a vast majority of soils in the world are low in phosphorus and the most effective time is near the seed or transplant at planting.

Studies conducted in Iowa have shown that phosphorus plowed down in the fall was more effective on corn than when disked in in the spring. The advantage of the plowdown application was thought to be related to deeper placement and less mixing with the soil. Spring plowdown was not compared.

Heavy initial applications at planting of forage crops are sometimes suggested. Generally the yields are high in the beginning but gradually decrease when compared with annual application. This is the result of fixation as well as crop removal.

NITROGEN. In contrast to phosphorus, the possibility of nitrogen losses must be considered in selecting the time at which it may be applied. Theoretically it would be most desirable to add nitrogen as close as possible to the peak requirement of the crop. In addition to time and amount needed by plants, climate and soil type influence the time of application.

The amount and distribution of rainfall is important. The map in Figure 13-16 shows the relation of annual water surplus to geographical

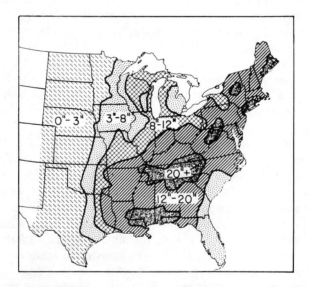

Figure 13–16. Average annual water surplus in inches. The surplus is the amount precipitation exceeds evapotranspiration. [Nelson et al., *SSSA Proc.*, **19**:492 (1955).]

area. To obtain the annual surplus, the potential evapotranspiration (after Thornthwaite) is subtracted from the annual precipitation. A 4-inch waterholding capacity is assumed. The greater the surplus the greater is the possibility of loss.

Another aspect of climate is temperature. As one goes further south the temperatures are more nearly optimum for nitrification during a greater portion of the year. Ammoniacal nitrogen applied before planting would thus be more subject to nitrification and leaching. Nitrification takes place at lower temperatures than was once assumed, for slow nitrification has been observed at temperatures close to freezing. The effect of temperature on nitrification was covered in Chapter 5.

In actual practice ammoniacal forms of nitrogen are generally recommended in the fall in North Central United States when the temperature drops to 45–50°F at four inches except on sandy or organic soil. Eighteen experiments in Illinois have shown that fall applications on corn are 92 per cent as effective as spring applications. Or putting it another way, for each 100 lb. nitrogen in the spring, 110 lb. would be needed in the fall.

The effect of time of nitrogen application on irrigated corn in Nebraska is shown in Figure 13-17. Summer side-dress was clearly superior to spring or fall applications. Eighty pounds side-dressed was equivalent to about 160 lb. applied in the fall or spring. As yields increase and nitrate accumulates in the profile and in the residues, response to nitrogen decreases and time of application becomes of lesser importance. Also, at higher than optimum rates time of application would not be so important. The grower must weigh all of these factors in relation to convenience, timeliness, and ease of earlier applications.

A delay in time of application beyond the 2-ft.-high stage may result in decreased utilization of nitrogen applied to corn. It is interesting to note, however, that nitrogen applications at the pretasseling stage, when deficiency was rather marked, were effective in increasing yields.

On fall-planted small grain on heavy soils in a cool climate application of all or most of the nitrogen in the fall has possibilities under many conditions. In warmer humid regions yields will be somewhat below those obtained by top-dressing nitrogen in late winter because of leaching or gaseous losses. However, there are several important advantages to fall applications on small grains. In late winter the ground may be too wet for machinery to be operated, it may be difficult to convince the farmer of the importance of early applications, and there is usually a saving in labor. On fall-planted small grain a delay in the late winter or early spring nitrogen application reduces the yields.

The following information from Nebraska shows that time of application of 40 lb. N/A. had little effect on yield, but the spring application

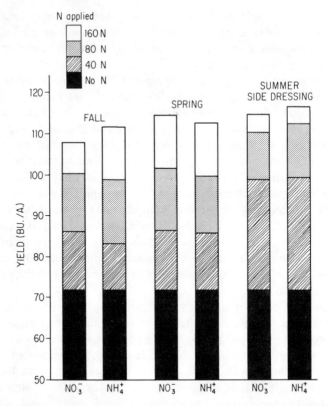

Figure 13–17. Yields related to kind and rate of nitrogen and time of application of nitrogen on irrigated corn in Nebraska. Average of fourteen experiments. Nitrate supplied by $Ca(NO_3)_2$ and ammonium by $(NH_4)_2SO_4$. Medium-to-fine-textured soils of the Crete, Hastings, and Hall series. [Olson et al., *Univ. of Neb. Bul. SB479 (1964)*.]

increased the protein content to some extent (86 experiments, cont. wheat).

	Yield	Protein
No nitrogen	27.0 (bu.)	10.7 (%)
Fall	34.6	11.2
Spring	34.5	11.6
Split (10 fall + 30 spring)	35.0	11.3

The NO_3-N in the 180-cm. soil profile had a marked effect on response of continuous hard red winter wheat to nitrogen in Nebraska (Figure 13-18). The per cent response was even greater with fallow because of more accumulation of NO_3-N and water. It should be kept in

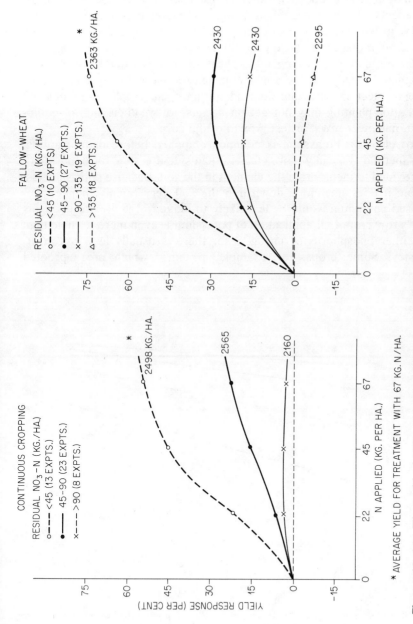

Figure 13–18. The residual NO_3–N in the 180 cm. soil profile affected response of continuous and fallow wheat to nitrogen in Nebraska. (Olson et al., *International Winter Wheat Conf. Proc.*, p. 179. Ankara, Turkey: USDA and University of Nebraska, 1972.)

mind that although some nitrogen may move down through the profile and enter the ground water, some will move down but perhaps not out of reach of next year's deeper-rooted crops. There is evidence of the residual effects of high rates of nitrogen. An example is given in Figure 13-19, in which the residual effect of 150 lb. of nitrogen in 1960–1961 was about equal to the yield of 50 lb. in 1962.

The trend is toward preplant application of ammonium nitrogen. This treatment spreads out the application season and helps to make it possible to meet the growing demand for nitrogen. Applications delayed until after planting may not get on because of weather, lack of equipment, narrower rows, or fast growth of the corn.

POTASSIUM. Potassium is commonly applied before or at planting. This method is usually more efficient than side-dressing, for opportunity is provided to incorporate the element in the soil. Because potassium is a cation, it does not move down into the soil very rapidly; hence side-dressed potassium would be less likely to move to the root zone of this year's row crop. Fall application of potassium is even more feasible than for either phosphorus or nitrogen, for there is usually less loss in efficiency. Some crops—for example, peanuts, which are taprooted

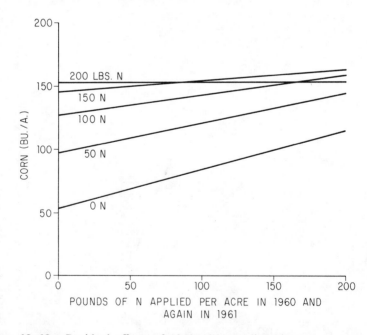

Figure 13–19. Residual effects of 1960–1961 applied nitrogen on corn were quite marked in 1962 on this black prairie soil in Indiana. (Courtesy of S. A. Barber, Purdue University.)

plants—respond better to applications made the year before than to direct applications. This is believed to be the result in part of movement of some of the potassium into the lower soil zones and to its more uniform distribution throughout the absorbing zone of the roots.

Under some cropping practices fertilizers high in potassium may be broadcast once or twice in the rotation. Fall plowdown is a practical approach. These applications are usually made before the more responsive crops such as corn and legumes. Potassium may or may not be included in the starter or planting fertilizer. When 30 to 50 lb. are being applied, all of it is usually most conveniently added to the starter fertilizer. The effects of broadcast and row applications are given in Table 13-8.

Maintenance application on forage crops can be made at almost any time. Fall applications are generally desirable, for the potassium will have had time to move down into the root zone. On hay crops an application made after the first cutting is desirable, for the first crop will have had the benefit of any potassium made available over the winter. There is some evidence that increased yields and efficiency are obtained by top-dressing alfalfa after each cutting. However, the grower may weigh the inconvenience against the value of the yield increase and prefer one large application.

FERTILIZATION OF THE ROTATION

In many respects the problem of distribution of fertilizer in the rotation is essentially one of time of application. When only moderate amounts of fertilizer are being used, the usual practice is to apply small quantities to each crop. Results of only a few comprehensive experiments are available in which comparisons have been made between splitting the application of fertilizer among all the crops in the rotation, applying all at one time, or treating only a few specific crops. Any added effectiveness from small frequent applications must be balanced against the extra time and cost of making them. Actually there are many times at which to fertilize in the rotation (Figure 13-20).

Results are reported on the effect of place in the rotation of the application of 53 lb. of phosphorus in a corn, oats, alfalfa-brome rotation on a calcareous Ida silt loam in Iowa. Splitting the phosphorus between oats and second-year meadow instead of applying it all to oats reduced oat yields by 7 bu. but increased corn yields by 9 bu./A. Splitting the phosphorus between oats and corn instead of applying it all to oats decreased oat yields 8 bu. and forage yields ½ ton but increased corn yields 19 bu. Splitting the application was desirable on this soil to obtain the greatest efficiency of the applied phosphorus. It is interesting to note that 106 lb. of phosphorus on oats gave the highest yield of all crops,

MANY POSSIBLE TIMES TO APPLY FERTILIZERS IN ROTATION

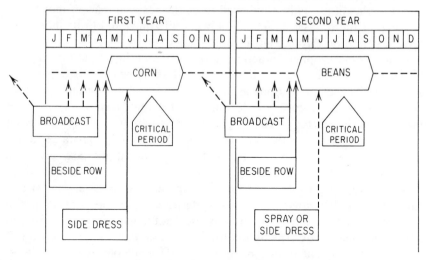

CONSIDER SAFETY, EFFICIENCY, AND LABOR

Figure 13–20. There are many times when fertilizer can be applied in a rotation. An example of a corn-soybean rotation is shown.

regardless of the placement in the rotation. *This illustrates the point that as higher rates of fertilizer are used and soil fertility increases the problem of time of fertilizer application becomes less important.*

Corn-soybeans is a common rotation and sufficient quantities of phosphorus and potassium may be applied to corn to take care of nutrient requirements of both crops. However, soybeans respond to these nutrients to about the same extent percentage-wise. In Iowa, 0-60-60 on a low-fertility soil for each crop increased yields of soybeans 5.5 bu. and corn 14 bu. Formerly, soybeans were considered as scavengers or "eating at the second table."

Peanuts usually do not respond to direct fertilization unless on a low-fertility soil. They are usually grown in rotation with highly fertilized crops such as cotton or corn and respond to increased soil fertility levels.

Forage crops, such as alfalfa, were formerly considered to be luxury consumers of potassium and recommendations were made that the element be applied to other crops in the rotation. However, greater yields are accompanied by increased amounts of potassium in the plant. More frequent cutting implies that younger plants are utilized and they are higher in potassium. The amounts of nutrients in eight tons by cuttings is shown in Table 13-10. The percentage of potassium in the four cuttings ranged from 2.25 to 2.50.

In tropical soils with insufficient water, nitrogen, and phosphate the K

TABLE 13-10. Pounds of Plant Nutrients in Four Cuttings of Alfalfa*

	1st cut	2nd cut	3rd cut	4th cut	Total
N	136 lbs.	111 lbs.	93 lbs.	75 lbs.	415 lbs.
P_2O_5	31	24	22	17	94
K_2O	124	107	98	72	401
Ca	50	41	36	24	151
Mg	13	9	7	7	36
S	6	8	7	5	26
Yield—T.	2.35	2.10	2.03	1.52	8 T.

* Flannery, *Better Crops with Plant Food*, **57**(2):1. Copyright 1973 by the Potash Institute of North America.

released by soil minerals is sufficient for low yields of rice. However, with the use of irrigation, nitrogen, phosphorus and new high-yielding rice varieties response to potassium increases as illustrated by the following, the response to potassium being 6, 12, and 33 per cent, respectively.*

N	P	K	Yield of paddy rice (kg./ha.)		
(kg./ha.)			1967	1968	1969
120	80	0	4.98	4.45	4.02
120	80	80	5.29	4.97	5.35

High Acre-Value Crops. In crops such as vegetables, fruits, and tobacco the acre value of the produce is quite high. Potatoes, for example, may be worth $1000/A. or more, and the cost of the fertilizer is a relatively small item. In such situations large quantities of fertilizer may be applied as an insurance measure, with the thought that even in a good growing season nutrients will not be a limiting factor. This is particularly true of phosphorus and potassium. Rates of 65 to 90 lb. P/A. may be applied to potatoes, whereas the crop may remove less than 35 lb. of this nutrient. These nutrients are not lost to any extent, for continued fertilization will build up the fertility level of the soil to the point at which crops of corn, soybeans, or small grain, with their more extensive root systems, will not respond to preplant applications of phosphorus and potassium. Hence it is possible to recover some of the nutrients and to get an additional return. Nitrogen, however, is usually needed on corn and small grains.

* Mengel, *Japanese Potash Symposium,* Yokendo Press, Tokyo: Kali Kenkyu Kai (1971).

Cover Crops. Wherever intensive vegetable growing is practiced there is a need for a well-developed system of crop management that will supply organic matter. Green-manure crops, such as small grains or crimson clover, are sometimes grown. In some areas intensive fertilization of these crops is practiced not only to produce large quantities of organic material but also to provide a source of nutrients for the next crop. The nutrients, in a sense, are stored in the growing crops and released on decomposition.

There are some advantages to this procedure. Large quantities of fertilizer are usually applied to truck crops, and there is a possibility of fertilizer injury. The cover crops will release the nutrients gradually for distribution throughout the root zone. Results of numerous studies with radioactive phosphorus have shown that this element in a green-manure crop may be more readily available than equivalent quantities supplied as superphosphate. If vegetable crops are being planted under cool temperatures, it may be necessary to use starter fertilizer in the row, for under such conditions the green manure may decompose too slowly.

CARRYOVER EFFECTS

Applications of nutrients to help to set the stage for maximum profits will result in a certain portion of these nutrients being left in the soil. The amounts remaining will depend on the amounts added, the yield, the method of harvesting, and the soil effects. On the average, a crop will remove about one half to three quarters of the nitrogen and potassium and less than one fourth of the phosphorus. What is left, however, is not all carryover because of leaching losses, fixation, and surface erosion.

Examples. Rotational fertilization or soil buildup programs are based on carryover effects, examples of which have been cited earlier in the chapter. Another good example of carryover nitrogen is shown in Figure 13-19. Examples of carryover nitrogen equal to one fourth of the preceding year's application are quite common. In corn, for example, if the desired yield level of 160 bu. was obtained last year and the goal is the same this year, 180 lb. of nitrogen might be recommended. However, if only 140 bu. was obtained an extra 20 to 25 lb. might be suggested.

Soybeans are thought to fix much of their nitrogen from the air. However, in Illinois soybeans removed a net of 1.7 lb. nitrogen per bu. from the soil.* On sandy or low organic matter soils following crops receiving little nitrogen, soybeans would probably fix the greatest share of their nitrogen.

The carryover effect of phosphorus is well known. An example can be cited on corn in Iowa, where 17.6 lb. P/A. increased yields by 18 bu. in the first year and with no further application yielded 15 bu. in the second, for a total 33-bu. gain.

---

* L. F. Welch, University of Illinois, personal communication.

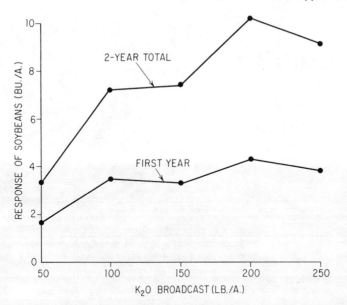

Figure 13–21. Response of soybeans to residual potassium on a soil low in that element. Note the marked response in the second year. [Miller et al., *Soybean Digest,* **21**(3):6 (1961).]

An example of residual effect of potassium on soybeans is shown in Figure 13-21. Actually response by soybeans to carryover fertilizer has been generally quite good provided enough is applied to the previous crop.

An example of carryover effect on soil test and soybean yield from potassium applied on corn over a four-year period with none on soybeans is shown in Table 13-11. This was on a heavy, dark, imperfectly drained soil in central Illinois. Adequate nitrogen and phosphorus was applied on corn.

In view of the many factors affecting carryover, periodic soil and plant tests offer a means of assessing the net effects. The soil may be consid-

TABLE 13–11. Effect of Potassium on Soil Test, Corn Yield, and Carryover Effect on Soybeans.*

Total K_2O on corn 1967–70 (lb./A.)	K soil test after 3rd yr. Mar. 1970 (lb./A.)	Corn yield 1967–69 ave. (bu./A.)	Corn yield (Southern leaf blight) 1970 (bu./A.)	K soil test Sept. 1971 (lb./A.)	Soybean yield 1971 (bu./A.)
0	280	149	104	272	50
1200	421	171	137	448	70

* L. F. Welch, University of Illinois, personal communication.

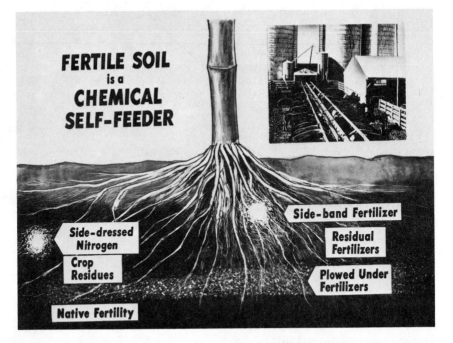

Figure 13–22. The soil might be considered a chemical self-feeder, and there are many possible sources of nutrients. Periodic soil tests are helpful in evaluating the amount in the soil. (Courtesy of the Potash Institute of North America.)

ered a chemical self-feeder (Figure 13-22), and soil tests will help to determine the amounts of nutrients present.

Significance. As fertilizer is applied in increasing quantities, it becomes apparent that increased attention must be given to the value of the carryover. In many cases the cost of fertilization is charged to the crop treated. However, carryover fertilizer is like money in the bank and is part of fertilizer economics. Hence it is apparent that if we are to make a critical economic evaluation of fertilizer use, the carryover value must be considered. This subject is covered in more detail in Chapter 15.

MICRONUTRIENTS

Approaches. Two approaches are commonly followed in supplying micronutrients to crop plants:

1. The addition of needed amounts of specific nutrients to take care of specific needs is applicable in areas known to be severely deficient in one or more of the micronutrients or to crops known to have especially high requirements. The micronutrient may be

added to a mixed fertilizer and, as in boron on alfalfa, in which 0-4.4-25 + B (0-10-30 + B) may be recommended generally over an entire state. In other instances the material may be applied separately such as a broadcast application of zinc or copper. This technique has been successful for many years and will continue to be used.

2. In a second approach small amounts of a micronutrient mixture are added to fertilizers intended for general use. The objective here is to add the material as insurance in forms and quantities that will not harm the more sensitive crops but that will take care of the needs of many crops under certain soil conditions. Justification is offered as follows: (a) it is impossible to determine needs on each field; (b) the absence of visual symptoms does not preclude hidden hunger; (c) it is better to anticipate needs rather than to wait and lose money; (d) low-cost insurance.

Many fertilizer companies offer a premium grade of fertilizer which contains micronutrients. In some areas of the country companies may add 0.5 lb. B/ton to all fertilizers.

Placement. Micronutrients are applied in small amounts and placement methods are directed toward securing the greatest efficiency. For example, to be most effective zinc is banded with the fertilizer near the seed but heavier rates may be broadcast. However, for corn a small amount of $ZnSO_4$ in contact with the seed was more effective than in a band. TVA has observed that Fe-EDDHA was more effective than $FeSO_4$ for grain sorghum. Location of the band near the seed row was important. Manganese and iron may be sprayed on the plants. However, although the grower has every intention of spraying his soybeans at the proper time, his schedule may not permit this, in which case addition to the planting fertilizer is most satisfactory. Broadcast applications allow for too much fixation of manganese or iron. Molybdenum may be applied on the seed or as a spray. Boron may be broadcast with the fertilizer for alfalfa or sidedressed or sprayed on row crops.

Identification of Needs. Some progress has been made in the development of diagnostic tools for identifying micronutrient deficiencies. Soil and plant analysis methods are available but generally satisfactory calibration of these tests is lacking. This is not surprising in view of the small amounts of elements involved and the marked effect of environmental conditions on responses.

In diagnosing a plant nutrient problem, the following steps should be considered:

1. Deficiency symptoms must be recognized.
2. Soil types or locations must be observed.
3. The crop response probability list must be checked.

4. A complete soil test must be made; *p*H is especially important.
5. An analysis must be made on the foliage.
6. Yield goal must be considered.

The micronutrient content of the seed of certain crops may be an indication of need, as determined with molybdenum on soybeans in Indiana. Yield increases from molybdenum occurred when the soybean seeds on the untreated plots contained 1.6 or less ppm. of this element.

Use. There are hazards in the indiscriminate use of minor elements. On an acid soil the addition of manganese would not as a rule be beneficial and might be detrimental. The following is an example of a response to copper but a depressing effect from additional micronutrients on wheat on a Hyde silt loam in North Carolina:

No micronutrients	15 bu./A.
5 lb./A. copper	53 bu./A.
5 lbs./A. each of Cu, Fe, Mn, Zn	36 bu./A.

The interaction of copper and lime on a Hyde soil containing 20.4 per cent organic matter and having a *p*H of 4.9 is shown in Figure 13-23. Wheat yields were increased by copper and lime but the response was greater to copper at the low lime levels.

It has been pointed out that the grower may add micronutrients even though experiments do not show much improvement in yield. The cost is not great and only a small increase will be required, though this increase may not be significant at 19:1 odds. However, the grower might be interested even if the odds were only 1:1.

In Chapter 10 the regulation of plant nutrients in addition to N-P-K was discussed. The amounts are considered to be too low in some states. Michigan, for example, has higher minima: 1 per cent manganese, ⅛ per cent boron, 0.1 per cent iron, ½ per cent copper, ½ per cent zinc, and 0.04 per cent molybdenum.

Michigan workers point out that a farmer may want to try an all-inclusive micronutrient mix on part of his crop. If a response is obtained, further tests must be carried out to determine the source of the response. They suggest a rate per acre of 5 lb. of manganese, 2 lb. of zinc, 1 lb. of copper, 0.5 lb. of boron, and 0.2 lb. of molybdenum per acre on soil with a *p*H of more than 6.0. This mixture should be added to N-P-K fertilizer and placed in bands about 2 in. away from the seed to prevent fertilizer injury. Boron should not be used in the mix for crops such as beans or small grain, which are easily injured. Rates should be doubled for peats or mucks.

As was pointed out in Chapter 8 the addition of certain acid-forming

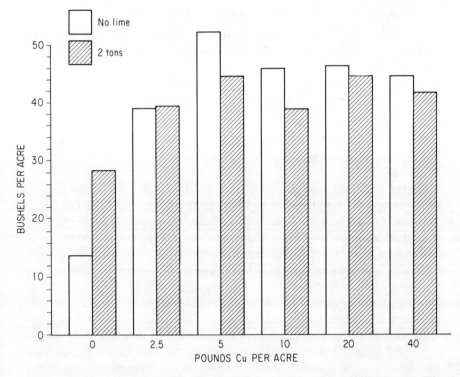

Figure 13–23. Wheat yields as affected by various copper–lime rates. The response to copper was greater at low lime levels. [Younts et al., *Agron. J.,* **56:**229 (1964).]

sulfur-containing fertilizers in a band near the seed or transplant may also serve to correct micronutrient deficiencies that are induced by high *p*H values. Elemental sulfur, ammonium polysulfide, ammonium thiosulfate and H_2SO_4 have been found to be effective under such conditions.

Trends. Micronutrient deficiencies are increasing and can be expected to continue. Higher and higher yields are being obtained and are putting a greater drain on all nutrients. Interaction among major and micronutrients will assume greater importance. As an example, it has been established that phosphorus fertilization may induce zinc deficiency with a high level of soil phosphorus. This is attributed, at least in part, to the formation of insoluble zinc phosphate. In potatoes healthy plants tend to have a P/Zn ratio of less than 400, and in deficient plants the ratio is generally greater than 400, all of which points to the fact that rapid advances in diagnostic techniques and a new philosophy of the use of micro- and macroelements are needed in the farming enterprise. This will take place in the next few years.

UTILIZATION OF NUTRIENTS FROM THE SUBSOIL

The utilization of nutrients from the subsoil has long been under investigation, but little information is available. Soil structure, aeration, pH, drainage, and root distribution are factors that are important with respect to physiological or positional availability but are beyond the scope of this chapter. Two general aspects of the problem are mentioned, however.

1. Utilization of the native nutrients from the subsoil.
2. Addition of nutrients to the subsoil.

Native Nutrients. The subsoils of most soils in the humid regions are generally acid and low in fertility. In the lower rainfall areas, however, the subsoil may be quite well supplied with certain nutrients.

On certain soils, particularly those derived from loess, the B and C horizons may be fairly high in phosphorus. The amount of phosphorus the plants can derive from these lower depths under field conditions has not been accurately determined. Results of greenhouse studies indicate that the phosphorus content of the B and C horizons of certain soils may be more available than that of the surface horizons. It has been suggested that the continued growing of sweet clover, a deep-rooted crop, will increase the available phosphorus in the surface by upward transfer from the subsoil. This improvement is brought about as the organic residues are returned to the surface soil and decomposed. The surface horizons of forest soils are commonly higher in nutrients than the subsoil horizons because of this upward transfer.

Certain soils, particularly loess or alluvial, may be uniformly high in potassium throughout the profile, and there is evidence that subsoil potassium can be utilized by plants. In some areas difficulty is experienced in correlating soil test results for potassium in the surface soil with potassium response. When the content of potassium in the subsoil is taken into consideration, the relation between exchangeable potassium and crop response may be considerably improved.

Some states have made a systematic analysis of the subsoils of major series. An example from Wisconsin is shown in Figure 13-24. Such data could be helpful in making more accurate fertilizer recommendations.

The absorption of micronutrients from the subsoil by crops has been considered by some to be one of the advantages of crop rotation. These elements are supposedly concentrated in the surface horizons as the organic residues from certain crops are turned in. This may be a reasonable assumption, but at present there is only limited experimental evidence to support this view.

Added Nutrients. Deeper application of nutrients may sometimes be desirable. It is known that the application of certain nutrients to a deficient soil zone will increase the concentration of roots in that zone.

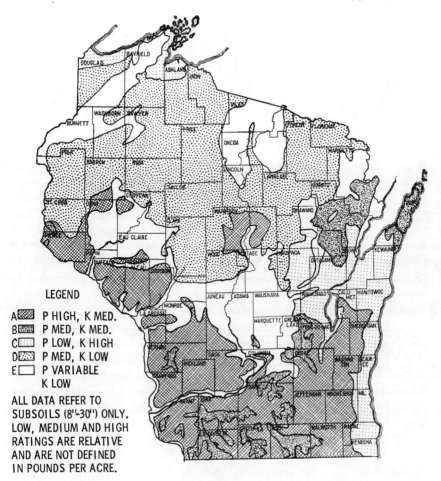

Figure 13–24. General subsoil fertility groups in Wisconsin have been established. (Courtesy of M. T. Beatty and R. B. Corey, University of Wisconsin.)

Lime added to an acid subsoil will not only supply calcium and magnesium but will also reduce the quantities of aluminum and iron in solution.

Much of the early work on subsoil fertilization involved the placement of fertilizer at a depth of 18 to 24 in. behind a subsoiler. Results of many field tests have failed to show consistent benefits, perhaps because the fertilizer was banded. The channel may have sealed off too quickly and the roots may have been unable to penetrate the low fertility zone to reach the fertilizer.

Profile modification, a unique approach, has received considerable attention in North Carolina. It was postulated that root penetration may be impeded for a number of reasons which include cementation of soil particles, compaction, poor aeration, soil acidity, toxicity, or nutrient deficiency. Accordingly, field trials were conducted on Norfolk soils which were plowed to a depth of 24 to 36 in. A portion of the lime and phosphorus was broadcast before plowing and the remainder after plowing. The cost of the operation was about 100 dollars/A. Results are reported to have been favorable.

In this situation the principal benefit may be related to the more efficient use of annual precipitation. More water should enter the soil and the roots should penetrate more deeply. If the plant is utilizing water effectively from a layer 24 to 36 in. thick, in contrast to one only 12 in. thick, there will be a greater reserve against drought. Under some conditions, turning up heavy clay subsoil material may cause the surface soil to seal off more rapidly and to decrease water intake.

Plowing a Palouse-area wheat soil to a depth of 36 in. has been beneficial. This soil has a dense clay subsoil but is permeable to 18 in. Summers are relatively dry in the Palouse area, but winter rainfall recharges the subsoil water. Hence with deep plowing the crops can absorb water from a 36-in. rather than an 18-in. soil layer.

In southern Illinois a low organic matter, low fertility claypan soil was tilled 9, 18, 27, and 36 inches deep and four rates of lime, phosphorus, and potassium were applied. Root penetration and growth during moisture stress were increased, but corn yields were not increased by deeper tillage and fertilization. Moisture penetration was reduced, probably due to the low organic matter, clay soil material being brought to the surface.

In Michigan a soil with a sand lens at 20 inches was tilled 26 inches deep. The sand lens interfered with downward movement of water and roots. Seven years after deep tillage, corn yields were 128 bu. as compared to 87 bu. for 10-inch depth tillage.

In the Solonetzic soils with sodic claypans, plowing 24 inches deep with addition of sulfur or gypsum increases water and root penetration. Wheat yields were increased from 19 bu. with 9-inch plowing depth to 34 bu. with 27-inch.

In somewhat the reverse of deep tillage, Michigan workers developed the idea of installing an asphalt barrier at 24 to 30 inches in very sandy soils in order to help hold the water. This approach has been quite effective for high-value crops.

From this discussion it is apparent that the effects of physical manipulation of the soil profile varies according to the soil. Much more needs to be learned. In the meantime deeper plowing or chiseling to break up plowpans and improved management practices to encourage deeper rooting are of help.

LIQUID FERTILIZERS

Surface or plowdown application of fluid fertilizers does not differ from the use of solid fertilizer but certain other possibilities will be considered here.

Fertigation. This is the application of fertilizer in irrigation water either in surface or overhead irrigation. The principal nutrient applied is nitrogen. The practice is used to some extent in the western United States and Hawaii. Various metering devices introduce the nitrogen into the irrigation system.

There are two main advantages. One, the nitrogen can be applied at the time it is needed. With corn, for example, the plant uses most of its nitrogen from the rapid growth stage to the milk stage. Applying some of the nitrogen at this time is more efficient on sandy soils and just as efficient on heavier soils. Second, one or more field operations are eliminated. Labor is saved, for it is much less time consuming to deposit a drum or cylinder of fertilizer at the end of a field and to let the water do the work. Sometimes the dealer will transport the drums directly to the field. With rank-growing, long-season crops such as sugar cane it is difficult for a man to get through the field to distribute fertilizer. Its application in the irrigation water is one answer.

The question of uniformity of application with surface irrigation is sometimes raised. This may be a real problem under some conditions and with low rates of application of plant nutrients. Under row irrigation a large proportion of the nutrients may be deposited near the inlet. With sprinkler irrigation the nutrients, of course, fall with the water.

Clear liquid mixed fertilizers can be applied as well as sulfur and certain micronutrients such as zinc. However, on annual crops in most instances sufficient phosphorus and potassium can be applied at planting time. The farmer must consider the economics in making his decision to use fertigation.

Foliar Applications. Certain of the fertilizer nutrients soluble in water may be applied directly to the aerial portion of plants. The nutrients must penetrate the cuticle of the leaf or the stomata and then enter the cells. This method provides for more rapid utilization of nutrients and permits the correction of observed deficiencies in less time than would be required by soil treatments. However, the response is often only temporary.

When problems of soil fixation of nutrients exist, foliar application constitutes the most effective means of fertilizer placement. Workers in the Soviet Union have emphasized the importance of foliar feeding in the Arctic regions, where permafrost retards nutrient release from plant residues, root growth, and nutrient uptake. Eventually the practice may be helpful to early-planted crops in the temperate zone, and it can always be combined with regular disease- and insect-spray programs, with a corresponding saving in labor.

So far the most important use of foliar sprays has been in the application of micronutrients. The greatest difficulty in supplying nitrogen, phosphorus, and potassium in foliar sprays is in the application of adequate amounts without severely burning the leaves and without an unduly large volume of solution or number of spraying operations. However, foliar sprays are an excellent supplement to soil applications of the major nutrients.

Foliar sprays in which urea supplies the nitrogen have been successful on apples, citrus, and a number of other crops. Spraying produces more rapid absorption of nitrogen than soil applications. Foliar applications, however, on small grains and wheat are no more efficient than soil applications. It has been found that greater absorption of urea nitrogen by tobacco occurs at night or when epidermal hairs are injured but that no difference exists between upper and lower leaf surfaces.

In Alaska potato yields were increased from weekly potassium sprays. Potassium-containing sprays have also been used on other crops, among which are apples, celery, and pineapple. As with phosphorus, however, the problem of adding sufficient amounts becomes critical.

Microelements lend themselves more readily to spray applications because of the small amounts required, and several have been supplied in this way. Zinc is a notable example, and leaf and dormant spray applications have been found to be many times more efficient than soil applications for fruit trees such as citrus or peach. On some soils difficulty has been experienced in obtaining much uptake from soil applications because of fixation of the zinc. Manganese and iron are other examples. Spray applications of 1 to 2 lb. Mn/A. a few weeks after planting soybeans are commonly recommended. Pineapples in Hawaii are regularly sprayed with ferrous sulfate because soil applications of iron are ineffective.

Efforts to correct iron chlorosis have not always been successful and more than one application may be needed on some crops. Magnesium has also been applied as a foliar spray on certain fruit trees and celery, and boron sprays on fruit trees have been quite successful.

To be most effective two or three spray applications repeated at short intervals may be needed, particularly if the deficiency has caused severe stunting. Care must be taken to identify the nutrient needed, or additional problems may develop. The microelements are usually required in only very small amounts, and too much of one element may be detrimental.

Much more needs to be learned about foliar application, and its value in supplementing standard soil fertility programs must be resolved. The approach to both problems should be the addition of plant nutrients, as revealed by soil and plant tests and crop performance, and the cost of the practice in relation to the increased value of the crop.

Starter Solutions. The application of a nutrient solution around the roots of vegetable crops has been widely used for transplants such as tomatoes and peppers. The principal response is from phosphorus, and such starter solutions are usually high in this element. Starter solutions, however, have generally proved to be of little economic value for flue-cured tobacco.

It has been observed that plants recover more rapidly from the shock of transplanting when starter solutions are used. Increased yields have also been obtained, and in several instances the crops matured at an earlier date. The reason for the observed responses is probably that the transplants have a limited root system and consequently a limited capacity for the absorption of water and nutrients. The addition of a dilute solution of these plant nutrients at the time of planting should make it possible for the plant to absorb these needed elements more readily. Starter solutions, however, must not have a high salt concentration if injury from plasmolysis is to be avoided.

Summary

1. Although nutrient uptake by crops cannot be used as an accurate guide for fertilizer recommendations, it does indicate differences among crops and gives an idea of the rate at which the reserves or storehouse nutrients in the soil are being depleted.

2. Root development helps to give an idea of the most effective placement of fertilizer. For example, potato roots are much less extensive than corn; hence potatoes can utilize nutrients closer to the plant more effectively. Cation exchange capacity of the roots influences ease of uptake of mono- and divalent cations.

3. Soil characteristics influence depth of rooting and yields may be directly proportional to the water available to the roots.

4. Root expansion is favored by a concentration of nutrients.

5. Proper fertilizer placement is important in the efficient use of nutrients from emergence to maturity, prevention of salt injury, and convenience to the grower.

6. Movement of some fertilizer salts in soils is appreciable. Nitrates move most freely, but ammoniacal nitrogen is adsorbed by the soil colloids and moves very little until converted to the nitrate. Potassium is also adsorbed and moves little except in sandy soils. Phosphorus for all practical purposes does not move.

7. The more concentrated materials have a lower salt index per unit of plant nutrient and when placed close to the roots have less salt effect on young plants.

8. Band application at planting is important in providing a rapid start and large healthy leaves. Under cool conditions nitrogen, phosphorus, and potassium are generally less available to the young plants, and band placement will enhance their adsorption.

9. Broadcast application is a means of applying large quantities of nutrients that cannot be conveniently added at planting. The nutrients

are generally mixed through the plow layer, which helps to provide a continuing source of nutrients later in the growing season. They are likely to be in a moist zone for a greater part of the year. Because elements such as phosphorus and potassium move to the roots by diffusion through water films, this point is important.

10. Phosphorus moves so little that it should be placed in the zone of root development. Band applications are most efficient, but for high yields of some crops it is essential to build up the soil level of this element. Forages are an exception in that they can use top-dressed phosphorus reasonably well by absorption through the crowns.

11. In no-till crop production where planting is done in killed sod or plant residues, surface application of phosphorus and potassium are available to the crop. The soil is moist under the residues and the roots develop near the surface in humid regions. If the soil is low in these elements, building the soil fertility of the plow layer is desirable before beginning no-till. In drier regions surface applications may not be as effective because of less moisture under the residues.

12. Small amounts of nitrogen in the planting fertilizer encourage absorption of phosphorus. Because nitrogen is mobile, or becomes mobile after nitrification, application of the major portion before planting or side-dress applications after planting are both effective. In general, the nearer the time of application to peak nitrogen demand, the more efficient the utilization. The amount and distribution of rainfall must be considered in connection with soil texture.

13. Leaching losses of potassium are insignificant except on sandy or organic soils under heavy rainfall. Hence band applications at planting and broadcast applications before planting or at some point in the rotation are effective. Starter responses from potassium similar to those from nitrogen and phosphorus are generally recognized on low potassium soils.

14. In determining the method and time of application, convenience to the grower must be considered with efficiency and safety.

15. As higher rates of fertilizer are used in conjunction with top management, more attention is given to fertilization for the entire rotation. Bulk applications of phosphorus and potassium may be applied once or twice in a four-year rotation in addition to starter applications at planting for the more responsive crops. Nitrogen is applied annually in bulk to the specific nonlegume crops.

16. Year-around fertilization implies fertilization anytime the soil, crop, or weather permits and is becoming a must with increased volume of fertilizer, declining transportation facilities, and a need to save labor. As soil fertility levels increase, the point and time of application of phosphorus and potassium declines in importance. The point is to apply adequate quantities for maximum economic yield.

17. Application of nutrients for the most profitable yield will result in a portion remaining in the soil. In many cases the cost of fertilization is charged to the crop treated. However, if a critical evaluation of fertilizer use is made the carryover value must be considered.

18. Two approaches are followed in the application of micronutrients: (a) addition to take care of specific needs; (b) addition to the fertilizer of a small amount of a mixture of micronutrients for general use. The latter is a method of insurance. Development of suitable diagnostic tools for the recognition of needs and to help predict responses is a major problem.

19. The theory of subsoil fertilization is the promotion of deeper rooting, greater water penetration, and more efficient use of water. So far fertilization with a subsoiler offers little promise on most soils.

20. With problems of soil fixation of nutrients, foliar application may constitute the most effective placement, particularly for certain micronutrients.

Questions

1. Why can phosphorus materials be placed close to the seed or plant? Why is it usually important that phosphorus be close to the seed or young plant? How do you account for the marked response of legumes to band seeding?

2. Why do roots expand in response to plant nutrients on an infertile soil?

3. What soil conditions might affect depth of crop rooting?

4. Under what soil texture conditions would ammoniacal nitrogen be more likely to move? Why? What soil environmental conditions would favor rapid transformation to nitrate forms?

5. What crops in your state are being underfertilized with phosphorus or potassium? Overfertilized?

6. What materials may be used in a low-analysis fertilizer such as 5-4.4-8.3 (5-10-10) to give it a lower salt index per pound of nutrients than a 10-8.8-16.6 (10-20-20)?

7. An experiment is being conducted on a sandy soil to determine the effects of nitrogen and potassium on snap-bean production. The fertilizer is placed in bands 2 in. to each side and 2 in. below the seed. In addition to the phosphorus, nitrogen and potassium are applied in quantities to furnish 50 lb. N/A. (ammoniacal) and 60 lb. K/A. On all plots in which the complete fertilizer was applied the stand was poor; when nitrogen was omitted, the stand was poor, but when potassium was omitted the stand was good. Explain just what happened. What would you do to avoid this trouble?

8. Explain specifically why crops are more likely to experience salt injury on a sandy soil than on a silt loam. Why does potassium not move appreciably in a silt loam?

9. Why might the nature of the root system of the crop being grown affect the decision to build up the fertility level of the soil versus applying fertilizer in the row? How would the economic status of the farmer affect the decision?

10. Explain how band and broadcast applications complement each other in encouraging efficient crop production.

11. You are planning to apply phosphorus broadcast. You had the choice of broadcasting and plowing down or broadcasting and disking in after plowing. Which procedure would be most desirable? Explain fully.

12. Under what conditions is surface broadcast phosphorus and potassium taken up by the plant? Explain.

13. What cropping systems exist in your state in which it might be desirable to apply all the phosphorus and potassium to one crop in the rotation?

14. Under what specific conditions in your state do you believe that all the nitrogen could be applied before planting? Under what conditions should all not be applied before planting?

15. Why does nitrogen applied with phosphorus cause more phosphorus to be absorbed by the plant?

16. Under what conditions would you advocate fall fertilization in your state?

17. What are the possibilities for summer, fall, winter, and spring application of fertilizer in your state? Why is there a need to spread the fertilizer season?

18. Calculate the removal of phosphorus and potassium in a corn-corn-soybean-wheat-alfalfa rotation and in a corn-soybean-wheat-alfalfa-alfalfa rotation. Assume yields given in Table 13-1 and corn, soybeans, and wheat for grain.

19. What is meant by carryover fertilizer? Why is there an appreciable amount in a properly fertilized rotation?

20. Give the pros and cons of the two approaches used in applying micronutrients.

Selected References

Barber, S. A., "Effect of tillage practice on corn root distribution and morphology." *Agron. J.*, **63**:724 (1971).

Barber, S. A., "The influence of the plant root system in the evaluation of soil fertility." *International Symposium on Soil Fertility Evaluation Proc.* **1**:249 New Delhi: *Indian Soc. of Soil Sci.* (1971).

Belcher, Chester R., and J. L. Ragland, "Phosphorus absorption by sod planted corn from surface applied phosphorus." *Agron. J.*, **63**:754 (1972).

Boawn, L. C., and G. E. Leggett, "Phosphorus and zinc concentrations in Russet Burbank potato tissues in relation to development of zinc deficiency symptoms." *SSSA Proc.*, **28**:229 (1964).

Cook, R. L., and J. F. Davis, "The residual effect of fertilizer," *Advan. Agron.*, **9**:205 (1957).

De Wit, C. T., "A physical theory on placement of fertilizers." *Staatsdrukkerij-Uitgeverijbedrijf*, Netherlands (1953).

Dumenil, L., J. Pesek, J. R. Webb, and J. J. Hanway, "Phosphorus and potassium fertilizer for corn: how to apply." *Iowa Farm Sci.*, **19**(10):11 (1965).

Fehrenbacher, J. B., and H. J. Snider, "Corn root penetration in Muscatine, Elliot and Cisne soils." *Soil Sci.*, **77**:281 (1954).

Fehrenbacher, J. B., P. R. Johnson, R. T. Odell, and P. E. Johnson, "Root penetration and development of some farm crops as related to soil physical and chemical properties." *Trans. 7th Intern. Cong. Soil Sci.*, **3**:243 (1960).

Fitts, J. W., and W. V. Bartholomew, "Modifying the soil profile for deeper root penetration." *Better Crops Plant Food* **44**(5):52 (1960).

Flannery, Roy, "Alfalfa absorbs much plant food." *Better Crops with Plant Food.* **57**(2):1 (1973).

Foth, H. D., K. L. Kinra, and J. N. Pratt, "Corn root development." *Michigan State Univ. Agr. Exp. Sta. Quart. Bull.*, **43**(1):2 (1960).

Fox, R. L., and R. C. Lipps, "Distribution and activity of roots in relation to soil properties." *Trans. 7th Intern. Cong. Soil Sci.*, **3**:260 (1960).

Grunes, D. L., "Effect of N on the availability of soil and fertilizer P to plants." *Advan. Agron.*, **11**:369 (1959).

Hall, N. S., W. V. Chandler, C. H. M. van Bavel, P. H. Reed, and J. H. Anderson, "A tracer technique to measure growth and activity of plant root systems." *North Carolina Agr. Exp. Sta. Tech. Bull. 101* (1953).

Hammes, J. K., and J. F. Bartz, "Root distribution and development of vegetable crops as measured by radioactive P injection technique." *Agron. J.*, **55**:329 (1963).

Hansen, C. M., L. S. Robertson, H. J. Retzer and H. M. Brown, "Grain drill design from an agronomic standpoint." *Trans. of ASAE.*, **5**(1):8 (1962).

Holt, E. C., and F. L. Fisher, "Root development of Coastal Bermuda grass with high N fertilization." *Agron. J.*, **52**:593 (1960).

Illinois Agronomy Handbook, Dept. of Agron., University of Illinois, 1973.

Krantz, B. A., W. L. Nelson, C. D. Welch, and N. S. Hall, "A comparison of phosphorus utilization by crops." *Soil Sci.*, **68**:11 (1949).

Lambeth, V., "Desirable soil nutrient level for commercial vegetables." *Better Crops Plant Food*, **42**:14 (1958).

Lavy, T. L., and S. A. Barber, "A relationship between the yield response of soybeans to molybdenum applications and the molybdenum content of the seed produced." *Agron. J.*, **55**:154 (1963).

Lawton, K., and J. F. Davis, "Influence of fertilizer analysis and placement on emergence, growth and nutrient absorption by wheat seedlings in the greenhouse." *Agron. J.*, **52**:326 (1960).

Longnecker, D., and F. G. Merkle, "Influence of placement of lime compounds on root development and soil characteristics." *Soil Sci.*, **72**:71 (1952).

Lucas, R. E., "Micronutrients for vegetables and field crops." Michigan State Univ. Ext. Bull. E-486 (1973).

Lutrick, M. C., "The downward movement of K in Eustis loamy fine sand." *Soil Crop Sci. Soc. Florida Proc.*, **18**:198 (1958).

Martens, D. C., G. W. Hawkins, and G. D. McCart, "Field response of corn to $ZnSO_4$ and Zn-EDTA placed with the seed." *Agron. J.*, **65**:135 (1973).

Mengel, K., "Potassium availability and its effect on crop production." *Japanese Potash Symposium Proc.*, Yokendo Press, Tokyo: Kali Kenkyu Kai (1971).

Miller, R. J., J. T. Pesek, and J. J. Hanway, "Soybean yield responses to fertilizer." *Soybean Digest*, **21**(3):6 (1961).

Munson, R. D., and W. L. Nelson, "Movement of applied K in soils." *Agr. Food Chem.*, **11**(3):193 (1963).

Nelson, L. B., and R. E. Uhland, "Factors that influence loss of fall applied fertilizers and probable importance in different sections of the United States." *Soil Sci. Soc. Am., Proc.*, **19**:492 (1955).

Olson, R. A., "N and Nebraska wheat." *Agr. Ammonia News*, **13**(4):37 (1963).

Olson, R. A., A. F. Dreier, C. Thompson, K. Frank, and P. H. Grabouski, "Using fertilizer nitrogen effectively on grain crops." *Nebraska Agr. Exp. Sta. Bull. SB479* (1964).

Olson, R. A., D. H. Sander, and A. F. Dreier, "Soil analyses—are they needed for nursery data interpretation?" *International Winter Wheat Conf. Proc.*, p. 179. Ankara, Turkey: USDA and University of Nebraska (1972).

Parks, W. L., and W. M. Walker, "Effect of soil potassium, fertilizer potassium and method of fertilizer placement upon corn yields." *SSSA Proc.*, **33**:427 (1969).

Peterson, L. A. and Dale Smith, "Recovery of K_2SO_4 by alfalfa after placement at different depths in a low fertility soil." *Agron. J.*, **65**:769 (1973).

Pumphrey, F. V., F. E. Koehler, R. R. Allmaras, and S. Roberts, "Method and rate of applying zinc sulfate for corn on zinc-deficient soil in western Nebraska." *Agron. J.*, **55**:235 (1963).

Rader, L. F., L. M. White, and C. W. Whittaker, "The salt index—a measure of the effects of fertilizers on the concentration of the soil solution." *Soil Sci.*, **55**:201 (1943).

Raper, C. D., Jr., and S. A. Barber, "Rooting systems of soybeans: I. Differences in root morphology among varieties." *Agron. J.*, **62**:581 (1970).

Runge, E. C. A. "How weather and soil moisture affect corn yield." *Get your answer from us.* Univ. of Illinois Agronomy Field Day (1973).

Sandoval, F. M., J. J. Bond, and G. A. Reichman, "Deep plowing and chemical amendment effect on a sodic claypan soil." *Trans. of ASAE*, **15**(4):68 (1972).

Shickluna, J. C., R. E. Lucas, and J. F. Davis, "The movement of potassium in organic soils." *Proceedings of the 4th International Peat Congress I-IV*, p. 131 (1972).

Terman, G. L., D. R. Bouldin, and J. R. Webb, "Evaluation of fertilizer by biological methods." *Advan. Agron.*, **14**:265 (1962).

Tidmore, J. W., and J. T. Williamson, "Experiments with commercial nitrogenous fertilizers." *Alabama Exp. Sta. Bull. 238* (1937).

Triplett, G. B., Jr. and D. M. VanDoren, Jr., "Nitrogen, phosphorus and potassium fertilization of non-tilled maize." *Agron. J.*, **61**:637 (1969).

Vavra, J. P., "The effect of deep tillage and subsoil fertilization on growth of corn." *Illinois. Fert. Conf. Proc.*, 60 (1966).

Wagner, R. E., and W. C. Hulburt, "Better forage stands." *Nat. Fertilizer Rev.,* **29:**13 (1954).

Wang, L. C., O. J. Attoe, and E. Truog, "Effects of lime and fertility level on the chemical composition and winter survival of alfalfa." *Agron. J.,* **45:**381 (1954).

Welch, L. F., D. L. Mulvaney, L. V. Boone, G. E. McKibben, and J. W. Pendleton, "Relative efficiency of broadcast versus banded phosphorus for corn." *Agron. J.,* **58:**283 (1966).

Welch, L. F., P. E. Johnson, G. E. McKibben, L. V. Boone, and J. W. Pendleton, "Relative efficiency of broadcast versus banded potassium for corn," *Agron. J.,* **58:**618 (1966).

Wilkinson, S. R., and A. J. Ohlrogge, "Fertilizer nutrient uptake as related to root development in the fertilizer band. Influence of N and P fertilizer on endogenous auxin content of soybean roots." *Trans. 7th Intern. Cong. Soil Sci.,* **3:**234 (1960).

Wilkinson, S. R., and A. J. Ohlrogge, "Principles of nutrient uptake from fertilizer bands: V. Mechanisms responsible for intensive root development in fertilized zones." *Agron. J.,* **54:**288 (1962).

Wittwer, S. H., M. J. Bukovac, and H. B. Tukey, "Advances in foliar feedings of plant nutrients." *Fertilizer Technology and Usage.* Soil Sci. Soc. Am., 1963, pp. 429–455.

Younts, S. E., and R. P. Patterson, "Copper-lime interactions in field experiments with wheat. Yield and chemical composition data." *Agron. J.,* **56:**229 (1964).

Younts, S. E., "Response of wheat to rates, dates of application, and sources of copper and to other micronutrients." *Agron. J.,* **56:**266 (1964).

Zubriski, J. C., E. H. Vasey, and E. B. Norum, "Influence of nitrogen and potassium fertilizers and dates of seeding on yield and quality of malting barley." *Agron. J.,* **62:**216 (1970).

14. CROPPING SYSTEMS AND SOIL MANAGEMENT

The principal objective of any cropping practice or soil management program is sustained profitable production. The Soil Conservation Service, begun in 1935 in the United States, has focused much attention on this aspect of agriculture. Soil conservation is essentially good soil management and embraces more than just the prevention of soil losses. Soil erosion is a *symptom* of poor soil management, whether it be inadequate plant nutrients or improper cropping systems. Scarseth (1962) said "Erosion is a symptom, not a primary cause of soil destruction. The primary cause of soil destruction by erosion is impoverishment of nutrients, especially N."

YIELD TRENDS

Crop yields in the United States have gradually increased, as shown in Table 14-1. The yields did not increase so rapidly up to the 1940's as might be expected. A partial explanation is given diagrammatically in Figure 14-1. According to estimates from Ohio, the capacity of the soils to produce decreased 40 per cent in the sixty-year period from 1870 to 1930. This corresponds in general to estimates made on the productivity of Iowa soils over the period 1890 to 1950 and is closely related to fer-

TABLE 14–1. Trends in Crop Yields in United States (1917–1973)

	1917–1926	1927–1936	1937–1946	1947–1956	1957–1963	1964–1970	1971–1973
Corn (bu.)	27	24	32	39	56	75.1	92.2
Wheat (bu.)	14	13	16	18	25	27.8	32.8
Cotton lint (lb.)	160	180	255	317	451	478	488

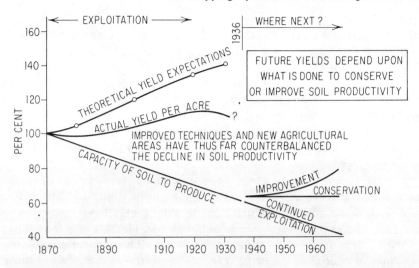

Figure 14–1. Improved practices in Ohio since 1870 should have resulted in yields 40 to 60 per cent higher per acre in 1930, but the aggregate yield increased less than 15 per cent. Improved practices only slightly more than counterbalanced the decline in the ability of the soil to produce. Yields can be increased only if proper soil-management programs are adopted. (*Ohio State Agr. Ext. Serv. Bull. 175, 1936.*)

tility levels. Soil organic matter, hence nitrogen supply, has decreased. The removal of such elements as phosphorus, potassium, calcium, magnesium, and sulfur has generally been greater than the amounts returned to the soil in the form of manure and commercial fertilizers.

The actual increase in crop yields in Ohio of about 15 per cent during the period 1870 to 1930 was the result of such factors as improved varieties, insect and disease control, cultural practices, machinery, drainage, and increased fertilization and liming. With these developments it was estimated that crop yields should have increased 40 to 60 per cent if soil fertility had been maintained. These data from Ohio illustrate the fallacy of using yield trends alone over a period of years as a measure of soil fertility. Yields of crops in Ohio improved considerably in the period from 1930 to 1972. Yields in the United States have increased greatly from 1962 to 1972. For example, corn averaged 64 bu./A. in 1962 and 97 bu./A. in 1972.

The marked increase in yields has been due to rapid advances in technology of which fertilizer is only one part. USDA estimated in 1964 that elimination of N and P would decrease corn yields in Illinois 37 per cent, grapefruit in Florida 94 per cent, and alfalfa in Arizona 34 per cent. The values would be greater now. Also, if fertilizer were not used on corn in Iowa 29 per cent more land would be needed. This would mean less suited land would go into production with greater opportunity of erosion. Much more dramatic results could be reported in developing countries. In Mexico wheat yields increased from 775 kg./ha. in 1943 to 2,700 kg./ha. in 1968.

In dry-land regions water is more of a limiting factor than fertility, and fertility has declined only slowly. In the Southeast most of the soils are inherently low in fertility and have been farmed intensively for 100 to 200 years. Much of the drop in productive capacity is probably more the result of losses of soil by erosion rather than the actual depletion of fertility by cropping.

The continued increase in yields on experiment station plots is important, for they serve as a guide as to what can be done. In those states and with those crops on which experimental yields have continued to increase, the average state yields, although considerably less than those of experiment stations, have continued to rise. This emphasizes the importance of continued intensive research to overcome the effect of those factors that might be limiting crop yields.

AIM OF CROP AND SOIL MANAGEMENT

The aim of all cropping systems should be sustained maximum profit from the farming operation. In evaluating a crop and soil management system for its effects on sustained high production there are several factors that must be kept in mind:

1. Organic matter and soil tilth.
2. Plant-nutrient supply.
3. Incidence of weeds, insects, and diseases.
4. Water intake and soil erosion.

Soils differ greatly in their characteristics, hence in their management requirements. For example, one soil, a silty clay loam high in organic matter, may have excellent tilth and could endure management practices that decrease tilth for some little time before problems are encountered. Another soil, a silt loam low in organic matter, might be in poor tilth, and the same management practice would cause trouble immediately. It is important that soils be evaluated with respect to their management requirements. Experimental data should be available to show the effects of management practices on different soils. In many instances there are insufficient data on management practices required for high yield levels.

ORGANIC MATTER IN THE SOIL

Much emphasis was once placed on the organic matter content of a soil as an indication of its productivity. With the increasing use of nitrogen it is not necessary nor even wise to rely on the soil organic matter for high yields of a crop such as corn.

Bradfield (1963) sums up the subject as follows:

> For most farmers, the only economical way to get more organic matter in their soil is to grow more organic matter on their own farms. Larger crops will mean more roots, more stalks and stubble, more food for livestock, and hence more manure to return to the soil. The cheapest way to grow these larger crops is by more liberal fertilization and by use of good soil-building rotations in which the soil is so handled that maximum efficiency is obtained from the fertilizers. This will require the best available seed, the best adapted cultivation practices, and the most efficient use of all organic residues. Organic farming with chemical fertilizers will result in even higher yields per acre and even more organic matter in our more productive soils.

Effect of Cropping Systems. Cropping systems affect the organic matter in the soil because of several factors:

1. Tillage of the soil produces greater aeration, thus stimulating more microbial activity, and increases the rate of disappearance of soil organic matter (Figure 14-2). A cropping system with a large proportion of row crops thus encourages a much more rapid loss of organic matter than a cropping system with a large proportion of close-growing or sod crops. Tillage in many instances has increased the yields of nonlegumes on soils containing a good supply of organic matter but to which insufficient nitrogen was applied. This is probably related in part to the effect of increased aeration on organic matter decomposition and subsequent release of nitrogen. Too, new surfaces are continually exposed, and there is greater opportunity for wetting and drying of the soil. Of course, all of this causes gradual depletion of the soil nitrogen.

2. Cropping systems differ in the amount of plant residues they con-

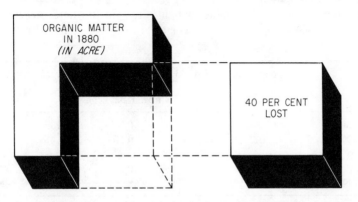

ORGANIC MATTER
IN 1880
(IN ACRE)

40 PER CENT
LOST

Figure 14-2. Loss of total organic matter from South Dakota soils has been heavy during 65 years of farming. (*South Dakota Agr. Expt. Sta. Circ. 92,* 1952.)

tribute. Corn for grain may add 3 to 5 tons/A. of stover. Roots may furnish another 1 to 3 tons. Corn for silage results in removal of the stover and grain. A grass-legume sod will produce about the same amount of residues, but most of it will be removed in hay and grazing. With a good growth of small grain, 2 to 3 tons of organic matter may be returned if the straw is left on the field, but only about ½ ton may be returned if the straw is removed. In peanuts the tops, nuts, and many of the roots are removed, a system that will encourage much more rapid loss of organic matter. In a livestock program, in which the grain and plant parts are consumed by the animal, only a fraction of the organic matter will be returned to the land, and then the manure will be applied to those fields closest to the barn.

3. Cropping systems vary in nitrogen content of the plant residues, and the accumulation of soil organic matter is in part related to this content. If residues low in nitrogen are turned under, much of the carbon will be evolved as carbon dioxide in decomposition before the ratio approaches 10 or 12:1.

Corn well fertilized with nitrogen may contain as much as 1.5 per cent in the stover. In contrast, corn receiving small amounts of nitrogen may have less than 0.5 per cent in the stover. Because of this and the much larger amount of organic matter produced, the stover from the well-fertilized corn will be more effective in maintaining organic matter than that from poorly fertilized corn. Most crop residues containing around 1.5 per cent nitrogen do not need additional amounts for rapid humus formation.

There has been much discussion of the effect of added nitrogen on accumulation of organic matter. The data in Table 14-2 indicate considerable gain in both carbon and nitrogen from the addition of nitrogen to mature millet residues.

Nitrogen applied to NuGaines wheat straw in the field increased the nitrogen percentage and gave a lower C/N ratio:*

	Sept. 7	Nov. 15	Mar. 22	May 22	Oct. 3
			Nitrogen, per cent		
N added	1.52	0.90	0.98	0.83	1.14
No N added	0.27	0.46	0.57	0.46	0.81
			C/N ratio		
N added	26	43	40	47	34
No N added	145	84	68	84	48

* Smith and Douglas, *SSSA Proc.*, **35**:269 (1971).

TABLE 14-2. Effect of Additions of Urea to 12.5 Tons per Acre of Mature Millet Residues on Gain in Carbon and Nitrogen on Evesboro Loamy Sand*

Nitrogen added (lb./A.)	Gain in carbon (tons/A.)	Gain in nitrogen (lb./A.)
0	1.25	171
200	1.82	276
400	2.20	330
800	2.89	402

* L. S. Pinck, F. E. Allison, and V. L. Gaddy, *J Am. Soc. Agron.*, **40**:237 (1948).

The application of nitrogen in the fall to crop residues low in this element is an interesting possibility. With a low nitrogen supply it is possible to find undecomposed plant residues such as corn stalks one or two years after plowing down. Under these conditions, a growing corn stalk may be pulled up and the roots may be matted in the undecomposed residue. The effect on the nitrogen supply to the growing plant is obvious.

To bring a 0.5 per cent nitrogen material to 1.5 per cent would require 20 lb. N/ton of material, which might be applied when the residues are turned under. It should be mentioned, however, that nitrogen applied on plant residues early will generally produce a smaller increase in the yield of the succeeding crop than if it were applied to it directly. The nitrogen assimilated by the soil organisms is not immediately available, and there may be other losses including volatilization and leaching. Nitrogen in the soil humus is released at an average rate of about 2 per cent of the total each year, until, of course, the C:N ratio reaches a point at which further decomposition would be inhibited.

Under some conditions a lack of sulfur can retard the decomposition of organic residues. This was discussed in Chapter 8, where it was pointed out that the addition of fertilizer sulfur is necessary under such conditions to bring about decomposition of the added organic material.

SOIL ORGANIC MATTER. Much work has been done in many areas and on many different types of soil to determine the effect of cropping systems on soil organic matter. In general, in experiments started on virgin soils it has been difficult to maintain the soil organic matter even with the best cropping systems. On soils in which the organic matter had been depleted before experimentation began, cropping systems with a high proportion of sod or close-growing crops and a minimum of cultivated crops are likely to result in an increase in the organic matter and nitrogen content.

In areas of high mean annual temperatures, such as in the southern United States, decomposition continues over a considerable part of the year. It is difficult to increase the organic matter content under such conditions. In contrast, in the northern United States the soil organic matter content, if depleted, can be increased more readily by certain cropping systems.

Many factors determine whether the soil organic matter is increased or decreased by cropping systems. *The key point is to keep large amounts of crop residues (stover and roots) passing through the soil. Continued good management, including adequate fertilization, helps bring this about.*

Effect of Added Plant Nutrients. The amount of lime and fertilizer in the cropping system affects not only the yield of the harvested crop and its composition but also the amount of crop residues produced. This increased quantity of crop residues brought about by greater amounts of plant nutrients is important in organic matter maintenance. In addition, higher yields mean more extensive root systems which distribute organic matter deeper in the soil.

NITROGEN. The effect of nitrogen on corn grain and stover yields is illustrated by the data shown in Table 14-3. The higher rates of nitrogen not only raised grain yields but also increased stover yields 50 per cent.

The general relation that exists between additions of nitrogen and loss of soil nitrogen is shown in Figure 14-3. This indicates that the yearly losses of soil nitrogen, hence soil organic matter, decrease with larger additions of nitrogen. It would appear that if nitrogen additions were equal to or slightly greater than crop removals the losses in the soil would be reduced to a minimum.

On a Cisne silt loam in southern Illinois over a twelve-year period the addition of nitrogen to a corn-soybean-wheat (legume, grass) rotation already receiving lime, phosphorus, and potassium increased the nitrogen content of the soil 0.014 per cent and the carbon 0.15 per cent.

PHOSPHORUS. The effect of phosphorus in increasing the grain, straw, and lespedeza yields is shown in Table 14-4. Although it should be kept in mind that these soils were low in fertility, adequate fertilization

TABLE 14-3. Effect of Nitrogen on Corn Grain and Stover Yields*

Nitrogen (lb./A.)	Grain (bu./A.)	Stover (lb./A.)
0	22.8	3,614
40	52.1	3,356
80	81.7	5,078
120	100.5	5,459
160	102.3	5,328

* B. A. Krantz and W. V. Chandler, *North Carolina Agr. Exp. Sta. Bull. 366* (revised 1954). Adequate phosphorus and potassium were applied.

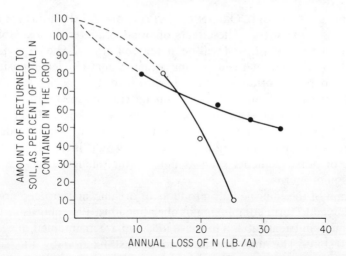

Figure 14–3. Yearly losses of N decrease as nitrogen, expressed as a percentage of that element contained in the crop, is returned to the soil in increasing amounts. ●—values for soils with more than 4500 lbs./A. of total nitrogen; o—values for less than 2500 lbs./A. of total nitrogen. (From Melsted, "New Concepts of Management of Corn Belt Soils," in A. G. Norman, Ed., *Advances in Agronomy,* Vol. VI. Copyright 1954 by Academic Press, Inc.)

greatly increased the amount of plant residues turned back to the soil. Such residues should be helpful in maintaining or even increasing soil organic matter.

OTHER NUTRIENTS. The effects of nitrogen and phosphorus are cited as examples. The addition of other nutrients such as potassium, lime, sulfur, or micronutrients that increase crop yields will also increase the amount of residues.

TABLE 14–4. Effect of Phosphorus on Yields of Oat Grain, Straw, and Lespedeza*

Phosphorus (lb./A.)	Grain (bu./A.)	Straw (lb./A.)	Lespedeza (lb./A.)	
			Same year	Next year
0	15.2	1,108	415	742
4.4	86.7	5,730	944	1,584
6.2	85.3	5,537	1,230	1,624
11.0	121.6	6,542	1,289	1,828
22.0	126.3	10,733	1,075	1,954
33.0	127.5	8,246	1,450	2,544

* Unpublished data, W. H. Rankin, North Carolina Agr. Exp. Sta. Adequate nitrogen and potassium applied.

EFFECTS OF ADDED ORGANIC MATTER ON AVAILABILITY OF NUTRIENTS IN THE SOIL. The effects of organic matter on availability of nutrients already in the soil will be mentioned briefly. The large quantity of carbon dioxide evolved during organic matter decomposition is thought to be important to the release of certain nutrients, inorganic phosphorus in particular. The carbon dioxide dissolves in water and forms carbonic acid. The result is a decrease in soil pH. This effect would be of greater importance on neutral or alkaline soils. Under such conditions the temporary reduction in pH would increase the rate of release of other elements such as boron, zinc, manganese, and iron as well as phosphorus.

Certain of the intermediate products of organic matter decomposition are believed to form complexing or chelating ions. Phosphorus or certain of the micronutrients attach to these ions and are maintained in a weakly ionized state. The ions are retained against fixation by the soil but remain in a form that can be utilized by plants.

SURFACE SOIL VERSUS SUBSOIL

Profitable crop production on eroded soils has been an important agricultural problem. The generally reduced crop yields of nonlegumes on subsoil is well known. On permeable soils this is largely the result of less organic matter and the subsequent lower release of nitrogen. In a Marshall soil in Iowa added nitrogen on the subsoil brought the corn yield up to about equal to that obtained on top soil (Table 14-5).

Research in Ohio on a permeable subsoil and on a tight subsoil high in clay from which the surface had been removed provides an interesting example. A corn, small-grain, and alfalfa rotation was used. On the permeable subsoil, with adequate lime, phosphorus, and potassium, corn yields were 95 per cent of those grown on the topsoil. Added nitrogen did not produce a notable response because the alfalfa supplied this element.

TABLE 14-5. **Effect of Nitrogen on Corn Yields from Subsoil and Normal Surface Soil**[*]

1958			1959		
Nitrogen (lb./A.)	Subsoil (bu./A.)	Normal soil (bu./A.)	Nitrogen (lb./A.)	Subsoil (bu./A.)	Normal soil (bu./A.)
0	31.4	77.3	0	30.3	82.0
67	83.5	100.5	60	69.6	105.4
133	101.3	103.5	120	106.6	122.3
200	107.3	104.8	180	122.4	122.6
267	105.1	96.1	240	122.8	116.7

[*] Adequate phosphorus and potassium added. O. P. Englestad and W. D. Schrader, *SSSA Proc.*, **25**:497 (1961).

On the tight soil considerable difficulty was met in getting stands of corn and legume seedings and yields of nonlegumes were generally low. Once a stand of alfalfa was obtained, however, yields of hay were reasonably good.

These data illustrate that on permeable subsoils high yields of non-legume crops can be had with added nitrogen or in rotation with alfalfa. In some years, however, moisture may be limiting because of less water entrance and less available waterholding capacity in the subsoil.

Under unfavorable physical conditions yields of nonlegumes compara-ble to those grown on topsoil are not generally attainable. However, once a stand of a legume seeding has been produced yields will be satis-factory. At such times management for high yields and longevity of stand reduces the frequency of growing uncertain grain crops and the difficulties of frequent establishment of meadows. Unfavorable moisture is a constant threat.

An additional aspect of this problem deals with the availability of zinc and sometimes sulfur in exposed subsoils. In subsoils exposed by lev-eling for irrigation zinc deficiency may become acute on corn. Defi-ciencies of sulfur have been reported also on crops planted on soils leveled for irrigation. This is related to a lower content of organic matter. The zinc deficiency is accentuated by high phosphorus, high lime, and soil compaction.

SHOULD ORGANIC MATTER BE MAINTAINED OR INCREASED?

To answer this question it is necessary to review some of the func-tions of organic matter:

1. It acts as a storehouse of nutrients – nitrogen, phosphorus, sulfur, and so on.
2. It increases exchange capacity.
3. It provides energy for microorganism activity.
4. It releases carbon dioxide.
5. It stabilizes structure and improves tilth.
6. It provides surface protection and thus increases infiltration.

It is interesting to note that all of these functions (except surface pro-tection) depend on decomposition. Hence the production of large quan-tities of residues, and their subsequent decay, is necessary to good crop and soil management. It is significant that one of the plant-nutrient problems of the Arctic is the resistance to decay of organic matter under low temperatures, and organic matter accumulates even on gravel ridges. In contrast, in the subtropics and tropics, although much organic matter is produced, it decays very rapidly.

Maintenance of organic matter for the sake of maintenance alone is not a practical approach to farming. It is more realistic to use a management system that will give sustained top profitable production. The greatest source of soil organic matter is the residue contributed by current crops. Consequently, the selection of the cropping system and method of handling the residues are equally important. Proper management and fertilization will produce high yields, the principal point in which farmers are interested. A by-product of these high yields is the organic residue. Therefore those soils that are being managed to produce large yields are being improved at the same time.

Workers at Michigan State University state: "The practices we perform to grow our top yields do the best job in conserving and building our soils."

LEGUMES IN THE ROTATION

Legumes were a mainstay of some rotations for many years. Their main purpose now is to supply large amounts of high-quality forage, whether hay or pasture. An additional and valuable benefit has been the nitrogen supplied to companion or succeeding crops. In a very few instances legumes are not desirable in rotations. A rotation that includes flue-cured tobacco, a crop demanding a low nitrogen level in the soil, is an example.

Nitrogen Fixation by Legumes. Elemental nitrogen occupies about 78 per cent of the air by volume. A group of bacteria (*Rhizobia*), called nodule or symbiotic bacteria, utilizes this free nitrogen by attaching to the roots of legumes and causing nodules to form. This mutually beneficial relationship is known as symbiosis.

Nodule formation is accomplished by bacteria entering a single-celled root hair. The bacteria then rapidly increase in number, grow toward the base of the root hair, and pierce the cortex of the root. As a result of this penetration, marked cell proliferation takes place, and a nodule, which is a mass of root tissue containing millions of bacteria, is formed. Nodules should not be confused with certain nematode infections which are merely thickened roots.

Nodule bacteria use both the carbohydrates and minerals contained in the host plant and in turn fix atmospheric nitrogen. This nitrogen may be utilized by the host plant, it may be excreted from the nodule into the soil and be used by other plants growing nearby, or it may be released by decomposition of the nodules or legume residues after the legume plant dies or is plowed down.

Amounts of Nitrogen Fixed. The amount of nitrogen fixed by *Rhizobia* varies with the yield level, the effectiveness of inoculation, the nitrogen obtained from the soil, either from decomposition of organic matter or residual nitrogen, and environmental conditions. A good legume crop such as soybeans, alfalfa, or peanuts contains large amounts of nitrogen (Table 13-1). In general about 50 to 80 per cent of the total nitrogen is fixed by the nodule bacteria.

Welch estimates that in Illinois soils each bushel of soybeans removes 1.7 lb. of nitrogen from the soil, and soybeans fix 45 per cent or more of the total nitrogen in the plant.* However, on lighter colored soils soybeans might fix 80 per cent or more. Other legumes would behave much the same way.

Legumes in combination with grasses for forage generally supply nitrogen for both crops. However, in the South, with a longer growing season, rotational grazing, and more complete utilization of forage, NPK is applied to grass-legume mixtures.

Much remains to be learned regarding nitrogen needs of very high-yielding legumes. Cooper (1972) reported increased yields of soybeans after nonlegume residues from 200 lb. of nitrogen. However, 80 bu./A. yields have been reported without extra nitrogen additions. Generally legumes such as alfalfa and soybeans have shown no response to added nitrogen. Large seeded legumes such as snapbeans and lima beans constitute an exception. They are short season crops in which the bacteria have insufficient time to fix adequate nitrogen.

Legumes versus Commercial Nitrogen. Formerly, one of the reasons for including legumes in a rotation was to supply nitrogen, but with the development of the synthetic nitrogen industry and the resulting availability of inexpensive nitrogen, agriculture is no longer dependent on legumes for this element. The selection of the program that the farmer should follow becomes a matter of economics; he should choose the program yielding the greatest net return on his investment.

The supply of fertilizer nitrogen in 1975–1980 is uncertain. Shortage of natural gas to manufacture ammonia is the main problem. Hence, the farmer may need to consider legumes as a possible supplement until the nitrogen supply problem is resolved.

In certain areas, particularly in some tropical countries, commercial nitrogen may not be available or the grower has no money to pay for it. Therefore a well-planned cropping system which includes legumes is essential to help to supply the nitrogen needed for the growth of nonlegumes. The main drawback, however, is often the lack of adapted legumes.

When legumes are used for forage in a livestock system of farming, the problem is different. The legumes serve the dual purpose of feed for livestock and some of the nitrogen for the grain crops. In this system legumes are essential to furnish at least part of the feed. The alternative is to grow only grass and to make heavy applications of nitrogen. Feeding trials generally demonstrate the superiority of legumes over nitrogen-fertilized grass.

Adequacy of Legume Nitrogen for Corn. It is generally recognized that in corn production fertilizer-nitrogen must supplement the amount

---

* L. F. Welch, University of Illinois, personal communication.

fixed by legumes. A good growth of a legume such as alfalfa will furnish nitrogen for a good but not a top yield of corn. The results from Illinois experimental fields indicate that, with proper treatment with lime, phosphorus, and potassium as needed, grass-legume mixtures have done reasonably well in supplying nitrogen for first-year corn. However, as much as 75–125 lb. N/A. may be recommended under these conditions.

The quantity of nitrogen that will be produced is often uncertain. Legumes have been relied on to furnish nitrogen for crops in rotation but have fallen short of supplying adequate quantities. This may have been caused by stand failure of the legumes, improper inoculation, or inadequate fertility. When interpreting experimental results, it is sometimes difficult to determine whether plant nutrients have been present in sufficient quantities for adequate growth of the legume. A good legume plowed down may supply 60 to 80 lb. nitrogen per acre. However, the farmer often overestimates the quality of the legume sod and only half or less of this amount may be furnished.

Two-year average results from three locations in Iowa showed yields of corn of 83 and 91 bu. after Madrid sweet clover and Ladino clover, respectively. However, nitrogen alone at the rate of 51 and 95 lb. gave yields of 90 and 98 bu., respectively.

ROTATIONS VERSUS CONTINUOUS CROPPING

Continuous cropping, or monoculture, is not new. There are examples from all over the world—rice in the Far East, wheat in the subhumid areas of the United States, cotton in the South. Although monoculture was once considered a sign of poor farming, the greatly increased supply of chemical nitrogen in the 1950's prompted much interest in continuous corn on soils on which erosion was not a serious problem. Information obtained indicated that the value of different rotations should be reexamined under conditions in which crop yields would not be restricted by an inadequate supply of plant nutrients. There is nothing inherent in the corn plant that makes it hard on the soil. Rather, inadequate fertilization and excessive cultivation may be responsible for any deleterious soil effects associated with corn production.

Until about 1950 continuous corn plots were used as examples to demonstrate the undesirability of the system. Most of these plots did not receive adequate fertilization, particularly nitrogen, and were compared with rotations that included legumes. Hence continuous corn showed up poorly. Since that time more logical comparisons have been set up. In general under high-yield conditions, results show continuous corn has a yield 15 per cent lower than the yield of corn in rotations. In some periods of limiting moisture continuous corn may be superior to corn after alfalfa. Alfalfa takes up water deep in the profile and a subsequent corn crop may be short of water.

The conditions under which continuous corn may be grown at a sustained high production level have not been clearly defined. Experience so far indicates that continuous corn is feasible over a wide range of soil conditions, provided that the soil is reasonably level. In Iowa it has been suggested that 0 to 2 per cent slopes are suitable and 2 to 5 per cent slopes are possible if careful attention is given to the erosion hazard and its control. On slopes greater than 5 per cent continuous cropping should not be attempted.

The possibility that monoculture may encourage certain diseases, weeds, or insects has been considered. In such cases the continuity may be interrupted and other cropping practices initiated.

Continuous corn does not mean that the whole farm is in corn. Rather, corn may be grown in the more adapted fields, and sod cropping confined to the other areas. For example, if a farm has both level and hilly land, the corn needs of the farmer and the cropping needs of the soils may be met by growing corn on the level land and leaving the hilly land in sod.

Calculations may be made to compare the returns from a rotation with those of continuous corn plus the cost of commercial nitrogen. The grower may accept a somewhat lower yield with continuous corn but still be ahead financially.

Effect on Soil Tilth. Well-fertilized thick stands of corn produce more above-ground residues than will a legume. The tillage practices commonly carried out in corn production, however, favor more rapid decomposition of organic matter. The rooting system is different from that of a legume such as alfalfa. The effect of the alfalfa roots in penetrating a fine-textured, compact subsoil would be advantageous.

Continuous row crops mean more tractor traffic and more soil stirring, and these result in compaction, breakdown of structure, decomposition of organic matter, and surface erosion. Some of these problems will be balanced off by thick stands of properly fertilized crops. Minimum tillage helps. Fall plowing may be more desirable than spring plowing to help maintain tilth on level dark soils, for freezing and thawing are quite effective in this respect.

More information is required concerning the soil conditions under which deep-rooted legumes are needed in the rotation. There is some evidence that roots or decomposing residues of a species are toxic to the same species. Workers in Indiana and Illinois have observed higher yields of corn in a corn-soybean rotation than in continuous corn. Additional information is also required on the conditions of the soil under which continuous production of a highly fertilized row crop such as corn will result in increased yields over the years. In this competitive period it is not sufficient just to maintain yields.

Double cropping such as small grain-soybeans, small grain-corn, triple cropping or even four crops of rice in areas with long growing seasons and possibilities of irrigation are being considered to a greater extent. With 4 crops a year 27 tons/ha. of rice is a real possibility. This means utilizing soil, solar, and water resources to the maximum. If adequate fertility and pest control are provided, soil productivity should gradually increase. Thus more attention will be directed toward measuring yield per acre per year.

Advantages of Both Systems

ROTATIONS
1. Deep-rooted legumes may be grown periodically over all fields.
2. There is more continuous vegetative cover with less erosion and water loss.
3. Tilth of the soil may be superior.
4. Crops vary in feeding range of roots and nutrient requirements: deep-rooted versus shallow-rooted, strong feeder versus weak feeder, and nitrogen fixer versus nonlegume.
5. Weed and insect control are favored, although chemicals are becoming increasingly effective.
6. Disease control is favored. Changing the crop residues fosters competition among soil organisms and may help reduce the pathogens.
7. Broader distribution of labor and diversification of income are effected.

CONTINUOUS CROPPING OR MONOCULTURE
1. Profit may be greater.
2. A soil may be especially adapted to one crop; for example, corn, rice, or forage.
3. The climate may favor one crop; for example, corn is more suitable than oats in the Corn Belt.
4. Machinery and building costs are probably lower.
5. The grower may prefer a single crop and become a specialist. Few people can become well enough informed to do an expert job of growing a large number of crops and also produce livestock. Monoculture demands greater skills, including pest, erosion, and fertilization control.
6. The grower may not wish to be fully occupied with farming the year around.

OTHER FERTILITY EFFECTS OF CROPPING SYSTEMS

Concentration of Nutrients in Surface. Crop plants vary considerably in their content of primary, secondary, and micronutrients. In addition, crops may absorb nutrients from different soil zones, thus making

TABLE 14-6. Minor Element Composition of Certain Crop Plants*

	Parts per million					
Crop plants	Boron	Molybdenum	Copper	Manganese	Zinc	Cobalt
Grown on Sassafras loam						
Bean tops	50	1.6	20	50	112	0.64
Carrot tops	30	0.9	18	120	163	0.38
Rye	15	1.5	12	48	93	0.30
Rye and vetch	90	3.4	17	90	263	0.72
Rye grass	15	2.9	19	80	123	0.76
Grown on Collington loam						
Bean tops	75	2.2	19	40	551	0.24
Carrot tops	56	0.8	18	160	460	0.28
Rye	30	1.1	15	32	456	0.20
Rye and vetch	100	11.1	16	80	465	1.32
Rye grass	30	2.4	20	80	175	0.30

* F. E. Bear, *SSSA Proc.*, **13**:380 (1948).

the choice of cropping sequences important to plant nutrition. Deep-rooted crops absorb certain nutrients from the subsoil. As their residues decompose in the surface soil, shallow-rooted crops may benefit from the remaining nutrients.

On a soil close to the borderline in its content of a particular micronutrient it is entirely possible that the preceding crop could have a considerable effect on the supply of this element to the current crop.

For example, rye on a Collington loam contained 1.1 ppm. of molybdenum, and rye and vetch contained 11.1 ppm. (Table 14-6). It appears that as the rye and vetch residues decompose in the surface soil a considerably greater amount of molybdenum will become available than after rye alone. It is possible that this effect may be one of the important benefits of a cropping system or rotation, particularly for the elements that are not contained in fertilizers.

Forests in the tropics and subtropics are an excellent example of a crop that transports nutrients to the surface. The fallen leaves, twigs, and stems decompose, but the nutrients remain to form the basis for the shifting cultivation in which the inhabitants clean, burn, and then farm their fields for two to five years. After this period the nutrients in the surface are almost entirely exhausted and the land is allowed to go back to trees for rejuvenation.

Effect on Phosphorus and Potassium in the Soil. The net effect of cropping practices on phosphorus and potassium levels depends on the removal of nutrients by the harvested portion of the crop, nutrients supplied by the soil, and supplemental fertilization. The variations in nutrient content of crops as well as those resulting from rotation fertilization were discussed in Chapter 13. Flue-cured tobacco may be used as an illustration:

	Pounds per acre		
	N	P_2O_5	K_2O
5-10-15, 1500 lb./A.	75	150	225
13-0-44, 200 lb./A.	26	0	88
Leaves removed, 3000 lb./A.	85	15	155
Net gain, lb./A.	16	135	158

In contrast, peanuts receive little or no fertilization, and the nutrient-element removal by a good crop (2 tons of nuts and 2.5 tons of hay) is about 240 lb. N, 39 lb. P_2O_5, and 185 lb. K_2O/A. Considerable difference between soils on which tobacco and peanuts are grown might be expected. These differences tend to be minimized, however, because crops in rotation with tobacco receive small quantities of fertilizer, whereas those in rotation with peanuts are heavily fertilized.

The differences in the phosphorus level of soils under various cropping systems were covered in Table 12-7. The relatively high phosphorus level in tobacco and potato soils is the result of high phosphorus fertilization coupled with low removal by leaching and cropping. The low phosphorus level of the Ladino-grass soils is a reflection of the low-fertility status of soils usually chosen for forage production plus the relatively high proportion of the added phosphorus which is removed in the harvested crops.

When little or no fertilizer is added, the removal of nutrients by crops is reflected in fertility levels. Many examples may be cited. The potassium content of the soil is often depleted after several years of alfalfa or other legume-hay crops or after corn and soybeans in rotation for many years.

Periodic soil tests serve to point out changes in fertility levels and a knowledge of removals and additions helps to explain trends. Unfortunately, less than 10 per cent of the fields may be tested, but, as pointed out in an earlier chapter, soil-test summaries are effective in revealing the influence of cropping systems on the status of soil fertility. The use of these summaries in research and extension programs should be helpful in formulating an over-all approach to the problem of cropping, as related to fertility requirements and maintenance.

Effect of Rotation on Soil and Water Losses. The erosion equation is $A = RLSKCP$. The soil loss per unit area-tons per acre annually (A)

is the product of factors for rainfall (*R*), slope length (*L*), slope steepness (*S*), soil erodibility (*K*), cropping and management (*C*), and conservation practices (*P*).

In this discussion primary attention will be directed to cropping and management of which soil fertility is an important part. The higher the crop yield the less runoff (Figure 14-4).

Erosion is a symptom, not a primary cause of soil destruction. The primary causes are impoverishment of nutrients, especially nitrogen, and inadequate plant population. The higher the proportion of water entering the soil the smaller the runoff.

Some of the characteristics of cropping systems and/or fertility management related to soil losses are the following:

1. The denseness of cover or canopy. This affects the amount of protection from impact of rain and evaporation. The amount of transpiration of water, thus making room for more water, is another factor. Residues and stems reduce velocity of water, and evaporation, and residues when turned back to the soil make it more permeable to water. Research has revealed the tremendous kinetic energy of the raindrop as it hits the soil with a bomblike blast and has helped to bring about a better understanding of the vital importance of plant cover in intercepting this force.
2. The proportion of time that the soil is in a cultivated crop versus the amount of time in a close-growing crop such as small grains or forage.
3. The time that the crop grows in relation to the distribution and

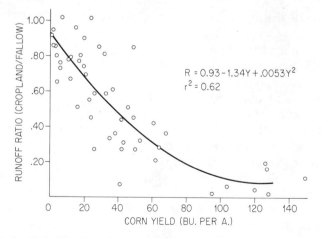

$$R = 0.93 - 1.34Y + .0053Y^2$$
$$r^2 = 0.62$$

Figure 14–4. Relation of surface runoff from growing or maturing corn to crop yield (Bethany and McCredie, Mo.; Clarinda, Castana, and Beaconsfield, Iowa; LaCrosse, Wis.; and Zanesville, Ohio). [Wischmeier, *SSSA Proc.*, **30**:22 (1966).]

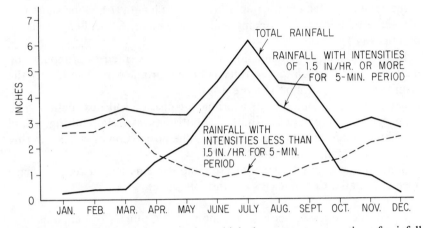

Figure 14-5. The time of year during which the greatest proportion of rainfall and most intense rains occur is important in determining the value of a crop used for controlling erosion. (*North Carolina Agr. Expt. Sta. Bull. 392*, 1954.)

intensity of rainfall (Figure 14-5). The May to September period would be most vulnerable.

4. The type and amount of root system.
5. The amount of residues returned. Points 4 and 5 will affect the soil structure.

There are two important ways in which the interception of rain by vegetative canopies may reduce erosion:

1. Some of the intercepted water evaporates before it reaches the soil.
2. Some of the impact force of the raindrops is absorbed by the vegetation.

The canopy effect is well illustrated by Wollny, who as early as 1880 showed the effect of cover and number of plants per unit area on the percentage of the total rainfall reaching the soil. For example, with corn at a population of 12,700 plants per acre, 60.7 per cent of the total rainfall penetrated the vegetative canopy. With 28,500 plants only 44.5 per cent penetrated. Similar figures for soybeans are 78.2 and 64.3 per cent, respectively.

The effect of the yield level of corn on soil loss has been well established by Wischmeier. For example, when corn followed grain and one-year hay, a decrease in the yield from 75 bu. to 25 bu. increased soil loss about three times. It would be interesting to know the effect of 150-bu. yields.

Results from Missouri indicate that fertilizer applied to small grains reduced erosion:

Crop	Fertilizer	Soil loss (tons/A.)
Wheat after soybeans	No fertilizer	4.3
Wheat after soybeans	0-18-17	2.7
Oats after corn	No fertilizer	3.7
Oats after corn	0-18-33	1.8

These effects are largely the result of a denser cover and a more extensive root system of the growing crop when fertilizer was applied.

The effect of improved practices including approximately tripling the rate of fertilizer and manure and increasing soil pH from 5.4 to 6.8 in a corn-wheat-meadow-meadow rotation in a small watershed in the hilly area of eastern Ohio is shown in Table 14-7. Yields were increased about 50 per cent and runoff and erosion drastically reduced. Management practice did not cause strong differences among soil characteristics. The effects on runoff and erosion were in part due to better surface protection through a quicker cover in the spring, a denser cover throughout the season and a more extensive root system of the growing crop. Contouring rather than straight rows across the slope was used.

Erosion causes considerable fertility loss, for when the surface soil is eroded the nutrients accumulated in the surface soil are lost. These nutrients include nitrogen and sulfur in the organic matter as well as elements such as phosphorus, potassium, calcium, and magnesium which may have accumulated in the surface soil.

The amounts of available nutrients lost are influenced by texture, organic matter content, and fertility level. The nutrients are held in the

TABLE 14–7. Effect of Management Level on Crop Yields, Runoff, and Erosion (1945–1968)*

	Prevailing practices	Improved practices
Corn (metric ton/ha.)	5.1	7.3
Wheat (metric ton/ha.)	1.5	2.3
Hay (metric ton/ha.)	4.3	7.8
Runoff, growing season (cm.)	1.9	1.0
Peak runoff rate (cm./hour)	2.3	1.5
Erosion from corn (metric ton/ha./yr.)	10.6	3.1

* Edwards et al., *SSSA Proc.*, **37**:927 (1973).

clay and organic matter particles. Hence, the key point in reducing nutrient losses to the streams is control of erosion.

WINTER COVER CROPS

Winter cover crops are planted in the fall and plowed down in the spring. A nonlegume or a nonlegume and a legume such as oats and crimson clover are often grown together. There are several advantages to the latter practice. A greater amount of organic matter is produced, the nonlegume can benefit from the nitrogen fixation, and because the nonlegume is usually more easily established a stand of at least one crop is assured.

Nitrogen Added. One important reason for using green-manure legume crops is that they supply additional nitrogen. The amount of nitrogen supplied depends on the amount of root and top growth turned under, so a large amount of plant growth, of course, is desirable.

When a nonlegume is turned under, only the nitrogen from the soil or that supplied in the fertilizer is returned. Although there are certain special cases in which a nonleguminous crop is needed, for example, before flue-cured tobacco, a legume is most often preferred in order that the nitrogen supplied may help to justify the practice.

In a winter cover crop experiment on Marlboro fine sandy loam in North Carolina the beneficial effects of cover crops on corn and cotton production were related to the nitrogen supplied (Table 14-8). Vetch, for example, furnished adequate nitrogen for corn or cotton, but the yields were no higher than when commercial nitrogen alone was used. It is significant to note that vetch increased corn yields 52 bu./A. in this experi-

TABLE 14-8. Effect of Winter Cover Crops and Nitrogen on Yields of Cotton, Corn, and Peanuts (5 yr. ave.)[*]

Crop	Rate of nitrogen (lb./A.)	Seed cotton (lb./A.)	Rate of nitrogen (lb./A.)	Corn (bu./A.)	Peanuts (lb./A.)
No cover	0	1,256	0	26	1,203
	80	1,635	160	80	
Vetch	0	1,674	0	78	1,425
	20	1,690	160	78	
Oats-vetch	0	1,754	0	56	
	20	1,676	160	79	
Crimson clover					1,528

[*] Adequate phosphorus and potassium applied. C. D. Welch, W. L. Nelson, and B. A. Kratz, *SSSA Proc.*, **15**:229 (1951).

ment. Peanut yields tended to be higher after a cover crop. Results of penetrometer measurements indicated that the soil was less compact after vetch than it was with no cover. Peanuts are produced underground, and the less compact soil was apparently helpful in encouraging larger yields.

Organic Matter Added. One of the benefits attributed to winter cover crops is the organic matter supplied to the soil. Organic matter is certainly a factor affecting the tilth of very fine sandy loams and heavier textured soils. Green manures will help maintain the soil organic matter or will sometimes even increase it. In North Carolina organic matter was increased by vetch cover crops plus high nitrogen on the corn on soils low in organic matter and was maintained on soils with relatively high contents of organic matter. Water infiltration was improved.

In rotations in which the crops return little residue, maintenance of soil productivity may be particularly difficult. The lengthening of the rotation to include crops such as well-fertilized small grains (for grain), corn, and green-manure crops would appear to be a step in the right direction. This has been found to be a must in vegetable-growing areas. The acreage of corn and sorghum silage is increasing in some areas in order to use the whole plant. This leaves the soil with almost no surface residues. Oats or rye seeded immediately after harvest or seeded by airplane before harvest will help to protect the soil.

Supply of Nutrients. The phosphorus in green manure may be even more effective than fertilizer phosphorus, probably because of its gradual release during decomposition, localized placement, the presence of organic acids to maintain availability, and the formation of certain complex ions.

An interesting green-manure program has been carried on by the Seabrook Farming Corporation of New Jersey. This concern, which grows large quantities of vegetables, has found it necessary to obtain large quantities of organic matter from cover crops in order to produce top vegetable yields. Most vegetable crops return only small amounts of plant residue to the soil. Small grains such as barley or small grains plus crimson clover are thus effectively used. The cover crops are planted on time, and yields of 3 to 5 tons of dry matter per acre are plowed down. With the extremely high fertilization remaining from previous vegetable crops, the cover crops grow well and produce high quantities of organic matter containing as much as 100 lb. of nitrogen, 30 lb. of phosphorus, and 250 lb. of potassium per acre.

Nitrogen is added at turning to ensure rapid decomposition of the organic matter, and more fertilizer is applied directly to the vegetable crops. The larger share of nutrients, however, comes from the decomposing organic matter. This means that less fertilizer is needed for the vegetable crop and that there is less chance of fertilizer injury. In addition, the nutrients are distributed through the plow layer, which has certain advantages.

Protection of the Soil Against Erosion. Protection against erosion has been suggested as one of the most important reasons for winter cover crops. Benefits, however, should be related to the distribution of rain during the year. In the South the most intense rains occur in the summer months. In North Carolina the greatest soil loss by water comes in June, July, and August (Figure 14-5).

Critical studies of the problem indicate that the effect of cover crops on soil loss is generally rather small when winter cover crops are turned under in the usual manner in early spring. A mulch balk treatment in which rye residues were left on the surface between the rows was much more effective than plant residues turned under.

Surface residues from summer crops and weed growth, if left undisturbed, may provide more protection than cover crops seeded in the fall. The greater the percentage of the soil surface covered by mulch the less the soil loss (Figure 14-6). Freshly prepared land is quite susceptible to erosion, and considerable time is required before the crop can prevent soil loss. For example, rye cover has been considered after corn but, because of the large corn residues from heavy fertilization and planting, it has not been generally beneficial.

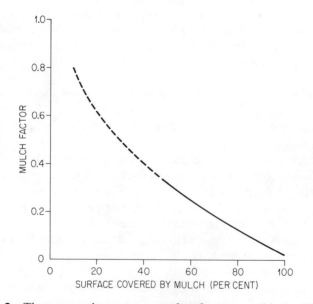

Figure 14–6. The greater the percentage of surface covered by mulch, the less the soil loss. The mulch factor is the ratio of soil loss with a given percentage of mulch cover to corresponding losses with no mulch. (Wischmeier, *Conservation Tillage, Proc. of National Conf.,* Soil Cons. Soc. Am., 133, 1973.)

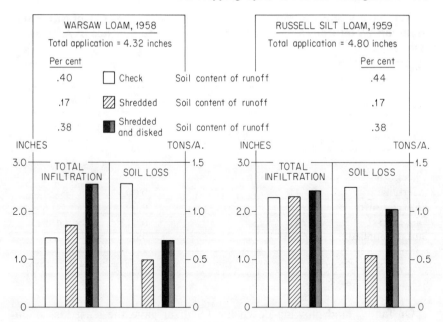

Figure 14–7. Effect of cornstalk residue management on infiltration and soil loss under simulated rainfall with an intensity of 2.4 in./hr. Warsaw 4–4.5 per cent slope, Russell 3–3.5 per cent slope. Check is cornstalks left by picker. [Mannering et al., *Soil Sci. Soc. of Am. Proc.*, **25**:506 (1961).]

Methods of handling cornstalks in the fall have a marked effect on soil loss. Shredding reduced soil losses to about half that obtained from cornstalks as left by the cornpickers (Figure 14-7). The shredded cornstalks provide a cushion which protects the soil from the impact of raindrops. Disking shredded cornstalks is undesirable from the standpoint of soil conservation. Therefore when land is not fall-plowed, shredded cornstalks represent a good possibility.

Ordinarily, winter cover crops should be concentrated on those fields on which peanuts, sweet potatoes, certain vegetables, or silage have been grown and on which little surface residue has been left. Under such circumstances small grains appear to fit in well, and rather than as cover crops *per se* small grains can be fall-seeded and harvested for grain the following summer.

For perennial crops, such as peaches and apples planted on steep slopes, continuous cover is helpful in reducing erosion. Since the trees and the cover crops occupy the land simultaneously, care must be taken, particularly in young orchards, to prevent competition for water and nitrogen. In some of the muck soils suitable for vegetables a strip of small grain or a row of trees at intervals helps to reduce losses from wind erosion.

TABLE 14–9. Effect of Method of Incorporation on the Loss of Nutrients in Runoff Water from a Silt Loam Soil Subjected to 5 Inches of Artificial Rainfall*

	Previous crop			
	Fallow		Bluegrass sod	
Method of incorporation	Soluble P	Soluble $(NH_4 + NO_3) - N$	Soluble P	Soluble $(NH_4 + NO_3) - N$
Check — no fertilizer	0.041	1.21	0.002	0.41
Plowed down	0.056	10.36	0.002	0.69
Disced in	0.281	28.59	0.011	0.81
Surface applied	0.627	26.11	0.071	2.96

* Nelson, D. W., reprinted from *Fertilizer Solutions*, **17**:10 (May–June 1973). Fallow soil treated with 150 lb. ammonium nitrate-N and 50 lb. superphosphate-P per acre. Area previously in sod was treated with 200 lb. of N and 50 lb. of P per acre in liquid suspension form.

On fallow land plowing down the fertilizer gave the least loss of nitrogen and phosphorus (Table 14-9). This was on a silt loam with 6 per cent slope and the rainfall was applied at the rate of 2.5 inches in one hour two weeks apart. Hence, with no cover, opportunity for loss of nutrients is great if they are not plowed down.

Residual Effects. Decomposition of green-manure crops is rapid, but the residual effects are well recognized. The smallest residual effects generally are expected in areas in which the mean annual temperature is high and the soil is sandy. However, data obtained in the South by using oats as an indicator crop show surprising results (Table 14-10). On Nor-

TABLE 14–10. Residual Effects of Vetch and Nitrogen on Yield of Oats After Vetch as a Cover in Continuous Corn for Seven Years*

	Nitrogen on corn (lb./A.)	Norfolk sandy loam one year after cover crop with one crop of corn between Yield of oats (bu./A.)	Norfolk loamy sand two years after cover crop with two crops of corn in between Yield of oats (bu./A.)
No cover	0	28.9	13.3
	90	54.5	23.6
	180	71.1	23.9
Vetch	0	79.5	45.6
	90	70.4	34.4
	180	62.2	44.2

* B. A. Krantz and W. V. Chandler, *North Carolina Agr. Exp. Sta. Res. Bull. 366* (Revised, 1954).

folk sandy loam one year after the last vetch crop the residual effect produced 50 bu. of oats. On Norfolk loamy sand the residual effect was more than 30 bu. two years after the last vetch crop.

In Norfolk sandy loam the residual effect of the high rate of nitrogen on corn was considerable. This effect presumably came from the stover as it decomposed as well as from residual nitrogen in the soil. In Norfolk loamy sand the residual effect from the high rate of nitrogen was much less than that from vetch.

Grazing. Small grain or other crops can be grazed in late fall and winter when the amount of growth and soil conditions permit. Adequate fertility, either residual or added, is a must and extra nitrogen is needed. Grazing supplies tangible evidence of profit from cover crops.

FARM MANURE

Farm manure is a by-product of the livestock industry. In any successful business full use must be made of by-products and farm manure is no exception. The nutrient content, losses by volatilization and leaching, and cost of handling must all be considered. Economics of using farm manure are discussed in Chapter 15.

Nutrient Content. Although there is much variation among animals and feed, in general about three fourths of the nitrogen, four fifths of the phosphorus, nine tenths of the potassium, and one half of the organic matter are recovered in the voided excrement. Because of losses by volatization and leaching, however, only one third to one half of the value of manure is actually realized in crop production.

Average amounts of nitrogen, phosphorus, and potassium in manures from different animals are shown in Table 14-11. A figure of 10-2.2-8.3 (10–5–10) in a ton of manure is often used as a guide. On the basis of readily available nutrients in a ton, however, the values may be about 5-0.5-4 (5–1–5). The relatively low content of phosphorus is thus emphasized.

Organic Matter. Solid manure contains 50 to 80 per cent water. Thus an application of 10 tons/A. would supply 2 to 5 tons of organic matter, which would help to maintain the soil in better tilth, improve water intake, increase release of carbon dioxide, and increase organic matter content in some soils.

Many comparisons of manure and commercial fertilizers have been made. On a fine sandy loam at Broadbalk field at Rothamsted, chemical fertilizers used for 100 years have been just as effective as manure for continuous wheat production. However, for some crops on heavy soils the additional organic matter supplied from manure could well improve crop production.

Manure in a Complete Fertility System. Manure is still a mainstay in the fertility programs on many farms. However, the rapid increase in use of commercial fertilizer and the greater interest in the most profit-

TABLE 14–11. Average Amounts of Nitrogen, Phosphorus, and Potassium in Manures from Different Farm Animals*

	Dairy cattle	Beef cattle	Poultry	Swine	Sheep
Animal size, lb.	1000	1000	5	100	100
Wet manure, tons/yr.	11.86	10.95	0.046	1.46	0.73
Moisture, %	85	85	72	82	77
Nutrients			*Pounds per ton*		
Nitrogen (N)	10.0	14.0	25.0	10.0	28.0
Phosphorus (P)	2.0	4.0	11.0	2.8	4.2
Potassium (K)	8.0	9.0	10.0	7.6	20.0
Sulfur (S)	1.5	1.7	3.2	2.7	1.8
Calcium (Ca)	5.0	2.4	36.0	11.4	11.7
Iron (Fe)	0.1	0.1	2.3	0.6	0.3
Magnesium (Mg)	2.0	2.0	6.0	1.6	3.7
Boron (B)	0.01	0.03	0.01	0.09	—
Copper (Cu)	0.01	0.01	0.01	0.04	—
Manganese (Mn)	0.03	—	—	—	—
Zinc (Zn)	0.04	0.03	0.01	0.12	—
Approximate value per ton	$2.00	$2.44	$6.30	$2.38	$4.70

* Walsh and Hensler, *Wisconsin Ext. Circ. 550* (1971).

able production has relegated manure to a lesser role. Unless an unusually large quantity of legumes were produced on the farm and considerable commercial feed were consumed, thus bringing in fertility from other farms, a livestock program by itself would tend to result in gradual depletion of soil fertility.

Manure must be considered primarily as a nitrogen fertilizer and, to a lesser extent, one of potassium. Losses of nutrients from manure are serious. For example, if fermented manure is left to dry on the soil surface after being spread and before plowing under, as much as 25 per cent of the nitrogen may be lost by volatilization in one day and as much as 50 per cent in four days. For most efficient use manure should be plowed under the same day it is spread. Weather or labor conditions may not permit it, however. Spreading 15 tons of manure on frozen ground on an 11 per cent slope in Wisconsin gave 11 lb. of nitrogen loss in the runoff compared to 3 lb. plowed down in spring.

Following crops such as corn silage and vegetables, from which little plant residue is returned, manure will be an especially important addition to the soil management program.

Only when the use of manure is coordinated with that of lime, commercial fertilizers, legumes, and other soil management practices will its full benefit as a by-product be realized.

In large feeding operations, in which the animals are confined to feed lots, drying and bagging the manure for use on turf and gardens results in a valuable by-product.

Distribution of Manure by Grazing Animals. One aspect of the farm manure problem which has received little attention has been the question of distribution of manure by grazing animals. This presents a problem in the maintenance fertilization of pastures. A study was conducted in North Carolina over a period of years in which the pattern of distribution of manure was determined. One beef animal unit per acre was grazed, and the distribution of the manure over a given area in relation to time was observed. It was found that for elements such as nitrogen, which do not remain in effective concentrations for more than a year, about 10 per cent of the grazing area would be effectively covered in one year. On the other hand, with elements such as phosphorus, which are not leached or luxury-consumed, some effect might be obtained from a given application as many as ten years later. The data showed that nearly all of the pasture area would have received deposits of manure in a ten-year period. Potassium is intermediate between nitrogen and phosphorus in retention in the soil, and manure-deposited potassium would be effective to some degree for at least five years. During this period about 60 per cent of the area would have been covered.

For a bunch type of forage growth with little lateral feeding the deposits of nitrogen from the manure would be of little consideration in a pasture fertilization program. With potassium, even though 60 per cent of the area would be covered in a five-year period, it would be necessary to fertilize the entire area to ensure good growth on the other 40 per cent. With phosphorus the deposition from manure should also be taken into consideration in fertilization, but it should not replace supplemental applications if the soil is low in this element. Mechanical spreading of feces has not been found to be beneficial since it causes more plants to be fouled and rejected by grazing animals.

Studies in Indiana revealed that with low stocking rates, animal excreta will essentially have no effect on soil fertility. On highly productive pastures with a high carrying capacity, excreta could have a beneficial effect on soil fertility over a period of time. Grain feeding on pastures had considerable effect on soil fertility and each increase of 4.5 tons of grain fed per acre resulted in an increase of 53 steer days grazing.

Animal excreta have other important effects on pastures. In the case of legumes the high nitrogen in the excreta would encourage competition from weeds and grasses. Also, selective grazing takes place and feces result in almost complete refusal of forage in the area.

The effectiveness of manure deposits for forage plants with lateral feeding roots, such as Ladino or Coastal Bermuda grass, should be somewhat different. Even if a given area of soil were not covered with excreta, the plants growing on that area might obtain nutrients from one adjacent. However, further investigations are needed on this problem to help answer many of the questions being raised with respect to differences among soils and forage plants. Growers may have a false sense of security concerning the adequacy of animal excreta on pastures.

Liquid Manure. In some beef, hog and dairy operations solid and liquid manure is collected in a lagoon. With concrete floors and little bedding, 10 to 25 per cent water may be used to help carry the manure to the lagoon. Such lagoons may hold 100,000 gallons or more. Periodically the slurry is agitated and is spread as a liquid with sprinkler wagons, sprinkler irrigation, or by direct injection in the soil. The latter helps to avoid odor and insect problems.

With sprinkler irrigation the soil should have a high intake rate, water holding capacity, and exchange capacity. The soil should be able to support high yields for maximum nutrient uptake.

This method of handling manure saves a high percentage of the plant nutrients, particularly if it can be mixed with the soil immediately after spreading. Another feature is the saving in labor.

Commercial Feedlot. Drylot dairies, swine units, beef feedlots, and poultry ranches often face a unique problem in the disposal of manure. There may be little or no cropland associated with the operations. The disposal of manure is based on the same operations that sanitary engineers consider in the handling and treatment of municipal and industrial wastes:

1. Collection—once or twice a day cleaning may be practiced.
2. Processing—drying is a possibility.
3. Storage—huge piles are used, but insects and odor may be a problem. Manure lagoons under anaerobic conditions demand a minimum of attention and sludge builds up slowly.
4. Utilization or disposal—the livestock operator would like to find a market in which he might break even or at least have little cost in disposing of the manure. Finding such a market becomes a problem in highly urbanized areas such as California.

Human Wastes. Disposal of the waste or sludge from municipal treatment plants is an increasing problem. For each 1,000,000 people daily in the United States, 1400 dry tons of sludge will be generated. The sludge is beneficial as a soil amendment and fertilizer.

Workers in Illinois report that the approximate pounds of major nutrients and secondary nutrients supplied when 250 lb. of NH_4-N is supplied in 1.4 acre inches of nonlagooned liquid digested sludge is approximately as follows:

P	236 lb.	Mg	138 lb.	Zn	28 lb.	B	1.6 lb.
K	48	S	92	Cu	10	Mn	5.0
Ca	390	Fe	470	Mo	0.02		

One half of the total nitrogen is in organic form and is released slowly to crops when applied to the land. Also, not all of the P is in available form. One of the questions still unanswered is the amount of heavy metals contained in the sludge and whether this will cause undesirable accumulations in the soil. This may vary with the soil exchange capacity and other characteristics. Eventually municipal sewage treatment facilities may be required to remove heavy metals and toxic substances from the sludge.

In some areas the effluent or liquid portion of the sewage is also being utilized. This is mostly on an experimental basis as yet. The material contains a high amount of the soluble nutrients such as potassium. A permeable soil is required in order that the large volume of liquid will move into the soil.

USE. Responses on crops have been similar to those obtained with chemical fertilizer. Reclaiming sand dunes and strip mine areas represent a real possibility. A main problem to solve for large cities is transportation to and spreading the sludge on areas where it can be used. However, many smaller municipalities have been returning sewage waste to the land for many years. One problem to overcome is the idea that the sludge may contain harmful bacteria. However, after the sewage has gone through the sewage digestion system this has not been a problem.

MINIMUM TILLAGE

Minimum tillage, which is a principle and not a practice, helps to create the row and interrow zones. It ranges from one less cultivation or discing than is usually performed to the use of a no-till planter.

With a row crop such as corn, the field can logically be divided into two zones, the soil immediately around the seed and the soil between rows. The first might be called the *seedling zone* (row zone) and the second the *water management zone* (interrow zone) (Figure 14-8).

The two zones have drastically different requirements in regard to the physical condition of the soil. In the row zone the soil should be firmed around the seed for quick germination and rapid seedling growth. In the interrow it should be loose and rough to discourage germination of weed seeds and encourage water intake.

When a layer of soil is loosened by tillage, it expands, or fluffs up. For example, if a soil is plowed 9 in. deep, the layer of soil fluffs up to 11 or 12 in. With this increased amount of pore space the soil can hold more

Figure 14-8. The soil in a cornfield can be divided into two zones, the seedling (row) zone and the water management zone (inter-row) zone. (Larson, and John Deere and Co., 1963.)

rainwater than it could before plowing. Repeated tillage reduces this temporary water-storage capacity. The effect of minimum tillage on water intake is well illustrated in Figure 14-9. The same effect on increasing infiltration and decreasing soil loss is shown in Table 14-12. Minimum tillage with cultivation gave greater water intake and less soil loss than minimum tillage alone.

The increasing effectiveness of chemical weed control gives minimum tillage an additional impetus. Growers who formerly cultivated a row crop three times now do so only once or not at all. On soils that do not crack, one cultivation may be helpful to increase water penetration.

Mulches. Plant residues applied to the soil surface or left on the surface from the preceding crop are used to reduce soil loss and weed growth and to conserve moisture. The effect of mulches on soil loss was shown in Figure 14-6. These residues slow down the water and allow more penetration. In Iowa early growth of corn was progressively retarded by increasing the amount of mulch (Figure 14-10), which decreased soil temperature. This practice, which may or may not be a disadvantage later in the growing season, would be outweighed by im-

Figure 14–9. Minimum tillage on the right half of this field has kept the soil loose and porous, so that it can quickly absorb rain water. Conventional tillage on the left half has caused so much compaction that water cannot easily soak into the ground. [Bateman et al., *Illinois Agr. Ext. Circ. 846* (1962).]

TABLE 14–12. Minimum Tillage Increases Infiltration and Decreases Soil Loss on Russell Silt Loam, Slope 4.5–5 Per Cent*

	Infiltration (in.)	Final infiltration rate (in./hr.)	Soil loss (Tons/A.)
Shortly after planting			
Conventional	1.28	1.10	7.58
Minimum tillage	1.95	1.63	3.25
After first cultivation			
Conventional-cultivated	1.91	1.36	2.44
Minimum tillage-cultivated	2.17	1.60	1.41
Minimum tillage-uncultivated	1.28	0.99	6.72

* L. D. Mayer and J. V. Mannering, *Agr. Eng.*, **42**:72 (1961).

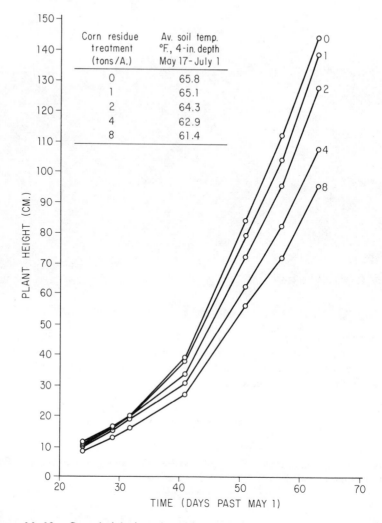

Corn residue treatment (tons/A.)	Av. soil temp. °F., 4-in. depth May 17-July 1
0	65.8
1	65.1
2	64.3
4	62.9
8	61.4

Figure 14–10. Corn height is reduced by increasing amounts of mulch. This is related to the lower temperatures. [Burrows et al., *Agron. J.*, **54**:19 (1962).]

proved moisture penetration in some years. Manure as a mulch versus manure plowed down produced a 4.8 bu. increase in the yield of corn in Ohio (seven-year average).

Chiseling may leave one half of the previous crop residues on the surface. Hence the chisel plow may be a good substitute for the mold-board plow on some soils to reduce soil and water losses.

No-Tillage. Two recent developments have made possible a revolution in tillage practices and form the basis of no-tillage:

One is the development of herbicides that effectively maintain season-long weed control or if planting in sod, kill the sod.

Figure 14–11. No-till planting of corn in cornstalk residue. (Courtesy of S. W. Melsted, Illinois Agricultural Experiment Station.)

The second is the development of corn planters that can plant in almost any kind of soil cover. An example of no-tillage in corn is shown in Figure 14-11.

In addition to saving energy, time and machinery in the many operations of plowing, discing, chiseling, and cultivation in conventional planting, row crops can be planted on sloping land. The crop residues are left largely on the surface and no-tillage reduces water runoff and soil losses. This means less nutrients are lost and more water is available to grow the crop (Figure 14-12). Also, if need for greater grain production arises, the need can be met without damaging the soil. In one county in a rolling area it was estimated that instead of the present 5000 acres of corn 40,000 acres could be grown with no-tillage.

With a row crop such as corn, no tillage is often equal to or better than plowdown, discing, and planting on well-drained soils. On heavy, poorly drained soils where fall plowing is practiced so that freezing and thawing can improve structure or on soils where drainage is a problem no-tillage is usually not desirable. The residues intensify the cooler and more moist conditions in early spring.

In Missouri 6 to 8 row strips of corn are alternated with 25 to 30 ft. of meadow. Small grain is drilled in the corn land after harvest and the corn planted no-till.

Double-crop systems where a row crop such as soybeans, corn, or sorghum is planted right after small grain harvest are well suited to no-till planting. With the exception of the nitrogen for the nonlegume, suf-

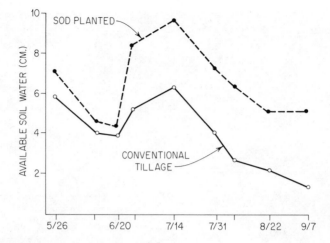

Figure 14–12. Available soil water in 0-60 cm. profile as affected by tillage practice for corn with orchardgrass. [Bennett et al., *Agron. J.*, **65**:488 (1973).]

ficient phosphorus and potassium is applied on the small grain to take care of the needs of both crops.

Seeding legumes and/or nonlegumes in rundown pastures may be accomplished by the same principle. A slit is worked up and fertilizer dropped beneath the seed and/or broadcast. A herbicide kills or retards the existing grasses. Again, this technique is useful on sloping lands which should not be otherwise tilled as well as on more level soils.

FERTILITY. An important key to the success of no-till is adequate fertility. The fertility requirements are affected by amounts of surface residues through effects on soil moisture, temperature, aeration, and biological activity. Hence, it is difficult to set hard, fast rules for given nutrient additions.

Broadcast application on the surface is often practiced. Hence, phosphorus, potassium, and lime accumulate in the surface inch or two. In low-fertility soils, building soil test levels to medium or high before starting no-till is desirable. Periodic plowing every four or five years, if terrain permits, may be required to obtain the top yields needed in the future.

Broadcast applications of phosphorus and potassium are fairly effective because with a good surface cover, as after sod, the soil is more moist than without cover and the roots grow closer to the surface. However, even in such situations, applications of phosphorus are often more effective if placed in a band beside the row at planting, particularly if the fertility is not high. After corn or other crops, the effectiveness of broadcast application depends greatly on the amount of the surface covered with residues. The greater the proportion of the surface covered with residues the more effective the broadcast application. Potassium uptake by the plant is often not as great as with plowdown. This may be due to low aeration or positional unavailability.

In drier areas uptake of nutrient from broadcast application with no-till may be somewhat erratic because of the shortage of moisture under the mulch.

Work in Kentucky indicates that a higher rate of nitrogen may be needed in no-tillage corn. The explanation is that nitrates leach to a greater depth because of a lower evaporation loss and greater penetration of water. However, studies in Indiana indicate that nitrogen deep placed was superior to surface-applied nitrogen on a poorly drained soil.

While no-tillage performs well in many areas, it is difficult to write the entire program for all soils as yet.

Plowing with Immediate Planting. Ordinarily the grower will disk and harrow once or twice before planting, with the result that the soil tends to compact and dry out. There are several methods, including plowing and planting at the same time, to help prevent these undesirable effects. Wheel-track planting, in which the seeding shoe operates in a track made by the tractor or other equipment, is also useful on freshly plowed soil.

A promising variation in plowed ground is the use of a three-wheel rotary hoe 5 to 6 in. deep just in the row. Ohio has measured corn yield increases of 5 bu./A. in favor of this strip processing. Advantages were even greater in years of lower rainfall.

The principles on which minimum tillage is based are sound, and commercial growers are adopting the practice rapidly. So far wheel-track planting has resulted in about the same yields as conventional tillage. As other limiting factors are taken care of, the yields obtained with minimum tillage should exceed those obtained with conventional tillage practices.

PRODUCTIVE CAPACITY

Bray of Illinois stressed the importance of determining the productive capacity of the soil. If it is known, the grower will then have an idea of the best yield levels to shoot for in planning the fertilizer program. With adequate supplies of plant nutrients, this capacity will be related largely to the available water that crops can utilize from the soil. Bray suggests that the grower select an area of representative land and make sure that all nutrient deficiencies are overcome. The soil should be limed in accordance with soil-test results and 200–250 lb. N, 65 lb. P, and 250 lb. K/A. should be broadcast and plowed down. Since corn is a good indicator crop, a variety that will respond to a high population is planted with a medium stand on one half the area and at a high stand on the other half. The amount of rainfall will affect results, but such trials over a period of years will give an indication of the average yield capacity of the soil.

Although the appropriate rates of nutrients will vary according to crop and soil conditions, the foregoing approach has much merit. It is important, however, to apply nutrients to the point at which fertility is not a limiting factor.

Bray suggested that lime, phosphorus, and potassium be added to the rest of the fields in accordance with the results of a soil test. The nitrogen is applied on the basis of the average yield possibility and should not be less than crop removal from the field. If 150 bu./A. appears to be a reasonable corn yield possibility, at least 150 lb. of nitrogen should be added annually. In order to provide for even higher yields an extra 50 to 100 lb. of nitrogen might well be tried.

Summary

1. Marked increases in yields of crops have taken place in the past ten years as a result of improved management practices and particularly of more adequate fertilization. Before that time yields increased slowly.
2. The aim of a crop and soil-management program should be to realize sustained maximum profit from the cropping program. Effects on tilth, water intake and erosion, plant nutrient supply, and pest control must be considered.
3. Cropping systems affect the amount of soil organic matter largely by the amounts of residue produced. As added plant nutrients increase yields, larger amounts of residues are obtained.
4. Organic matter serves as a storehouse for nutrients, increases exchange capacity, provides energy for microorganisms, releases carbon dioxide, improves tilth, and provides surface protection. All of these functions except the last depend on decomposition. Hence organic matter accumulation in the soil is not an end in itself. The important point is the production of high quantities of organic matter and its subsequent decay.
5. One of the problems in soil erosion is related to the lower supply of nutrients, nitrogen in particular. On permeable subsoils adequate fertilization and careful timing of tillage operations in regard to soil moisture content can go a long way toward producing satisfactory yields when combined with other good management practices.
6. The main function of legumes in a rotation, with the exception of legumes for grain, has been to furnish large quantities of high-quality forage. An additional benefit has been the nitrogen supplied on some soils. With the development of the commercial nitrogen industry growing legumes for nitrogen alone is not economical in most instances.

7. It is recognized that supplemental nitrogen must be applied for top profitable production of first-year corn after a legume.
8. Legumes or a nonlegume-legume mixture in forages reduces nitrogen requirement and generally results in superior animal performance.
9. The amount of nitrogen fixed by a legume is dependent on the amount of nitrogen the soil or residual fertilizer will supply. When the soil supplies a high amount of nitrogen the legume may not fix over 50 per cent of its needs.
10. Where soil, solar, and water resources permit, two to four crops may be grown in a single year. Double cropping is common in the United States. In some tropical areas three or four crops may be possible.
11. Monoculture or continuous cropping to one crop is practiced in many parts of the world. Continuous corn may yield somewhat less than corn in rotation, but economics may favor the former when production costs and the value of other crops in the rotation are considered.
12. Sod crops, including legumes, in rotation improve tilth and water intake, and continuous row cropping may favor compacting and breakdown of soil structure. Much information is needed on long-time effects on top production.
13. Deep-rooted crops may bring nutrients to the surface. The net effect of cropping practices on fertility levels depends on removal of nutrients by the harvested portion, amount supplied by the soil, and the amount of supplemental fertilization.
14. The primary cause of *erosion is impoverishment of nutrients.* The effects of cropping systems on soil losses can be drastically altered by the denseness of the cover, which in turn can be increased by more adequate fertilization and thicker planting.
15. Winter cover crops will supply residues, and nitrogen if they are legumes. They are most useful in cropping systems in which few residues remain on the land over the winter. One of the uncertainties is the amount of growth produced. Early planting of these crops and late plowing to turn them under are helpful in increasing amount of growth.
16. Farm manure is a valuable by-product of the livestock industry. Because of losses by volatilization and leaching, generally not more than one third to one half of the value is realized. Manure alone on a farm will result in a gradual depletion of soil fertility, but it is a valuable supplement to a well-designed lime, fertilization, and soil-management program.
17. Disposal of human waste or sludge from municipal treatment plants is receiving increasing attention. The sludge will supply large

amounts of most essential elements. The possible toxic effects of heavy metals is still to be answered.

18. Minimum tillage is a principle and not a practice. It is based on fewer trips over the field, results in better soil tilth, lower production costs, and generally higher yields, if other limiting factors are taken care of. This makes for more efficient use of fertilizers.

19. No-till planting directly in sod or crop residues involves more herbicides but saves energy, water, and soil. It is particularly important on lighter soils and on sloping lands not suitable to other tillage because of erosion losses.

Questions

1. Why are state yield trends over a period of years likely to be misleading as a measure of soil productivity? What might happen to yield trends if breeding studies ceased?

2. Why should a continued increase in ceiling yields in experiment station plots be conducive to increased state yields?

3. What is the aim of a crop and soil-management program?

4. Explain why additions of nitrogen equal to crop removal help to reduce loss of organic matter.

5. Under what soil condition may a corn-small grain (alfalfa) rotation be preferable to corn plus commercial nitrogen as well as other nutrients each year? Under what conditions may the latter cropping system be preferable?

6. What influences how much of its total nitrogen a legume will fix?

7. On what soils in your state could organic matter be increased? Under what soil conditions in your state would additions of organic matter be beneficial other than by the nutrients supplied?

8. What nutrient may be the most likely one to give a marked yield response on corn grown on an eroded soil? Why?

9. Loss of surface soil is serious, but the seriousness varies considerably with the soil. In what soils in your state is the loss likely to be the most serious?

10. Why is it important that plant residues decompose? What functions do they serve in the undecomposed state?

11. List the advantages of rotations and monoculture.

12. What cropping systems have depleted phosphorus and potassium in soils in your state? Explain. In what cropping systems have these elements been increased? Explain.

13. In what ways may nitrogen, phosphorus, and potassium be lost other than by crop removal? In what other ways than fertilization may the supplies be increased?

14. Are there examples in your area in which the farmer is relying too heavily on a legume to furnish the nitrogen needs of a cropping system? Explain.

15. Explain the statement that the primary cause of erosion is depletion of plant nutrients.
16. In what cropping systems may winter cover crops fit? Why?
17. Why will fertility level on a given farm gradually decrease if manure is the only carrier of plant nutrients used? Explain the fertility distribution problem in a pasture.
18. What is minimum tillage? What are the advantages?
19. Would no-till fit in your state? If so, where? Why?
20. In which of the major nutrients is sludge from sewage plants usually low? Is sludge being used in your state? If so, where?

Selected References

Backtell, M. A., C. J. Willard, and G. T. Taylor, "Building fertility in exposed subsoil." *Ohio Agr. Exp. Sta. Res. Bull. 782* (1956).

Bateman, H. P., and W. Bowers, "Planning a minimum tillage system for corn." *Illinois Agr. Ext. Cir. 846* (1962).

Bear, F. E., "Variation in mineral composition of vegetables." *SSSA Proc.*, **13**:380 (1948).

Benacchio, S. S., G. O. Mott, D. A. Huber, and M. F. Baumgartner, "Residual effect of feeding grain to grazing steers upon the productivity of pasture." *Agron. J.*, **61**:271 (1969).

Bennett, O. L., E. L. Mathais, and P. E. Lundberg, "Crop responses to no-till management practices on hilly terrain." *Agron J.*, **65**:488 (1973).

Bradfield, R., in W. H. Garman, Ed., *The Fertilizer Handbook*. National Plant Food Institute. 1963, p. 120.

Christensen, R. P., and R. O. Aines, "Economic efforts of acreage control programs in the 1950's." *ERS, USDA, Agr. Econ. Rpt. No. 18.*, p. 23, (1962).

Cooper, R. L., "Nitrogen response of soybeans with different crop residues in the absence of moisture and lodging stress." *Agronomy Abstracts*, Am. Soc. of Agron. (1972).

Copley, T. L., L. A. Forest, and W. G. Woltz, "Soil management of bright tobacco in lower Piedmont." *North Carolina Agr. Exp. Sta. Bull. 392* (1954).

Duncan, E. R., and F. W. Schaller, "Continuous corn." *Plant Food Rev.* **8**(4):2–5 (1962).

Edwards, W. M., J. L. McGuinness, D. M. Van Doren, Jr., G. F. Hall, and G. E. Kelley, "Effect of long-term management on physical and chemical properties of the Coshocton watershed soils." *SSSA Proc.*, **37**:927 (1973).

Engelstad, O. P., and W. D. Shrader, "The effect of surface soil thickness on corn yields: II. As determined by an experiment using normal surface soil and artificially exposed subsoil." *SSSA Proc.*, **25**:497 (1961).

Fenster, C. R., and T. M. McCalla, "Tillage practices in Western Nebraska with a wheat-sorghum-fallow rotation." *Nebraska Agr. Exp. Sta. SB 515* (1971).

Hart, S. A., "Manure management." *Calif. Agr.*, **18**(12):5 (1964).

Ibach, D. B., and J. R. Adams, "Crop yield response to fertilizer in the United States," *U.S. Dep. Agr. Stat. Bull. 431* (1968).

Kamprath, E. J., W. V. Chandler, and B. A. Krantz, "Winter cover crops—their effects on corn yields and soil properties." *North Carolina Agr. Exp. Sta. Tech. Bull. 129* (1958).

Krantz, B. A., and W. V. Chandler, "Fertilize corn for higher yields." *North Carolina Agr. Exp. Sta. Bull. 366* (revised 1954).

Larson, W. E., and W. C. Burrows, "Effect of amount of mulch on soil temperature and early growth of corn." *Agron. J.*, **54:**19 (1962).

Larson, W. E., *Zone System for Corn.* Moline, Illinois: John Deere, 1963.

Melsted, S. W., "Sewage sludges and effluents: effect on soils, plants and fertilizer markets." *Illinois Fertilizer Conf. Proc.* (1973).

Meyer, L. D., and J. V. Mannering, "Minimum tillage for corn: Its effect on infiltration and erosion," *Agr. Eng.*, **42**(2):72 (1961).

Meyer, L. D., and J. V. Mannering, "Tillage and land modification for water erosion control." *Tillage for greater crop production,* ASAE (1967).

Nelson, Darrell W., "Losses of fertilizer nutrients in surface runoff." *Fertilizer Solutions*, **17:**10 (1973).

No-Tillage Systems Proc. Sponsored by Ohio State Univ., OARDC and Chevron Chemical Co., Columbus, Ohio, 1972.

Parker, D. T., and W. E. Larson, "Effect of tillage on corn nutrition." *Crops Soils*, **7**(4):15 (1965).

Peterson, J. B., "The relation of soil fertility to soil erosion." *J. Soil Water Cons.*, **19**(1):15 (1964).

Peterson, J. R., T. M. McCalla, and George E. Smith, "Human and animal wastes as fertilizers," in R. A. Olson, T. J. Army, J. J. Hanway, and V. J. Kilmer, Eds., *Fertilizer Technology and Use,* second edition. Madison, Wisconsin: Soil Sci. Soc. Am., 1971.

Peterson, R. G., W. W. Woodhouse, Jr., and H. L. Lucas, "The distribution of excreta by freely grazing animals and its effect on pasture fertility: 1. Excretal distribution." *Agron. J.*, **48:**440 (1956).

Phillips, J. A., "No tillage fertilization principles." No-tillage Research Conference, Univ. of Kentucky, 1970.

Pinck, L. A., F. E. Allison, and V. L. Gaddy, "The effect of green manure crops of varying carbon-nitrogen ratios upon nitrogen availability and soil organic matter content." *J. Am. Soc. Agron.*, **40:**237 (1948).

Puhr, Leo F., and W. W. Worzella, "Fertility maintenance and management of South Dakota Soils." *South Dakota Agr. Exp. Sta. Cir. 92* (1952).

Rohweder, Dwayne A., and Richard Powell, "Grow legumes for green manure." *Wis. Ext. Fact Sheet A2477* (1973).

Salter, R. M., R. D. Lewis, and J. A. Slipher, "Our heritage—the soil." *Ohio State Agr. Ext. Serv. Bul. 175* (1936).

Scarseth, G. D., *Man and His Earth.* Ames: Iowa State University Press, 1962.

Smith, J. H., and C. L. Douglas, "Wheat straw decomposition in the field." *SSSA Proc.*, **35:**269 (1971).

Triplett, G. B., Jr., and D. M. van Doren, Jr., "Chemicals make possible no-tilage corn." *Ohio Farm Home Res.*, **48**(1):6 (1963).

Triplett, G. B., C. A. Osmond, and P. Sutton, "Fertilizer application methods for no-till corn." *Ohio Report,* OARDC, **57**(3):39 (1972).

van Doren, D. M., Jr., and G. J. Ryder, "Factors affecting use of minimum tillage for corn." *Agron. J.*, **54:**447 (1962).

van Doren, D. M., Jr., and G. B. Triplett, Jr., "Mulch and tillage relationships in corn culture." *SSSA Proc.*, **37:**766 (1973).

Walsh, L. M., and R. F. Hensler, "Manage manure for its value." *Wisconsin Ext. Circ. 550* (1971).

Welch, C. D., W. L. Nelson, and B. A. Krantz, "Effects of winter cover crops on soil properties and yields in a cotton-corn and in a cotton-peanut rotation." *SSSA Proc.,* **15:** 29 (1951).

Whitaker, F. D., V. C. Jamison, and J. F. Thornton, "Runoff and erosion losses from Mexico silt loam in relation to fertilization and other management practices." *SSSA Proc.,* **25:** 401 (1961).

Wischmeier, W. H., "Relation of field plot runoff to management and physical factors." *SSSA Proc.,* **30:** 272 (1966).

Wischmeier, W. H., "The erosion equation—a tool for conservation planning." *Proc. of 26th Ann. Meeting of Soil Cons. Soc. Am.* (1971).

Wischmeier, W. H., "Conservation tillage to control water erosion." *Conservation Tillage Proc. of a National Conference.* Soil Cons. Soc. Am., 1973.

Yoshida, S., F. T. Parao, and H. M. Beachell, "A maximum annual rice production trial in the tropics." *International Rice Com. Newsletter,* **21**(3): 27 (1972).

15. ECONOMICS OF LIME AND FERTILIZER USE

The commercial farmer is interested in maximum profit per acre. This is the point at which the last dollar spent to produce these yields returns just a dollar. These yields are higher than most people think and it means pushing all good management practices to the limit.

Higher crop yields continue to offer the greatest opportunity for reducing per-bushel or per-ton production costs. The modern grower has the attitude of any other good businessman. He would like to know if a given expenditure is the best he can make with his available funds and if his production costs are properly distributed.

The grower realizes that he must spend money to make money. This is certainly true of expenditures for lime and fertilizer. The greater demand for fertilizer in the United States and in other parts of the world during the last few years is evidence that growers have recognized the returns from added plant nutrients.

In spite of the rapid rise in the use of plant nutrients, reliable estimates show that much land is still underfertilized and that plant nutrient additions could be profitably increased. Indiana applied more plant nutrients per acre in 1973 than any other state in the North Central region. Yet when one considers the yield potentials of row crops, small grain, hay, and pasture the state is still at less than three-fourths potential even with present technology. As one moves West, South, or East potentials are even greater. Lime use in the United States is about 30 per cent of the need.

For the developing countries in Asia and the Far East, the Near East, Africa, and Latin America, N-P-K use in 1972 is estimated at about one half of the 1980 demand. Economic development must give high priority to agriculture. Rostow says, "Put another way, the rate of increase in output in agriculture may set the limit within which modernization proceeds." Use of fertilizers is an index of the use of modern agricultural methods. The close correlation between average fertilizer use and value index of crop production has been pointed out in the Introduction.

As higher rates of plant nutrients are required, it becomes more and more important that the nutrients be applied so that they will be most efficiently utilized. Returns from added nutrients are generally related inversely to the available supply of these nutrients in the soil.

It is recognized, of course, that lime and fertilizer are only two of the many factors needed for high production. Attention must be given to water control, seedbed, variety, date and rate of seeding, stands, fertilizer placement, cultivation, weed, insect and disease control, and harvesting practices. In this discussion it is assumed that these factors are supplied in line with the best available information. Proper use of fertilizers complements the effects of other management practices.

PRICES OF FERTILIZERS COMPARED
WITH OTHER FARM PRICES

To obtain a given level of production, farmers can vary the inputs of land, fertilizer, labor, machinery, etc. The actual use of each depends on relative costs and returns. It is interesting to compare the costs of certain farm inputs shown in Figure 15-1. In the period 1950–1972 there was little change in fertilizer price. On the other hand, machinery, wages, and real estate increased rapidly. Hence, the farmers have tended to substitute fertilizer and lime, as well as machinery and other items, for labor (Figure 15-2). The output per man-hour increased 375 per cent from 1950–1970. While the price of fertilizers and lime will rise in the next few years, the price may not rise as fast as other inputs.

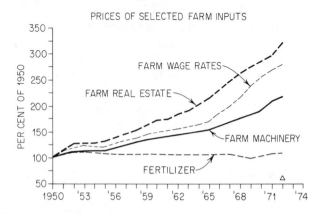

Figure 15–1. Prices of selected farm inputs 1950–1972. (*Farm Cost Situation*, Feb. 1973, ERS, USDA.)

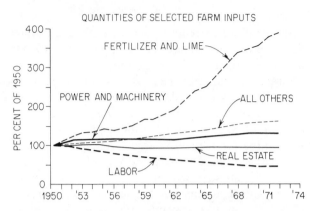

Figure 15–2. Quantities of selected farm inputs, 1950–1972. (*Farm Cost Situation,* Feb. 1973, ERS, USDA.)

FERTILIZER SALES AND FARM INCOME

Sales of fertilizers in a given year once tended to follow farm income of the year before. If the grower had a good income in one year, he would be more than likely to increase his fertilizer use in the next. On the other hand, if his income were low, he would reduce his expenditure for fertilizer. However, because of increased emphasis on research and education in fertilizer use the grower is now spending a higher proportion of his income on plant nutrients. The price of produce and availability of markets are powerful incentives for improved practices by farmers the world over.

YIELD LEVEL AND UNIT COST OF PRODUCTION

Practices that will increase yield per acre usually lower the cost of producing a bushel or ton of the crop (Table 15-1), for it costs just as much to prepare the land, plant, and cultivate a low-yielding field as it does a high-yielding field. Land, buildings, machinery, labor, and seed will be essentially the same, whether the production is high or low. These and other costs are called *fixed* and must be paid regardless of yield. *Variable costs* are those that vary with the magnitude of the total yield, such as the quantity of fertilizer applied, pesticides, harvesting, and handling.

The effect of fertilizer in increasing yields and decreasing the cost per unit of production, in this case a quintal, is well illustrated in Table 15-1. Machinery, labor, management, interest, and other costs remained the same at all levels of fertilizer. However the operating costs including fertilizer, lime, machinery operation, and seed and pest control increased. The lowest cost per unit and the greatest return were obtained at the

TABLE 15–1. Effect of Rate of N–P$_2$O$_5$–K$_2$O on Corn Yields and Profits[*]

Corn production details	0-0-0	67-22-22	134-45-45	270-90-90
			kg./ha.	
Corn yield q./ha.	48	70	85	98
Gross income per ha.	$188	$274	$336	$390
Operating costs ([1])	56	69	82	111
Other cash costs ([2])	25	25	25	25
Fixed costs (machinery, labor, and management)	88	88	88	88
Interest ([3])	82	82	82	82
Total costs	251	264	277	306
Cost per q.	5.3	3.8	3.3	3.1
Return to land, labor & mgt.	−63	+10	+59	+84

[*] Sullivan, *Phosphorus in Agriculture,* **61**:19 (June 1973).
[1] Fertilizer, lime, machinery operation, seed, pest control.
[2] Real estate taxes, land maintenance, and overhead.
[3] Interest costs of land.

highest rate of fertilizer. While cost of inputs and price of product will vary, this principle remains the same the world over.

The total cost of growing an acre of corn in the Midwest was estimated to be $135 per acre in 1973 and is climbing each year. The average yield in the United States in 1973 was 97 bushels per acre. This means that the farmer must receive $1.40 per bushel to break even.

Level of Management. An important key in obtaining the most efficient use of the land and inputs is the weather and the ability of the farmer. In humid areas the weather cannot be controlled. In Chapter 16 the relationships between soil fertility and efficiency of water use will be discussed.

Timeliness of operations and precision in seed selection, planting, pest control application, and timely harvesting are the attributes of a good farmer. Delaying planting soybeans one month may reduce soybean yields 10 to 20 bushels in some areas. Delaying planting corn one day may cut yield one to two bushels or more a day after a given date. Ten per cent of the corn or soybean crop may be left in the field. Fifty per cent of the forage may be wasted in a pasture. When only the corn grain is harvested, about half of the total digestible nutrients may be left in the field in the form of stover.

In Ohio the average cost of production for corn farms grouped by yield into three categories was about the same, yet the profit ranged from $10 to $86 per acre. The same was true on cost of production for three groups of soybean farmers, yet the profit ranged from $20 to $90.[*]

[*] J. E. Beuerlein, Ohio State University, personal communication.

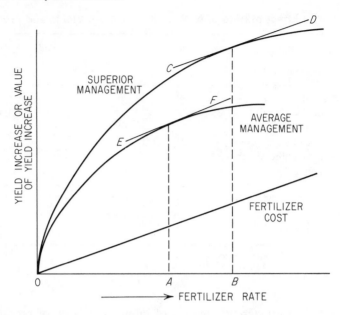

Figure 15–3. With superior management, higher rates of fertilizer can be profitably used than with average management. With average management Rate A is optimum, whereas with superior management Rate B is optimum. [Stritzel, *Iowa Farm Sci.*, **17**(12):14 (1963).]

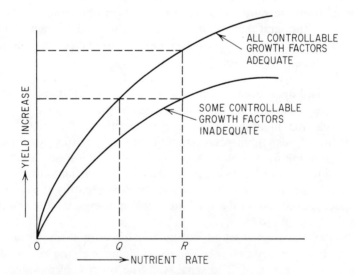

Figure 15–4. With a given rate of fertilizer (Q or R), a higher yield is obtained when all controllable growth factors are adequate. [Stritzel, *Iowa Farm Sci.*, **17**(12):14 (1963).]

The soil or the weather might be blamed but most of the difference was probably due to the management ability of the farmer.

With superior management higher rates of fertilizer can and must be used. This general principle is illustrated in Figure 15-3, in which A is the most profitable rate with average management and B, the most profitable rate with superior management. In Figure 15-4 it is seen that with a given rate of fertilizer Q or R a higher yield is obtained when all controllable growth factors are adequate.

Superior management, including the use of more lime and fertilizer, reduced the cost of production of irrigated corn in Kansas from 85 cents/bu. at the 90-bu. yield level to 68 cents/bu. at a 150-bu. yield level (Table 15-2). Similarly, superior management gives higher yields of soybeans (50 bu./A.) and reduced the cost from $2.10 to $1.11/bu. in North Carolina (Table 15-3).

In Pennsylvania an extra expenditure of $11.16 for lime and fertilizer on alfalfa decreased the unit cost of production from $34.07 to $18.95/ton. In Colorado the expenditure of $28.80 for fertilizers on sugar beets decreased the cost of production from $10.40 to $8.60/ton. In Ohio the production cost of hay was $20 per ton with a 3-ton yield and $15 with a 5-ton yield. With wheat the production cost was $1.71 per bu. at 35-bu. yield and $1.20 at 50 bu.

Unit Production Cost of Prime Importance in Agriculture. As prices decrease, it is particularly important that the unit cost of production be kept at a minimum. With average management (90 bu./A. yield level) in the studies reported in Table 15-2 the price of corn must be

TABLE 15–2. Influence of Improved Management Including Adequate lime and Fertilizer on Profits from Irrigated Corn*

	Average management	Superior management
Yield per acre	90 bu.	150 bu.
Fertilizer cost	$8.30	$23.20
Irrigation cost	8.00	8.00
Total costs	76.80	102.20
Cost per bushel	0.85	0.68
Profit per acre	13.20	47.80
Profit per bushel	0.15	0.32
Acres for $1000 profit	75.7 A.	20.9 A.
Bushels for $1,000 profit	6,666 bu.	3,125 bu.

* Kansas State University, cited by W. H. Garman. *The Fertilizer Handbook.* Copyright 1963 by the National Plant Food Institute.

TABLE 15–3. Influence of Level of Management on Returns from Soybeans in North Carolina[*]

Soybeans	Level of management		
	Current	Good	Superior
Yield per acre (bu.)	20.0	34.0	50.0
Price per unit	$ 2.00	$ 2.00	$ 2.00
Value per acre	40.00	68.00	100.00
Costs per acre	41.98	53.72	55.75
Cost per unit	2.10	1.56	1.11
Return over cash operating expenses per acre	− 1.98	14.28	44.25

[*] W. H. Garman, *The Fertilizer Handbook*. Copyright 1963 by the National Plant Food Institute.

above 85 cents/bu. for the farmer to make money, for 85 cents is the cost of production. With superior management the price of corn would need to be above only 68 cents/bu. for the farmer to make money. This general relationship is rather difficult to make clear. In periods of declining prices the farmers may be hesitant to spend the extra money for fertilizer or for any other practice that might increase yields. However, if the farmer has the capital or can secure the necessary credit, it is obvious that this outlay would enable him to farm profitably at lower produce prices, even though he had to spend more money.

RETURNS PER DOLLAR SPENT ON FERTILIZER OR LIME

Some farmers must consider the returns on each dollar spent for fertilizer, lime, or any other farming practice. They often have only a limited amount of cash or can obtain only a limited amount of credit. They must decide whether to apply nitrogen, phosphorus, potassium, and/or lime, in what quantities and on what fields, and whether to buy extra feed for

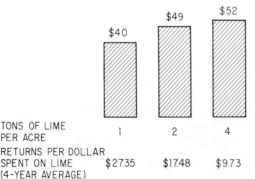

Figure 15–5. The value per acre (above the cost of the lime) of the response to lime of Ladino clover grown on an acid Bladen silt loam. (*North Carolina Agr. Expt. Sta. Bull. 385*, 1953.)

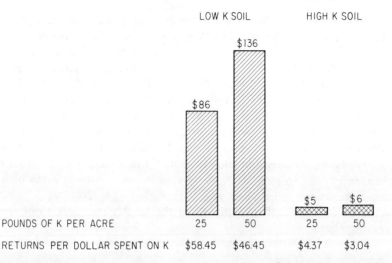

LOW K SOIL   HIGH K SOIL

POUNDS OF K PER ACRE	25	50	25	50
RETURNS PER DOLLAR SPENT ON K	$58.45	$46.45	$4.37	$3.04

Figure 15–6. The value (above the cost of potassium) of the response to added potassium of cotton grown on soils low and high in potassium. Adequate lime, nitrogen, and phosphorus applied. Three-year average on Norfolk sandy loam and Cecil loam, respectively. (*North Carolina Agr. Expt. Sta. Bull. 385*, 1953.)

their livestock or improved seed for their crops. In each operation they are interested in the return that can be obtained per dollar spent.

In general, as the rate of a particular nutrient is increased, the returns per dollar spent decrease. This is illustrated in Figures 15-5 and 15-6. This decrease is the result of a reduced response for each successive increment of any given treatment. Eventually the point is reached at

TABLE 15–4. Effect of Rate of Nitrogen on Net Return per Added Dollar Invested[*]

Nitrogen rate (lb./A.)	Added input (lb. N/A.)	Net return per added dollar invested
20	20	$7.25
40	20	5.75
60	20	5.00
80	20	3.87
100	20	2.38
120	20	1.63
140	20	0.88
160	20	0.50
180	20	0.12
200	20	−0.62

[*] P. R. Robbins and S. A. Barber, Purdue University, private communication.

which there is no further response to increasing amounts of a particular element, which is a corollary of the growth response curves discussed in Chapter 2. Other information on the effect of rate of nitrogen on net return per added dollar invested is given in Table 15-4. On this particular soil the extra 20 lb. N/A. over 160 lb., to give a total of 180 lb., returned 12 cents over cost for each dollar spent on this 20-lb. increment.

Credit agencies are very much interested in the returns the farmer realizes for each dollar spent on fertilizer or lime. As a general rule, 2 to 3 dollars is the expected return for each dollar spent. With this information available, the credit agencies would be in a more favorable position to extend credit to farm patrons. Certainly, with interest rates of 8 per cent, investment in fertilizer and lime with 100 to 200 per cent return would be an excellent investment for the farmer and a safe one for the credit agency.

PROFIT PER ACRE

Progressive growers recognize that although returns per dollar spent are important the significant figure is the profit per acre (Figure 15-7). With adequate cash or credit, the farmer must select the treatment that will earn the greatest net return per acre. The effect of rate of fertilizer,

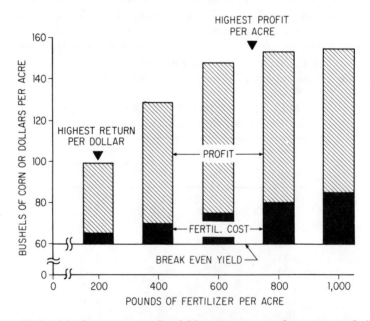

Figure 15-7. Maximum economic yields mean top profit per acre. Only the man with very limited capital can afford to stop fertilizing at the rate which produces the highest return per dollar invested. (Courtesy of the Potash Institute of North America.)

TABLE 15–5. Economic Aspects of Corn Yield Response to Nitrogen Rates*

Nitrogen per acre (lb.)	Yield increase per acre (bu.)	Marginal yield increase[1] (bu.)	Marginal cost[2]	Marginal return[3] Price of corn per bushel			Gross profit per acre[4] Price of corn per bushel		
				$1.25	$1.00	$0.75	$1.25	$1.00	$0.75
20	8	8	$2.40	$4.17	$3.33	$2.50	$ 7.60	$ 5.60	$ 3.60
40	15	7	2.40	3.65	2.92	2.19	13.95	10.20	6.45
60	21	6	2.40	3.13	2.50	1.88	19.05	13.80	8.55
80	26	5	2.40	2.60	2.08	1.56	22.90	16.40	9.90
100	30	4	2.40	2.08	1.67	1.25	25.50	18.00	10.50
120	32	2	2.40	1.04	0.83	0.63	25.60	17.60	9.60

* J. A. Stritzel. *Iowa Farm Sci.*, **17**(12):14 (1963).
[1] Marginal yield increase is the additional increase obtained for each successive 20-lb. nitrogen addition.
[2] Cost for each successive 20-lb. addition of nitrogen at 12 cents/lb.
[3] Marginal return is the return per dollar invested for each successive $2.40 spent for nitrogen in this example.
[4] Phosphorus and potassium costs have not been subtracted from the gross cost. Phosphorus and potassium rates are constant for all nitrogen rates in this example.

lime, and potassium on this net return above the cost of the nutrients is shown in Table 15-1, Figures 15-5 and 15-6, respectively. For example, for Ladino clover the net return per acre from lime was greatest at the 4-ton rate, even though the return per dollar was considerably below that obtained at the 1-ton rate.

Maximum profit is obtained when the added return in yield just equals the cost of the last increment of fertilizer. For example, in Table 15-5 the marginal return (returns per dollar) with $1.25 corn is $1.04 at the 120-lb. rate, so it is still profitable. With one-dollar corn the return is only 83 cents at the 120-lb. rate.

Credit agencies may tend to encourage farmers to apply fertilizer and lime at the lower rates. They feel that there is less risk and that there may be more of a chance that the farmer will be able to pay for the fertilizer or lime in spite of yields below the top profitable point. However, according to the discussion on the importance of cost per unit, a low rate of fertilizer may result in a high cost of production.

WHAT ARE THE MOST PROFITABLE RATES OF PLANT NUTRIENTS?

More and more growers are asking this question, but the answer is difficult because of several factors.

1. The expected increase in yield from each increment.
2. The level of management.
3. The price of fertilizer.
4. The price the farmer expects to receive for his crops.
5. Additional harvesting and marketing costs.
6. Residual effects.
7. Levels of other nutrients in the soil or fertilizer.

Obviously, the most profitable rate of fertilization will vary some from year to year because of the variation in prices of farm products and fertilizer. Agronomists and economists thus are faced with a sizable task of calculation if the growers' questions are to be answered each year. They must work with expected yields, prices, and costs.

There are risks in using fertilizers because of the uncertainties of yields and prices. However, the farmer may not know what represents attainable levels of yields on his own soil. A program of education to establish realistic goals on given soil associations would help to establish the possibilities.

Many universities have worked out yield potentials with average and good management for the major soil areas or types. A check list has been prepared to outline those practices that must be considered in good management. Such potentials were extrapolated from field trials.

In a more specific approach the grower himself may conduct simple trials over the years on a few acres of his own fields, as suggested in Chapter 14. One and one half or double rates of recommended nutrients may be tried on a limited scale in conjunction with two plant populations and two or three varieties. Careful observations can make these few acres into a field laboratory from which the more profitable practices for the entire farm can be taken.

Once the goals are worked out, the farmer can examine the profitable responses from adequate nutrients over a period of years. He can then minimize risks during dry or unfavorable years which occur only occasionally.

Expected Increase in Yield from Each Increment. This prediction must be based on the best experimental evidence available and for the environmental conditions under which the question is raised. Experiment stations have these data for some soil conditions in the state, but by and large experimental results are limited at high yield levels. The possibility of unfavorable climatic conditions or the incidence of disease and insects must also be considered.

Yield responses to nitrogen on corn over a series of years at eight experiment stations in Illinois indicated that economically optimal rates ranged from 100 lb. to 240 lb./A. (Figure 15-8).

An important factor is the amount of nutrients present in the soil. The use of soil tests to monitor the levels helps to predict the need for lime and fertilizer. Many more calibration studies will be needed in the future at high yield levels to characterize crop responses with modern technology and the multiplicity of soil, management, and environmental conditions.

Examples of the effect of soil fertility level on the response to lime and potassium are shown in Figures 15-5 and 15-6. Cotton responds to a greater extent on a soil low in potassium than on a soil high in this element (Figure 15-6). It should be noted on the high potassium soil that although 25 lb. of potassium is essentially adequate and the return from the 50-lb. rate over the 25-lb. rate is very small, the return is sufficient to more than pay for the potassium.

The level of phosphorus and potassium affects response of Coastal Bermuda grass to nitrogen (Figure 15-9). With low levels, the maximum profit rate was 200 lb. of nitrogen and only $40/A. With high levels, however, the optimum profit rate was 310 lb., giving a profit of $79.55. This illustrates the importance of balanced soil fertility to obtain maximum profit.

The curves on nitrogen in Figure 15-9 illustrate the law of diminishing returns. This means that the first increment gives the highest increase and succeeding increments give progressively lower increases.

The data in Figure 15-6 indicate that if the farmer has a limited

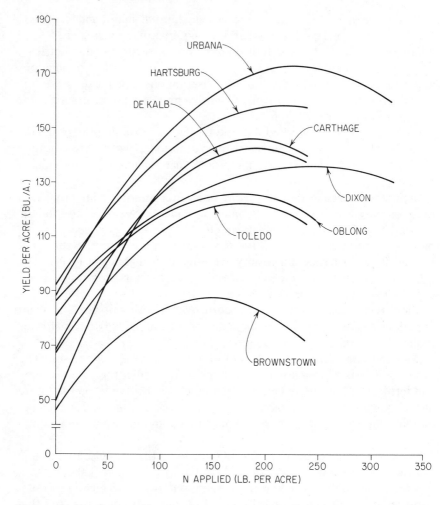

Figure 15–8. Corn yield responses to nitrogen applied in the spring, Browns-town, 1967–1970; Carthage, Hartsburg, and Toledo, 1967–1971; DeKalb, 1966–1971; Dixon, 1970–1971; Oblong, 1967–1971; and Urbana, 1968–1971. [Swanson et al., *Illinois Agr. Econ.*, **13**(2):16 (1973).]

amount of money to spend on fertilizer he would realize the greatest re-turn per dollar spent by putting it on the soils with the lowest fertility level.

The inability to predict weather conditions for the coming season is one of the main deterrents to writing out an exact formula for fertilizer needs. The effect of moisture on the response of corn to nitrogen illus-trates the importance of water (Figure 15-10). In a total of fifty-four experiments, fourteen (26 per cent) were conducted under conditions classified as dry. Under such dry-weather conditions on these soils there

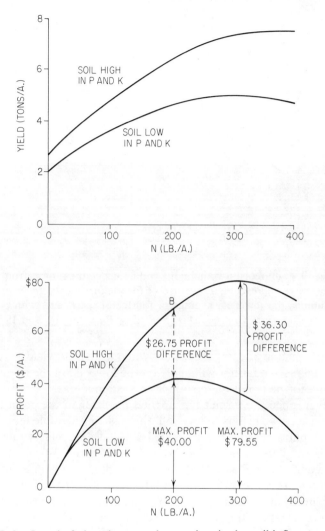

Figure 15–9. Level of phosphorus and potassium in the soil influences response of Coastal Bermudagrass to nitrogen. (L. F. Welch cited by J. E. Engibous, *The Fertilizer Handbook*. Copyright 1972 by the Fertilizer Institute, Washington, D.C.)

is considerable risk in the application of more than 80 lb. N/A. Under conditions of adequate moisture, however, 160 lb. of nitrogen would be profitable. The farmer is usually a rather optimistic individual by nature, and the more aggressive ones may fertilize for a good year most of the time. They will consider that there will be more good years than bad. In the good years they will stand to gain considerably from high fertilization, whereas in the poor years they will not lose much. In addition,

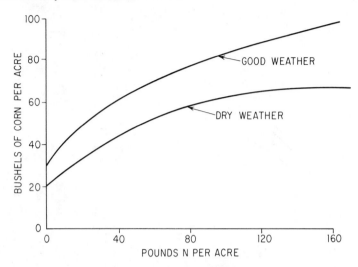

Figure 15–10. Soil moisture conditions will influence the returns from nitrogen. The upper curve represents an average of forty experiments under good moisture conditions, the lower an average of fourteen experiments with poor moisture conditions. (*North Carolina Agr. Expt. Sta. Bull. 366,* revised 1954.)

there is the matter of carryover effect when little crop yield is removed. The less aggressive farmer will apply on the low side of the optimum fertilizer rate, for he is not quite so able to sustain a loss.

However, once the farmer has applied fertilizer to the point at which the response curve begins to level off, there is considerable leeway in the amount of nutrients that can be added without appreciably affecting returns. This is illustrated by the example in Figure 15-6 on the high-K soil. In addition, in Table 15-5 the gross profit with one-dollar corn ranged from $16.40 to $18/A., with 80 to 100 lb. N/A. This means that in this example once the grower has reached the optimum rate of 100 lb., 20 lb. more or less will have little effect on profit per acre. This is fortunate in view of uncertain environmental conditions and the lack of specific experimental information.

When too much fertilizer is added the economic loss is not as great as when a crop is underfertilized by the same proportion (Table 15-6). The residual effect must be considered and helps to compensate for the interest on the extra fertilizer. Thus it appears more profitable to use the optimum amount recommended, even if the rate would be more than optimum in unfavorable years. This of course applies to nutrients that do not leach from the soil.

Price of Fertilizer in Relation to Value of Crop. Although the cost per pound of nutrients in a given fertilizer may fluctuate, these variations are much less than the fluctuations in crop prices. For some crops, how-

TABLE 15-6.   Effect of Underfertilizing vs. Overfertilizing on the Net Return from Added Fertilizer.*

	Yield of corn (bu./A.)	Cost of nutrients in terms of yield (bu./A.)	Net return from fertilizer (bu./A.)	Difference from optimum (bu./A.)
¼ less	120.5	8.1	46.3	−4.3
Optimum	127.5	10.8	50.7	0
¼ more	129.0	13.5	49.5	−1.2
None	66.0	0		

* Walsh, L. M., and James D. Beaton, Eds., *Soil Testing and Plant Analysis*, p. 204, Madison, Wisconsin, Soil Sci. Soc. of Am., 1973.

ever, government regulations have stabilized the floor price, and the grower may use this figure best as a basis for calculations. When these prices have not been regulated, careful consideration of trends and outlook will be helpful in establishing a profitable rate of fertilization.

There are several methods that take into consideration both the cost of the nutrient and the price of the crop. In any case, the response of the crop in question to increasing rates of a nutrient must be known for the general soil condition. These data are obtained by research and are available from the agricultural experiment stations.

One approach worked out by agronomists and economists (Figure 15-11), charts the corn response curve in bushels for a particular element, in this case nitrogen (line *Y*). The cost of the nitrogen plus application is plotted in terms of bushels of corn (*C*). For example, if corn were $1/bu., fifteen dollars worth of fertilizer would be equal to 15 bu. of corn. The advantage of this method is that the fertilizer cost can be shown on the same chart with the yield response.

The area between the two curves in Figure 15-11 represents the extra bushels of corn obtained from the fertilizer. The point at which the difference between the two curves is the greatest represents the point of maximum profit. The exact point (*P*) can be determined by drawing a line tangent to *Y* and parallel to *C*. In this particular example 87 lb. of nitrogen gave the greatest return. Extreme accuracy is not necessary to locate the point of greatest profit, because the variation that would result from applying a little more or a little less fertilizer than is indicated by the most profitable point would be negligible. A satisfactory determination of the most profitable rate can usually be made visually from the graph.

If the price of fertilizer or the expected crop changes, the line representing the cost of fertilizer can be changed (Figure 15-11). For example, if the price of corn went up to $3.00, it would take only 5 bu. to

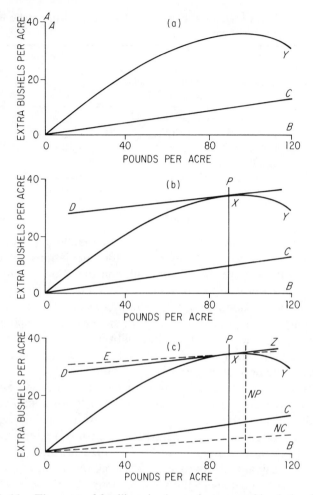

Figure 15–11. The cost of fertilizer is shown in terms of bushels of corn. The point at which there is the greatest difference between the yield curve (Y) and the cost-of-fertilizer curve (C) is the point of greatest profit. If the price of corn goes up, the new cost line (NC) for fertilizer drops. (From Allstetter, *Natl. Fertilizer Rev.*, January–February–March 1953, a publication of the National Plant Food Institute, Washington, D.C.)

pay for fifteen dollars worth of fertilizer; *NC* is the next cost line calculated on the basis of a higher price for corn, and because it is lower there is a greater spread between the *Y* and the *NC* lines. A higher rate of nitrogen can then be used.

Additional Application, Harvesting, and Marketing Costs. These additional costs are rather hard to estimate. If the farmer has been applying 200 lb. of superphosphate, it will cost little more for labor and machinery to apply 400 lb. If a farmer is combining 50 bu. of wheat per acre, it will cost little more to combine 80 bu., but transportation costs are greater.

Residual Effects of Fertilizers. This subject has been treated in considerable detail in preceding chapters and is mentioned here because residual effects are definitely part of fertilizer economics. In general, the entire cost of moderate soil treatment with nitrogen, phosphorus, and potassium is charged to the crop being treated, whereas lime is charged off over a period of five to ten years. With high rates of fertilizer, however, residual effects are striking.

At optimum fertilization rates it has been shown on some soils that about one third of the nitrogen may be residual for next year's crop. Although the carryover of phosphorus varies greatly with the soil, it usually amounts to 40 to 60 per cent on well-limed soils. The residual effects of potassium depend on the soil and how the crops are handled but vary from 25 to 60 per cent. The lower figure would apply when the hay, straw, and stover are also removed from the land.

The increased yield of next year's crop, due to residual effects, may be sufficient to pay for the fertilizer application on this year's crop. This is illustrated by the data in Table 15-7. At the 80-lb. rate of nitrogen the 11-bu. increase in the following corn crop is sufficient to pay for the nitrogen. These data were obtained under irrigated conditions and little leaching occurred. However, the information agrees with that found in many other states.

The carryover effects of phosphorus and potassium for two or more years are also well known, and many examples could be cited. In Indiana the first increment of 50 lb. K/A. produced an increased corn yield in the first year of 8.6 bu., but the four-year total was 22 bu. Application of 1200 lb. K_2O/A. over a four-year period on corn in Illinois increased the soybean yield the next year 20 bu. as well as increasing the soil potassium level 175 lb.*

TABLE 15–7. Residual Effect of Nitrogen on the Next Corn Crop — Nebraska*

Nitrogen applied (lb./A.)	Corn yield same year (bu./A.)	Increase on following corn crop (bu./A.)
0	39	—
40	72	5
80	93	11
120	110	17

* F. V. Pumphrey and L. Harris, *Nebraska Agr. Exp. Sta. Quart.* (Spring 1955).

* L. F. Welch, University of Illinois, personal communication.

Important factors affecting carryover are soil characteristics, weather, rate of applied nutrients, time and method of application, yield increases, crops fertilized, and how these crops were handled. More information is needed on these problems to determine more precisely the carryover effects.

Effect of Fertilizer on Value per Ton or Bushel. One factor often overlooked is the effect of fertilizers on the value of the product. This would apply particularly to forages consumed by animals or even food consumed by man. This may be through increased content of protein or certain plant nutrients. For example, Indiana found the following effects of rate of nitrogen on value of orchardgrass hay per ton because of effect on protein:

		N lb./A.			
0	100	125	250	500	1000
$23.90	$23.82	$24.76	$27.15	$31.33	$33.37

Similar results might be reported for the effects of sulfur, magnesium, potassium, phosphorus, and other elements.

PRIORITY IN USE OF FUNDS

The grower is constantly faced with the task of investing his funds at the point in the farm business that will give the greatest return. In the earlier part of the chapter it was pointed out that between 1950 and 1972 fertilizer costs did not rise, whereas the cost of land, labor, and machinery rose markedly. The grower can well justify the use of more fertilizer per acre of land, per hour of labor, or per dollar of machinery and equipment expense than in 1950.

A second aspect is related to the profitability of the various practices. Although the importance of management will vary greatly from one situation to another, the following contributions to final corn yields are of interest (Iowa State University Agronomy Report, 1962):

	Yield increase (bu./A.)	Contribution (%)
Weed and insect control	3.7	11.4
Improved varieties	4.4	13.5
Planting rates	4.4	13.5
Rotations	7.2	22.2
Fertilizer and lime	12.8	39.4
Total	32.5	100.0

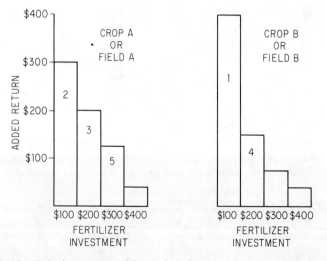

Figure 15–12. Priority on use of funds is influenced by many factors, including the crop and field, as illustrated here. The first 100 dollars should be spent on B, the next 100 dollars on A, the next on A, the next on B, and the next on A. (*Illinois Agr. Circ. 855, 1962.*)

It was emphasized that the rotation effect is primarily the result of nitrogen additions, hence may also be considered a fertilizer effect. Fertility could therefore be said to account for at least 60 per cent of the total potential yield increase.

A third aspect in the priority use of funds is related to the crops and fields that will receive various amounts of fertilizer, as illustrated in Figure 15-12. The first 100 dollars would be spent on B because it would return 400 dollars, the next 100 dollars on A because it would return 300 dollars, the third on A, the fourth on B, and the fifth on A.

Iowa research workers (Gibbons and Heady, 1960) state:

> . . . Fertilizing good cropland according to soil test recommendations should have a high priority for operating funds. This practice gives some of the highest returns on farm invested funds. It is generally more profitable and less risky to invest funds here than in nearly any other place in the farm business.

PRICE PER POUND OF NUTRIENTS

In a discussion of the economics of fertilizer use it is important that the cost per pound of nutrients be considered. The farmer, of course, is interested in the most economical source, but he may be accustomed to buying on the basis of cost per ton of fertilizer rather than on the cost per ton of plant nutrients. It is unfortunate that the profit a fertilizer dealer makes often depends on the number of tons of fertilizer rather than on the tons of nutrients sold.

In Materials. The variation that may occur in the cost of plant-nutrient elements in different carriers is shown in Table 15-8. The prices should be recognized as examples and will, of course, vary considerably from one area to another with time of year and discounts. In addition, bulk is usually about $6/ton less than bagged goods. Local prices can be used for more exact comparisons.

The wide variation in the cost per pound of nitrogen is significant. Of course, other factors such as cost of application, content of supplementary nutrients, and the effect on soil acidity must be taken into consideration.

With the advent of liquid fertilizers, such as anhydrous ammonia and nitrogen solutions, and bulk handling, different application equipment was needed. On small acreages custom application or rented equipment is generally advisable, and the extra charge for application must be considered. Usually the charge is on a per-acre basis. Typical charges might be $1.25/A. if the dealer spreads the fertilizer and $2.00 per ton rental on the equipment if the grower spreads it himself. If the charge is $1.25/A. for 40 lb. of nitrogen applied, the cost would be 3 cents/lb. of nitrogen in addition to the actual cost. If 125 lb. of nitrogen is being spread, the application cost would be 1 cent/lb.

The economy of ordinary and triple superphosphate will depend on location and the need for sulfur. With respect to potassium carriers, muriate is the most economical. Potassium from sulfate of potash or potassium nitrate is more expensive than that derived from muriate because of manufacturing costs, but these forms may be required for special-purpose fertilizers.

TABLE 15–8. Approximate Prices of Fertilizer Materials and Cost of Nutrients per Pound (for illustration only)

Material	Analysis	Price per ton	Cost of nitrogen P_2O_5 or K_2O (cents/lb.)
Sodium nitrate	16% N	$60.00	18.7
Ammonium sulfate	20.5% N	41.00	10.0
Ammonium nitrate	33.5% N	67.00	10.0
Urea	45% N	90.00	10.0
Anhydrous ammonia	82% N	89.00	5.5
Nitrogen solution	28% N	45.00	8.0
Superphosphate (ordinary)	20% P_2O_5	40.00	10.0
Superphosphate (triple)	48% P_2O_5	95.00	10.0
Muriate of potash	60% K_2O	54.00	4.5
Sulfate of potash	52% K_2O	85.00	8.3

In Mixed Fertilizers. The grower is often faced with the choice of using all mixed fertilizers, mixed and straight materials, or all straight materials. A knowledge of the method of calculation of costs is important.

For example, a farmer has a choice of 12-24-24 or 6-24-24 plus ammonium nitrate or anhydrous ammonia.

6-24-24 costs $74.50/ton

12-24-24 costs $80.50/ton

Assuming that the cost of the phosphorus and potassium is the same in both mixtures, the additional nitrogen in 12-24-24 amounts to $6.00 or 5 cents/lb. of nitrogen. The ammonium nitrate will furnish nitrogen for 10 cents/lb. and the anhydrous ammonia for 5.5 cents/lb.

Another calculation that farmers need to make relates to the economics of high-analysis fertilizer. In a comparison of 5-10-10 and 10-20-20, 2 tons of 5-10-10 are required to furnish the same amount of nutrients contained in 1 ton of a 10-20-20. If 5-10-10 costs $50.00 and 10-20-20 costs $90.00 per ton, the 10-20-20 will be $10 cheaper than 2 tons of 5-10-10.

Other Factors. In addition to the actual cost of the material, the farmer must consider the cost of transportation, storage, and labor used in applying the fertilizer. These costs may be difficult to evaluate, but if the actual price of the nutrients from one source is the same as another, growers will take the one requiring the least labor. The higher-analysis goods require less labor in handling. Time is also saved because fewer stops are made in applying the material.

The practice of custom application of bulk fertilizers is spreading rapidly. Fertilizers are often applied on the land for the same price as they are delivered to the farm in bags. The dealer can spread the fertilizer for the cost of the bags, and the farmer does not have to handle the fertilizer. The desirability of such a method of placement depends on a number of factors, as discussed in Chapter 13.

Service. The term *service* usually refers to some form of agronomic assistance to the grower. It may include sampling soils, testing the soils in the company laboratory or sending them to another laboratory, fertilizer and lime recommendations, maps of farm showing requirements, planning rotations, and advice on other agronomic practices, including land preparation, varieties, pest control, and harvesting. The agronomist or dealer may spend time with the grower in his fields during the summer to evaluate his program and suggest improvements. Such specialized advice from qualified individuals can be particularly helpful to the grower over a period of years.

Personalized service costs money and the grower must eventually pay for it. Actually, however, the investment per acre is not large. If the extra cost of the fertilizer is $5/ton, the additional yield of corn needed to pay for it is calculated as follows (price of corn is one dollar/bu.):

Rate of application (lb./A.)	Corn (bu./A.)
200	0.5
400	1.0
600	1.5

A sound advisory service program could well increase yields 10 to 20 bu./A.; therefore such a service will enter into the grower's decision when he is ready to make his choice of fertilizer.

LIME AND AN EFFICIENT RETURN FROM NITROGEN, PHOSPHORUS, AND POTASSIUM

Lime furnishes the least expensive nutrients, for the cost of limestone spread on the land ranges from about $3 to $6/ton (90 to 100 per cent calcium carbonate equivalent). The returns are usually much higher on lime when it is applied where needed than those obtained from other nutrients. Returns of 25 dollars or more per dollar spent on lime are common. It would be difficult to cite another agricultural practice that yields such a return on the investment.

An example of the cumulative returns from limestone in a corn-wheat-hay rotation is shown in Table 15-9. The three-ton initial rate with reapplication as needed appeared to be a satisfactory solution for the conditions under which this experiment was conducted and did not require a large expenditure at any one time.

In spite of a high return, however, lime is often neglected in the fertility programs in humid regions. This may be because visual responses to

TABLE 15–9. Cumulative Net Returns from Limestone Applications at End of Each Year (corn-wheat-hay rotation on an acid Indiana soil) *

Rate of application (tons/A.)	Accumulated net returns at end of								
	1st year	2nd year	3rd year	5th year	9th year	12th year	15th year	20th year	27th year
1.5	−0.71	6.64	16.25	38.46	75.70	107.82	146.46	185.14	223.93
3.0	−5.78	4.17	15.90	46.71	98.44	149.58	193.74	255.87	302.19
12.0	−29.42	−21.15	−7.76	24.95	76.97	135.92	190.96	286.82	373.04
5.0†	−6.25	2.02	13.78	40.36	75.48	129.86	183.92	273.12	360.78

* Limestone valued at $4/ton spread. Nitrogen, phosphorus, and potassium applied. S. A. Barber, *Purdue Univ. Agron. Dept. Mimeo. AY124* (1954).
† Three tons applied initially and two additional tons applied after a lapse of seven years.

lime are often not so spectacular as those obtained with such nutrients as nitrogen, phosphorus, or potassium, unless the soil is particularly acid.

A second reason for the limited use of lime may be that its effects last over a period of time and the returns are not all realized in the first year. A third reason may be the lack of a concerted promotion and sales campaign by the lime companies. Certainly, if the fertilizer industry left the promotional aspects of fertilization up to state extension workers, with all due respect to that group, there would not be nearly the demand for fertilizer that there is when supported by an extremely active group of salesmen and industry agronomists.

The importance of liming in accordance with soil and crop requirements to obtain efficient use of the primary nutrients applied should be stressed. There are many such examples, one of which is shown in Figure 15-13.

At the Brownstown Soil Experiment Field in Illinois the effect of lime in a corn-soybean-wheat-hay rotation was studied on an acid soil for many years. With adequate N-P-K over a period of eight years the unlimed treatment showed a net annual loss per acre of $13.64, whereas with lime the net increase was $19.75 per acre.

Lime is the first step in any sound soil management program, and broadcast applications in accordance with soil and plant requirements are essential for greatest returns from fertilizer.

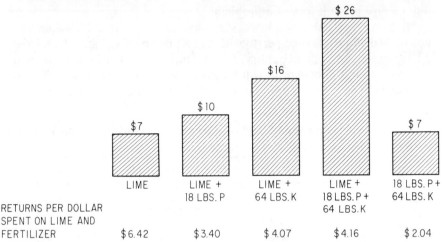

Figure 15-13. Lime alone or fertilizer alone when both are needed will not give maximum returns. Here is shown the value of the response of soybeans (above the cost of the nutrients) from lime and fertilizer applied singly and in combination. Average of nine experiments. (*North Carolina Agr. Expt. Sta. Bull. 385,* 1953.)

SUBSTITUTION CONCEPT

When an economist refers to two nutrients as substitutes, he means that both will have the same effect on yield. This is illustrated in Table 15-10. Either 53-lb. P or 200 lb. N/A. increased the yield to 46 bu. However, adequate quantities of both produced yields of 139 bu./A.

The concept of substitution refers to economic substitutes (i.e., they both increase yield) and not physiological substitution. The fact that either nitrogen or phosphorus increases yield may be the result of a number of factors. One nutrient may have an effect on the utilization of other nutrients; for example, phosphorus is utilized more efficiently in the presence of ammoniacal nitrogen. Increased efficiency of water utilization is another possibility.

Eventually nutrients not added will be depleted in the soil. For example, if nitrogen alone were applied without phosphorus over a period of years it is possible that the phosphorus level would be depleted to the point at which nitrogen alone would not produce even a small increase unless phosphorus were added. Hence the substitution concept is not a sound long-time approach.

ANIMAL WASTES

When the supply of commercial fertilizers is plentiful, the high cost of labor and equipment used in spreading animal manure makes it difficult to calculate its value. However, front-end loaders and large-capacity spreaders for solid wastes and improved techniques for distributing liquid manure make it less of a problem.

Michigan State University workers have calculated the following values of N-P-K in manure per ton: dairy, $2.00; beef cattle, $2.44; hog, $2.38; sheep, $4.70; and poultry, $6.30. Only a portion of nutrients such as nitrogen and phosphorus will be available the first year. In addition secondary and micronutrients and organic matter are supplied by

TABLE 15–10. Predicted Effect of Nitrogen and Phosphorus on Yields of Corn in Iowa*

Phosphorus applied (lb./A.)	Nitrogen applied (lb./A.)					
	0	40	80	120	160	200
0	—	13.3	29.2	39.8	45.5	46.1
18	16.2	38.3	55.4	67.4	74.4	76.3
35	34.1	57.5	75.9	89.2	97.5	100.7
53	46.3	71.0	90.7	105.3	114.9	119.4
70	53.0	78.7	99.8	115.6	126.6	132.3
88	53.4	80.7	103.0	120.2	132.2	139.4

* R. D. Munson and J. P. Doll, in A. G. Norman, Ed., *Advances in Agronomy*, Vol. XI. New York: Academic, 1959.

manure. The latter should be important in affecting soil structure on some soils. Too, solid manure may be applied on sloping land to help to reduce erosion, but in such cases much of the nitrogen would be lost. Profitable rates of application in Michigan were found to range from 10 to 30 tons per acre.

It is recognized, of course, that there is much variability in manure, depending on methods of storing and handling. However, it appears in general that with current prices of nutrients, labor, and equipment it is profitable for the grower to use livestock manure. Because it is largely a N-K fertilizer, best returns should be obtained by applying it to non-legumes. Hauling charges can be reduced by applying it on fields close to the barn and using commercial fertilizer on the more distant fields.

In some farm management programs it is imperative that the manure be disposed of, and its value may be balanced against equipment and hauling costs. The general feeling, however, is that it is worth a little more than cost of handling.

In areas or times when fertilizers are difficult or impossible to obtain, animal manures play an important role in supplying fertilizer needs.

MAINTAINING TOTAL PRODUCTION ON FEWER ACRES

In some areas and for some crops emphasis is being placed on controlled production. This has the effect of preventing a glutted market and thus maintaining a reasonable price structure. Many farmers question controlled acreage because of loss of profit. However, the history of any controlled crop has been that the yields per acre increase. This is due primarily to two factors; first, the less productive acres are abandoned or put in soil-improving crops, and, second, more attention is devoted to proper practices on the acres under cultivation.

As fertilizers increase yields, fewer acres are required to produce a given amount of production. For example, if fertilizer increased production 30 per cent, acreage could be reduced accordingly. Total profit would be greater because fixed costs would be saved on the reduced acreage.

In 1966 Ibach indicated that in the United States, with a 1960–64 crop production index of 100, cropland use was 451 million acres and total nutrient N-P-K use was 8.7 million tons. In 1980 with a crop production index of 150 and a cropland use of 301 million acres total nutrient use would be 26.6 million tons. This illustrates the part that superior management can play in increasing crop production and at the same time use fewer acres. These extra acres would be the marginal acres more subject to erosion and would essentially be put in reserve until needed.

The need for food in the 1970's over the world, however, necessitated that farmers put these unused acres back into production. In 1973 cropland use was 320 million acres, up 9 percent over 1972, and total nutrient use was 17.8 million tons. About 12 million additional acres were put into production in 1974. Careful land management will be needed to prevent excessive soil losses.

PLANT NUTRIENTS AS PART OF INCREASING LAND VALUE

When buying land, the farmer may be faced with the possibility of choosing high-priced or low-priced property. There is usually the question which is the better buy. The higher-priced land is generally in a better state of fertility and has better improvements in terms of buildings, drainage, and fences.

The lower-priced land may actually be a good buy, however, provided it has not been severely eroded or has no other physical limitations. Such land, however, is usually infertile and may need considerable lime, nitrogen, phosphorus, and/or potassium. Adequate liming and heavy fertilization, as indicated by soil tests, combined with other good practices can rapidly bring this land up to good production. Such expenditures to improve fertility may be considered as part of the cost of the land. If the problem is considered in this light, $40 to $50/A. for liming and heavy fertilization may not seem out of reason. Thus by proper management it is possible to increase land productivity and value.

Summary

1. Higher crop yields continue to offer the greatest opportunity for maximum net profit per acre and for reducing production costs per bushel or per ton. Adequate fertility is a major factor in obtaining high yields. Although fertilizer use has increased in the past few years, it is still below optimum in most areas.
2. Fixed costs are those that remain about the same, regardless of yield. Hence practices that increase yields usually lower the cost of production per unit.
3. In a typical response curve each successive increment of a plant nutrient gives a smaller yield increase in accordance with the law of diminishing returns. Hence the return per dollar spent decreases, an important consideration to the grower with a limited amount of capital.
4. Progressive growers recognize that profit per acre is more important than return per dollar spent. Maximum profit from fertilizer application is obtained when the added return in yield just equals the cost of the last increment of fertilizer.
5. The most profitable rate of plant nutrients is related to expected increases in yield from each increment, level of management, price of fertilizer, expected price of crop, additional harvesting and marketing costs, residual effects, and soil fertility level.

6. Management level is the degree to which all the factors affecting crop production are successfully controlled. At higher rates of fertilization, with expectation of higher yields, managerial ability becomes more and more important.
7. Price per pound of nutrients varies among sources of materials. Higher-analysis mixed fertilizers are generally the most reasonable in cost and in labor required for application.
8. Personalized agronomic service to the grower by the fertilizer industry will enter more and more into the grower's choice of a fertilizer.
9. Priority in use of funds is of prime importance to the grower. It is generally more profitable to invest funds in fertilizer and lime applied in accordance with soil-test level than in other parts of the farm business.
10. Lime usage is about 30 per cent of adequate in the United States. Where needed, the returns from lime are usually much higher than those obtained from other plant nutrients.
11. Although there is much variability in composition of manure, it represents a profitable by-product on livestock farms as a supplement to a complete fertility program.
12. Overproduction is a problem in developed countries in some periods. Studies clearly indicate that superior management on fewer acres maintains profits. Although yields per acre are higher, fewer total bushels or tons of product are produced because of the lower acreage.
13. In periods when increased production of food is needed, the unused or storage acres represent an important market for fertilizer and lime. These acres are usually the marginal acres and good management, including fertilizer and lime, can bring them back into production quickly.
14. Residual effects of fertilizers and lime are an important part of fertilizer economics. With the increasing amounts of fertilizers being applied the value of the residual fertilizer must be considered.

Questions

1. Explain why high yields are a necessity in periods of low prices.
2. What are fixed costs? Variable costs? How do they affect unit cost of production?
3. When a farmer has a limited amount of capital for fertilizer, is he going to be more interested in the rate that will give him the greatest return per dollar spent or the rate that will give him the greatest net profit? How would soil tests help him in his decision? Would the situation be changed if the farmer had unlimited capital? Explain.
4. What are the banks in your area doing in regard to credit for lime and/or fertilizer?

5. How does use of irrigation make it easier to decide what rate of fertilizer to apply?

6. Discuss the factors that determine the most profitable rate of plant nutrients.

7. Why does level of management affect the return from a given level of fertilization?

8. Explain why there is considerable leeway in the amount of nutrients that can be added after a reasonable level of application is reached. Explain how the following affect degree of carryover of fertilizer: soils, weather, application rate, yield increases, crop fertilized and harvested, and nutrient considered.

9. What nitrogen carrier would you choose to fertilize corn? Why? You are quoted the following prices: 3-12-12, $47; 4-16-16, $61; 5-20-20, $76; and 6-24-24, $87. Which would you choose and why?

10. Explain why funds invested in plant nutrients usually return more profit than investments in other phases of the farm business.

11. Obtain the average yield of corn in your state and the number of acres used for this production. What would a reasonable yield of corn per acre be if recommended practices were followed, and how many acres would be required to maintain present total production? What would you estimate the profits to be under each system?

12. What is holding back lime use in your state?

13. Is farm manure a valuable source of plant nutrients in your state? Why or why not.

14. How would you evaluate the residual nutrients in your area? Considering present crop production levels, are the soil test levels suggested in the 1960's adequate now?

Selected References

Abbring, F. T., R. W. Taylor, L. H. Smith, K. L. Washburn, C. L. Rhykerd, and C. H. Noller. "Economics of nitrogen fertilization of orchardgrass." *Agronomy Guide ID86*, Purdue University (1972).

Ahrens, C. L., and E. R. Swanson, "Choosing the most profitable fertilizer program." *Illinois Agr. Ext. Circ. 855* (1962).

Allstetter, W. R., "Can fertilizer recommendations be geared to crop prices? The answer is 'Yes.'" *National Fertilizer Rev.*, 6 (January, February, March 1953).

Barber, S. A., "Liming Indiana soils." *Purdue Agron. Dept. Mimeo AY124* (1954).

Engibous, J. E., "Economics of fertilizer use." Chap. 8, *The Fertilizer Handbook*, Washington, D.C.: The Fertilizer Institute, 1972.

Farm Cost Situation, ERS USDA (Feb. 1973).

Garman, W. H., Ed., *The Fertilizer Handbook*. Washington, D. C.: National Plant Food Institute, 1963.

Gibbons, J., and E. O. Heady, "More about choosing a hog system." *Iowa Farm Sci.*, **14**(8):9–11 (1960).

Ibach, Donald B., *Agr. Econ. Report No. 92.*, USDA (1966).

Illinois Department of Agron. Mimeo AG-1870. "Brownstown – celebrating a quarter century of soil and crop research 1937–1961" (1962).

Krantz, B. A., and W. V. Chandler, "Fertilize corn for higher yields." *North Carolina Agr. Exp. Sta. Bull. 366* (Revised, 1954).

Munson, R. D., and J. P. Doll, "The economics of fertilizer use in crop production." *Advan. Agron.* **11**:133–170 (1959).

Nelson, W. L., "Lime and fertilizer pay off." *North Carolina Agr. Exp. Sta. Bull. 385* (1953).

Parker, F. W., "Fertilizer and economic development," Chapter 1. *Fertilizer Technology and Use.* Madison, Wisconsin: Soil Science Society of America, 1963.

Pesek, J. T., E. O. Heady, and L. C. Dumenil, "Influence of residual fertilizer effects and discounting upon optimum fertilizer rates." *Proc. 7th Intern. Congr. Soil Sci.*, **III**, 220–227 (1960).

Pumphrey, F. V., and L. Harris, "Nitrogen boosts corn yields; increased fertilizer will aid next year's crop." *Nebraska Exp. Sta. Quart.* (Spring 1955).

Rostow, W. W., *The Stages of Economic Growth.* New York: Cambridge University Press, 1960.

Shaudys, E. T., and G. R. Prigge, "The costs of producing major field crops in Ohio." *Ohio Res. Bull. 1051* (1972).

Stritzel, J. A., "Narrative guide for slide sets to be used in county intensified soil fertility programs." *Iowa State University Agron. Rept. A-543* (July 1962).

Stritzel, J. A., "You have a choice of fertilizer rates." *Iowa Farm Sci.*, **17**:14–16 (1963).

Sullivan, L. J., "What motivates U.S. farmers in use of fertilizer." *Phosphorus in Agriculture*, **61**:19 (June 1973).

Swanson, E. R., C. R. Taylor, and L. F. Welch, "Economically optional levels of nitrogen fertilizer for corn: An analysis based on experimental data, 1966–1971." *Illinois Agr. Econ.* **13**(2):16 (1973).

Vitosh, M. L., J. F. Davis, and B. D. Knezek, "Long-term effects of fertilizer, manure and plowing depth on corn." *Research Report 198*, Michigan State Univ. (1972).

Wagner, R. E., and W. K. Griffith, "Maximum economic yields are cost cutters." *Better Crops with Plant Food*, **47**(3):34–39 (1964).

Walsh, L. M., and James D. Beaton, Eds., *Soil Testing and Plant Analysis*, p. 204, Madison, Wisconsin: Soil Sci. Soc. of Am., 1973.

Walsh, L. M., and R. F. Hensler, "Manage manure for its value." *Wisconsin Circ. 550* (1971).

16. FERTILIZERS AND EFFICIENT USE OF WATER

Water stress has often been a convenient scapegoat on which to blame any poorly growing crop, even though nutrient deficiency, pests, and other factors were fullfledged accomplices. There is much talk about good and poor seasons. Most of the time a good season in unirrigated regions is interpreted as one which has an ample amount of rain.

Agriculture so far has had first call on water supplies. This may not continue indefinitely as pressures grow for increased industrial, recreational, and urban use. Agriculture will have to justify the water it uses, and agricultural needs will be balanced against other demands.

All of this means that water must be used as efficiently as possible. Finding ways of raising this efficiency is a prime challenge to agriculture. It is estimated that on the basis of what is now known over-all efficiency of water in irrigation farming is only about 50 per cent. It is probably much less than that in unirrigated areas. In general, any growth factor that increases yield will improve the efficiency of water use. These factors include variety, plant spacing, pest control, time of planting, and plant nutrient supply.

The importance of the effect of fertilization on water requirement of plants has been recognized for many years. In 1913 Briggs and Shantz, USDA scientists, made this statement:

> Almost without exception the experiments herein cited show a reduction in the water requirement accompanying the use of fertilizer. In highly productive soils this reduction amounts to only a small percentage. In poor soils the water requirement may be reduced one-half or even two-thirds by the addition of fertilizers. Often the high water requirement is due to a deficiency of a single plant-food element. As the supply of such an element approaches exhaustion, the rate of growth as measured by the assimilation of carbon dioxide is greatly reduced but no corresponding change occurs in the transpiration. The result is inevitably a high water requirement.

About 50 years later another USDA scientist (Viets, 1962) said:

> Whether fertilizers increase consumptive use not at all or only slightly, all evidence indicates that water-use efficiency, or dry matter produced per unit of water used, can be greatly increased if fertilizer increases yield. So fertilization for the adequate nutrition of all crops plays a major role in the efficient use and conservation of water resources. Fertilizers may also increase root development within the soil so that soil water is used to higher tensions and at greater depths. This effect is important in dryland agriculture and even in farming in humid areas during periods of drought.

In the following discussion the influence of plant nutrient supply on efficient use of water is the major factor covered, for this book deals with soil fertility and fertilizers.

Level of fertility influences:

1. The amount of the crop that will be produced from each inch of water. With a higher yield more crop will be produced per inch of water.
2. The amount of water that will be used from the soil. This is related to depth of rooting and amount of water held per foot of soil. As fertility increases the extent of the root system, the plant can forage more extensively for water.
3. The fullness of the crop canopy. With a more complete canopy infiltration of water is increased.
4. The speed of maturity of many crops. If the period of crop growth is shortened more favorable weather may be encountered.

SOIL MOISTURE LEVEL AND NUTRIENT ABSORPTION

Water is a key factor in all three mechanisms of nutrient uptake. Roots intercept more nutrient ions when growing in a moist soil than when growing in a drier one because growth is more extensive. This is especially important for calcium and magnesium.

Mass flow of soil water to supply the transpiration stream transports most of the nitrate, sulfate, calcium, and magnesium to roots.

Roots usually will not receive enough phosphorus and potassium by these two methods. A third method, diffusion, is important. The plant absorbs nutrients adjacent to the roots and a concentration gradient is established. Nutrients then diffuse slowly from areas of higher concentration to areas of lower concentration but at distances no greater than $1/4$ in. As this occurs through the water films, the rate of diffusion depends partly on the water content of the soil. An example of the effects of water level on rubidium diffusion, an element similar to potassium, is shown in Figure 16-1. With thicker water films or with a higher nutrient content in the soil the elements can diffuse more readily.

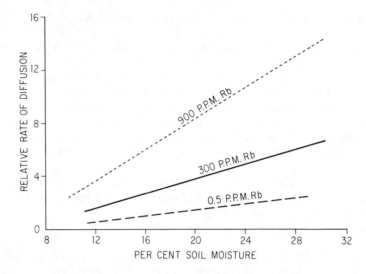

Figure 16–1. The effect of soil-moisture level on rate of diffusion of rubidium (an element similar to potassium) at three levels of applied rubidium. [Barber, reprinted from *Plant Food Review*, **10**(2):5 (1964). A publication of the National Plant Food Institute, Washington, D.C.]

Considerable work has been done to establish the relationship between soil moisture levels and nutrient absorption by the plant. Nutrient absorption is affected directly by level of soil moisture as well as indirectly by the effect of water on the metabolic activity of the plant, soil aeration, and the salt concentration of the soil solution.

In Iowa as moisture stress increased, the concentration of N, P, and K in the corn leaves decreased (Table 12-2). While application of N, P, and K increased the nutrient content of the plants, concentrations were still below the optimum range.

An example of the direct effect of moisture tension is shown in Figure 16-2. Uptake of phosphorus decreases to 80 per cent at 1 bar and to 50 per cent at 3 bars moisture tension in relation to uptake at ⅓ bar. Water films between the roots and soil particles become thinner and the path length of ion movement increases as tension increases. This reduces the rate of ion diffusion to the roots.

The effect of decreasing soil H_2O (higher pF) on decreasing corn growth is shown in Figure 16-3. However, as per cent K saturation is increased growth increased at all three moisture levels. Hence, an adequate amount of nutrient, in this instance potassium, helps to overcome some of the stress associated with a low supply of water.

Summer rainfall (June, July, and August) is related to the response of corn to potassium (Figure 16-4). The lower the rainfall, the greater the

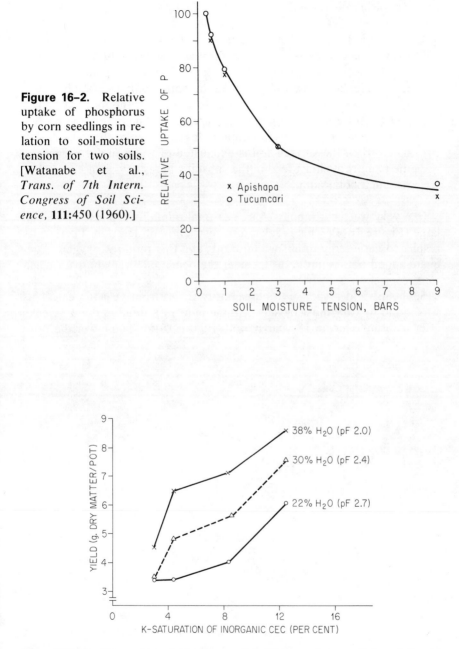

Figure 16–2. Relative uptake of phosphorus by corn seedlings in relation to soil-moisture tension for two soils. [Watanabe et al., *Trans. of 7th Intern. Congress of Soil Science,* **111**:450 (1960).]

Figure 16–3. Dry matter yield of corn after three weeks growth in relation to potassium saturation and water supply. [Grimme et al., *International Symposium on Soil Fertility Evaluation Proc.,* **1**:33. New Delhi: Ind. Soc. Soil Sci. (1971).]

response, which is related to at least two factors:

1. Most of the potassium absorbed must move to the plant root from adjacent areas. It passes through the water films, and the lower the water content the harder it is for the potassium to move. Hence at low soil moisture the potassium absorbed by the plant may originate in the immediate vicinity of the root. At high soil moisture it may originate at a greater distance from the root.
2. In some soils the subsoil contains less potassium than the surface. When the surface soil is exhausted of water in dry periods, the plant roots must feed in the subsoil where they cannot absorb so much potassium.

In a very wet year (about 25 in. seasonal rainfall) response was again large (Figure 16-4). This is related to restriction of aeration. Plant roots respire to get energy to absorb nutrients and this requires oxygen. Adequate added potassium helps to meet the needs of the plant even when root respiration is restricted.

Soybean yield response to phosphorus varies from year to year. This was found to vary most closely with amount of rainfall in the 12-weeks after planting over an 18-year period (Figure 16-5). The lower the rain-

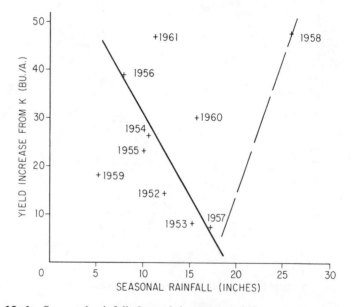

Figure 16–4. Seasonal rainfall (June, July, August) influences corn response to potassium. Low or very high rainfall produces greater responses. [Barber, *Better Crops with Plant Food,* **47**(1):16. Copyright 1963 by the Potash Institute of North America.]

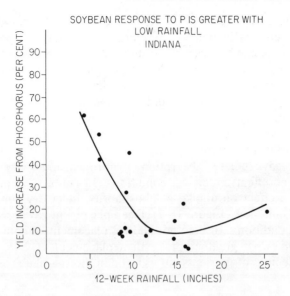

Figure 16–5. The less the rainfall for 12 weeks after planting, the greater the percentage yield response of soybeans to phosphorus. [Barber, *Better Crops with Plant Food*, **55**(2):9. Copyright 1971 by the Potash Institute of North America.]

fall the greater the percentage response to phosphorus. The same relationship was found for potassium.

Although absorption of nitrogen is definitely reduced on dry soils, it is not reduced nearly as much as that of phosphorus and potassium. Ammoniacal nitrogen does not move readily, but nitrate-nitrogen is an anion and moves in and with the soil water. In heavy rains nitrates move downward in the soil profile and provide a storehouse of nitrogen for later use, provided that they do not move out of the root zone.

Another aspect is that under drought conditions organic matter decompositon, hence nitrogen release, is slowed.

Much more information is needed about the effect of soil moisture levels on uptake of micronutrients. Temporary boron deficiency during periods of dry weather is quite common. Explanations include (1) a good part of the boron is in the organic matter, and under dry conditions organic matter decomposition and release of boron is slowed down; (2) in some areas the lower soil horizons are lower in boron content than the surface. Under dry conditions the roots are less active in the surface horizons and the plants tend to take up less boron. On the other hand, in quite sandy soils excessive rainfall and leaching may remove some of the available soil boron.

Increased soil moisture has been shown to result in greater amounts of molybdenum and cobalt uptake by alsike clover. There was more of these two elements in the soil solution, and the larger amounts in the plant were explained by greater transpiration and the sweeping of the enriched solution into the plant. Moisture level did not affect the uptake of copper. Manganese becomes more available under moist conditions because of conversion to reduced, more soluble forms.

Placement.   Under drought conditions it is generally accepted that it is best to place the fertilizer in the zone of the soil that retains water for a greater part of the season. As shown in Figure 16-6, deeper placement of fertilizer gave greater absorption of nitrogen under dry conditions. There was no effect under wet conditions. In contrast, depth of placement had no effect on uptake of phosphorus under dry conditions and uptake was quite low (Figure 16-7). Deep placement gave greater uptake under wet conditions. In data not shown, placement of potassium at 12 in. gave somewhat greater uptake than placement at 2 in. In general deeper placement of nutrients, so as to be in moist soil a greater portion

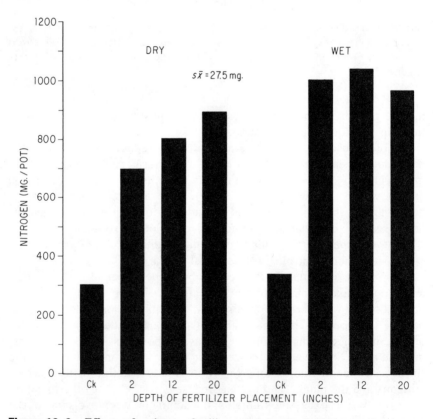

Figure 16-6. Effects of moisture, fertilizer, and depth of placement on nitrogen uptake by grain sorghum. [Eck et al., *Agron. J.,* **53**:335 (1961).]

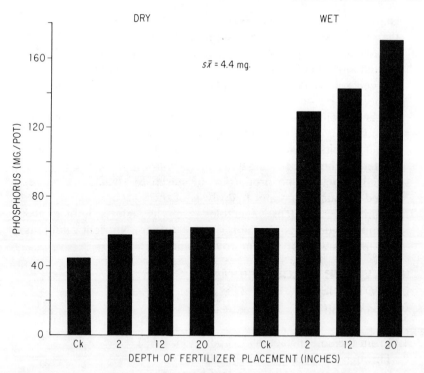

Figure 16-7. Effects of moisture, fertilizer, and depth of placement on phosphorus uptake by grain sorghum. [Eck et al., *Agron J.*, **53**:335 (1961).]

of the season, will result in more efficient utilization. Of course if a soil is very dry deeper placement will not be effective.

PHYSIOLOGICAL EFFECT OF PLANT NUTRIENTS RELATED TO WATER ECONOMY

There are numerous nutrient-water interactions in plants, some of which are the following:

1. On a P-deficient soil added phosphorus speeds up maturity. The plant thus grows for a shorter period and, other things being equal, uses less water. For example, if maturity is advanced seven days and the plant is using 0.2 in. of water a day, 1.4 in. of water will be saved.
2. A K-deficient plant is flaccid. Added potassium increases turgidity and helps to maintain internal water balance and the hydration of the protoplasm. A reduction in turgidity is accompanied by a decrease in stomatal openings to conserve water.
3. Stomatal opening is controlled by active transport of K into the

guard cells. The guard cells of the closed stomate contain less K than those cells of the open stomate. Hence, uptake of CO_2 and photosynthesis by K-deficient leaves is less.

4. A high ion concentration in the cell increases osmotic pressure of the cell solute and consequently the plant's ability to withstand high water tension in the soil.

5. Water deficits in the plant affect all processes of cell growth, including division, enlargement, and maturation.

6. Water deficits cause a decreased rate of photosynthesis, which is related to the decreased supply of carbon dioxide caused by the closed stomata. The proportion of starch to sugar is often reduced because of greater hydrolysis of the starch.

7. Quality of succulent vegetables such as celery, lettuce and cucumbers depends on their state of turgor. Nutrient additions that increase turgor will improve quality.

HOW WATER IS LOST FROM THE SOIL EVAPOTRANSPIRATION

Water in a soil is lost in three ways:

1. From the soil surface by evaporation.
2. Through the plant by transpiration.
3. By percolation beyond the rooting zone.

The sum of the water used in transpiration and evaporation from soil plus intercepted precipitation is called evapotranspiration. With a more complete cover less water evaporates from the soil directly and more goes through the plant. Fertilizer and adequate stands are among those factors that help to provide more plant cover rapidly and thus get more benefit from the water.

Evapotranspiration can best be described in terms of the net radiation. This is the difference between the incoming radiation and radiation losses from soil and crop surfaces. Net radiation is used (1) to evaporate water from the soil or plant, (2) to heat the air, the soil, or the plant, or (3) in photosynthesis.

With a sparse stand or growth much sunlight will reach the soil, and considerable water may be evaporated directly from a moist soil without passing through the crop. With a heavy crop canopy a blanket of green is presented to the sun and less energy reaches the soil. The soil is kept cooler, the crop provides insulation to maintain a higher humidity just above the soil, and there is less air movement. These three effects are among those helping to reduce evaporation from the soil. It should be kept in mind, however, that even with a heavy canopy a considerable amount of energy still reaches the soil.

Fertilizer affects plant size, total leaf area, and often the color of the foliage. Close rows and adequate stands, along with adequate fertilization, help to provide a heavy crop canopy quickly. For example, in Iowa water use by corn was less in 21-in. rows than in 42-in. rows. Differences in evapotranspiration among crops may be small once a complete cover is developed. Daily use of water with a growing crop on the soil varies greatly from one day to another, depending on soil and atmospheric environmental conditions such as temperature, moisture present, and wind. However, losses of 0.1 to 0.3 in. of water daily per acre are common.

Evaporation from the soil may account for one third to two thirds of the total disappearance of water in a crop year in humid areas or where the soil is wet. Experiments in Illinois have shown that when the soil was covered with plastic to reduce evaporation from the soil corn used 6.1 in. during the growing season in contrast to 13.4 in. when grown normally. Under these conditions a heavy crop canopy would increase water efficiency.

With local droughts or in arid regions the soil surface is dry, and very little water is lost from the soil. The moisture films between the particles are thin, and little water is transported to the soil surface by capillarity or diffusion of water vapor. Hence on dry soils most of the water use is by transpiration, and low plant spacings are generally employed to reduce the amount of water used by the plants.

Heat advection, in which there is horizontal and vertical movement of air in a turbulent fashion, brings in more heat. In a hot dry area with a strong wind the heat from the air may contribute to 25 to 50 per cent of the total evapotranspiration. In arid and semiarid areas advection is great, and thus quite variable evapotranspiration may be obtained.

EFFECT OF PLANT NUTRIENTS ON WATER REQUIREMENT

Water requirement is defined as the ratio of the weight of water absorbed by the plant during the growing season to the weight of dry matter produced by the plant. As far back as 1912 workers in Nebraska showed that manure greatly improved effectiveness of water use by corn.

	Pounds of water per pound of dry ears	
	No manure	Manure
Infertile	2136	692
Medium fertility	1160	679
Fertile	799	682

TABLE 16–1. **Effect of Nitrogen on Water Used per Pound of Dry Matter**[*]

	Water used per pound of dry matter		
	50 lb. N/A.	100 lb. N/A.	200 lb. N/A.
Common Bermuda	8,275	3,962	2,941
Coastal Bermuda	2,012	1,206	722
Suwanee Bermuda	1,515	915	572
Pensacola Bahia	2,652	1,633	1,054
Pangola	2,546	2,049	2,625

[*] G. W. Burton, G. M. Prine, and J. E. Jackson, *Agron. J.*, **48**:498 (1957).

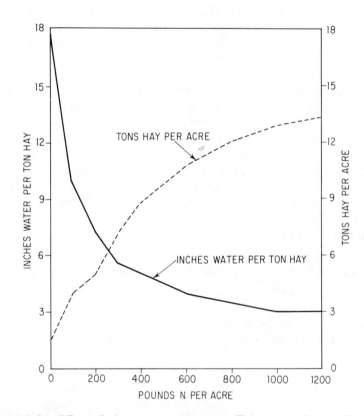

Figure 16–8. Effect of nitrogen on water-use efficiency by Coastal Bermuda hay. Phosphorus and potassium were applied in liberal amounts. [*Texas A and M College Progr. Rept. 2193* (1961), given on page 192 of *Fertilizer Salesmen's Handbook,* published and copyrighted by National Plant Food Institute, 1963, by permission of copyright owners.]

Adequate nitrogen decreases the water used per pound of dry matter in grasses (Table 16-1). With the exception of Pangola grass, in which there was a stand problem at the 200-lb. rate of nitrogen, the grass receiving the 200-lb. rate used only 35 to 40 per cent as much water per pound of dry matter as was used at the 50-lb. rate. The wide difference among the grasses is significant, for Common Bermuda uses five times as much as Suwanee.

$$\frac{\text{Dry matter production}}{\text{water used}} = \text{water-use efficiency is a fundamental relation}$$

in agriculture. Any practice that promotes plant growth and the more efficient use of sunlight in photosynthesis to increase dry matter production will also increase water-use efficiency.

Irrigated Soils. Fertility is one of the important controllable factors affecting plant growth. This principle is well illustrated in Figure 16-8. Adequate nitrogen markedly increased the yield of Coastal Bermuda hay. At the same time the amount of water per ton of hay decreased from 18 in./ton with no applied nitrogen to 3 in./ton with 1,000 lb. N/A.

On a low-P soil phosphorus banded before seeding increased the amount of alfalfa hay obtained with each inch of water (Table 16-2). At each water level the 264 lb. P/A. treatment increased the tons per inch of water 35 to 40 per cent over the 44 lb. of phosphorus treatment.

A diagrammatic presentation of these data on alfalfa, as well as the effect of nitrogen on cotton, related to soil-water stress, is shown in Figure 16-9. Increased water stress reduced the alfalfa yield about the same proportion under high and low phosphorus treatment. With low nitrogen on cotton the yields were apparently already so low that increased water stress had little effect.

TABLE 16-2. Effect of Phosphorus and Irrigation on Total Yield of Alfalfa and Tons of Hay per Acre-Inch — Arizona, 1950–1952*

Phosphorus applied (lb./A.)	Irrigation treatment					
	Dry		Medium		Wet	
	Yield (Tons/A.)	Tons per acre-inch	Yield (Tons/A.)	Tons per acre-inch	Yield (Tons/A.)	Tons per acre-inch
44	25.8	0.120	28.7	0.129	33.3	0.130
88	28.6	0.133	30.7	0.138	37.7	0.148
132	34.5	0.160	37.7	0.169	44.8	0.175
264	34.9	0.162	40.6	0.182	47.2	0.185

* C. O. Stanberry, C. D. Converse, H. R. Haise, O. J. Kelley, *SSSA Proc.*, **19**:303 (1955).

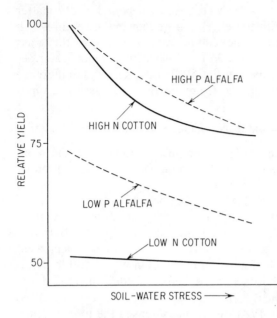

Figure 16–9. Fertility-irrigation relationships for cotton and alfalfa on a sandy Arizona soil. [Stanberry et al., *SSSA Proc.*, **19**:303 (1955).]

Results from the Colorado Experiment Station show that 150 lb. N/A. on irrigated corn drastically increased yields and water efficiency:

	No nitrogen	150 lb. nitrogen
Yield of corn	91 bu.	147 bu.
Bushels per inch of water	4.9 bu.	7.4 bu.

In many parts of the world large sums of money are being spent for irrigation without proper attention to cultural practices. After the lack of moisture is eliminated by irrigation, a great number of things may limit yields (Figure 16-10). Because of these other factors, there can be many disappointments. It must be kept in mind that if yields of 180 bu. rather than 90 bu. of corn or 8 tons rather than 4 tons of alfalfa are to be obtained, the nutrient removal is at least doubled. This means that the crop must obtain more of these elements from some source, whether from the native supply, manures, or fertilizers.

Unirrigated Soils. On cereal crops in Nebraska nitrogen increased water use efficiency as compared with no added nitrogen. Water use was taken to be the difference in soil moisture content to a depth of 6 ft. between planting and harvest plus 90 per cent of the precipitation. The other 10 per cent was assumed to have run off.

| Crop | Number of experiments | Yield | | Increased water use (inches) | Increased water use efficiency (%) |
		No nitrogen (bu./A.)	Nitrogen (bu./A.)		
Corn	12	70	112	1.3	44
Wheat	29	31	37	0.9	12
Oats	16	46	66	0.8	38
Sorghum	9	60	75	1.4	11

In other Nebraska studies, 121 field experiments on four crops were evaluated. Optimum fertilizer treatment increased water use efficiency of grain crops an average of 29 per cent. This increase was almost in proportion to yield response to fertilizer. It was emphasized that it represents a major economy in crop water use and is significant in view of shrinking water resources.

In Montana nitrogen increased wheat yield, evapotranspiration, and bushels per inch of water. Without added nitrogen water extraction was

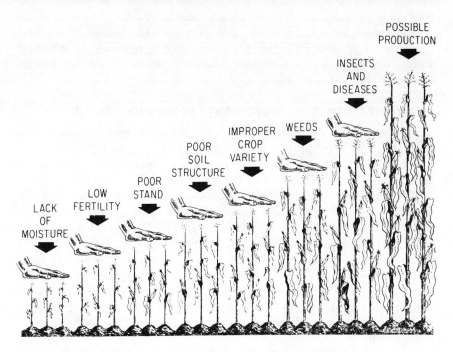

Figure 16–10. After the lack of moisture is eliminated by irrigation, a great number of things may limit yield. Careful attention is needed to get the most return out of the water and fertilizer. (*USDA, SCS Bul. 199.*)

largely limited to the upper 3 feet. With nitrogen, water was extracted to 6 feet.

	0 N	60 lb. N	240 lb. N
Yield of wheat, bu./A.	24	46	54
Evapotranspiration, in.	8.7	10.7	12.4
Water remaining in 7 ft. profile, in.	7.1	5.2	3.8
Bushels per acre-inch of H_2O	5.1	6.7	6.4

In North Dakota the same trends were observed for wheat but fertilizer had little effect on the amount of water remaining in the profile. Yield, rainfall, and soil would affect the latter. In one experiment on corn with no nitrogen, the yield was 107 bu./A. and with 200 lb. nitrogen 177 bu. Bushels per inch of water were 3.7 and 6.1, respectively.

Much additional information could be cited, including sugar beets in Colorado, cotton in California, cotton in Alabama, potatoes in Idaho, grain sorghum in Texas, hay in South Dakota, and corn in Wisconsin. However, the data shown serve to illustrate the effect of adequate nutrients on obtaining greater efficiency from water.

Water and fertilizer make a good team. When water is limiting, response to applied fertilizer is dwarfed but increases with adequate water. When fertility is limiting, response to water is reduced but increases with adequate fertility.

FERTILIZATION AND WATER EXTRACTION BY ROOTS

Most crops use water more slowly from the lower root zone than from the upper soil. The top quarter is the first to be exhausted of available water (Figure 16-11). In periods of stress the plant must draw water from the lower three quarters of the root depth.

The favorable effects of applied plant nutrients on the mass and distribution of roots when soils are deficient is well known. This subject was discussed in Chapter 13 and illustrated in Figures 13-6 and 13-7. With soils requiring the addition of plant nutrients, the plant may extract water from a depth of only 3 to 4 ft. With fertilization the plant roots may be effective to a depth of 5 to 7 ft. or more. Hence the effective depth of the reservoir from which the plant can draw water is increased. If the plant can utilize an extra 4 to 6 in. of water from the lower depths, the crop can endure droughts for a longer period of time without disastrous results.

Studies at the Midwest Claypan Soil Conservation Experiment Farm at McCredie, Missouri, showed that on August 17 there was 1.04 in. of available water in the top 42 in. of soil under well-fertilized corn (Figure 16-12). When no fertilizer was applied there were 4.5 in. On the well-

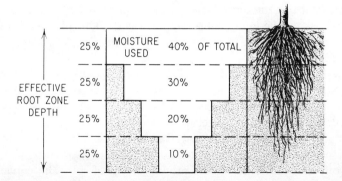

Figure 16–11. The top quarter of the root zone is first to be exhausted of available moisture. Certain management practices, including adequate fertilization, help to develop a deeper root system to use the moisture from the lower root zone. (*USDA, SCS. Bul. 199.*)

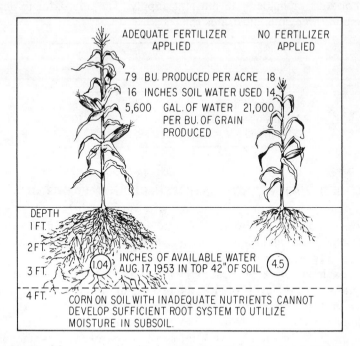

Figure 16–12. Corn on soil with inadequate nutrients cannot develop a sufficient root system to use subsoil moisture effectively. [Smith, *Missouri Farmers Assoc. Bul.* (1963).]

fertilized plot the amount of water needed per bushel of corn was 5,600 gal., whereas on the low-fertility area it was 21,000 gal. Can growers afford to be this wasteful of water? Can growers complain about the lack of water?

Many more examples of the effects of fertilization on the extent of root systems could be cited. It should be emphasized that in areas in which the subsoil is dry increased fertilization will not help crops to penetrate the soil further to get more water.

Soils Vary in Water-Holding Capacity. Texture, structure, and organic matter influence the water-holding capacity of soils, which may vary from less than 1 in. to more than 2 in./ft. of soil. The capacities in the surface 5 ft. of representative soils in Illinois were reported to be as follows:

Oquawka sand	5 in.
Ridgeville fine sandy loam	7 in.
Swygert silt loam	9 in.
Muscatine silt loam	12 in.

However, crops root differently in different soils because of compact soil horizons or inadequate nutrient supply. The approximate depth of rooting in three soils illustrates this:

	Depth of roots (feet)	Water available (inches)
Clarence silt loam	3	6½
Saybrook silt loam	4.5	10½
Muscatine silt loam	5+	14

STORED WATER AND FERTILIZER RECOMMENDATIONS

The relation of stored soil water to crop response to fertilizer has received a good bit of attention in semihumid and low rainfall regions. In such areas the rainfall during the growing season is low and stored soil water is of great importance.

An example of responses to N-P in 66 trials on wheat in North Dakota with three levels of stored moisture and three levels of rainfall is shown in Table 16-3. It is of interest to note that even with the lowest levels of moisture yield responses were obtained.

On the basis of such types of information, nitrogen recommendations in some states were made according to stored soil moisture. However, most of these states now base their recommendations on the residual NO_3-N in the soil profile or fertility level, as related to yield potential.

TABLE 16–3. Maximum Increase in Yield of Wheat from N-P on Nonfallow Land as Related to Stored Moisture at Seeding and to Rainfall During the Growing Season— 66 Trials [*]

Rainfall from seeding to 20 days before harvest (inches)	Available soil moisture inches/4 ft. of soil		
	0–2 (low) (bu./A.)	2–4 (med.) (bu./A.)	4–6 (high) (bu./A.)
>8	7.1	10.0	15.0
6 to 8	5.0	9.5	16.4
<6	2.4	5.9	10.5

[*] E. B. Norum, *Better Crops Plant Food*, **47**(1):40 (1963). Copyright 1963 by the Potash Institute of North America.

This approach is based on the concept that adequate fertility will help to make better use of the water that is available.

In some states a systematic survey of the moisture in the soil profile to rooting depth, 4 to 6 ft., is made in late fall or early spring. This information must then be weighed against the probability of summer rainfall.

Fertilizers have an indirect effect on the amount of stored water in the soil profile. When there is a response to fertilizer, an increased amount of vegetative cover in the growing crop is produced. Runoff of intense rains is retarded and infiltration increased. In the nongrowing season the greater amount of residues on the soil surface act in a similar manner.

In some irrigated regions fall irrigation after crops are harvested is practiced as an insurance policy. On deep soils up to 12 in. of water can be stored in next year's root zone. This has several advantages. Water is usually lower in price in the fall. Wetting and drying and freezing and thawing over winter may improve soil structure. Sometimes water is short during the growing season.

FERTILITY PAYS IN DROUGHT OR FLOOD

Adequate fertility helps to minimize the dips in production due to a poor season. An example of the effect of L-P-K on corn grown in a corn-soybean-wheat-hay rotation over a thirteen-year period is shown in Figure 16-13. The soil was an acid planosol low in fertility in southern Illinois. It is evident that fertility was beneficial in both very dry and very wet seasons. The following ranges are of interest during the thirteen-year period:

June-July-August rainfall	4.4 to 17.0 in.
Increase over check for L-P-K	17.8 to 92.9 bu./A.
Average increase	58.6 bu.

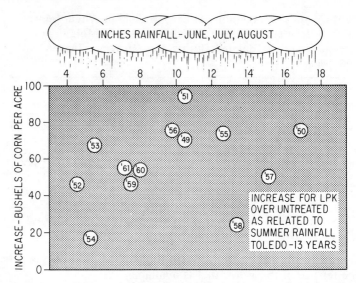

Figure 16-13. Increase in corn yield from L-P-K, as related to June, July, August rainfall—thirteen years in southern Illinois. Corn grown in rotation with legumes. [Lang et al., *Better Crops with Plant Food*, **47**(1):16. Copyright 1963 by the Potash Institute of North America.]

The range in yield of corn on the check plot was 0.3 to 49 bu./A. and on the fertilized plot, 23 to 113 bu.

An indirect effect of fertility was that the limed plots in this experiment could usually be plowed one to two weeks before the unlimed plots were dry enough to support a tractor. This faster drying is probably the result of improved physical structure:

1. More organic residues.
2. Greater root volumes.
3. Deeper root penetration.

Such effects improve water penetration and drainage. In poorly drained soils excess water may be a problem and tile drainage is essential. In northwest Ohio on above-average drained land 450 lb. of fertilizer per acre gave a 15-bu. increase, whereas the increase with average drainage was only 3 bu. Corn grown on a poorly drained silty clay soil in Ohio yielded 79 and 96 bu./A. with surface and tile drains, respectively. However, the variability in yield was 30 per cent with surface and 17 per cent with tile drains. This variability is important for the farmer to know in deciding on inputs to use.

Periodic fluctuations in crop yields related to rainfall have been noted in the Netherlands. Average yields of wheat and rye after some dry

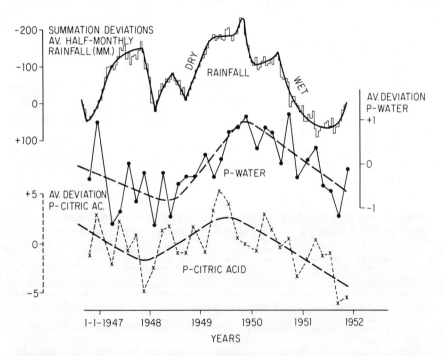

Figure 16-14. Trends of phosphorus soluble in water and in citric acid, compared with summation curve of deviations from average half-monthly rainfall in the Netherlands. [van der Paauw, *Plant Soil,* **17**(2). Copyright 1962 by Martinus Nijhoff, The Hague.]

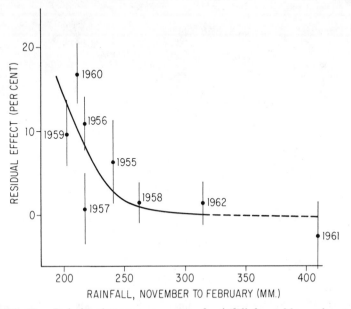

Figure 16-15. Relation between amount of rainfall from November to February and average residual effect of nitrogen in different years in the Netherlands. Vertical lines through dots indicate standard errors. [van der Paauw, *Plant Soil,* **19**(3). Copyright 1953 by Martinus Nijhoff, The Hague.]

641

years amount to 1.5 times those obtained after a succession of wet years. Pea yields are about three times higher. Water-soluble phosphorus and exchangeable potassium rise in dry periods and fall in wet periods, whereas *p*H is the reverse. The phosphorus-moisture relationship is shown in Figure 16-14.

Although the amount of available nitrogen is affected by alternating rainfall periods, the effect of winter rainfall on nitrogen response is paramount in most cases. The effect of increasing winter rainfall on decreasing carryover of nitrogen is shown in Figure 16-15. It should be theoretically possible to eliminate some of the yield fluctuations at a high level of production by appropriate fertilization and management.

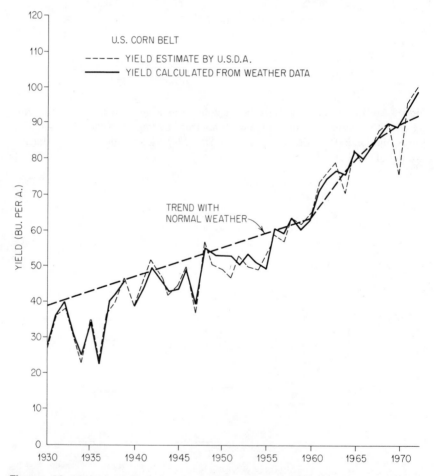

Figure 16-16. Actual and calculated yields of corn for the Corn Belt, 1930–1972. (Unpublished data, L. M. Thompson, Iowa State University.)

The effects of weather and of technology on corn yields have been studied in several Corn Belt states. Calculated yields were compared with actual yields and were determined by considering weather factors, along with the technological changes. Eighty-six per cent of the yield variation was accounted for. It is of interest to note the increases from 1930 to 1972 (Figure 16-16). Many factors, including adequate nitrogen and other plant nutrients, have helped to bring this about. Similar studies have been made on soybeans, wheat, and grain sorghum.

Summary

1. A good or a poor season is often related to the amount of rain received during the growing period.
2. It has been known for many years that plant nutrients applied on deficient soils will increase the crop yield per inch of water.
3. Increased moisture tension which reduces the percentage of many elements in plants is related to thinner water films around the soil particles, hence less diffusion to the roots.
4. Response to potassium is greater in dry years and in very wet years.
5. A high ion concentration in the cell increases the osmotic pressure of the cell solute and the ability of the plant to withstand high water tensions in the soil.
6. Water is lost from the soil by evaporation from the soil surface and from the plant by transpiration. The sum of these two losses plus the evaporation of intercepted precipitation held on the plant parts is called evapotranspiration.
7. On moist soils a heavy crop canopy helps to reduce direct evaporation from soils and a higher percentage of the water is used by the plant. Many factors including heavier plant stands and fertility bring about a heavier crop canopy.
8. On dry soils little water is evaporated directly from the surface. Low plant spacings are generally employed to reduce the amount of water used by the plants.
9. The relation $\dfrac{\text{dry matter production}}{\text{water used}}$ = water-use efficiency is fundamental in agriculture. Any practice that promotes plant growth and the more efficient use of sunlight in photosynthesis to increase dry-matter production will increase water-use efficiency.
10. Many examples are available to show the effect of fertilizer on increased yield of crops per acre-inch of water.
11. Large sums of money are being spent for irrigation in many parts of the world. Limiting factors, including lack of fertility, are often holding down yields and disappointing results are obtained.

12. Adequate fertility promotes a more extensive and deeper root system. The effective depth of the reservoir from which plants can draw water is thus increased. Compaction and low fertility discourage deep rooting.
13. Texture, structure, and organic matter content are factors that cause soils to vary in water-holding capacity. Capacity may vary from less than 1 in. to more than 2 in./ft. of soil.
14. In semi-humid areas stored water in the soil is a significant factor in crop yields and influences the amount of fertilizer recommended.
15. Adequate fertility helps to avoid the drastic dips in crop yields caused by inadequate rainfall or even excessive water.
16. Growers in the past have worried about having enough moisture to get the most out of the fertilizer. In the future they will worry about having enough fertilizer to get the most out of the moisture.

Questions

1. Describe a good or a poor season in an irrigated area and in an unirrigated area.
2. In a given soil volume why is absorption of many nutrients decreased as soil moisture tension increases?
3. Explain the greater response to potassium in dry years and in wet years. Why is boron generally less available under dry conditions?
4. Why does a higher ion concentration in the cell increase drought resistance?
5. What is evapotranspiration? How does a heavy crop cover affect the losses in the various components of evapotranspiration? Is there more total water loss with a greater yield?
6. Explain the difference in evaporation losses from a moist soil surface and a dry soil surface.
7. Define water-use efficiency. Why is it so important in agriculture? List factors that affect water-use efficiency.
8. What is the effect of adequate plant nutrients on water-use efficiency? Why?
9. Explain the effect of adequate nutrients in increasing the extent of the root system. Why is this important in drought periods?
10. On what soils in your state will root penetration be limited by lack of fertility and by physical condition in the lower soil horizons?
11. What soils in your state have a low water-holding capacity? Why?
12. Why is stored water of importance in dry regions? What advantages are there in irrigation in the fall after crops are harvested?
13. Are there irrigated farms in your region in which full returns are not being obtained from an investment in irrigation? Why?

Selected References

American Society of Agronomy, "Research on water." *ASA Spec. Pub. Series No. 4* (1964).

Barber, S. A., "Rainfall and response." *Better Crops with Plant Food*, 47(1):6 (1963).

Barber, S. A., "Water essential to nutrient uptake." *Plant Food Rev.*, 10(2):5 (1964).

Barber, S. A., "Soybeans do respond to fertilizer in a rotation." *Better Crops with Plant Food*, 55(2):9 (1971).

Bauer, A., "Fertilized wheat uses water more efficiently." *Farm Research*, 24(3):4, North Dakota Agr. Exp. Sta. (1966).

Blosser, R. H., "Land management practices affecting income on Hoytville soil." *Ohio Agr. Exp. Sta. Res. Bull. 911* (1962).

Briggs, L. J., and H. L. Shantz, *USDA Bur. Plant Ind. Bull. 285* (1913).

Brown, Paul L., "Water use and soil water depletion by dryland winter wheat as affected by nitrogen fertilization." *Agron. J.*, 63:43 (1971).

Burton, G. W., G. M. Prine, and J. E. Jackson, "Studies of drouth tolerance and water use of several southern grasses." *Agron. J.*, 48:498 (1957).

Eck, H. V., and C. Fanning, "Placement of fertilizer in relation to soil moisture supply." *Agron. J.*, 53:335 (1961).

Garman, W. H., *The Fertilizer Handbook*. Washington, D.C.: National Plant Food Institute, 1963.

Grimme, H., K. Nemeth, and L. C. v. Braunschweig, "Some factors controlling potassium availability in soils." *International Symposium on Soil Fertility Evaluation Proc.*, 1:33. New Delhi: Ind. Soc. Soil Sci. (1971).

Henderson, D. W., R. M. Hagan, and D. S. Mikkelsen, "Water use efficiency in irrigation agriculture." *Better Crops Plant Food*, 47(1):46 (1963).

Kelly, O. J., "Requirement and availability of soil water." *Advan. Agron.*, 6:67 (1954).

Kubota, J., E. R. Lemon, and W. H. Allaway, "The effect of soil moisture content on the uptake of Mo, Cu, and Co by alsike clover." *SSSA Proc.*, 27:679 (1963).

Lang, A. L., L. E. Miller and P. E. Johnson, "Fertility pays in flood or drouth." *Better Crops Plant Food*, 47(1):16 (1963).

Murick, J. T., D. W. Grimes, and G. M. Herron, "Water management consumptive use, and N fertilization of irrigated winter wheat in western Kansas." *USDA, ARS Prod. Res. Rept. No. 75* (1963).

Odell, R. T., *Univ. of Illinois Agron. Facts SP-16* (1956).

Olson, R. A., D. G. Hanway, and A. F. Dreier, "Fertilizer, limited rainfall do mix." *Nebraska Agr. Exp. Sta. Quart.*, 7(3):7 (1960).

Olson, R. A., C. A. Thompson, P. H. Grabouski, D. D. Stukenholtz, K. D. Frank, and A. F. Dreier, "Water requirement of grain crops as modified by fertilizer use." *Agron. J.*, 56:427 (1964).

Sawhney, B. L. and I. Zelitch, "Big pump for stomata." *Better Crops With Plant Food*, 55 (3):3 (1971).

Schwab, G. O., N. R. Fausey, and D. W. Michener, "Comparison of drainage methods in a heavy-textured soil." 72–727. *ASAE*. St. Joseph, Mich. (1972).

Smith, G. E., "Soil fertility—basis for high crop production." *Missouri Farmers Assoc. Bull.* (1953).

Stanberry, C. O., C. D. Converse, H. R. Haise, and O. J. Kelley, "Effect of moisture and P variables on alfalfa hay production on the Yuma Mesa." *SSSA Proc.*, **19:**303 (1955).

United States Department of the Interior, Bureau of Reclamation and United States Department of Agriculture, Soil Conservation Service, "Irrigation on western farms." *Agr. Inf. Bull. 199* (1959).

van der Paauw, F., "Periodic fluctuations of soil fertility, crop yields and of responses to fertilization as affected by alternating periods of low or high rainfall." *Plant Soil,* **17**(2):155 (1962).

van der Paauw, F., "Residual effect of N fertilizer on succeeding crops in a moderate marine climate." *Plant Soil,* **19**(3):324 (1963).

Viets, F. G., Jr., "Fertilizers and the efficient use of water." *Advan. Agron.,* **14:**233 (1962).

Watanabe, F. S., S. R. Olsen, and R. E. Danielson, "P availability as related to soil moisture." *Trans. 7th Intern. Cong. Soil Sci.,* **III:**450 (1960).

Yao, A. Y. M., and R. H. Shaw, "Effect of plant population and planting pattern of corn on water use and yield." *Agron. J.,* **56:**147 (1964).

17. ATTACKING SOIL FERTILITY PROBLEMS

The objective of a sound soil management program is increased and sustained efficiency in the production of agricultural crops.

Until 1900 a large part of the increase in agricultural production was effected by bringing more land under cultivation. Very little unused land remains today in many developed countries, and it appears that future increases in production will come about largely by improvement in acre yields. Consequently, raising yields by better fertilizer and management practices is of paramount importance.

NEW PROBLEMS ARISE

New problems in soil fertility are continually arising. Some require refinement of present recommendations, whereas others present entirely new questions.

The nutrient requirements of new crop varieties, which are continually being released, may be quite different from those currently in use. For example, certain new corn hybrids require greater amounts of magnesium than some old-line varieties, and some of the new tobacco varieties have a higher potassium requirement than those currently grown. High-oil, high-protein, and high-amylase corn hybrids are being developed, although little is known about their fertility requirements. Opaque-2 or high-lysine corn has been found to contain higher K in the kernels. A good example of the examination of the requirements of a new crop was provided during World War II, when numerous studies were made of the needs of guayule, a plant used as a source of rubber.

Changes have been made in the materials used as sources of plant nutrients, particularly for nitrogen and phosphorus. Such changes require that the nutrient availability of these materials be determined. Certain of the water-insoluble phosphorus sources may be more available if al-

→ changes in techniques of applic.

lowed to come in contact with a limited amount of soil than if placed in a band. The nature of these interrelationships is investigated and the behavior of these ions in the soil is determined.

Higher crop yields result in greater removal of mineral nutrients. Although this creates problems with the major nutrient elements, the amounts of these elements in soil can be readily increased. A more important question deals with removal of those elements that are not being added in fertilizers. As yields increase, it may be expected that secondary and micronutrient deficiencies will gradually become more common. An excellent example is found in the increasing number of sulfur deficiencies in North America. These have resulted from the greater use of high-analysis sulfur-free fertilizers. Soil fertility levels change with continued fertilization, necessitating a re-evaluation of placement and rates of fertilizer. There is considerable interest in adding nutrients to the lower horizons of some soils.

More and more soil series and soil types are being separated by soil surveys. Where one soil type was mapped forty years ago, six may be mapped today, and ten tomorrow. The agronomic significance of these separations should be determined by studying the responses to added fertilizers at different levels of soil fertility and with different soil management practices. This is a very large undertaking, however, and only the major soil series can be studied initially. The basic differences among soil types are reflected in their capacities for retaining plant nutrients, in their moisture relations, and in their structure. Remote sensing may be of real value in rapidly assessing some of the broad groupings of soils in states or countries.

New problems continually arise as specific crop production programs are initiated in different parts of the country. Obviously, the experimental work on which a program is based cannot cover all of the expected soil and environmental conditions. As a result problems peculiar to an area will develop.

PREPARING TO ATTACK A PROBLEM

Every research worker is confronted with a wide range of problems, all of which he cannot attack. He must select the most pertinent. It is occasionally difficult for agricultural workers in related fields to understand why the research worker cannot concentrate on every problem presented to him.

Before an investigation is made, a project outline is prepared. This should include the following:

1. Title.
2. Objectives.
3. Reasons for undertaking investigations, including economic evaluation.

4. Previous work and present status of investigations.
5. Outline of procedure.
6. Duration.
7. Personnel.

The title and objectives are reasonably broad to permit a number of investigations. The research worker will usually not write up a project every year or two or every time a new segment of the problem is uncovered. The reason for undertaking these investigations is the basis of the objectives, and the economic justification for the project is usually written up in this section. Extension and industry people are important sources of information in developing a project because of their broad contact with what the problems really are out in the field.

A knowledge of previous work is of first importance in developing a project and should be the point from which the investigator launches his attack. Full knowledge of the present status of the problem is essential. This type of information is obtained from the library and from personal contact with research workers.

Usually only the broad plan of the proposed work is given in the project outline. Detailed work plans are submitted annually. Each problem requires a different type of experimental approach, and a given problem may demand a wide range of experimentation. Various experimental techniques that involve both high and low forms of plant life were discussed in Chapter 12.

The duration of the project depends on the nature of the problem. Generally such projects are planned for limited time periods to encourage frequent re-evaluations based on the recently collected data. Generally rewriting or closing out a project within five years is desirable to keep abreast of results and changing conditions.

With the advent of large grants from special granting agencies, writing of proposals has taken on a new dimension. The project writeup will depend greatly on requirements of the agency and the proposal involved.

Selection of Variables in Field Studies. The selection of the treatments to be used depends, of course, on the objectives of the experiment. An experimenter with vision is not limited in his selection of the range of variables by what is currently considered to be a practical farm practice. Too often in the past such thinking has led to the stagnation of progress. Sound fundamental research in soil fertility should be free of such shackles. The human mind as a rule is not able to foresee the limits of practicality twenty or thirty years hence. To impose such thinking on a research program is to limit its value and to curtail progress.

Trials to Determine Top Yields. Problems at a 175-bu./A. yield level of corn or an 8-ton yield level of alfalfa are different from the problems at 100 bu. or 4 tons, respectively. The higher yield level puts an undetermined stress on balance among nutrients, nutrient behavior in the soil, and plant physiological processes. The results of studies at the lower yield levels may be different and quite meaningless when compared with results at the higher yield levels.

Field trials are needed to determine the top yields in major cropping areas. Such studies will serve as a continuous yardstick for other experiments and may actually result in revision or discard. Following are some of the items to consider in establishing such trials:

1. A cropping system to run several years.
2. Flexible treatments on most plots.
3. Practices according to need and new technology.
4. Bases for comparing best varieties.
5. Bases for coordinating the efforts of many specialists and departments.
6. Best leads for the laboratory, growth chamber, or greenhouse.

Several states have established such trials with impressive results. An important by-product is Item No. 6 in that the most pertinent problems are being identified.

Importance of Other Factors. It is necessary that all factors affecting growth, other than the variable under study, be supplied according to the best information available. The variety or the hybrid used should be the best one that can be obtained. Adequate quantities of the plant nutrients other than the one under study must be applied. For example, the soil may not be adequately limed, and a low-lime level will limit the effectiveness of other growth factors. The fertilizer must also be properly placed. Attention must be given to the control of pests as well as to other cultural practices. Many examples of the limiting effect of insects and weeds in experimental work could be cited.

Certain cultural practices have a pronounced effect on yields and may interact with nutrient requirements. For example, in the Midwest yield increases of 10 to 30 bu. of corn per acre have been obtained by planting by May 1 rather than in the middle of May. In addition, the plants are shorter with the earlier planting. Such effects have a real impact on plant spacing and selection of varieties as well as on nutrient responses. Thus the change of one management practice sets off a chain reaction.

Water as a growth factor may also be limiting. Whenever possible this limitation should be removed by providing for an irrigation variable, unless the study is to run for a long time and the results correlated not only with the imposed treatments but with climatic factors as well.

Characterization of Experimental Conditions. Necessary to the establishment of any good field study is provision for the adequate definition of the experimental environment. With long-time field studies the yield from a given treatment may vary from one year to the next by as much as 200 to 300 per cent. This may be the result of rainfall variation; consequently, a complete record of amount and distribution of rainfall during the growing season is needed for proper interpretation of the data. A study of the rainfall pattern covering a long period may establish the probability of occurrence of a given moisture pattern for a given locality, which, related to the response under varying moisture conditions, would be helpful in formulating recommendations.

A continuous measurement of temperature and radiation is becoming more important in interpreting variability in experiments. These data may be of value in devising treatments to obtain higher yield levels, and careful notes must be made of unusual conditions such as disease or insect infestation. These data are useful in that an indication of the value of plant nutrients under conditions of such infestations is needed to encourage insect control before adequate fertilization can be attempted.

Characterization of the soil is essential in interpreting the yield data. In many instances no attempt has been made to describe the physical, chemical, and biological properties of the surface and subsoils on which the experiments were conducted. Crop response to applications of a given nutrient is related to the amount of the available form of that nutrient in the soil as well as to the physical and biological conditions in the soil.

Samples of soil from the experimental area should be saved for future studies. New techniques which may help to interpret past results are continually being developed. Many countries, among them the Netherlands and Denmark, have thousands of standard samples of soils on which yield responses are known. When new extractants are to be tested, large numbers of these samples are analyzed. The results are then compared and calibrated with the yield responses and the most efficient extractant is determined.

The plant is the final product and the one in which the grower is interested. It integrates all the factors in the environment. Hence a careful study of the plant itself is essential. This may be in the form of periodic growth or biochemical and chemical measurements on the living plant or on the dried tissue in the laboratory. Measurements on the final product to evaluate quality are becoming increasingly important. Results of all such studies aid in interpreting effects of treatments even in the absence of response data. The nature and importance of these basic relationships were discussed in Chapters 2 and 3.

Regional Experiments. A problem common to a rather wide area may be attacked by a uniform experimental setup, often called a regional experiment, at many locations over the country. It permits the gathering of a large amount of data within a relatively short time over a variety of soil and climatic conditions. Soils which have a range in content of the nutrient under study and also other characteristics are selected as sites for the experiments. Analysis of the data may establish relationships that would be difficult to obtain by a more limited approach.

An example of the regional approach was the uniform phosphorus study in 1951 carried out at 74 locations in the United States. Several rates of phosphorus were applied on soils varying widely in their P-content. A-values and crop response data were obtained in the field. Bulk samples were sent to a central location and *A*-values and crop response data were obtained in the greenhouse. These studies played a significant role in determining the relationship among *A*-values, yields, and phosphorus uptake by crops. The soil samples were helpful in establishing more effective extractants for phosphorus tests and for calibration studies.

Regional studies on potassium, carried out in the United States in the 1950's, involved 83 locations for alfalfa and 51 locations for corn and were supplemented with greenhouse studies on soils from these experiments. Extensive soil and plant analyses were made. Regional studies on nitrogen have also been conducted on corn.

Basic Soil Studies. A full discussion of this subject is beyond the scope of this chapter. However, laboratory studies under controlled conditions are essential in establishing fundamental nutrient-element behavior. Certain soil factors affecting retention of ions such as nitrate, phosphate, ammonium, sulfate, or potassium must be studied under laboratory conditions. This type of investigation is basic in predicting responses of plants under given soil conditions and in helping to determine the treatments to be utilized in the greenhouse or in the field.

SIGNIFICANCE OF RESEARCH DATA

Variability of Data. An invariable property of biological material is its variability, or the failure of one organism to be completely similar in all respects to another organism of the same genetic makeup. If any particular property, say height, were measured for a large number of corn plants, all of the same variety and all growing under seemingly identical conditions, it would be found that a few plants would be short, a few quite tall, with the remaining plants at heights intermediate between these two extremes. Further, the majority of the plant heights would tend to cluster around some mean value about half way between the highest and lowest values observed.

If the number of plants falling within certain size limits were plotted for the entire range of values, a bell-shaped figure, known as a normal curve of distribution, would be obtained. This type of distribution is characteristic of a number of properties exhibited by living organisms and forms the basis of a field of study known as experimental statistics. The application of this science to agricultural research has been of tremendous value in enabling the research worker to interpret his data more adequately. In addition to the variability inherent in plants themselves, environmental and soil conditions in the field are variable even though they may give the appearance of being perfectly uniform.

If it is desired to determine the effect of some fertilizer treatment on the yield of corn, for example, two plots could be set out side by side. The particular treatment in question could be applied to one plot with the other receiving some standard or no treatment at all. The difference in yield between the two plots would give a measure of the effectiveness of the fertilizer treatment in question. According to the normal curve of distribution, the yields of these two plots could be different even if they were treated alike. Conversely, they could be alike even though they received different treatments. Of course, the yields may be different when different treatments are applied. The question is how to determine whether the differences observed between treatments are the result of chance or of the experimental variables imposed. It is here that the tool of statistics is so valuable to the agronomist.

Populations and Samples. The normal curve of distribution is based on measurements made on large numbers of individuals. The totality of individuals exhibiting any particular property is said to be a *population*. Populations may be finite or infinite in number. It is not necessary to measure a property of each individual within a population in order to make some statement about the distribution of that property. Measurements may be made on a *sample* drawn from the population. If the sample is taken in such a way that each individual in the population has an equal chance of being included, the sample is said to be *random*. If the sample drawn constitutes only one individual, the probability of this one individual exhibiting the mean characteristics of the population is small. If 1,000 individuals constituted the sample, however, the probability would be considerably greater that the mean of the sample of 1,000 would describe the mean of the population fairly well.

Replication and Randomization. So with the two corn plots mentioned, if only one plot of each treatment is included, the probability is small that the differences observed are the result of the treatment imposed. If each treatment were repeated several times, however, and in each instance the treated plot outyielded the untreated plot, the odds would be much greater that this difference in yield was the result of the fertilizer treatment and not of some chance variation. The repetition of each treatment several times is known as *replication* and is one of the hallmarks of a properly conducted field experiment.

It should be emphasized here that the difficulty of determining whether the cause of treatment difference is due to chance or the variables imposed is greatest when the differences are small, say less than 10 per cent. There is somewhat greater justification, however, for attaching significance to a difference of 40 to 50 per cent between the yields of two differently treated unreplicated plots.

Not only is replication essential, but randomness of the plots within a replication is necessary for a good experiment. By randomness in the physical placement of treatments is meant allowing each treatment to have an equal opportunity of being assigned some particular space in the experimental area. By so doing the opportunity for introducing a bias is minimized.

STATISTICAL TREATMENT OF DATA AND STATEMENTS OF CONFIDENCE

Data collected from replicated and randomized experimental plots can be subjected to various types of *statistical analysis,* depending on the objective. For example, suppose that it is desired to know whether there is any difference in the yields of wheat when different sources of phosphorus are included in the fertilizer. If only one rate of phosphorus were employed and the only imposed experimental variable was the source of phosphorus, the statistical treatment given would probably be one known as an *analysis of variance.* The results of this analysis permits the investigator to state within certain probability limits which of the materials, if any, was better than the others. Frequently a statistic known as the *LSD value* (least significant difference) accompanies data of this type. It enables the person reading the research report to have some measure of the differences among the treatment means that are necessary before they can be considered as the result of the imposed experimental variables rather than of random variation. Data reported in research publications are of considerably greater value to the reader and constitute a more worthwhile contribution to our knowledge when they are accompanied by some *statement of significance* that enables us to determine whether the differences are likely to be the result of chance or of the experimental variables employed.

If an experimenter wished to determine the yield response of wheat to increasing rates of nitrogen, the type of statistical analysis employed would be that of determining the *coefficient of regression* or the *regression equation* relating the yield to the amounts of applied fertilizer. Such information makes it possible to calculate within limits the yields that would be expected from the application of a given quantity of fertilizer when applied under similar conditions of soil and environment.

The use of regression techniques is particularly helpful in interpretating data that are nonlinear. Examples of a nonlinear relationship will be found in the growth response curves shown in Chapter 2. When growth is functionally related to increasing amounts of fertilizer inputs, the curve is said to be nonlinear, or curvilinear, because each successive input gives yield increases that were smaller than the preceding one. This response gives rise to the type of curves shown in Chapter 2.

Another statistic employed in agricultural research is the *correlation coefficient*. When one is interested in determining which of a number of suspected factors is most closely related to some particular plant property, measurements of these variables are made on a number of samples and the relation of each to the property in question is determined by correlation methods.

There are numerous other statistics encountered in research reports. Each has its place and each contributes its part to the sum total of our knowledge. For the student interested in pursuing this important subject a few of many excellent references are listed at the end of this chapter. The important point in this discussion of statistics is that the person working with biological media, and the agricultural research worker in particular, should recognize the importance of the statement of confidence concerning the data being reported. Without a confidence statement, the reader of the report has little way of knowing if the results represent a significant difference among the various treatments or if they are merely due to random variation.

Probability Levels. The various statistics, such as LSD values, regression coefficients, and correlation coefficients, are usually computed at some stated probability level, such as the 5 per cent level or the 1 per cent level. These probability levels simply mean that, in the case of the 5 per cent level, the odds are 19 to 1 that the observed behavior is due to something other than chance. In other words, the differences are due to the imposed treatment variables. At the 1 per cent level, the odds are 99 to 1 that the differences observed are due to something other than chance.

Odds such as these are justified in the interpretation of research results. The investigator wants a high degree of assurance that the relationships indicated by his data are due to the experimental variables involved and are not the result of chance. Frequently randomized fertility trials are put out with the sole objective of determining under farm conditions responses to inputs of various plant nutrients. The data collected from these tests are usually the basis for making fertilizer recommendations to the grower. The question is whether there is a need for odds such as 19:1.

With the fast-changing picture in agriculture, progressive growers are willing to take greater risks for high yields. This is particularly true when the treatment involved is low in cost in relation to the value of the increased yield; for example a dollar's worth of boron may produce a ton of hay worth 20 dollars or more. With the possibility of a reward such as this, the grower may be interested in significance at odds as little as 50:50, or the 50 per cent level.

Another aspect is frequency of response. Results may show that a treatment will give a response only once in five years but in that one year the response more than pays for the cost of treatment for the five years. The response may not be *statistically* significant over the five-year period or at the 5 per cent level but progressive growers will be interested in using the treatment. A profit will be obtained.

Another important point is that with yields of 150 bu. of corn, 8 tons of alfalfa, 60 bu. of wheat, or 50 bu. of soybeans these crops are fast becoming high-value crops. The grower will want to protect his investment in other production costs, hence will be willing to take risks or odds considerably less than 19:1. This point is mentioned because in experimental work, even though the results show trends, if the data are not significant at the 5 per cent level they may be filed and forgotten.

APPRECIATION OF STATISTICS AT THE GRASS-ROOTS LEVEL

Research workers are by no means the only persons who should be interested in and have an appreciation of such a valuable tool as the confidence limit offered by statistics. The farmer, the orchardist, and other persons vitally concerned with the merits and demerits of agricultural inputs are continually confronted with the problem of deciding on which of several is the best. This decision occasionally has to be made with no more information available than the claims made by a salesman interested largely in the sale. A demand by the grower to see *reliable* data concerning the product should be a practice more widely employed. Too often glowing claims, supported by photographs, have been made. The photos, usually purporting to show plants that have received a certain fertilizer next to those that have received some standard brand, may be misleading. The person seeing the pictures has no way of knowing the soil treatment imposed on the plots, whether the fertilizers were applied at the same rate, whether similar varieties were used, and a host of other questions, the answers to which would be necessary for an intelligent interpretation of the picture in question. However, a common-sense appreciation of the role of these factors in plant growth should prompt him to ask such questions concerning the validity of the evidence offered.

An understanding of the technical details of statistics is not needed for the interpretation of many pieces of information; all that is required is an appreciation of the inherent variability of plants and the fact that this variability can be estimated at certain probability levels if a set of confidence limits is given. If greater cognizance were taken of these simple principles by a greater segment of the farm population, it would be reasonable to suppose that even further improvements in the efficiency of the farming operation would not be long in coming.

REPORTS OF RESEARCH WORK

The director of the experiment station or research group has the responsibility of making certain that the results of research work are reported to the tax-paying public. The files of many research men in all countries are filled with data that merit summary and publication. Some of these data are released each year. However, many more need to be made available, not only so that they may be used in the area in which they were gathered, but so that other investigators interested in the same general field will not duplicate the work.

Experiment Station Bulletins. This type of report contains results of a complete series of investigations and usually represents distinct progress in the field. Some bulletins are in popular style and can be utilized by agricultural leaders in the counties. Others are of a more technical nature and may be useful only to individuals connected with research work. Experiment stations release lists of these bulletins and publications periodically.

Journal Articles. There are numerous publications which have articles on plant-nutrient studies. These articles are usually rather short and contain the results of a limited amount of work. In most instances these publications are read only by agronomists and soil scientists. Some of the journals commonly used in the United States are *Agronomy Journal, Proceedings of the Soil Science Society of America, Soil Science, Plant Physiology, Proceedings of the American Horticultural Society,* and the British abstracting journal, *Soils and Fertilizers.* Others are given in the reference lists at the end of each of the preceding chapters.

Reports of the Experiment Station or Research Group. Many experimental groups publish a periodical which contains short reports of significant segments of work in all fields. This merely calls attention to the development of given lines of research and presents little detail.

Extension Circulars or Folders. The recommendations in extension publications are based on research data and other information. This is the fruition of the research. How well the recommendations will work and how widely they will apply depends on the quality and quantity of the research. Adjustments in recommendations must be made as conditions outside the scope of the original research work are encountered.

One point should be mentioned in connection with the publication of results of research. With the increase in numbers of scientists more research is being done and more papers are being published. This means that although periodicals are being expanded to take care of the increased load more attention may be given to the quality of a paper before it is accepted.

Although it is more difficult to assess the quality of laboratory and greenhouse work, a measure in field trials is the level of yields obtained. Depending on the environment, various standards might be set up. Perhaps in the Corn Belt yields below 175 bu. of corn, 60 bu. of soybeans, 8 tons of alfalfa, and 70 bu. of wheat should not be reported unless the reasons for the lower yields are known, documented, and outlined. The very study of these reasons should contribute much to a better understanding of factors limiting crop production.

FERTILIZER PRACTICES FOR TOMORROW'S AGRICULTURE

Fertilizer practices in the future will center around the adequacy of the level of nutrients in plants. At present many recommendations take crops out of the deficiency symptom stage but leave them in the "hidden hunger" zone where they would respond to more fertilizer. An eight-year summary of corn plant samples in Ohio showed that while 83 per cent were classified as normal by the sender, 44 per cent were short in one or more nutrients.

It is important to consider the needs and goals of the growers. The number of farms is decreasing and the size and gross income is increasing. In Indiana it is estimated that by 1980 there will be 10,000 farms with a gross income over $40,000, or an average of about 100 farms per county. This is mentioned because industry and extension could actually give personal attention to these farmers. This group of farmers will be using a high percentage of the purchased farm inputs.

As the size of the farm increases the goal of the farmer increases at about the same proportion or a little faster. The single most important

Figure 17–1. Set high goals for commercial growers and leaders. Goals such as two more tons of alfalfa, ten more bushels of soybeans, and thirty more bushels of corn will attract their attention.

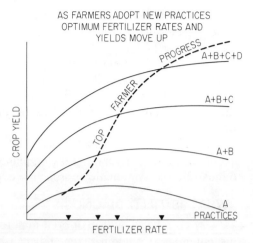

Figure 17-2. As top farmers adopt improved practices, B, C, and D, yields move up and optimum fertilizer rate increases. (Potash Institute of North America, **M-152,** 1968).

factor related to net profit is a high yield. Hence setting yield goals high becomes of prime importance (Figure 17-1).

The growers will have a higher and higher capital investment, hence will find it necessary to control as many of the limiting factors as possible in order to get highest net returns (Figure 17-2). Among other things, they will want especially to eliminate low fertility as a limiting

Figure 17-3. For many years livestock growers have recognized the importance of full feed for their livestock to get top profits. It is equally important to plan to full-feed the crops. (Courtesy of the Potash Institute of North America.)

factor. Growers have been indoctrinated with the idea of full feed for cattle. They will become increasingly interested in full feeding their crops (Figure 17-3).

The commercial grower will be persistent and give careful attention to details. Above all he will use a diagnostic approach; he will study the growing plants to figure out how he can do even better. Today's pressures demand precision farming and this means doing things right and on time.

A brief summary of the approach used in identifying fertility problems and other limiting factors in growing plants follows. This summary helps to illustrate the systematic approach needed in field diagnosis.

WHAT IS FIELD DIAGNOSIS?

Every crop has a built-in or bred-in potential of production. In farming this potential is seldom reached because there may be one or more limiting factors holding back production. Identifying these factors on the spot is field diagnosis. It is a study of the adequacy of what was done to grow the crop and of new factors that may be appearing on the scene, such as a new disease, insect damage, or nutrient deficiency. It is done when the plants are growing so that the appearance of the plant can be observed when the chemical tests and observations are being made.

The same thorough diagnostic approach should be used whether in experiments, demonstrations, or farm fields. There is no simple method. All known factors must be examined and weighed before arriving at conclusions.

Use

To Adjust the Management on This Year's Crop. Diagnosis must necessarily be done early in the growing season. It may involve control of insects or weeds or the addition of nitrogen, potassium, manganese, or some other nutrient.

To Evaluate the Current Program as a Guide for a Better One Next Year. This evaluation is most effective if a diagnosis can be made four or five times during the season. If the field is checked only once, it is best done just as the plant is preparing to make seed (at the early flowering stage). These two uses are part of a routine management program.

Trouble Shooting. With this use an effort is made to identify a critical problem in plant growth which may have appeared unexpectedly.

Who Can Do It? It is essential to know a few fundamentals and to apply a few important rules. A medical doctor's interpretation of data from questions and tests is based on an understanding of the human body. Similarly, a field diagnostician depends on his understanding of a few principles of soil, fertilizer, and plant behavior.

Such understanding can be obtained only by a study of the soil and plants in the field after learning basic principles by personal study. Initial field observations are best made when treatment differences exist so that the diagnostician can gain confidence in himself and in the tests. Proficiency is gained by continued practice. In unusual cases samples of plant material in plastic bags can be taken to specialists for verification of the diagnosis or for plant analysis in the laboratory.

Limiting Factors.   Man can control many of the factors that limit production but he has little control over weather factors. More efficient use of the rain which does fall can be made. Practices include surface residues, contour rows, cover on the soil as much of the year as possible, high yields in order to return large amounts of residues, and cultivation on noncracking soils.

Look for these controllable factors:
Shortages of nutrients or water.
Imbalance of nutrients N and K, P and Zn, Mg and K, N and S.
Improperly placed fertilizer.
Acid soil.
Insect or disease damage.
Improper variety.
Incorrect plant spacing.
Weeds.
Poor soil management—compaction, low residues.
Drainage.
Poor crop management—late planting, harvesting.
Cultivation damage.

Note that fertility is only one of the many possible factors affecting yield. However, many of the other factors may affect nutrient supply in the plant. Insects may cut off roots. Corn borers may cut the vascular system or pipe lines in the stalk. Compact soils will prevent root penetration. Drought will slow nutrient uptake.

There is usually one factor more limiting than others. When it is identified and corrected, another one may appear. A diagnostic approach is thus a continuous process and must consider all known growth factors. It takes an extremely alert diagnostician to identify specific needs and to anticipate others.

The Tools Needed.   (Figure 17-4)

A shovel—to examine the root system and the fertilizer band.
A soil sampling tube—to examine the soil for a tight or a sandy horizon.
A soil acidity kit—to check soil *p*H and placement of lime.
A knife—to probe the plant for insect damage, disease damage, and accumulation in nodes.

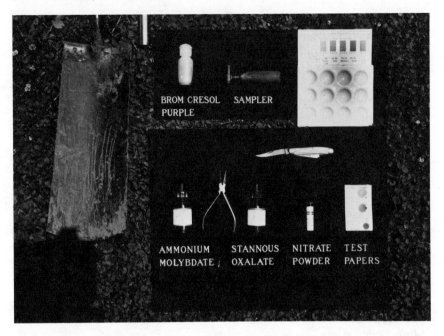

BROM CRESOL   SAMPLER
PURPLE

AMMONIUM   STANNOUS   NITRATE   TEST
MOLYBDATE   OXALATE   POWDER   PAPERS

Figure 17-4.   Many different tools may be used to examine the soil and the growing crop. Above are some examples of the possibilities. (Courtesy of the Potash Institute of North America.)

A tissue test kit — to test the plants for N-P-K and other nutrients. *Sample bags* — for detailed soil and plant analyses in the laboratory.

APPROACH IN THE FIELD

Field Background.   Much of this background must be obtained from the grower. Some will have good records but many will not. It is important for the grower to accompany the diagnostician while the fields are being studied. Points will come out that were not made around the kitchen table.

1. *Check soil test reports.* Who took the samples and who made the analyses and recommendations?
2. *Check lime, fertilizer, and manure applied.* What was applied, when, where and how during the last two years?
3. *Check cropping history.* Watch for crops with heavy removal characteristics such as those in which the entire aboveground portion is removed. Corn or beans after sugar beets may be low in zinc. Limited root system or short season crops such as cabbage or radishes, respectively, after an extensive root system crop, such as corn or tomatoes, may be in trouble.

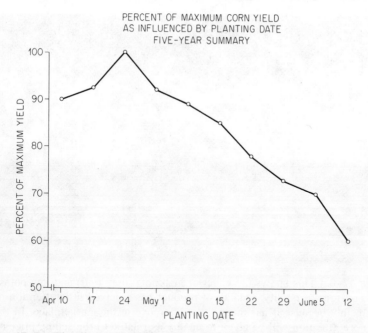

PERCENT OF MAXIMUM CORN YIELD
AS INFLUENCED BY PLANTING DATE
FIVE-YEAR SUMMARY

Figure 17–5. Timeliness of planting is perhaps the single most important practice a farmer can follow—and it doesn't cost him anything. Corn is shorter in
height with early planting. In the Midwest for each day's delay in planting corn
after the first of May yields will be reduced about 1 to 2 bu./A. (Potash Institute
of North America, **N-9, 1972.**)

4. *Check variety used.* Many have been developed under inadequate fertility and wide spacing. Progressive growers are conducting their own population trials with existing varieties.
5. *Check current management.* This includes land leveling, time of
   plowing, number of trips over the field, time of planting, rate and
   depth of planting, and chemicals for weed and insect control.
   Timeliness is the least exact and most difficult to describe but is
   perhaps the most important of all factors at high-yield levels
   (Figure 17-5.)
6. *Check rainfall records at the farm.* Those at the county seat may
   be of little value. What were the temperatures?
7. *Yields.* What yields has the grower been getting? What are his
   goals?
8. *Quality of product.* This measurement is difficult to make except
   in cases of extreme deficiency or pest damage. More attention
   must be given to this point however. (Figure 17-6.)

It is recognized that the grower will not have much of this information
at his fingertips. However, the very fact the questions were asked in-

Figure 17-6. This shows the effect of high potassium (*left*) and low potassium (*right*) on tomato quality. Quality of product has received little attention in the past, mainly because it has made little difference in the marketplace. In the future more attention will be given to quality as related to the selling price of the product. (Courtesy of G. E. Wilcox, Purdue University.)

Figure 17-7. One of the major limiting factors on most fields is wide rows. Narrower rows help make maximum use of sunlight, water, and also the other growth factors.

dicates the importance of the information. The progressive grower will get it and have it for the diagnostician next year, which brings out the point that the diagnostician can be most effective by working with a grower over a period of years.

Field Observations. If the field is to be checked only once a season, the ideal time is when the crop is preparing to make seed (flowering time). If time permits, monthly checks starting shortly after crop emergence are desirable.

1. *Check the plants visually.* This test includes plant spacing (Figure 17-7), deficiency symptoms, insect damage, disease damage, and weeds. Study the roots (Figure 17-8). Use a knife to cut open the plants and check for unusual conditions.
2. *Check the soil.* Use a shovel to examine fertilizer band, compaction, hardpan, and texture and structure (Figure 17-9). Observe drainage.

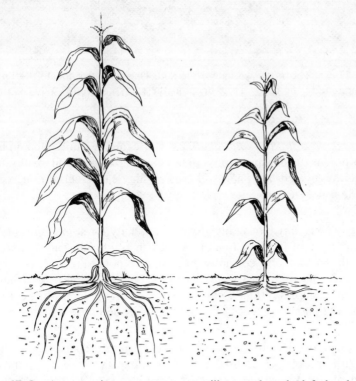

Figure 17–8. An extensive corn root system illustrated on the left develops in a fertile soil which has a good physical structure throughout the profile. On the right is a restricted root system that would develop on a soil low in fertility with a tough impervious subsoil or plowpan. [Ohlrogge, *Purdue Agr. Exp. Sta. Bul. 635* (1962).]

Figure 17–9. Too often we ignore the soil. Use a shovel, locate the fertilizer band, and study the roots. How deep do the roots penetrate? What is the soil type?

3. *Observe soil types and locations.* For example, molybdenum is a problem on organic soils containing bog iron. Spoil banks, recently leveled land, and soil erosion may cause certain micronutrient deficiencies.

Certain soils may have a high fixation capacity for some nutrients. For example, soils of the San Joaquin Valley in California may require 300 to 500 lb. K/A. before cotton will respond to potassium.

Chemical Tests. Tests on the plant help to confirm a deficiency symptom diagnosis. *Hidden hunger* is the major problem, however, and tests on the plant are necessary to determine the nutrient status.

1. *Test tissue of plants on the spot.* Test eight to ten average plants. With trouble spots in the field, compare with test results on plants from good areas. It must be kept in mind that the plants in the good area may be on the borderline and in the hidden hunger zone. If the plants can be tested several times during the season, watch for starter effects, as it is possible for the plant to run out

of certain nutrients later if basic fertility is not adequate. Tissue tests help to indicate levels of N-P-K in the plant at the time tests are made. It is important to keep in mind, however, that the plant is a dynamic biological system.

2. *Sample plants for laboratory analysis.* Detailed analysis on the dried samples will be helpful for all nutrients. Spectrographic analyses are available at a reasonable cost and as many as sixteen elements can be reported.

3. *Test soil p*H. Soil acidity is a major problem and a colorimetric test will give an approximation of this value in the field. To determine lime needs a regular soil sample should be taken to the laboratory for a lime requirement test. The problem is usually not whether the soil is *p*H 6.8 or 6.5 but rather *p*H 6.5 or 5.2. Such tests are helpful in locating the position of preceding lime applications and often reveal some surprising information.

4. *Sample soil.* This has the advantage of getting a representative sample and also of getting a sample under the growing crop. Nutrient drawdown during periods of peak demand by the crop may be a limiting factor with some nutrients on some soils.

CHECK LIST

A check list is helpful in formulating a systematic approach and for preserving observations. An example of one type of list follows:

Date _____

Field designation

Field background

Soil tests _____ date_____ laboratory _____

This Year   Last Year

Lime rate _____   _____

Fertilizer_____   _____

Placement_____   _____

Manure_____   _____

Land leveling _____   _____

Crop yield goal_____   Yield _____

Crop quality_____   _____

Rainfall_____   _____

Insecticide_____   _____

Herbicide _____   _____

Time of plowing _____   No. of trips over field_____

this year

Time of planting_____ Variety_____ Rate_____

this year

Moisture and Temperature (Circle)

Moisture summary			Temperature summary		
Early	*Middle*	*Late*	*Early*	*Middle*	*Late*
V. dry	V. dry	V. dry	V. hot	V. hot	V. hot
Dry	Dry	Dry	Hot	Hot	Hot
Ideal	Ideal	Ideal	Ideal	Ideal	Ideal
Wet	Wet	Wet	Cool	Cool	Cool
V. wet	V. wet	V. wet	Cold	Cold	Cold

Obvious limiting factors (severe-S, moderate-M, light-L, none-O)

N_____   Fertilizer placement _____
P_____   Soil compaction_____
K_____   Hardpan_____
Zn_____   Root insects_____
S_____   Plant insects _____
Lime_____   Low stand _____
Drainage _____   High stand_____
Soil type_____   Variety_____
Disturbed soil _____   Disease_____
                                Weeds_____
                                Cultivation damage _____
Other limiting factors   _____

Tissue tests (field average of 1 sample per acre)

	Good area	Poor area
NO$_3$	_____	_____
PO$_4$	_____	_____
K	_____	_____
Other	_____	_____

Leaf analysis results

Soil tests (field average of one core per acre)

	Good area	Poor area
Quick pH		
Depth 1 in.	_____	_____
4 in.	_____	_____
8 in.	_____	_____
12 in.	_____	_____

Soil tests on sample taken under growing crop

_____   _____
_____   _____
_____   _____
_____   _____

Yield goal_____   Final yield_____

Conclusion

1.   (Most limiting factor)
2.   (Next limiting factor)
3.   (Next limiting factor)

Recommendations

_____

_____

Fitting the Puzzle Parts Together. A thorough study as just outlined will suggest limiting factors. It will be necessary to obtain resource information from references or specialists. A key factor is the managerial ability of the farmer.

The more background knowledge and experience the diagnostician can obtain, the more proficient he will be in identifying limiting factors. An alert individual well grounded in the basic fundamentals of plant physiology, soil chemistry, and fertilizer technology and use will succeed rapidly.

Diagnostic Clinics or Workshops. Once the diagnostician has gained proficiency it is important to train future leaders. Two half-day sessions on experimental plots, demonstrations, or grower fields will help to acquaint the students with the problems, the significance of the information desired, and the concept of limiting factors (Figure 17-10). Similar sessions with research personnel, extension workers, vocational teachers, and industry representatives are of real educational value. Periodic review sessions to acquaint all concerned with new information are essential.

The response to increased amounts of nutrients is the net sum of the potential response and the depression response (Figure 17-11). The depression response may be due to a number of factors including nutrient imbalance or other interactions, salt damage, variety, and effect of pests. Researchers, extension, and industry have a real challenge to identify and correct depression responses. Careful study of the growing plants in the field is an important part in improving yields. Computer

Figure 17–10. Diagnostic clinics or workshops out in the field are helpful in training leaders to recognize limiting factors in crop production. Periodic review sessions are essential in order to keep abreast of recent developments.

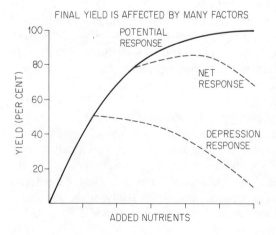

Figure 17–11. Net response to fertilizer is the sum of the potential and depression responses. [Nelson, L. B., et al., Eds., *Changing patterns in fertilizer use,* p. 117 (1968). Reprinted with permission of Soil Science Society of America, Madison, Wisconsin.]

programs to predict yields from production practices are being developed. The purpose is to teach farmers how cultural practices work together and interact to affect yields.

Summary

1. New questions are constantly arising as varieties, management practices, and yield levels change.
2. The research workers must select the most pertinent problems on which to work. Project outlines are then prepared to make a systematic study. A thorough investigation of previous work is of first importance.
3. Well-conducted field experiments using the best known practices and cooperation of people from various related disciplines are essential in order that top yields for the area are obtained. Characterization of environmental factors and careful observations of the plant and soil help to lead to pertinent follow-up laboratory and greenhouse studies.
4. Data from plot to plot vary and hence, replication and randomization are necessary. Various types of statistical treatment are made on the data including analysis of variance, least significant difference, regression techniques, and correlation coefficient.
5. Probability levels are often calculated at the 5 or 1 percent levels in order to determine if differences are real. However, with a low cost treatment, a good opportunity for higher yield, or a high crop price the farmer may be interested in the 20 or even the 50 percent level.
6. The researcher has a real responsibility to report his data so that the results might be used in his area and also be available to other investigators. However, more attention will be given to the quality of the results and of the reports in the future.
7. As yield goals increase there is a need to know the next limiting factor in a farmer's field. This is where field diagnosis comes in. Field diagnosis can be used to adjust management on some crops the current year, improve management the next year, and in trouble shooting.
8. There are many possible limiting factors in crop production of which fertility is only one. Timeliness, pests, variety, drainage, drought, and soil compaction are a few of these factors.
9. It is essential to obtain the background on what has been done in the current and in past years. Such may be obtained from the grower but often memory is short. However, by working with a grower year after year information can be accumulated. Timeliness is perhaps the most important of all factors at high yield levels.
10. Observations in the field are of prime importance and if checked only once, flowering is the ideal time. Once a month is best, how-

ever. Plants should be cut open with a knife and a shovel should be used to examine the roots and the soil. Tissue tests on the plants, samples for analysis in the laboratory, a quick soil pH and a regular soil test in the laboratory are among the approaches.

11. A check list should be used in order to develop a systematic approach and for preserving observations. Once the information is obtained the whole picture can be seen and suggestions can be made. No matter how high a field is yielding there is always a "next limiting factor".

12. The response to increased amounts of nutrients is the net sum of the potential response and depression response. Investigators have a real challenge to identify and correct the depression responses.

Questions

1. Why might fertility needs change as variety is changed?
2. What are the main parts of a project outline? Why is a thorough study of previous work needed?
3. What are some of the items to consider in designing field trials? Why is it important to identify environmental conditions under which an experiment is conducted?
4. Why are regional experiments particularly useful?
5. Data often fall into a normal curve of distribution. Why? How does replication and randomization help?
6. What does LSD mean? Explain why it is calculated. Why might a farmer be interested in odds less than 5 percent? Why is a farmer interested in reliable data?
7. Is there hidden hunger on crops in your state? On what crops and for what elements? Compare full feeding cattle and full feeding crops.
8. For what purpose is field diagnosis used? Who can do it?
9. How would you go about diagnosing a field? Why is a visit with the farmer of first importance? Why is a check list essential?
10. In one field the corn plants test low in K, yet the soil test is high. What might be the reason for the low test? In another field cotton plants tests low in NO_3 but the plant shows no symptoms. The area agent says this may be hidden hunger. What does he mean? Will symptoms appear soon?
11. Explain Figure 17-11.

Selected References

American Potash Institute, "Know your limiting factors in crop production." *Better Crops with Plant Food,* **44**(1):1 (1960).

American Potash Institute, "Fighting hidden hunger with chemistry." *Better Crops with Plant Food,* **47**(3):1 (1964).

Cochran, W. G., and G. M. Cox, *Experimental Designs,* 2nd ed. New York: Wiley, 1957.

Evans, Clyde E., W. R. Thompson, Jr., H. T. Rogers, and D. L. Thurlow, "Relationships between total phosphorus and potassium in soybean petioles and tissue test ratings for these elements." *Agronomy Abstracts,* p. 105 (1972).

Fisher, R. H., *Statistical Methods for Research Workers,* 14th ed. New York: Hafner, 1973.

LeClurg, E. L., H. W. Leonard, and H. G. Clark, *Field Plot Technique,* 2nd ed. Minneapolis, Minnesota: Burgess, 1962.

Munson, R. D., "Influence of nutrient balance and other factors on corn maturity." *Agr. Amm. News,* **12**(4):34 (1962).

Nelson, L. B., et al., Eds., *Changing patterns in fertilizer use,* p. 47. Madison, Wisconsin: Soil Science Society of America, (1968).

Ohlrogge, A. J., "The Purdue soil and plant tissue tests." *Purdue Agr. Exp. Sta. Bull. 635* (revised 1962).

Potash Institute of North America, "Do you use sure practices?" *M 152* (1968).

Potash Institute of North America, "It pays to time things right." *N 9* (1972).

Snedecor, G. W., and W. G. Cochran, *Statistical Methods,* 6th ed. Ames, Iowa: The Iowa State University Press, 1967.

Steel, R. G. D., and J. H. Torrie, *Principles and Procedures of Statistics.* New York: McGraw-Hill, 1960.

Wickstrom, G. A., "Ask the plant about NPK needs." Fertilizer Solutions, **5**(3):36 (1963).

Index

A-value technique, diagnostic tool, discussed, 457–459
Absorption (*see also* Ion absorption; Nutrient absorption; Nutrient uptake) active, in plant roots, 114, 115; passive, in plant roots, 116, 117; plants, of sulfur, 297; of water (*see* Water absorption)
Absorption sites, on plant roots, for phosphorus, 71
Acid fertilizers, discussed, 367
Acid-forming fertilizers, *defined*, 366
Acidity (*see also* pH; Soil acidity; Soil pH), active, discussed, 398, 399; buffering of, discussed, 400, 401; *defined*, 398 (*see also* Soil acidity; pH); neutralization, discussed, 400; of nitrogen fertilizers, 176–179; potential, defined, 398, 399
Acids, strong, discussed, 399, 400; weak, discussed, 399, 400
Acidulating, soil, discussed, 430–432 (*see also* Soil acidulation)
Acinum, use of as green manure crop, in ancient agriculture, 7
Active absorption (*see* Absorption, active)
Active acidity (*see also* Acidity, active)
Activity index, of urea-formaldehyde, 171
Adenosine diphosphate, in plants, 73, 74
Adenosine monophosphate, in plants, 73, 74
Adenosine triphosphate, in plants, 73, 74
Adsorption, cations, in soil, discussed, 244–265; micronutrients, by soil, discussed, 302–332; phosphate, soil, discussed, 196–202; sulfates, soil, mechanisms of, 287–291
Aeration, soil (*see also* Soil aeration; Soil oxygen; oxygen); effect on nitrification, 136, 137
Agriculture, legendary, beginnings, 5–10

Agrobiology (*see* O. W. Wilcox)
Akiochi, disease of rice, 164; effect of soil iron on, 316
Algae, blue-green, nitrogen fixation by, 125
Alkali disease of cattle (*see* Selenosis)
Alkali spots, in soil, 266
Aluminosilicate clays, soil acidity, contribution of, 401–404
Aluminum, role, in soil acidity, 401–406; soil, effect of liming on, 425; reaction with phosphate, 199, 200; root growth, 425, 426; sulfate, adsorption, 289; leaching, 286, 287; reaction in soil, 432; use of as soil acidulent, 432
Amendments, soil, use of in ancient agriculture, 8
American Association of Plant Food Control Officials (AAPFCO), *defined*, 366
Aminization, in soil, 133
Amino acids, synthesis in plants, 68
Ammonia, anhydrous (*see also* Nitrogen fertilizers), 155–159; defoliant, 159; effect of, placement on plant growth, 158, 159; soil pH, 156, 157; fixation of in soil, 157, 158; nitrification of in soils, 156, 157; retention of in soil, 155–159; storage and handling of, 155; with sulfur (*see* Anhydrous ammonia-sulfur); toxicity of, 156
Ammonia, conversion to nitrate, 133, 134; effect of, pH on losses, from soil, 150, 151; losses of, from fertilizers, 367; soil, 150, 151; urea, 165
Ammonia oxidation, in fertilizer manufacture, 347
Ammonia-sulfur, fertilizer, 300
Ammonia, synthetic production of, 345, 346
Ammoniated superphosphate, 215